ENCYCLOPEDIA OF BUILDING TECHNOLOGY

About the Editor

Henry J. Cowan obtained Ph.D. and Doctor of Engineering degrees from the University of Sheffield, England, and an Honorary Doctorate in Architecture from the University of Sydney, Australia. He was dean of architecture of the University of Sydney (1966–1967) and head of its Department of Architectural Science (1953–1984). He was visiting professor at Cornell University in 1962.

Professor Cowan is a fellow of the American Society of Civil Engineers, the Institution of Structural Engineers, the Royal Society of Arts, the Institution of Engineers Australia, and the Accademia Pontaniana in Naples. He is an honorary fellow of the Royal Australian Institute of Architects and a past president of the Building Science Forum of Australia. He was awarded the Chapman Medal of the IEAust in 1956 and was made an Officer of the Order of Australia in 1983. He is the editor of the *Architectural Science Review* and the author of 16 books and more than 200 papers in various journals.

ENCYCLOPEDIA OF BUILDING TECHNOLOGY

edited by

Henry J. Cowan

PRENTICE HALL
Englewood Cliffs, New Jersey 07632

Library of Congress Cataloging-in-Publication Data

Encyclopedia of building technology.

Includes bibliographies and index.
1. Building—Dictionaries. 2. Engineering—Dictionaries. 3. Architecture—Dictionaries.
I. Cowan, Henry J.
TH9.E48 1988 690'.03'21 87-2556
ISBN 0-13-275520-3 (casebound)

Editorial/production supervision and interior design: Reynold Rieger
Cover design: Edsal Enterprises
Manufacturing buyers: Carol Bystrom and Lorraine Fumoso

Printed in the United States of America

10 9 8 7 6 5 4 3 2 1

ISBN 0-13-275520-3 025

Prentice-Hall International (UK) Limited, *London*
Prentice-Hall of Australia Pty. Limited, *Sydney*
Prentice-Hall Canada Inc., *Toronto*
Prentice-Hall Hispanoamericana, S.A., *Mexico*
Prentice-Hall of India Private Limited, *New Delhi*
Prentice-Hall of Japan, Inc., *Tokyo*
Simon & Schuster Asia Pte. Ltd., *Singapore*
Editora Prentice-Hall do Brasil, Ltda., *Rio de Janeiro*

To Renate

Contents

Preface

This encyclopedia is intended to provide information for architects, engineers, and others interested in building technology, particularly on subjects outside their field of specialization. Most articles are written in simple language, and all specialist terms and abbreviations, except those in common use, are explained. Common abbreviations are listed on pages xx–xxi.

Articles on wide-ranging topics contain cross references, denoted by ROMAN CAPITAL LETTERS, to other more specialized articles; for example, the article on TALL BUILDINGS contains cross references to 31 other articles. In addition, there is a detailed index at the end of this volume.

The volume contains 210 articles by 161 contributors, including some of the most distinguished authorities in the field of building technology. Most of these authors live in Australia, Britain, Canada, or the United States, but there are a few from 11 other countries. The contributors are identified at the end of each article by their initials; these initials, together with their full names and addresses, are listed on pages xi–xix.

Each article is followed by references to books or papers containing further information to assist those who wish to delve more deeply into the subject.

Sydney, Australia **H. J. C.**

Contributors

This list is arranged in alphabetical order by initials, with the initial of the surname dominant. Thus H.G.P. (H. G. Poulos) is printed after Z.S.M. (Z. S. Makowski) and before R.P. (R. Park).

H.O.A. Howard O. Aaronson, Partner, Jaros, Baum, and Bolles, Consulting Engineers, 345 Park Avenue, New York NY 10154, U.S.A. ELECTRICAL SERVICES.

J.M.A. J. M. Antill, formerly Professor of Civil Engineering, University of New South Wales, Kensington NSW 2033, Australia. NETWORK ANALYSIS.

L.D.A. L. D. Armstrong, formerly of the Division of Building Research, Commonwealth Scientific and Industrial Research Organization, Highett VIC 3190, Australia. DAMP-PROOF COURSES.

L.E.A. Lewis E. Akers, Technical Consultant, Winchcombe GL54 5LG, England. WOOD-BASED SHEET MATERIALS.

P.V.A.-L. P. V. Angus-Leppan, Professor of Surveying, University of New South Wales, Kensington NSW 2033, Australia. SURVEYING AND SETTING OUT OF BUILDINGS.

R.M.A. R. M. Aynsley, Professor of Architecture, Georgia Institute of Technology, Atlanta, GA 30332, U.S.A. SANITATION FOR DEVELOPING COUNTRIES.

S.A. Samuel Aroni, Professor, Graduate School of Architecture and Urban Planning, University of California, Los Angeles CA 90024, U.S.A. CONCRETE ADMIXTURES.

A.Bi. Art Bijl, Edinburgh Computer-Aided Architectural Design Research Unit, School of Architecture, University of Edinburgh, Edinburgh EH1 1JZ, Scotland. KNOWLEDGE ENGINEERING.

A.Bl. A. Blaga, Building Materials Section, Institute for Research in Construction, National Research Council, Ottawa K1A OR6, Canada. PLASTICS.

A.D.B. Arthur D. Bernhardt, President, Program of Industrialization in the Housing Sector, P.O. Box 303, Cambridge, MA 02141, U.S.A. MOBILE/MANUFACTURED HOUSING.

D.A.B. David Alan Bies, Reader in Mechanical Engineering, University of Adelaide, Adelaide SA 5000, Australia. SOUND ABSORPTION (ATTENUATION OF SOUND IN ROOMS).

D.B. Don Bartlett, Central Investigation and Research Laboratory, Department of Housing and Construction, 177 Salmon Street, Port Melbourne VIC 3207, Australia. PAINT.

F.A.B. F. A. Blakey, Chief of the Division of Building Research, Commonwealth Scientific and Industrial Research Organization, Highett VIC 3190, Australia. CELLULAR CONCRETE; LIGHTWEIGHT CONCRETE.

G.A.B. G. A. Brown, Research Manager, Australian Gypsum Industries, 350 Latrobe Street, Melbourne VIC 3000, Australia. GYPSUM.

J.B. John Brebner, Reader in Psychology, University of Adelaide, Adelaide SA 5000, Australia. ERGONOMICS.

J.E.B. J. E. Benson, Electro-Acoustic Consultant, 4 Beaumont Street, Denistone NSW 2114, Australia. SOUND-REINFORCEMENT SYSTEMS.

J.J.B. James Beaudoin, Building Materials Section, Institute for Research in Construction, National Research Council, Ottawa K1A OR6, Canada. CLEANING OF BRICKWORK AND MASONRY.

L.S.B. Lynn S. Beedle, Distinguished Professor of Civil Engineering, Lehigh University, Bethlehem, PA 18015, U.S.A. TALL BUILDINGS.

M.Ba. Michael Barron, Martin Centre for Architectural and Urban Studies, Department of Architecture, University of Cambridge, Cambridge CB2 2EB, England. ACOUSTIC MODELS.

M.Bu. M. Burt, Professor of Architecture, Israel Institute of Technology, Technion City, Haifa 32000, Israel. POLYHEDRA.

M.C.B. Maxwell C. Baker, Head of the British Columbia Regional Station, Institute for Research in Construction, National Research Council, Ottawa K1A OR6, Canada. FLASHINGS; FLAT ROOFS; ROOF SURFACES.

N.G.B. N. G. Brown, Life Cycle Performance Research Program, Division of Building Research, Commonwealth Scientific and Industrial Research Organization, Highett VIC 3190, Australia. SEALANTS AND JOINTING.

P.B. Paul Brown, Leader of the Cement Hydration Group, U.S. National Bureau of Standards, Washington, DC 20234, U.S.A. PORTLAND CEMENT.

P.J.B. Peter J. Bliss, School of Civil Engineering, University of New South Wales, Kensington, N.S.W. 2033, Australia. REFUSE DISPOSAL; SANITARY APPLIANCES; SEWAGE DISPOSAL.

R.B. Richard Bender, Dean of the College of Environmental Design, University of California, Berkeley, CA 94720, U.S.A. SYSTEM BUILDING.

S.A.B. Sydney Baggs, formerly Senior Lecturer in Landscape Architecture, University of New South Wales, Kensington NSW 2033, Australia. EARTH-COVERED ARCHITECTURE.

S.B. S. Blencowe, Director, Douglas Lift Slab, 330 Edgar Street, Bankstown NSW 2200, Australia. LIFT SLAB CONSTRUCTION.

S.K.B. Steve Brown, Division of Building Research, Commonwealth Scientific and Industrial Research Organization, Highett VIC 3190, Australia. ASBESTOS IN BUILDINGS.

D.C. David Canter, Professor of Applied Psychology, University of Surrey, Guildford GU2 5XH, England. HUMAN BEHAVIOR IN FIRES.

D.C.-A. Denison Campbell-Allen, Professor Emeritus of Civil Engineering, University of Sydney, Sydney NSW 2006, Australia. QUALITY CONTROL OF MATERIALS.

D.J.C. David Cook, Associate Professor of Civil Engineering, University of New South Wales, Kensington NSW 2033, Australia. TESTING OF MATERIALS FOR STRENGTH, DUCTILITY, AND HARDNESS.

F.R.S.C. F. R. S. Clark, Institute for Research in Construction, National Research Council, Ottawa K1A OR6, Canada. SMOKE AND TOXIC GASES.

G.M.E.C. Gordon M. E. Cooke, Head of the Structural Safety in Fires Section, Fire Research Station of the Building Research Establishment, Borehamwood WD6 2BL, England. FIRE-RESISTING CONSTRUCTION.

H.J.C. Henry J. Cowan, Professor Emeritus of Architectural Science, University of Sydney, Sydney NSW, 2006, Australia. ANCIENT ROMAN CONCRETE; BUCKLING; CATENARY; CONIC SECTIONS; DOMES; HELIODON; HYPOCAUST; IRON; METRIC AND CONVENTIONAL (BRITISH) AMERICAN SYSTEMS OF MEASUREMENT; MODELS; PROJECTIONS; PSYCHROMETRIC CHART; SPAN; STRUCTURAL MODEL ANALYSIS; SUN; TEMPERATURE AND HUMIDITY MEASUREMENT; THERMAL CONDUCTION, CONVECTION AND RADIATION; VERMICULITE.

I.J.C. I. J. Calderone, Manager of the Technical Advisory Service, Pilkington ACI, Dandenong VIC 3175, Australia. GLASS and GLAZING.

J.C. John Chalmers, Bensasson and Chalmers Ltd., 6 Kings Parade, Cambridge CB2 1SJ, England. DIGITAL COMPUTERS.

J.E.C. J. E. Cermak, Director of the Fluid Dynamics and Diffusion Laboratory, Colorado State University, Fort Collins, CO 80523, U.S.A. WIND LOADS; WIND TUNNELS.

J.P.C. John P. Cook, Jacob Lichter Professor of Engineering Construction, University of Cincinnati, Cincinnati OH 45221, U.S.A. ADHESIVES.

M.J.C. Malcolm J. Crocker, Professor of Mechanical Engineering, Auburn University, Auburn, AL 36830, U.S.A. SOUND-LEVEL METER.

R.G.Co. R. G. Coggins, Peddle Thorp and Walker, Architects, 50 Bridge Street, Sydney NSW 2000, Australia. CEILING; PARTITIONS.

R.G.Cr. Robert G. Crossman, Department of Applied Mathematics, University of Sydney, Sydney NSW 2006, Australia. STATISTICS.

T.M.C. T. M. Charlton, Professor Emeritus of Engineering, University of Aberdeen, Aberdeen AB9 1FX, Scotland. ELASTIC DESIGN OF STATICALLY INDETERMINATE STRUCTURES; STATICALLY DETERMINATE STRUCTURES.

W.W.S.C. W. W. S. Charters, Reader in Mechanical Engineering, University of Melbourne, Parkville VIC 3052, Australia. SOLAR COLLECTORS FOR SOLAR THERMAL SYSTEMS.

C.M.D. Charles M. Dabney, Principal, Charles M. Dabney Associates, 1005 W. Bay Avenue, Newport Beach, CA, 92661, U.S.A. COLORED CONCRETE.

J.B.D. J. Dalton, Technical Manager, S. G. Brooker Pty, Limited, 8 Help Street, Chatswood NSW 2067, Australia. FORM WORK; SCAFFOLDING.

J.B.d.B. J. B. de Boer, Professor Emeritus of Electrotechnics, Technische Hogeschool, 5600 MN Eindhoven, The Netherlands. GLARE.

L.A.D. L. A. Drake, School of Earth Sciences, Macquarie University, North Ryde NSW 2113, Australia. EARTHQUAKE.

P.D. Philip Drew, Architectural Historian, 35 Lorne Street, Killara NSW 2071, Australia. TENSILE ROOFS.

R.J.D. Richard J. Dinham, Chief Architect, Leighton Contractors, 472 Pacific Highway, St. Leonards NSW 2065, Australia. VALUE ENGINEERING.

R.M.E.D. R. E. Diamant, Engineering Consultant, 7 Goodwood Avenue, Manchester M23 9JQ, England. THERMAL INSULATION.

B.J.F. B. J. Ferguson, Chief Civil Engineer, Humes A.R.C., World Trade Centre, Flinders Street, Melbourne VIC 3005, Australia. REINFORCEMENT FOR CONCRETE.

D.F. D. Fischer, Professor Emeritus of Electrotechnics, Technische Hoogeschool, 5600 MN Eindhoven, The Netherlands. DISCHARGE LAMPS; INCANDESCENT LAMPS; LIGHTING LEVELS APPROPRIATE TO VARIOUS TASKS; LUMINAIRES.

F.F. Fergus Fricke, Senior Lecturer in Architectural Science, University of Sydney, Sydney NSW 2006, Australia. ACOUSTIC TERMINOLOGY AND MEASUREMENTS.

J.W.F. John W. Fisher, Professor of Civil Engineering, Lehigh University, Bethlehem, PA 18015, U.S.A. JOINTS FOR STEEL STRUCTURES.

M.F. Mark Fintel, formerly Director of Advanced Engineering Services, Portland Cement Association, Skokie, IL 60077, U.S.A. EARTHQUAKE RESISTANT DESIGN.

N.H.F. N. H. Fletcher, Director of the Institute of Physical Sciences, Commonwealth Scientific and Industrial Research Organization, Canberra ACT 2600, Australia. SOUND.

N.O.F. Nancy O. Feldman, Associate, Program of Industrialization in the Housing Sector, P.O. Box 303, Cambridge, MA 02141, U.S.A. MOBILE/MANUFACTURED HOMES.

B.G. Baruch Givoni, Professor of Architecture, University of California, Los Angeles, CA 90024, U.S.A. PASSIVE SOLAR DESIGN.

B.W.G. B. W. Gould, formerly Associate Professor of Civil Engineering, University of New South Wales, Kensington NSW 2033, Australia. WATER QUALITY AND TREATMENT.

J.S.G. J. S. Gero, Director of the Computer Application Research Unit and Professor of Architectural Science, University of Sydney, Sydney NSW 2006, Australia. COMPUTER GRAPHICS; DRAFTING; OPTIMIZATION.

M.A.G. M. A. Green, Professor of Solid-State Electronics, University of New South Wales, Kensington NWS 2033, Australia. PHOTOVOLTAIC CELLS.

P.U.A.G. P. U. A. Grossman, Principal Research Scientist, Division of Chemical and Wood Technology, Commonwealth Scientific and Industrial Research Organization, Clayton VIC 3168, Australia. RHEOLOGY.

R.E.G. Richard E. Gobee, Wormald Data Systems, 16 Giffnock Street, North Ryde NSW 2113, Australia. SECURITY IN BUILDINGS.

R.G.G. Richard G. Gewain, Chief Fire Protection Engineer, American Iron and Steel Institute, 1000 16th Street, Washington, DC 20036, U.S.A. FIRE PROTECTION OF STEEL STRUCTURES.

T.V.G. T. V. Galambos, Professor of Civil Engineering, University of Minnesota, Minneapolis, MN 55455, U.S.A. SAFETY FACTORS IN STRUCTURAL DESIGN.

A.W.H. A. W. Hendry, Professor of Civil Engineering and Building Science, University of Edinburgh, Edinburgh EH9 3JL, Scotland. PHOTOELASTICITY.

E.B.H. Edwin B. Huddleston, Consulting Engineer, 185 Condamine Street, Balgowlah NSW 2093, Australia. PRESERVATION OF TIMBER.

E.H. Edmund Happold, Professor of Building Engineering, University of Bath, Bath BA2 7AY, England. PNEUMATIC STRUCTURES.

G.E.H. Graham E. Holland, Senior Lecturer in Architecture, University of Sydney, Sydney NSW 2006, Australia. FLOOR FINISHES.

G.H. Gilbert Herbert, Mary Hill Swope Professor of Architecture, Israel Institute of Technology, Haifa, Israel. PREFABRICATION: ORIGINS AND DEVELOPMENTS.

H.H. Howard Harrison, Associate Professor of Civil Engineering, University of Sydney, Sydney NSW 2006, Australia. COMPUTER METHODS IN STRUCTURAL ANALYSIS.

J.H. Jacques Heyman, Professor of Engineering, University of Cambridge, Cambridge CB1 1PZ, England. MASONRY ARCHES, VAULTS, AND DOMES.

J.L.H. J. L. Heiman, National Building Technology Centre, Department of Housing and Construction, P.O. Box 30, Chatswood NSW 2067, Australia. DRAINAGE, DAMP-PROOFING, AND WATERPROOFING OF BASEMENTS.

K.H. K. Harper, Advanced Building Systems, 90b Hassall Street, Wetherill Park NSW 2164, Australia. TILT-UP CONSTRUCTION.

T.Z.H. Tibor Z. Harmathy, Head of the Fire Research Section, Institute for Research in Construction, National Research Council, Ottawa K1A OR6, Canada. FIRE RESISTANCE OF BUILDING ELEMENTS; FIRE-RESISTANT COMPARTMENTATION.

W.L.H. William L. Harris, formerly Chief Engineer, Services, Metropolitan Water, Sewerage and Drainage Board, Sydney NSW 2000, Australia. DRAINAGE INSTALLATIONS; WATER SUPPLY.

H.M.I. Max Irvine, Professor of Civil Engineering, University of New South Wales, Kensington NSW 2033, Australia. CABLE AND SUSPENSION STRUCTURES, THEORY OF.

A.E.J. Alexander E. Jenkins, Professor Emeritus of Materials and Mining Engineering, University of Sydney, Sydney NSW 2006, Australia. ALUMINUM; METALLURGY; NONFERROUS METALS.

B.K.J. Brian K. Jones, Building Performance Section, Institute for Research in Construction, National Research Council, Ottawa K1A OR6, Canada. EVACUATION OF BUILDINGS IN CASE OF FIRE.

G.H.J. G. Henry Johnston, Geotechnical Section, Institute for Research in Construction, National Research Council, Ottawa K1A OR6, Canada. PERMAFROST.

R.P.J. R. P. Johnson, Professor of Engineering, University of Warwick, Coventry CV4 7AL, England. LIMIT STATES DESIGN.

T.J. Tom Jumikis, Director, Woolacott, Hale, Corlett and Jumikis, Consulting Engineers, 140 Pacific Highway, North Sydney NSW 2060, Australia. EXPANSION AND CONTRACTION JOINTS.

W.G.J. Warren G. Julian, Dean of the Faculty of Architectures, University of Sydney, Sydney NSW 2006, Australia. COLOR MEASUREMENT SYSTEMS; ELECTRICAL MEASUREMENTS; LIGHTING (ILLUMINATION), UNITS AND MEASUREMENT OF.

A.K. A. Klarich, Peddle Thorp and Walker, Architects, 50 Bridge Street, Sydney NSW 2000, Australia. CURTAIN WALLS.

F.K.K. F. K. Kong, Professor of Structural Engineering, University of Newcastle-upon-Tyne, Newcastle-upon-Tyne NE1 7RU, England. PRESTRESSED CONCRETE.

K.M.K. Kevin M. Kelly, President, Jay-K Independent Lumber Corporation, Seneca Turnpike, New Hertford, NY 13413, U.S.A. VAPOR BARRIERS.

R.K. Richard Kittler, Institute of Construction and Architecture of the Slovak Academy of Sciences, Bratislava, Czechoslavakia. ARTIFICIAL SKY; DAYLIGHT DESIGN.

S.J.K. Stephen J. Kirk, Associate, Smith Hinchman and Grylls Associates, 1050 17th Street NW, Washington, DC 20036, U.S.A. VALUE ENGINEERING.

S.K. Steve King, Lecturer in Architecture, Canberra College of Advanced Education, Belconnen ACT 2616, Australia. PASSIVE SOLAR DESIGN.

T.K. T. Koncz, Consulting Engineer, 297 Wittkowerstrasse, CH 8053 Zürich, Switzerland. PRECAST CONCRETE.

W.G.K. W. G. Keating, Division of Chemical and Wood Technology, Commonwealth Scientific and Industrial Research Organization, Highett VIC 3190, Australia. TIMBER.

A.L. Asher Lazer, Chairman, Rezal Industries, 99 Rookwood Road, Yagoona NSW 2199, Australia. FIRE AND SMOKE DETECTORS; SPRINKLER SYSTEMS.

B.M.L. Barry M. Lee, Wormald International, Alexander Street, Crows Nest NSW 2065, Australia. FIREFIGHTING EQUIPMENT.

B.P.L. B. P. Lim, Dean of the Faculty of Architecture and Building, National University of Singapore, Singapore 0511. FENESTRATION.

E.T.L. Edward T. Linacre, formerly Associate Professor of Earth Sciences, Macquarie University, North Ryde NSW 2113, Australia. CLIMATE AND BUILDING.

J.B.L. J. B. Large, Professor, Institute of Sound and Vibration Research, University of Southampton, Southampton SO9 5NH, England. NOISE.

J.K.L. J. Kenneth Latta, formerly of the Energy Conservation Research and Development Section, Institute for Research in Construction, National Research Council, Ottawa K1A OR, Canada. CONDENSATION, FORMATION OF.

K.J.L. K. J. Lyngcolm, Engineer, Plywood Association of Australia, P.O. Box 8, Newstead QLD 4006, Australia. PLYWOOD.

O.G.L. O. G. Löf, Professor of Civil Engineering, Colorado State University, Fort Collins, CO 80523, U.S.A. SOLAR ENERGY.

R.E.L. R. E. Luxton, Professor of Mechanical Engineering, University of Adelaide, Adelaide SA 5000, Australia. HEAT ENGINES AND HEAT PUMPS.

R.H.L. R. H. Leicester, Head of the Safety and Risk Section, Division of Building Research, Commonwealth Scientific and Industrial Research Organization, Highett VIC 3190, Australia. LAMINATED TIMBER.

T.Y.L. T. Y. Lin, T. Y. Lin International, 315 Bay Street, San Francisco, CA 94133, U.S.A. LOAD BALANCING IN PRESTRESSED CONCRETE.

W.S.L. William S. Lewis, Partner, Jaros Baum and Bolles, Consulting Engineers, 345 Park Avenue, New York, NY 10151, U.S.A. ELEVATORS; ESCALATORS AND MOVING WALKS.

A.H.M. A. Harold Marshall, Professor and Director of the Acoustics Institute, University of Auckland, Auckland, New Zealand. ACOUSTICS OF AUDITORIUMS.

C.M. Charles Massonet, Professor Emeritus of the Strength of Materials and the Stability of Structures, University of Liège, 6 Quai Banning, Liège, Belgium. LIMIT DESIGN OF STEEL STRUCTURES.

H.J.M. H. J. Milton, Technical Consultant, 10 Janari Place, Aranda ACT 2614, Australia. DIMENSIONAL COORDINATION; TOLERANCES.

J.E.M. Joseph E. Minor, Professor of Civil Engineering, Texas Tech University, Lubbock, Texas 79407, U.S.A. HURRICANE RESISTANT DESIGN.

K.G.M. Keith G. Martin, Division of Building Research, Commonwealth Scientific and Industrial Research Organization, Highett VIC 3190, Australia. ACCELERATED WEATHERING; DURABILITY.

R.C.M. R. C. Mangal, Deputy Director, Central Building Research Institute, Roorkee 247667, India. BAMBOO AS A BUILDING MATERIAL.

R.E.M. Richard E. Master, Partner, Jaros Baum and Bolles, Consulting Engineers, 345 Park Avenue, New York, NY 10154, U.S.A. AIR CONDITIONING AND HEATING.

R.K.Ma. R. K. Macpherson, Professor Emeritus of Environmental Health, University of Sydney, Sydney NSW 2006, Australia. THERMAL COMFORT CRITERIA.

R.K.Mü. R. K. Müller, Professor and Director of the Institute for Model Analysis, University of Stuttgart, 7000 Stuttgart 80, German Federal Republic. MEASUREMENT OF STRAIN.

T.M. Tom McNeilly, formerly Director of the Brick Development Research Institute, University of Melbourne, Parkville VIC 3052, Australia. BRICKS AND BLOCKS.

W.H.Me. W. H. Melbourne, Professor of Fluid Mechanics, Monash University, Clayton VIC 3168, Australia. ANEMOMETERS.

W.H.Mi. William J. Mitchell, Professor of Architecture, Harvard University, Cambridge, MA 02318, U.S.A. COMPUTER-AIDED ARCHITECTURAL DESIGN.

Y.W.M. Y. W. Mai, Associate Professor of Mechanical Engineering, University of Sydney, Sydney NSW 2006, Australia. CRACKS IN BRITTLE MATERIALS.

Z.S.M. Z. S. Makowski, Professor of Civil Engineering, University of Surrey, Guildford GU2 5XH, England. SPACE FRAMES.

A.N. Adam Neville, Principal of the University of Dundee, Dundee DD1 4HN, Scotland. SHRINKAGE AND CREEP.

F.N. Frank Newby, Partner, F. J. Samuely and Partners, Consulting Engineers, 231N. Gower Street, London NW12 NS, England. CONCRETE STRUCTURES.

H.G.P. H. G. Poulos, Professor of Civil Engineering, University of Sydney, Sydney NSW 2006, Australia. FOUNDATIONS FOR BUILDINGS; PILES AND PILING; SOIL CLASSIFICATION; SOIL MECHANICS.

R.P. R. Park, Professor of Civil Engineering, University of Canterbury, Christchurch, New Zealand. REINFORCED CONCRETE SLABS.

S.J.P. S. J. Perrens, Senior Lecturer in Resource Engineering, University of New England, Armidale NSW 2351, Australia. HYDROLOGY OF ROOF WATER COLLECTION.

A.S.R. A. S. Rakhra, Building Performance Section, Institute for Research in Construction, National Research Council, Ottawa K1A OR6, Canada. LIFE-CYCLE COSTING.

C.B.R. C. B. Rolfe, Consulting Engineer, 16 Arcadia Road, Glebe NSW 2037, Australia. STAINLESS STEEL.

J.R. Jack Rose, formerly of the National Acoustic Laboratories, 50 Hickson Road, Sydney NSW 200, Australia. ACOUSTICAL TEST FACILITIES; HELMHOLTZ RESONATORS.

L.E.R. Leslie E. Robertson, Robertson, Fowler and Associates, Consulting Structural Engineers, 211 E. 46th Street, New York, NY 10017, U.S.A. VIBRATIONS IN TALL BUILDINGS.

S.M.R. Sidney M. Renof, Consultant, 32 Russell Street, Vaucluse, NSW 2030, Australia. CORROSION OF METALS; CORROSION PROTECTION.

V.S.R. Vangipuram S. Ramachandran, Head of the Building Materials Section, Institute for Research in Construction, National Research Council, Ottawa K1A OR6, Canada. CLEANING OF BRICKWORK AND MASONRY.

W.G.J.R. W. G. Ryan, BMI Ltd., P.O. Box 42, Wentworthville NSW 2145, Australia. CONCRETE MIX PROPORTIONING; PLACING OF CONCRETE.

A.H.S. Alan M. Spry, Senior Consultant, Australian Mineral Development Laboratories, Frewville SA 5063, Australia. NATURAL BUILDING STONE.

A.M.S. A. Monem Saleh, Consultant, P.O. Box 439, Strathfield NSW 2135, Australia. SUN SHADING AND SUN-CONTROL DEVICES.

A.S. Allan Shaw, formerly of the Department of Mechanical Engineering, University of Adelaide, Adelaide SA 5000, Australia. REFRIGERATION CYCLES.

B.S.S. Balwint Singh Saini, Professor of Architecture, University of Queensland, St. Lucia QLD 4067, Australia. VERNACULAR ARCHITECTURE.

E.S. Edward Steinfeld, Professor of Architecture, State University of New York, Buffalo, NY 14214, U.S.A. BARRIER-FREE DESIGN.

J.C.S. John C. Scrivener, Professor of Building, University of Melbourne, Parkville VIC 3052, Australia. BRICKWORK AND BLOCKWORK.

J.W.S. J. W. Sloman, Manager Building Design Standards, Telecom Australia Headquarters, Melbourne 3000, Australia. TELEPHONE SYSTEMS.

K.S. K. Sumi, Institute for Research in Construction, National Research Council, Ottawa K1A OR6, Canada. SMOKE AND TOXIC GASES.

P.E.S. Paul E. Sprague, Associate Professor of Art History, University of Wisconsin, Milwaukee, WI 53201, U.S.A. BALLOON FRAME.

P.R.S. Peter R. Smith, Associate Professor and Head of the Department of Architectural Science, University of Sydney, Sydney NSW 2006, Australia. PLENUM SYSTEMS AND DUCTS.

R.G.Sl. Roger G. Slutter, Professor of Civil Engineering and Director of Operations Division, Fritz Engineering Laboratory, Lehigh University, Bethlehem, PA 18015, U.S.A. COMPOSITE CONSTRUCTION.

R.G.St. Richard G. Stein, The Stein Partnership, Architects, 20 West 20th Street, New York, NY 10011, U.S.A. CONSERVATION OF ENERGY IN BUILDINGS.

R.J.S. R. Standen, Peddle Thorp and Walker, Architects, 50 Bridge Street, Sydney NSW 2000, Australia. DOORS.

T.J.S. Theodore J. Schultz, Consultant in Acoustics, 7 Rutland Square, Boston, MA 02118, U.S.A. IMPACT AND VIBRATION ISOLATION; SOUND INSULATION AND ISOLATION.

T.L.S. Thomas L. Saaty, University Professor, University of Pittsburgh, Pittsburgh, PA 15260, U.S.A. FUZZY SETS, ANALYTICAL HIERARCHIES, AND DECISIONS.

W.E.S. W. E. Scholes, Consultant in Building Acoustics, 117 Kindersley Way, Abbots Langley WD4 ODG, England. ACOUSTICAL REQUIREMENTS FOR BUILDINGS.

W.R.S. W. R. Sharman, Materials Division, Building Research Association of New Zealand, Porirua, New Zealand. ABRASION RESISTANCE.

C.P.T. C. P. Thorne, Coffey and Partners, Consulting Engineers in the Geotechnical Sciences, 12 Waterloo Road, North Ryde 2113, Australia. SUBSURFACE EXPLORATION.

D.A.T. Donald A. Taylor, Building Structures Section, Institute for Research in Construction, National Research Council of Canada, Ottawa K1A OR6, Canada. SNOW LOADS.

E.T. E. Tauber, Consulting Ceramic Engineer, 16 Malua Street, Ormond VIC 3204, Australia. CERAMIC FLOOR AND WALL TILES.

L.T. Lambert Tall, Professor of Civil Engineering, Florida International University, Miami, FL 33199, U.S.A. STEEL; WELDING, BRAZING, AND SOLDERING.

N.S.T. N. S. Trahair, Professor of Civil Engineering, University of Sydney, Sydney NSW 2006, Australia. STEEL STRUCTURES.

P.R.T. Peter R. Tregenza, Department of Architecture, University of Nottingham, Nottingham NG7 2RD, England. STAIRS.

G.V. George Vasilareas, Department of Architectural Science, University of Sydney, Sydney NSW 2006, Australia. DETAILING OF CONCRETE STRUCTURES; TIMBER JOINTS FOR STRUCTURES AND JOINERY.

J.F.v.S. J. F. van Straaten, formerly Director of the National Building Research Institute, Pretoria 0001, South Africa. NATURAL VENTILATION.

J.J.J.v.R. J. J. J. van Rensburg, Cement Technologist, National Building Research Institute, Council for Scientific and Industrial Research, Pretoria 0001, South Africa. PLASTERING.

A.W. Alexander Wargon, Director, Wargon Chapman Partners, Consulting Engineers, 51 Bathurst Street, Sydney NSW 2000, Australia. ROOF STRUCTURES.

B.W. Brian Woodward, Architect, Earthways Farm, Wollombi NSW 2325, Australia. ADOBE, PISE, AND PRESSED EARTH BLOCKS.

C.B.W. C. B. Wilby, Visiting Professor of Civil and Structural Engineering, University of Bradford, Bradford BD7 1DP, England. FOLDED PLATES; SHELL STRUCTURES.

G.P.W. Gaynor P. Williams, Head of the Technical Information Section, Institute for Research in Construction, National Research Council, Ottawa K1A OR6, Canada. FLOOD-PROOFING OF BUILDINGS.

R.F.W. Robert F. Warner, Professor of Civil Engineering, University of Adelaide, Adelaide SA 5000, Australia. REINFORCED CONCRETE DESIGN.

W.Z. William Zuk, Professor of Architecture and Director of Architectural Technology, University of Virginia, Charlottesville, VA 22803, U.S.A TIMBER STRUCTURES.

Abbreviations

Abbreviations that occur in only one article and are explained in the text are not included in this list.

a	annum (year)
Al	aluminum
amp	ampere
Btu	British thermal unit
C	Celsius, carbon
c ft	cubic foot
CIE	Commission Internationale de l'Éclairage
cm	centimeter
cosh	hyperbolic cosine
Cr	chromium
Cu	copper
cu ft	cubic foot
dB	decibel
dB(A)	decibel (A scale)
deg	degree
e	base of the natural logarithm, electron
F	Fahrenheit, fluorine
Fe	iron
fps system	the conventional American system of units (foot, pound, second)
ft	foot
ft^2	square foot
ft^3	cubic foot
g	gram
gal	gallon
GPa	gigapascal (= 1,000,000,000 Pa)
h	hour
H	hydrogen
Hz	hertz (= cycles per second)
in.	inch
J	joule
K	kelvin, degree Kelvin
kg	kilogram

kHz	kilohertz
km	kilometer
ksi	kilopound per square inch
kV	kilovolt
kW	kilowatt
kWh	kilowatt-hour
L	span
l, L	liter
lb	pound
lm	lumen
m	meter
m^2	square meter
m^3	cubic meter
M	mega (= 1,000,000)
Mg	magnesium
min	minute
mm	millimeter
Mn	manganese
MN	meganewton
MPa	megapascal
ms	millisecond
mw	meter width
μ (mu)	micro (= 0.000 001)
μm	micrometer
μPa	micropascal
N	newton, nitrogen
Na	sodium
Ni	nickel
nm	nanometer (= 0.000 000 001 m)
O	oxygen
P	person
Pa	pascal
psi	pounds per square inch
rms, RMS	root mean square
s	second
S	sulfur
Si	silicon
SI	SI metric system (Le Système International d'Unités)
Sn	tin
s^2	second, squared
sq ft	square foot
t	ton, tonne
V	volt
W	watt
Zn	zinc
°	degree
′	foot, minute of arc
″	inch
$	U.S. dollars
#	number

Conversion factors between metric and conventional American units are given in the article METRIC AND CONVENTIONAL BRITISH AND AMERICAN SYSTEMS OF MEASUREMENT

Note on Typography

Cross references to other articles are printed in ROMAN CAPITALS.

The title of each article is printed in **BOLD CAPITALS.**

Subheadings are printed in **Bold Capital and Lower-Case Letters.**

Minor subheadings, foreign words, and words that require special emphasis are printed in *italics.*

The initials of the author are given at the end of each article. These initials are listed in alphabetical order on pages xi–xix, followed by the name and address of the author.

A

ABRASION RESISTANCE. Ideally, abrasion resistance should be measured as the weight or thickness loss of flooring materials exposed under actual working conditions, but this takes too long to be of practical value. Instead, two distinctive types of wear are measured by abrasion resistance tests: (1) wear arising from foot traffic, and (2) wear arising from industrial wheeled traffic.

Abrasive grits such as silicon carbide in loose or wheel form are used in accelerated tests to reproduce the grinding and cutting effects of foot traffic. An international committee (Ref. 1) has reviewed a number of test methods. The Taber abraser is the most common testing machine. Frick (Ref. 2) has suggested modifications to improve its reproducibility.

Accelerated tests employing steel wheels or rolling ball bearings are considered to best reproduce the effects of industrial traffic, although abrasives and dressing wheels have also been used. Test methods have been reviewed by Sharman (Ref. 3).

See also DURABILITY.

References

1. Review of test methods for abrasion resistance of flooring. *Wear*, Vol. 4 (1961), pp. 479–494.
2. O. F. V. FRICK: Studies of wear on flooring materials. *Wear*, Vol. 14 (1969), pp. 119–131.
3. W. R. SHARMAN: *Abrasion resistance of freezing works flooring materials*. Research Report R36, Building Research Association of New Zealand, Wellington, 1982.

W.R.S.

ACCELERATED WEATHERING is a term used to describe the exposure of materials to controlled conditions in a laboratory with the aim of accelerating degradation mechanisms associated with natural weathering. Such laboratory exposures are used in attempts to predict the durability of building materials (see DURABILITY) and are usually performed in chambers serviced with controls of heating and cooling, relative humidity, waterspray, and light to simulate solar radiation.

History. Much of the development of accelerated weathering early in this century concerned asphalt; it followed from the findings that its degradation was due to oxidation by air, promoted by sunlight, and that it occurred at different rates behind different-colored glasses. Understanding of the chemical action of ultraviolet (UV) radiation and the simulation of this radiation by artificial lamps followed. Virtually all organic materials from textiles to paints were found to be susceptible to this radiation. Such materials were known to have less durability than inorganic materials and to be influenced by small changes in composition due to either geographically different sources of supply or the addition of other chemicals. An industry developed to produce chemical additives for UV, heat and antioxidant stabilization for oils, resins, rubbers, and, more recently, plastics.

Procedures. In accelerated weathering chambers, materials were traditionally exposed to arbitrary combinations of climatic factors in cycles in an attempt to simulate the cycle of sunshine and rain that constitutes the weather. Despite decades of use and the development of sophisticated devices, these methods have not provided useful quantitative information on service life, because each material has different degradation mechanisms during weathering, and the degradation rates do not have a common time scale. However, these procedures have been useful for preliminary assessment and for quality-control testing. Both of these uses depend on establishing correlations between the changes induced by natural weathering. Certain climatic factors have been artificially enhanced during outdoor exposure trials, either by mirrors to extend the periods of sunlight or by water sprays to extend the periods of wetness.

Recent Developments. Recently, a systematic approach and a standard nomenclature for service-life prediction have been developed. The degradation mechanisms have been separated and accelerated tests developed for each to determine critical performance conditions, including stress factors. Data obtained from accelerated tests are used to establish mathematical models relating degrading factors to degradation rates. These models are applied both to service conditions and failure criteria to establish predicted service levels for each degradation mechanism. They may include the influence of air pollutants on the atmospheric corrosion of metals, as well as the climatic factors.

References

1. C. ELLIS, A. E. WELLS, and F. F. HEYROTH: *The Chemical Action of Ultraviolet Rays*. Reinhold, New York, 1941.
2. AMERICAN SOCIETY FOR TESTING AND MATERIALS: *Performance Concept in Building* (proceedings of symposium). Special Technical Publication 361, ASTM, Philadelphia, 1972.
3. AMERICAN SOCIETY FOR TESTING AND MATERIALS: *Standard Practice for Developing Accelerated Tests to Aid Prediction of the Service Life of Building Components and Materials*. Standard E632-81. ASTM, Philadelphia, 1981.

K.G.M.

ACOUSTICAL REQUIREMENTS FOR BUILDINGS. The acoustical environment within a room is determined by many factors, such as the reflecting properties and arrangement of internal surfaces, noise levels outside, the sound isolation outside to inside, and the noise generated within the room. In design, the usual aim is to provide conditions that allow occupants to carry out their particular activities, such as listening, conversing, or relaxing, without undue interference or annoyance by unwanted sounds. Designing to meet these requirements is a very uncertain process because of individual differences; the best that can be done is to reduce, to acceptable proportions, *the risk* of a person being annoyed, dissatisfied, or awakened by the acoustical climate.

Requirements may be divided into two types: (1) those particularly relevant to listening to music or formal speech, as influenced by the properties of the room itself, and (2) those concerned with the effects of noise, internally or externally generated, on particular activities.

Room Acoustics. Listening conditions are largely influenced by reverberation time (RT), which is defined as the time taken for a sound to decay to 1 millionth of its original intensity (i.e., by 60 dB). Too long an RT, as found for example in large rooms with reflecting boundaries, can lead to the masking of current sounds by previous sounds, which, although dying away, may still be audible. Too short an RT can give an impression of "deadness" in a room.

For speech, an RT of from 1.0 to about 1.3 s, with audience present, is desirable. However, it is possible to provide adequate listening conditions for speech, even in rooms with much longer RTs, by reinforcing the direct and early reflected sounds by the use of directional loudspeakers, as, for example, in Westminster Abbey (see SOUND REINFORCEMENT SYSTEMS).

For music, the optimum RT at mid-frequencies lies in the range from 1.7 to 2.2 s, depending on the type of music, and an increase of up to 50% is desirable at lower frequencies (around 125 Hz) (Ref. 1).

Multipurpose rooms used for both speech and music pose a particular problem, but good listening conditions overall may be achieved by using an assisted resonance system. With these systems the natural RT is designed to be suitable for speech, and the electroacoustic system is used to lengthen the RT so as to be appropriate to music (see also ACOUSTICS OF AUDITORIUMS).

Noise Levels. Many studies of response to NOISE have been carried out, and these have yielded a multiplicity of scales, indexes, and rating procedures for quantifying noise in a way intended to indicate its effects on people (Ref. 2). Each rating procedure takes into account, to a greater or lesser extent, the frequency content of the noise, its temporal characteristics, the noise source, and the noise exposure history and activity of the people involved. For design purposes it is desirable to have a simple scale that is generally applicable over a wide range of noise exposure situations, and the equivalent noise level (L_{eq}) seems to provide the closest approach to such a scale. It is defined as the level of a continuous sound having the same energy content as the noise in question. Table 1 (based on Ref. 3) gives equivalent A weighted noise levels, L_{eq} dB(A), appropriate for a range of activities and spaces.

Sound Isolation. Legislation or guidelines exist in many countries to ensure that the transmission of noise

TABLE 1
Recommended Maximum Noise Levels

Area	dB(A)
Concert halls, opera houses, large theaters and auditoriums, broadcasting and recording studios (for very good listening conditions)	20–30
Small auditoriums and theaters, music rehearsal rooms, large courtrooms, conference and lecture rooms (for very good listening conditions)	30–35
Bedrooms, sleeping quarters, hospitals, and hotels (for sleeping, resting, and relaxing)	30–40
Private offices, small conference rooms, small courtrooms, classrooms (for good listening conditions); living areas (for conversation and listening to radio and television)	40–45
Large offices, retail shops, cafeterias and restaurants (for moderate listening/noise conditions)	45–50
Laboratory work spaces, drawing offices, computer rooms, laundry rooms, accounting machine rooms, typing pools	50–55
Intrusive noise levels higher than 55 dB(A) may be justified in areas where there are high levels of internally generated noise, as in some workshops, machine rooms, or plant rooms	Over 55

between dwellings is kept down to reasonable proportions. The International Standard Methods (Ref. 4) for rating physical sound isolation performance have not yet been adopted in all countries, and so reference must be made to the rating methods and requirements in force locally. However, a standard of airborne sound isolation broadly equivalent to that provided by a solid masonry wall, nominally some 230 mm (9 in.) thick, is generally regarded as the minimum desirable between dwellings (see SOUND INSULATION).

References

1. P. H. PARKIN et al.: *Acoustics, Noise, and Buildings*. Faber and Faber, London, 1979.
2. T. J. SCHULTZ: *Community Noise Rating*. Applied Science Publishers, London and New York, 1982.
3. *Thermal, Visual and Acoustic Requirements in Buildings*. Building Research Establishment Digest 226. Her Majesty's Stationery Office, London, 1979.
4. *Acoustics—Rating of Sound Insulation in Buildings and of Building Elements*. I.S.O.717. International Organization for Standardization, Geneva, 1982.

W.E.S.

The **ACOUSTICAL TEST FACILITIES** with the widest application have diametrically opposite characteristics; they are free-field (anechoic) and diffuse-field (reverberation) rooms. When contemplating the acquisition of one or more, all possible uses should be considered at the planning stage and weighed against their cost implications, because such facilities are generally expensive and difficult to modify.

The most critical factors are the following:

1. Range of frequencies to be covered
2. Minimum background noise requirements
3. Working space needed
4. Site characteristics
5. Mechanical characteristics of items to be tested

Design considerations also include access for people, test items, and services; test equipment support; ventilation; lighting; vibration and noise isolation; and (often overlooked) safety factors.

The design of **anechoic rooms** is well treated by Duda (Ref. 1) and also specified by the ISO (Ref. 2). In simple terms, the internal dimensions wall to wall approximate one wavelength (6.8 m (22 ft) at 50 Hz or 3.7 m (12 ft) at 90 Hz) of the lowest test frequency plus working space; costs escalate rapidly with size. Working space should include the largest specimen and the maximum measuring distances in three dimensions.

Acoustic linings are designed to absorb more than 99% of incident sound down to the cutoff frequency of the test. Materials commonly used are fiber glass, rockwool, and various foams, with a thickness approximately one-quarter the wavelength of the cutoff frequency. Graduated-density linings can be used; but generally wedges are preferred, arranged in patterns to minimize reflections of grazing sound. The technology of wedge design is not simple; experiments with a plane-wave tube do not always accurately predict full-scale performance. Some anechoic rooms have not met their specification on commissioning and can only be put to limited use.

As far as possible, all internal surfaces, especially the door, should be covered with acoustical absorption or reduced in size so that they do not reflect sound within the frequency range of interest. Trampoline floors of fine wire provide the greatest reduction in size, but removable perforated flooring can support heavier loads.

Site selection is particularly important for large rooms. There must be sufficient space to accommodate the room, its control room, the specimen preparation area, and any isolating walls and plenum spaces. Soil stability and load-bearing capacity are important because the room is very heavy; in addition there must be freedom from seepage.

Low background noise limits may require a site remote from major noise sources and/or double-wall construction, plenum spaces, vibration isolation, and noise-reduced ventilation. Access doors and service ducts must match wall insulation. It may be necessary to completely shield the room for sensitive electrical measurements. Some forms of lighting also interfere with acoustical measurements.

Safety needs consideration. The design should specify linings that do not permit free fibers to contaminate lungs and equipment or to give off noxious fumes under heat. Power-operated doors and hoists must have effective safety interlocks. Provision should be made for access, light, communication, and observation during power failures. Severe injuries can result from falls onto trampoline or grid floor; equipment should be at this floor level. Appropriate footwear should be used at all times.

Semi-Anechoic Rooms. In many situations, a fully anechoic room is not needed, and a much less expensive semi-anechoic room is sufficient. This utilizes a hard floor, usually of concrete, and the upper half of an anechoic room. Heavy equipment, such as dynamometers, can be built into this concrete floor.

Reverberation Rooms come in two main categories: (1) rectangular spaces with built-in diffusing elements, and (2) spaces with nonparallel surfaces for wall, floor, and ceiling, with less reliance on diffusers. In both, rotating blades are sometimes used to improve the even distribution of low frequency sound. As with anechoic rooms, larger rooms are needed for low-frequency testing. The ISO (Ref. 3) outlines the dimensions best suited for undertaking standardized tests down to 100 Hz ($\frac{1}{3}$ octave band).

Since the aim is to achieve as even a distribution of sound as possible, combined with a long reverberation time and a smooth rate of sound decay, all surfaces must be hard and smooth and resonate beyond the range of test frequencies. This typically calls for thick concrete walls with very smooth and hard internal surfaces. All openings must be regarded as areas of 100% absorption. They are sealed with doors that match the walls as closely as possible.

As the sound field is dependent on air temperature and humidity in the enclosed space, a method for stabilizing these is advantageous. Very careful attention must be given to both air- and solid-borne noise reductions if a low background noise level is needed. This includes good site selection, vibration isolation, the use of massive double walls with an allowance for the accommodation of plenum spaces, and the elimination of ventilation during tests.

Whereas single rooms are useful for evaluating sound power and acoustical absorption, rooms in coupled pairs are required for measuring transmission loss of walls and floors. The test specimens are mounted in a separate wall or floor between the two rooms to reduce the effect of flanking transmission. Injury can occur in these rooms from falls due to the very smooth hard floor, particularly when it is wet. Attention to correct footwear is important. A means for communicating with and observing those inside should be provided.

The **Utilization** of most test rooms is limited by the time taken for setting up experiments and dismantling them. The locating and design of the control room, together with ease of interconnection of equipment, can improve usage rate. Computer control of testing also helps.

References

1. J. DUDA: Basic design considerations for anechoic rooms. *Noise Control Engineering*, Vol. 9, No. 2 (Sept./Oct. 1977), pp. 60–67.
2. *Acoustics. Determination of Sound Power Levels of Noise Sources—Precision Methods for Anechoic and Semi-anechoic Rooms. ISO 3475*. International Organization for Standardization, Geneva, 1977.
3. *Acoustics. Determination of Sound Power Levels of Noise Sources—Precision Methods for Broad-band Sources in Reverberation Rooms. ISO 3741*. International Organization for Standardization, Geneva, 1975.

J.R.

ACOUSTIC MODELING can take several forms: computer modeling, two-dimensional wavefront photography with air waves or water waves, light modeling, and finally acoustic scale modeling. The latter involves the fewest restrictions and has been the most widely used to investigate sound propagation where sound meets obstacles. Since the wavelengths of sound are comparable to the dimensions of obstacles (walls, building elements, etc.), mathematical prediction is extremely difficult and often impossible. Acoustic scale modeling offers an economical technique for predicting acoustic behavior for sound in enclosures and for environmental noise propagation.

History. Most development work in acoustic modeling has been related to models of auditoriums (Ref. 1). The father of quantitative room acoustics, W. C. Sabine, produced in 1912 photographs of acoustic wavefronts in two-dimensional model sections and plans of theaters. In 1936, Vermeulen and De Boer published results of tests with three-dimensional light models. Light models continue to be used by certain designers, although they inevitably omit the elements of frequency and arrival time. Comprehensive acoustic scale modeling also dates from the 1930s, pioneered by Spandöck. The gradual improvement of transducers and associated equipment, such as tape recorders, led to continuing refinements, with many laboratories conducting model studies in the 1950s and 1960s. By 1967, Spandöck's group could demonstrate an impressively accurate acoustic reproduction with their auditorium models. This achievement was based on having developed model equivalents for all the acoustic absorbing materials found in typical auditoria.

The same basic techniques have also been used in open plan offices and factories.

Computer modeling was begun in the 1960s, but it is still unable to take account of complex wave behavior. Modeling for environmental noise problems also dates from the 1960s (e.g., Ref. 2) and has proved a valuable tool for transport and other environmental noise propagation.

Acoustic Scale Modeling. The theory is based on a fundamental relationship:

$$\text{speed of sound} = \text{frequency} \times \text{wavelength}$$

The model medium is nearly always air. For correct wave behavior, the ratio between wavelength and size has to be maintained. So in a 1 : 10 scale model, the wavelength is reduced by one-tenth and the frequency is increased by a factor of 10. The attractive feature about acoustic modeling is that it can offer a perfect modeling capability: the behavior in the model is related to full-sized behavior by simple ratios. There is, however, the complicating feature of absorption of sound at high frequencies as it propagates through the air. Air absorption at model frequencies can be substantially reduced by using dehumidified air (less than 5% relative humidity) or a nitrogen atmosphere. At scale factors of 1 : 8 or 1 : 10, by these means air absorption is nearly perfectly modeled. In other cases, excessive model absorption can often be corrected mathematically.

Auditorium Modeling. At the commonest scale factors of 1 : 8 or 1 : 10, model techniques allow both objective testing (to produce numerical results) and subjective testing, which involves playing music or speech through the model for subsequent subjective assessment, usually over headphones. However, the problem of incorporating an elaborate model exercise in the design program for an auditorium, as well as the expense involved, has led to smaller models at scale factors of between 1 : 20 and 1 : 50 (Ref. 4 and Fig. 1). Although in smaller models subjective testing is no longer possible, growing confidence in the objective measures of ACOUSTICS OF AUDITORIUMS make these models a valuable aid. They allow prediction of these quantities, which cannot otherwise be calculated in advance.

Figure 1 Acoustic model of the Theatre Royal, Plymouth, England, opened in 1982, to a scale of 1 : 50. The model is constructed of varnished timber, with accurately modeled seating.

References

1. L. CREMER and H. A. MÜLLER (translated by T. J. SCHULTZ): *Principles and Applications of Room Acoustics, Vol. 1*, Chapter I.7. Applied Science Publishers, London, 1982.
2. M. E. DELANY, A. J. RENNIE, and K. M. COLLINS: A scale model technique for investigating traffic noise propagation. *Journal of Sound and Vibration*, Vol. 56 (1978), pp. 325–340.
3. M. BARRON: Auditorium acoustic modeling now. *Applied Acoustics*, Vol. 16 (1983), pp. 279–290.
4. M. BARRON and C. B. CHINOY: 1 : 50 scale acoustic models for objective testing of auditoria. *Applied Acoustics*, Vol. 12 (1979), pp. 361–375.

M.B.

ACOUSTICS is the science of sound. In this encyclopedia it is discussed under the following headings: ACOUSTICAL REQUIREMENTS FOR BUILDINGS; ACOUSTICAL TERMINOLOGY AND MEASUREMENTS; ACOUSTIC MODELS; ACOUSTICS OF AUDITORIUMS; ACOUSTICAL TEST FACILITIES; HELMHOLTZ RESONATORS; IMPACT AND VIBRATION ISOLATION; NOISE; SOUND; SOUND ABSORPTION; SOUND INSULATION AND ISOLATION; SOUND-LEVEL METER; SOUND REINFORCEMENT SYSTEMS; and TELEPHONE SYSTEMS.

ACOUSTICS OF AUDITORIUMS considers the effects of rooms on audible communication taking place in them. Speech and music consist of sequences of transient sounds that vary rapidly in frequency content, directivity, loudness, and acoustical form (see SOUND). Human hearing is responsive to minute and subtle variations in the received signal. Because of their complexity and subtlety, these variations may well be unmeasurable, but they are perceived as essential elements in the appreciation of an artistic performance. This places great demands on the acoustical design of auditoriums. The larger the space, the greater the difficulty in providing a satisfactory acoustical design for the performing arts. For relatively simple sources, such as a person speaking, electronic reinforcement can greatly improve communication in large auditoriums. Severe difficulties remain in the use of SOUND REINFORCEMENT SYSTEMS for large and complex sources, such as orchestras.

The negative effects of auditorium acoustics introduce distortions, echoes, or blurring to the communication between source and listener; the positive effects reinforce and enhance both the signal at the listener's position and the sense of involvement and communication between artist and audience. The positive attributes contributed by the room are actively sought in acoustical design. Concert spaces in which they are insufficiently realized are disliked by their audiences.

Acoustics of performing arts auditoriums can seldom be discussed in isolation from other architectural attributes of the space in which the performance occurs. Sightlines, comfort, visual relationships between audience and performer, and background NOISE all contribute interactively to global assessments of the merits of an auditorium.

History. The effects of enclosure on communication have been known since antiquity (Ref. 1). Vitruvius, for example, about 2 B.C., noted not only the adverse effects of interference, reverberation, and echo on speech (although he used other terms), but also the reinforcing and beneficial effects of some spaces. He in turn quoted from Greek building theoreticians who preceded him. Technical advances in masonry construction, which permitted the enclosure of large volumes during the succeeding millennium, culminating in the Gothic cathedrals of the 11th to 15th centuries, exposed both the difficulties of verbal communication in such spaces, and also the positive contribution such extreme acoustical environments made to the religious milieu in which they were built.

Not until the late 19th century and the start of the 20th century, however, was a systematic attempt made to describe the acoustical properties of auditoriums with the intention of including specifically acoustical dimensions in the design of new spaces. Wallace Clement Sabine, then a junior physics professor at Harvard University, showed that the persistence of sound in a space, which he defined as *reverberation*, could be characterized by the time taken for a decay of sound intensity to 10^{-6} (60 dB) of its original intensity. He called this the *reverberation time* and showed it to be proportional to room volume and inversely proportional to the total amount of acoustical absorption present in the space:

$$T = \frac{kV}{A}$$

where T = reverberation time
V = room volume (m^3 or ft^3)
A = total absorption (m^2 or ft^2)

The constant of proportionality, k, was found to be 0.16 for the metric, and 0.05 for the fps system. So-called optimum reverberation times were derived for different functions and room volumes.

Sabine's work was reviewed, revised, and extensively applied, both in North America and Europe, during the succeeding 50 years. It retains extensive currency in the field of auditorium acoustics today. It has, however, three major drawbacks, so it must be used with caution. The *reverberation time* (RT) measured over 60 dB of decay correlates poorly with the subjective *reverberance* in a space. The absorbing power of the audience, usually the principal absorbent surface in an auditorium, varies from space to space, thus making calculated RT unreliable in rooms of original design. The third limitation, related to the second, is that Sabine's relationship takes no account of room shape. Recent computer techniques, in which the reverberation time is calculated by ray tracing, give promise of overcoming these defects.

An attempt was made in the 1930s and 1940s to apply wave theory to auditoriums. Conceptually, it is possible to obtain solutions to the general wave equation that, with appropriate boundary conditions, would characterize any transmission path in an auditorium. Practically, however, the complexities of both the source and realistic room boundaries have thus far made this problem intractable. Recent techniques of *finite element analysis* may yet succeed in producing an adequate mathematical model of an auditorium.

Since the 1950s, there has been a marked change of emphasis in research relating to auditorium acoustics. From Sabine in 1900 to about 1950, almost all the research effort had been directed toward description and prediction of the physical properties of sound fields in auditoriums. Since then, work has been directed toward the way human binaural hearing processes the reflections of sound in a room in relation to the direct sound. The development of electronic delay devices and accurate acoustical measuring systems has permitted realistic simulations of sound fields, and thus the understanding of the contribution the several components of such fields in auditoriums make to the listener's perception of both speech and music.

Speech Auditoriums require a high level of clarity in the received sound. The sound power of the unamplified human voice is relatively small (10^{-4} W). To provide an adequate precision of syllable recognition, a signal-to-noise ratio of at least 25 dB is required (see ACOUSTICAL TERMINOLOGY AND MEASUREMENTS). These facts place constraints on the number of people who can be addressed by the unaided human voice and also demand a low level of background noise in the auditorium as a first requirement for good acoustical conditions.

Since the 1950s, it has been known that human hearing

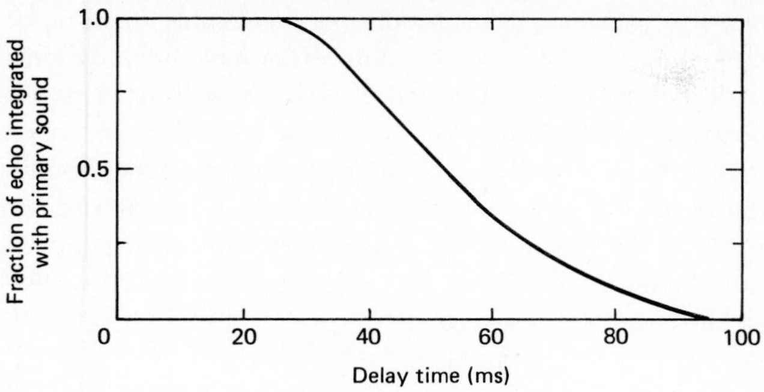

Figure 3 Fraction of reflection energy integrated with the direct sound as a function of delay. Reflection energy is equal to direct sound.

integrates the energy of early reflections of speech with the energy of the direct sound. There is a time limit on this process, however, which accounts for the perception of some late reflections as *echoes*; they are perceived by the listener as discrete acoustical events each having its own direction. Integrated reflections, on the other hand, merely alter the loudness, clarity, and timbre of the sound, all of which seem to come from the source.

Lochner and Burger (Ref. 2) made measurements of the time dependence of the echo threshold and the integration functions for speech (Fig. 3). Reflections delayed by about 35 ms after the direct sound are totally useful; after that time, progressively less energy from them is integrated, while the risk of their being audible as echoes increases.

Reverberant sound does not integrate with the direct sound. It lacks the discreteness of echoes, however, and may be considered in speech rooms as adding to the unwanted sound, as noise.

The objective measures available for speech rooms reflect these factors. Clarity is most usefully measured by the fraction of the total energy arriving at a listener position within 50 ms after the direct sound. The expression for this quantity, called *Deutlichkeit* in German, is

$$D = 10 \log \frac{\int_0^{50} p^2(t)\, dt}{\int_0^{\infty} p^2(t)\, dt}$$

where $t = 0$, the time of arrival of the direct sound
$p =$ instantaneous sound pressure

In rooms that are excellent for speech, D will have a value of -3 dB or more; that is, at least half the total energy will arrive within 50 ms of the direct sound. Such ratios seldom occur except by design. Rooms with relatively long reverberation times may still have adequate speech clarity as long as there is sufficient early reflected sound in them. In rooms without such early reflections, but with a long reverberation time, D will have a low value, and clarity will be poor. Too little reverberation is also acoustically adverse. Some reverberant feedback from the room is essential to support the person speaking.

Acoustics of Music Auditoriums is complicated by the variety of the musical signal, the complexity and spatial extent of the source on stage, and the fineness of judgments brought to bear by listeners in a music performance. Furthermore, music as an art form often extends the performer to the limits of human intellectual and motor skills. Acoustical support generated by the auditorium is essential to performers if these limits are to be realized in the performance. Performers' acoustical needs are quite different from those of the audience.

Audiences in the late 20th century have come to expect the enhancement of the musical experience provided by good auditoriums. While there has been agreement which auditoriums are best for symphonic music, it has been based on the pronouncements of perceptive individuals. The application of head-oriented stereophony and factor analysis in the early 1970s permitted statistically reliable statements about audience preference to be made after paired comparisons of a number of concert halls, and also identification of the factors (and their physical correlates) that give rise to these judgments.

There appear to be four major acoustical factors produced by an auditorium that positively enhance the musical experience for an audience. These are *loudness* (or strength), *spatial impression* (or envelopment), *clarity* (transparency), and *reverberance* (audible reverberation). Further factors proposed recently are *timbre* and *delay of the first major reflection* (as a function of the autocorrelation function of the music being heard). Negative factors, which acoustical designs seek to avoid, are echo, tonal distortion (related to timbre), unevenness throughout the seating area, unbalance between orchestral sections, and, of course, intrusive noise. The four principal factors are all functions of the (integrated) early reflected energy and/or its relationship to later arriving, unintegrated sound. It is not possible to give integration functions for music comparable with those for speech. It is, however, certain that music reflections are useful at much greater delays than those for speech. Integration times of 80 to 100 ms have been proposed.

Objective measures have been derived that correlate reasonably well with the four principal factors in audience preference. For *loudness*, the measure is sound pressure level (SPL) referred to the SPL of the direct sound at 10 m (33 ft). A tentative criterion is $+2$ dB; the range to be expected throughout an auditorium is about 5 dB, although it may be as great as 10 dB or more in poor spaces.

A number of measures have been proposed for *spatial impression*. These depend either on interaural cross-correlation measurements or on measurements of the relative amount of lateral reflected energy over the first 80 ms after the arrival of the direct sound. The latter is easier to measure and correlates reasonably with the subjective spatial impression. The expression derived by Barron and Marshall (Ref. 3) is

$$L = \frac{\int_0^{80} r \cos \theta \, dt}{\int_0^{80} r \, dt}$$

where r = arriving sound energy (including the direct sound)

θ = arrival angle relative to the axis through the listener's ears

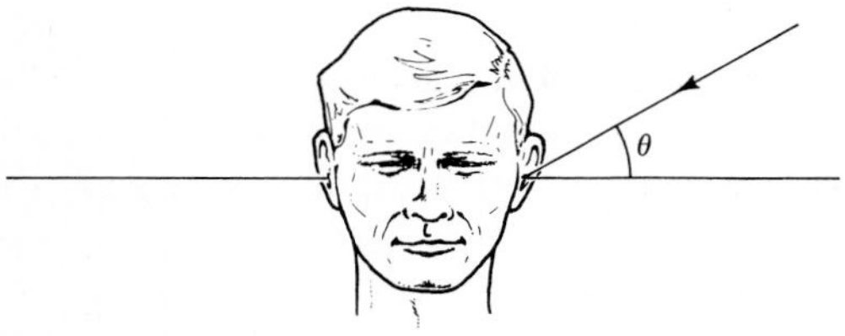

Usually expressed as a decimal fraction, a criterion for L is 0.12, with no upper limit.

Clarity for music is measured by the *Klarheitsmass*, derived by Reichardt, Alim, and Schmidt (Ref. 4). The expression is

$$C = 10 \log \frac{\int_0^{80} p^2(t) \, dt}{\int_{80}^{\infty} p^2(t) \, dt}$$

This compares the sound energy ($\propto p^2$) arriving in the first 80 ms of time (t), after and including the direct sound, with that arriving later. Satisfactory clarity is found to occur at values of C of about 0 dB and upward. The upward limit is determined by the audibility of the reverberant field.

Reverberance is measured by the reverberation time, described earlier, or one of its derivatives, such as *early decay time*. This is the time for 60 dB of decay, but with the slope measured only over the first 15 dB of decay. Atal, Schroeder, and Sessler showed that this quantity correlates better with subjective reverberance than RT (Ref. 5). All the foregoing quantities can be measured in accurate acoustical scale models during the design stages of an auditorium.

Performers' requirements of auditorium acoustics are quite different from those of the audience. Both speakers and musicians depend on feedback from the room, particularly in artistic presentations. Where it is lacking or inaudible, the quality of the performance is impaired. The acoustical conditions required for ensembles and soloists to hear themselves and each other and to enable them to interact acoustically with the auditorium have equal priority with the conditions preferred by audiences. Musicians depend on the audibility of the attack transients and other high-frequency components of instrumental sound, and require audible reflections of approximately equal strength between all players. This requirement limits the temporal range within which on-stage reflections are useful to musicians.

References

1. T. D. NORTHWOOD: *Benchmark Papers in Acoustics*, Vol. 10. Dowden, Hutchinson & Ross, New York, 1977.
2. J. P. A. LOCHNER and J. F. BURGER: The influence of reflections on auditorium acoustics. *Journal of Sound and Vibrations*, Vol. 1 (April 1964), p. 426.
3. M. F. E. BARRON and A. H. MARSHALL: Spatial impression due to early lateral reflections in concert halls: the derivation of a physical measure. *Journal of Sound and Vibrations*, Vol. 77, (July 1981), p. 211.
4. W. REICHARDT, ABDEL ALIM, and W. SCHMIDT: Definition und Messgrundlage eines objectiven Masses zur Ermittlung der Grenze zwischen brauchbarer und unbrauchbarer Durchsichtigkeit bei Musikdarbietung. *Acustica*, Vol. 32 (1975), p. 126.
5. B. S. ATAL, M. R. SCHROEDER, and G. M. SESSLER: Subjective responses to non-exponential reverberation. *Proceedings 5th ICA Liège*, Paper G32.

A.H.M.

ACOUSTIC TERMINOLOGY AND MEASUREMENTS. The acoustics of spaces is important for conversation, relaxation, work, and sleep. In most cases the effect of SOUND and NOISE on these activities depends on the *intensity*, *spectrum* (frequency content), and *duration* of the sound (Ref. 1).

Intensity, level, volume, and loudness are interchangeable terms when used loosely; however, loudness is a subjective assessment of the intensity of a sound. As intensity, in watts per square meter (W/m^2), is difficult to measure directly, it is estimated from the pressure variations in the air resulting from the passage of the sound. The root mean square (rms) value of the pressure fluctuations, expressed in pascals (Pa) or micropascals (μPa), is proportional to the intensity when there is no significant reflected sound and where the source is not close.

Because the ear is sensitive to logarithmic changes in intensity, rather than linear changes, and because the range of sound pressures is large (Table 1), the sound pressure is given as a *sound pressure level* (L_p) in *decibels* (dB). The

TABLE 1
Examples of Sound Pressures and Sound Pressure Levels

Situation	Sound Level (μPa)	Sound Pressure Level (dB re 2×10^{-5} Pa)
Jet take-off at 100 m	20,000,000	120
Discotheque/textile mill	2,000,000	100
Heavy truck at 20 m	200,000	80
Department store	20,000	60
Suburban living room	2,000	40
Recording studio	200	20
Threshold of hearing	20	0

sound pressure level or sound level or level, as it is often abbreviated, is given by

$$L_p = \frac{10 \log_{10} p^2}{p_0^2} \text{ dB}$$

where p = root mean square of the sound pressure (Pa)
p_0 = reference sound pressure, taken as 2×10^{-5} Pa

SOUND-LEVEL METERS indicate the sound pressure level in decibels. A simple sound-level reading is of little use, because it cannot be used to predict the acoustic performance of materials or spaces, nor can it be used to predict the reaction of people to sound. To do this, the sound-level meter signal must be conditioned by a frequency analysis.

Frequency Analysis is commonly undertaken with either a broadband filter or a proportional bandwidth filter. The main broadband filter used is the *A-weighting filter*, which gives the sound-level meter a frequency response similar to the human ear, and hence a reading, in dB(A), that corresponds to the loudness of the sound being measured.

Octave and *one-third octave band filters* (see also SOUND-LEVEL METER) are the proportional bandwidth filters commonly used. In conjunction with a sound-level meter, octave band filters indicate the sound level in a frequency band, which is 70% of the center frequency (Table 2). These proportional filters are more often used to determine the acoustic absorption and transmission of materials, but they can also be used to obtain a noise rating.

Duration is the third important characteristic of a sound. Some sounds are steady in level (e.g., a fan), whereas others vary relatively slowly with time (e.g., a passing vehicle); some *impulse* sounds vary rapidly with time (e.g., a hammer blow or gunshot). It is possible to measure the level of some sounds by using an appropriate sound-level meter response time (slow, fast, impulse, or peak). In other cases it is necessary to undertake a statistical analysis of the sound-level meter signal to obtain an *energy equivalent sound level*, L_{eq} dB(A); a *10 percentile level*, L_{10} dB(A); or some more complex value such as the *community noise rating* (CNR) or *sound exposure level* (L_{AX}) (Ref. 2).

L_{eq}, the most commonly used measure of the level of time-varying sound, is defined as

$$L_{eq} = \frac{1}{T} \int_0^T 10^{0.1 L_A} \, dt$$

where T = time period of interest
L_A = instantaneous sound level in dB(A)

TABLE 2
Octave Band Center Frequencies (Hz)

31.5,	63,	125,	250,	500,	1000,	2000,	4000,	8000,	16,000

Sound Power (W), unlike sound pressure level, is largely independent of the acoustical environment of the source. Like sound pressure, sound power can be quoted as a logarithmic ratio. The sound power level (L_W) in decibels is given by

$$L_W = \log \frac{W}{W_0}$$

where W = sound power in watts
W_0 = reference sound power, which is taken as 10^{-12} W (See Table 3.)

Absorption and Transmission are the two most important acoustic properties of materials. These properties, together with the size and shape of the room and the power of the source, determine how well sounds can be heard.

The ability of a material to absorb sound is given by its *absorption coefficient*, α, which is frequency dependent:

$$\alpha = 1 - \frac{I_r}{I_i}$$

where I_r = intensity of the sound reflected from a surface
I_i = intensity of the sound incident on the surface

The transmission of sound by a material is defined either by its *transmission coefficient*, τ, which, like the absorption coefficient, is frequency dependent,

TABLE 3
Examples of Sound Powers and Sound Power Levels

Source	Sound Power (W)	Sound Power Level (dB re 10^{-12} W)
Large jet aircraft	10,000	160
Large orchestra	10	130
Shouted speech	0.001	90
Conversational speech	0.00001	70
Whisper	0.000000001	30

$$\tau = \frac{I_t}{I_i}$$

where I_t = intensity of the sound transmitted through the material

or, more commonly, by the *transmission loss* (TL):

$$\text{TL} = 10 \log \frac{1}{\tau}$$

The transmission loss of a wall is often quoted as a weighted average value, for example, *sound transmission class* (STC) or *sound reduction index* (I_R). In practice, both the sound transmission class and sound reduction index are approximately equal to the arithmetic average of the one-third octave transmission losses between 100 and 4000 Hz.

Room Acoustics. The transmission coefficient or transmission loss determines the sound transmitted into or out of a space. The absorption coefficient determines the acoustic environment of a space, that is, its sound pressure level, L_p, in decibels:

$$L_p = L_W + 10 \log \left[\frac{Q}{4\pi r^2} + \frac{4(1 - \overline{\alpha})}{\overline{\alpha} s} \right]$$

where L_W = sound power (W)
Q = directivity of the source ($Q = 1$ for a source radiating omnidirectionally)
r = distance from the source (m)
$\overline{\alpha}$ = average absorption coefficient for the space
s = surface area of the space (m^2)

The free-field component of the sound field is given by the $Q/4\pi r^2$ term and the reverberant (reflected) component by the $4(1 - \overline{\alpha})/(\overline{\alpha} s)$ term.

The absorption coefficient also determines the *reverberation time* (RT) of the space, which is the time it takes the sound level to decay by 60 dB:

$$\text{RT} \simeq \frac{0.16V}{\overline{\alpha} s} \text{ seconds}$$

where V = room volume (m^3)

The main factors influencing the intelligibility of speech in a room are the reverberation time, the background noise level [expressed in dB(A), or as a noise rating (NR) or noise criterion (NC)] and the signal-to-noise ratio (the difference in decibels between the speech sound level and background noise level) (see also ACOUSTICS OF AUDITORIUMS).

The *articulation index* is an objective measure largely dependent on the signal-to-noise ratio over a number of frequency bands in the speech frequency range (Ref. 3). *Speech intelligibility*, the number of words correctly heard by subjects, can also be determined subjectively. *Sentence intelligibility* can be determined from speech intelligibility if a standard word list is used, for example, a phonetically balanced (PB) list or a rhyme list.

References

1. J. R. HASSALL and K. ZAVERI: *Acoustic Noise Measurement*, 4th edition. Bruel and Kjaer, Copenhagen, 1979.
2. T. J. SCHULTZ: *Community Noise Rating*, 2nd edition. Applied Science Publishers, London, 1982.
3. H. KUTTRUFF: *Room Acoustics*, 2nd edition. Applied Science Publishers, London 1979.

F.F.

ADHESIVES are substances used to bond the surfaces of materials together. They obtain their grab to a surface either through a chemical or electrostatic attraction between adhesive and substrate or by filling the gaps in the rough surface of the substrate for mechanical adhesion.

Adhesive bonding has many advantages to offer the construction industry. No other method of attachment is satisfactory for many applications. Examples would be the installation of CERAMIC WALL TILE and fastening the various plies together into a sheet of PLYWOOD.

Adhesives may also be called glues, pastes, mucilages, cements, or resins. They are generally furnished to the job in powder, liquid, or paste form. Some physical or chemical change is necessary to convert the adhesive to a solid. This change may be a hydration, as in a paste of PORTLAND CEMENT and water; it may be a polymerization; or it may be the simple evaporation of volatile materials, as in the various solvent-based adhesives. Other methods of cure are used for specific applications.

An adhesive is the only method of furnishing a continuous bonding plane between two parts that are to be joined. Bolts, nails, screws, and rivets all furnish attachment only at discrete points and consequently cause stress concentrations. Even welding furnishes only edge attachment.

Also, with the use of discrete fasteners, the great majority of all connections are some variation of a lap joint. Usually, only with welding is the butt joint connection possible, and this of course is only valid with metal substrates for structural uses. The adhesive, however, furnishes a continuous bond plane, which gives a more nearly uniform stress distribution.

History. Adhesives have been in use since the time of the early Egyptians and Assyrians. The Egyptians used adhesives for the bonding of papyrus, and pieces of veneered furniture have been found in the tombs of some of the Pharaohs. These adhesives were undoubtedly of animal origin. The Assyrians used a clay-based mortar to bond mud bricks together. The ancient Chinese and Japanese also used clay-based mortars in structural work and animal glues in their decorative building work. The Assyrians also discovered a natural bitumen near the Euphrates River and used this pitch as a cementing material.

In the Western world, mortars, bitumens, and starches have been used freely, but growth in the use of adhesives up to World War II had been mostly the increased use of animal and fish glues.

The history of the synthetic resins can be traced to the work of Faraday on natural rubber in 1826. An important step was the work of Bakeland in 1909 with phenol and formaldehyde. During World War I the Germans, from necessity, increased their research into synthetic rubber and gained a great deal of plastics technology. World War II forced the United States to develop a similar capability. Shortly after the war, the growth in the use of adhesives paralleled the growth of the petrochemical industry. Epoxy resins, although known before, began their real growth about 1960, as did urethane adhesives. Both thermoplastics and thermosets have shown a growth that roughly parallels the growth of plastics technology. The use of adhesives in the construction industry has grown more rapidly than the construction industry as a whole over the last 15 years.

Theories of Adhesion. Several different theories of adhesion are widely known and respected. Some of these are the theory of true chemical union, Zisman's theory of critical surface polarity, Bikerman's weak boundary layer theory, and the theory of fracture mechanics. Almost all the theories, with the exception of fracture mechanics, depend to some degree on a chemical or electrostatic union.

Almost all theories agree that the adhesive must wet the substrate to form a good union. Another point of agreement is glue line thickness. The joint must contain enough adhesive to wet both substrates. Any excess thickness may not only be wasteful, but may actually weaken the joint. This can easily be demonstrated by placing a drop of lubricating oil between two glass microscope slides and pressing them by hand into intimate contact. This thin liquid line forms a connection with fairly good tensile strength. It is obvious that a thicker glue line would weaken the joint.

Types of Adhesives. Animal and vegetable glues include the animal and fish glues, blood-albumin, casein, and the starch pastes. These are low- to moderate-strength adhesives (less than 7000 kPa or 1000 psi). They are used for installation of wallpaper and the manufacture of cheaper grades of plywood. They represent a very small percentage of the construction market.

The synthetic rubbers include both the solvent- and emulsion-based elastomers. They may be extended with fillers and furnished to the job as mastics for caulking gun or trowel application. The solvent-based neoprenes may be used as contact adhesives. Strengths of the materials in this group are variable. Some are well in excess of 7000 kPa (1000 psi). They constitute about 5% of the market.

Thermosetting adhesives include the ureas, melamines, epoxies, polyesters, and others. Some are polymerized by heat and pressure, whereas others require the addition of a catalyst, or curing agent, to effect a cure. These materials have excellent durability and high strength. They are used for structural and load-bearing applications. They account for about 70% of the market.

The ''white glues'' used in construction are formulated from polyvinyl acetate (PVA). They have moderate shear strength with wood substrates, but poor water immersion resistance. A companion product, polyvinyl butyral, has better immersion resistance.

The versatile thermoplastic ''hot melts'' occupy a small but rapidly growing segment of the market. They bond successfully to concrete, steel, wood, and structural plastic.

Other adhesives, such as pressure-sensitive tapes and the anaerobics, can be used for metal-to-metal bonding.

Testing of Adhesives. Adhesives are tested for tension, shear, peel, and cleavage, and sometimes for impact strength and creep. Some tests depend on the type of substrate to be used. For instance, wood shear specimens may be tested for both end grain shear and side grain shear. Of the different types of tests, the shear test is the most widely used and respected.

Testing procedures naturally vary to meet specification requirements. At least 50 specifications for adhesives are applicable to the construction industry. Fortunately, in the construction industry, virtually no specifications cross building trades. For example, wood specifications are determined by one group.

The tests for shear, tension, peel, and cleavage are relatively straightforward, but creep and rupture tests have not received the same attention. The creep test gives a plot of deformation versus time under constant load. The stress rupture test determines how much load a specimen can carry for an indefinite length of time.

Recent Research. Recent research has concentrated on the mechanics of the glued joint and on product development research. Intermediate between these two has been the work of organizations on systems, such as the glued wood–plywood floor system of the American Plywood Association and the work of the British Transport and Road Research Laboratory (TRRL).

Much of the work on the mechanics of the joint has been done at various academic institutions and at the Forest Products Laboratory of the U.S. Department of Agriculture. Much of this research has been published in the journals of the American Chemical Society and the bulletins of the Forest Products Laboratory. The work of Dietz, Bikerman, Zisman, Hata, Bergin, and many others has been notable.

Naturally, the bulk of product development research is carried on in-house by the various chemical companies. The results of this work show up in better products for the industry, but results are considered as proprietary information and are seldom published.

Adhesive-bonded composite members and the strengthening of concrete beams by externally bonded plates are being investigated at several universities and at TRRL. The greatest volume of adhesive research concerns itself with the development of stronger, more versatile synthetic resins. Although a host of synthetics are being studied, probably the most intensive work is being done on epoxies and urethanes.

References

1. J. J. BIKERMAN: Strength of Adhesive Joints. *American Chemical Society, Division of Organic Coatings and Plastics Chemistry, Paper V, No. 2, for Washington, D.C. Meeting, 12–17 Sep. 1971*, pp. 156–159.
2. J. P. COOK: *Construction Sealants and Adhesives*. Wiley, New York, 1970.
3. *Evaluation of Adhesives for Building Construction*. Forest Products Laboratory Report No. 172, Madison, Wis., 1972.
4. W. A. ZISMAN (ed., P. WEISS): *Adhesion and Cohesion*. Elsevier, Amsterdam, 1962.
5. M. D. MacDONALD: *The Flexural Performance of 3.5 m Concrete Beams with Various Bonded External Reinforcements*. TRRL Supplementary Report 728. Transport and Road Research Laboratory, Crawthorne, England, 1981.
6. D. J. McKEE and J. P. COOK: *A Study of Adhesive Bonded Concrete–Metal Deck Slabs*. Transportation Research Record No. 762. Transportation Research Board, Washington, D.C., 1980.

J.P.C.

ADOBE, PISÉ, AND PRESSED EARTH BLOCKS.

Earth has been used in numerous ways as a building material. Some of these date back into prehistory and some have been more recent innovations. Adobe, pisé, and pressed earth blocks are the most popular techniques in use today.

In the search for other materials to suit expanding requirements, people learned to hew stone and bake bricks; recently, manufactured materials have been used in home building. However, with the cost of labor continually rising, the shortage of building materials, and an increased concern for the environment, more people are turning again to earth as a domestic building material; and with the greater interest in the retention of our heritage, a larger number of old earth buildings are being considered for renovation (Ref. 1).

History. Earth is one of the oldest building materials. Historically, the first reference to its use dates from the time of Hannibal, during the second Punic War, when it was used to build watch towers. Investigators in recent years have vouched for the existence of these soil structures, twenty centuries old. Earth building has been carried out in almost every country of the world, and in many third world countries earth building in one form or another is still the major method of building domestic structures. Throughout the world, more people are housed in buildings constructed of earth than of all other materials.

Adobe is the most popular of techniques, due to its suitability to more varied circumstances. Adobe is mud or puddled earth, which is molded into blocks and then dried. Most earth, other than top soil, may be used for adobe. There must be sufficient clay to bind the particles together, but not too much so that cracking becomes a problem. The upper and lower limits of clay vary, but generally 25% clay is the minimum necessary to bind the soil together. Maximum percentages vary much more and are more affected by workability, but tend to be in the region of 70% to 80% clay.

The blocks are usually molded in steel or wooden molds, which vary in size depending on the wall thickness and method of bonding. Water is added to the earth until the resulting mix is able to stand up under its own weight, once the mold is removed, straight after forming the blocks. The blocks are left to dry flat until they are firm enough to be turned on their side to allow even drying. Some building codes require waterproofing additives, but these are not necessary if good design and constructional practices are followed. Building with adobe is similar to building with other masonry.

Pisé is an abbreviation of the French pisé-de-terre, meaning rammed earth. In the pisé method, moist earth is rammed between temporary formwork in the final wall position. Because of the large monolithic area of earth, shrinkage must be kept to a minimum. The soil range therefore is much more limited than with adobe, and is between 30% and 50% clay. The forms are moved along the wall after each section has been rammed to form courses that vary in height from 300 to 900 mm (12 to 36 in.). Corners and partitions are usually rammed first, followed by the sections in between. Openings are created by inserting bulkheads in the forms at appropriate positions.

Pressed Earth Blocks. Regular, sharp-edged blocks can be made using a manually operated or automatic dry press. Due to the nature of the compacting process, resulting in inadequate bonding between particles, pressed earth blocks often exhibit poor durability characteristics. Additives, such as cement, are therefore needed in most cases. These should be mixed dry to ensure even distribution. As the blocks need to be removed from the compacting device when wet, a minimum of 35% clay is required to hold the blocks together; 50% is usually the highest clay content.

Blocks are usually made in a proprietary compacting device; the CINVA ram is the most popular. The blocks need to be kept damp for a three- or four-day curing period and should not be stacked on top of each other during this time.

See also BRICKWORK AND BLOCKWORK.

References

1. B. WOODWARD: Restoration and Preventative Maintenance of Earth Walls. *Sixth Masonry Symposium*. NSW Institute of Technology, Sydney, 1982.
2. H. FATHY: *Architecture for the Poor*. University of Chicago Press, Chicago, 1973.
3. *Earth Wall Construction*. Bulletin No. 5, Experimental Building Station. Australian Government Publishing Service, Canberra, 1982.
4. B. WOODWARD and R. MITCHELL: *Mudbrick Notes*. Earthways, Wollombi, Australia, 1980.
5. *Soil Cement—Its Use in Buildings*. United Nations, New York, 1964.

B.W.

AIR CONDITIONING AND HEATING. Air conditioning is essentially a process for the treatment of air. Its objective is to deliver treated air to satisfy the temperature, humidity, and cleanliness requirements of the conditioned space. Heating, as it generally applies to a building, is a process of utilizing heat to raise the temperature of the occupied space to comfort level.

Since the introduction of air conditioning and heating into buildings, the various systems that have been developed to satisfy environmental needs have grown in number, as well as in complexity. Two major dimensions have been recently added to the ever-growing technology of air conditioning and heating systems:

1. The need for designs that are responsive to the diminishing supply and escalating cost of fossil fuels.
2. The increasing demand to provide safer building environments.

This article discusses the approach to the various problems associated with the design process.

Conceptual Phase. The design process usually begins with a conceptual phase, during which the air-conditioning components of a building are evaluated against stated or implied goals, such as the following:

1. Cost
2. Initial and future occupancy needs
3. Architectural, structural, and acoustical restraints and objectives
4. Internal and external environmental requirements
5. Applicable seismic restraints
6. Energy consumption
7. Operating cost
8. Smoke and fire management

Determination of Loads. The design of any system requires a determination of cooling and heating loads. The most significant reference sources for air conditioning and heating are the publications of the American Society of Heating, Refrigerating, and Air-Conditioning Engineers (ASHRAE), specifically, its *Handbook of Fundamentals* (Ref. 1), *Handbook and Product Directory* (Refs. 2, 3, and 4), and energy conservation standards (Refs. 5, 6, and 7).

The factors that should be considered in the determination of *heating loads* are as follows:

1. Transmission losses (exterior walls and fenestrations, doors, skylights, and roofs)
2. Ventilation loads (fresh air)
3. Infiltration losses (windows, doors, and walls)
4. Process requirements (appliances and humidification)
5. Domestic hot-water loads

The factors that should be considered in the determination of *cooling loads* are as follows:

1. Sensible heat gains
   - **(a)** People
   - **(b)** Electrical (lights and power)
   - **(c)** Transmission (walls, windows, roofs)
   - **(d)** Solar (direct and indirect)
   - **(e)** Outside and infiltration air
   - **(f)** Miscellaneous (computers and special requirements)
   - **(g)** Fan heat
2. Latent heat gains (people and outside air)

Air-conditioning Systems. Two major techniques are employed to condition air whenever refrigeration is required:

1. A mixture of fresh air and recirculated air is passed over coils through which refrigerant gas is circulated (Fig. 4).

2. A mixture of fresh air and recirculated air is passed over coils through which chilled water is circulated (Fig. 5).

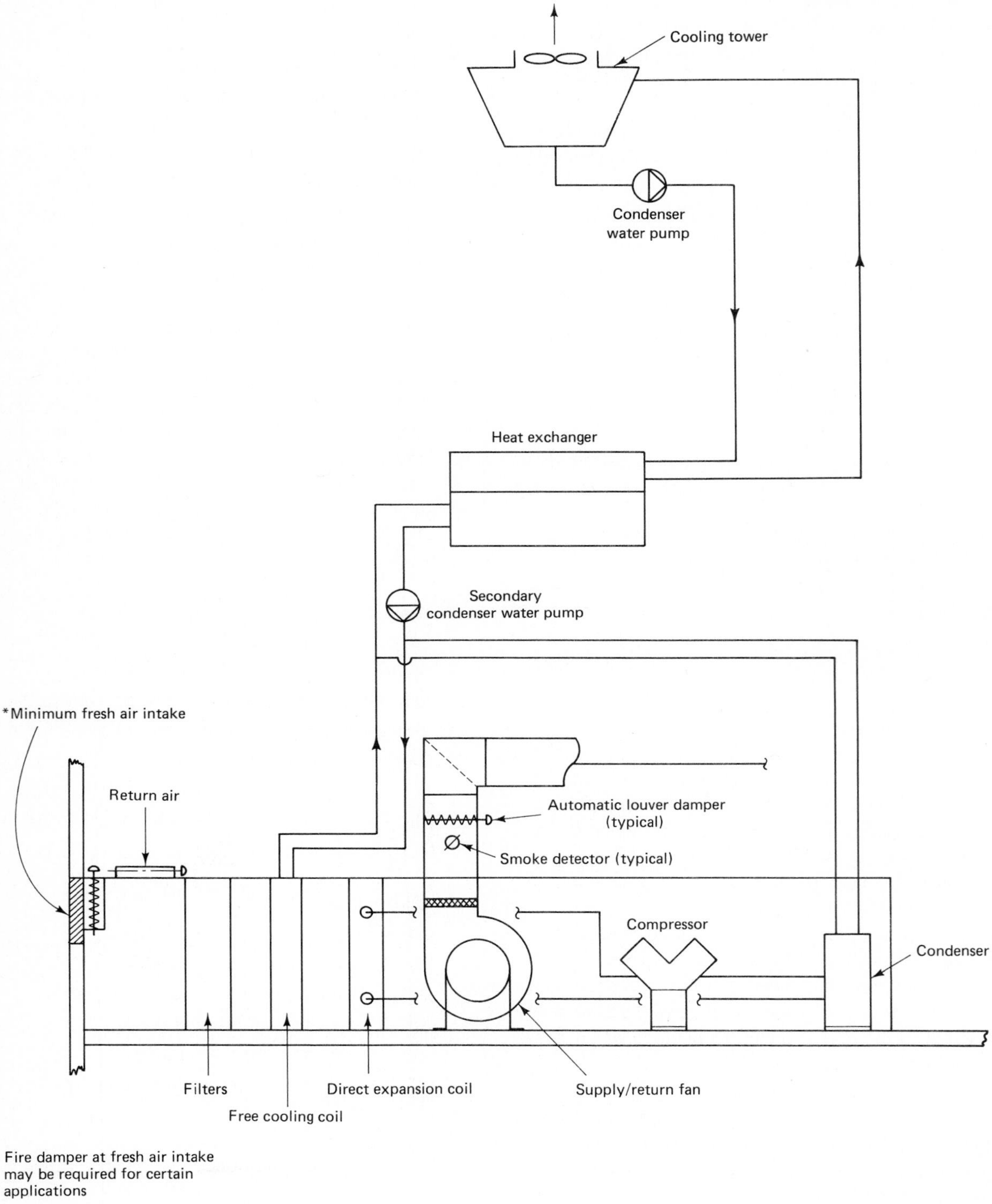

Figure 4 Elementary diagram for a local fan system. (For simplicity, detailed controllers and devices are not indicated.) *The quantity of fresh air drawn into the system is expelled through the toilet exhaust and building pressurization.

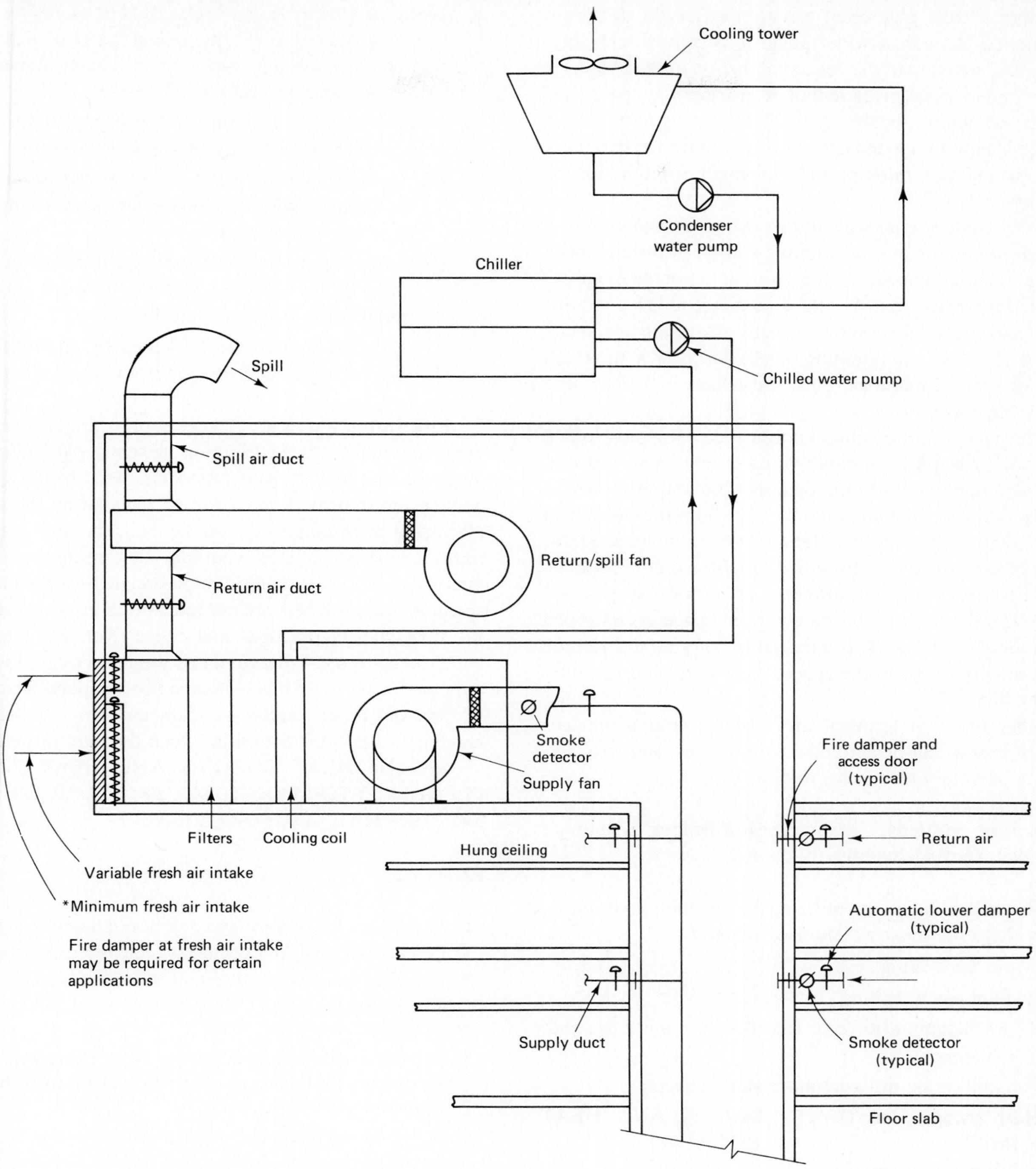

Figure 5 Elementary diagram of a central fan system. (For simplicity, detailed controllers and devices are not indicated.)

In appropriate climates, varying quantities of fresh air are mixed with recirculated air for a part of the year to ''manufacture'' conditioned air without the use of mechanical refrigeration. In certain applications, outdoor air may be utilized to chill the water and achieve free cooling without the use of mechanical refrigeration.

There are two principal types of air-handling systems for distributing the conditioned air to the occupied spaces. The *central station* variety serves more than one floor. The *local type* serves, in most cases, the floor or a portion of the floor upon which the equipment is located. Either type may be factory assembled or built in place.

Either system type may, through ductwork, distribute conditioned air with a wide range of velocities and pressures. The selection of the best combination of duct velocity and pressure for the given length of distribution, space limitations, economic factors, and the choice of terminal air devices is based on extensive study and design experience. No hard and fast rules provide a proper solution for all buildings.

Either system type may utilize *draw-through* or *blow-through* fan arrangements. Figures 4 and 5 illustrate draw-through configurations. A blow-through arrangement requires more space, but in cold climates it reduces stratification and produces a better mixture of outside and recirculated air. The configuration is chosen after a thorough study of space conditions, cost, local labor practices, and quality objectives.

The type of air terminal chosen today for most office occupancies employs variable-volume devices in order to conserve energy and to minimize distribution costs. These usually require medium-velocity, medium-pressure duct distribution for the supply air fans to variable-volume boxes. Low-pressure ductwork from the variable-volume boxes is connected to ceiling air diffusers of various shapes and types. The temperature of the occupied space is controlled by a room thermostat. It is adjusted by varying the volume of the supply of air to the space from its related variable-volume box.

Other types of terminal are constant-volume devices without reheat, fan-powered induction boxes, and constant-volume devices with reheat, to cite a few.

Heating Systems. Many types of heating system are available. They include the following:

1. Hot-water radiation, utilizing finned tube convectors, perhaps the most widely used method
2. Warm air heating
3. Radiant panel heating
4. Steam radiation utilizing finned tube convectors and/or radiators
5. Fan coil units, utilizing hot water or steam
6. Heat pumps (see HEAT ENGINES AND HEAT PUMPS)
7. Space heaters utilizing hot water or steam

A typical hot-water heating system may include the following components:

1. *Heat source:* On-site boilers that generate hot water or steam or purchased district steam generated by a public utility. The boilers may be gas-fired, oil-fired, or employ electric heating elements.
2. *Other equipment and distribution:* Hot water may be produced directly from a boiler plant or by adding steam to water in a heat exchanger. Hot water is distributed with pumps through pipes to hot-water finned-tube convectors located under windows.
3. *Controls:* The temperature of the occupied space can be controlled in a variety of ways, for example:
   - **(a)** The temperature of the hot water can be varied in accordance with the outside temperature by exposure to the outside air.
   - **(b)** Local thermostats can vary the quantity of water flowing through the convectors through automatic modulating valves within the piping.

   Either technique can be varied by controlling the flow of water and the overhead air distribution.

Life Safety Factors. The control of fire and smoke is very important. The distribution systems, particularly the ductwork and its various components, must be designed to prevent the transmission of fire and smoke by the correct placement of fire dampers, smoke dampers and smoke detectors, fresh-air intakes, and spill air terminations. Their interaction with fire- and/or smoke-related events must respond to standards that are not totally definitive and to local fire management practices and codes (Ref. 8). Figures 4 and 5 depict typical placement of these devices. However, with the exception of heat-actuated fire dampers, the control of fans and related automatic dampers varies widely, depending on the jurisdiction in which they are installed.

See also HEAT ENGINES AND HEAT PUMPS; PLENUM SYSTEMS AND DUCTS; SOLAR ENERGY; and THERMAL COMFORT CRITERIA.

References

1. *Handbook of Fundamentals*. American Society for Heating, Refrigerating, and Air Conditioning Engineers, Atlanta, 1981.
2. *Handbook and Product Directory, 1974, Applications*. American Society for Heating, Refrigerating, and Air Conditioning Engineers, New York, 1974.
3. *Handbook and Product Directory, 1975, Equipment*. American Society for Heating, Refrigerating, and Air Conditioning Engineers, New York, 1975.
4. *Handbook and Product Directory, 1976, Systems*. American Society for Heating, Refrigerating, and Air Conditioning Engineers, New York, 1976.
5. *Energy Conservation in New Building Design. Standard 90-75*. American Society for Heating, Refrigerating, and Air Conditioning Engineers, New York, 1975.
6. *Thermal Environmental Conditions for Human Occupancy. Standard 55-74*. American Society for Heating, Refrigerating, and Air Conditioning Engineers, New York, 1974.
7. *Standards for Natural and Mechanical Ventilation. Standard 62-73*. American Society for Heating, Refrigerating, and Air Conditioning Engineers, New York, 1973.

8. *Standard for the Installation of Air Conditioning and Ventilating Systems, Standard NFPA-90A*. National Fire Protection Association, Quincy, Mass., 1981.

9. *Mechanical and Service Systems*. Chapter SC-2 in Volume SC of *Monograph on Planning and Design of Tall Buildings*. American Society of Civil Engineers, New York, 1980.

R.E.M.

ALUMINUM's alloys are light, strong, and durable under severe service conditions and are also capable of being wrought, cast, or extruded into a wide range of useful forms. Aluminum may be joined by bolting, welding, riveting, and adhesive bonding. Although the material is frequently used with a bright, smooth, and natural surface that possesses a high reflectivity, low emissivity, and high thermal conductivity, it may also be subjected to a wide variety of mechanical surface finishing procedures, such as polishing, embossing, sandblasting, or wire brushing. Many colors are available in chemical, electrochemical, and paint finishes.

Aluminum and its alloys rely for surface protection on the existence of an ultrathin oxide film about 1×10^{-5} mm in thickness, which forms instantaneously on first exposure to the atmosphere. Atmospheres or substances that degrade this film are therefore to be avoided if possible (Fig. 6). Although resistance to oxidizing acids is excellent, aluminum should not be exposed to strong alkaline media, particularly cement or concrete in the setting stage.

The thickness of the oxide film can be increased and the corrosion resistance improved by an electrolytic process known as *anodizing* (see CORROSION OF METALS). Color can be added during the anodizing process, but mineral pigments should be used externally, as organic pigments fade in sunlight.

Figure 6 Deterioration of aluminum frame for screen door. [Reproduced from *Metals Australia*, Vol. 15, No. 5 (1983), by permission of the publishers.]

TABLE 1
Wrought Alloys

Code	Major Alloying Element	Example
1XXX	Essentially pure aluminum	1060 (99.6% Al)
2XXX	Cu (two phase)	2014 (4.5% Cu)
3XXX	Mn (one phase)	3003 (1.3% Mn)
4XXX	Si (two phase)	4032 (12.5% Si)
5XXX	Mg (one phase)	5050 (1.2% Mg)
6XXX	Mg and Si (two phase)	6063 (0.4% Si, 0.72% Mg)
7XXX	Zn (two phase)	7075 (5.6% Zn)

Alloy and Temper Designations. All major aluminum producers have adopted a common code for designating aluminum and its alloys; this code is incorporated within most national standards. Where a particular country for some reason varies the code in question, the numeral identifier is generally followed by a letter identifying the source of the variation. Aluminum and its alloys are characterized by a four-numeral code to identify wrought alloys (Table 1) and a three-numeral code if the material is a cast alloy (Table 2).

The wrought alloys are divided into two general classifications, single and multiphase alloys. Whereas the mechanical properties of the single-phase group can be varied only by a combination of cold working and annealing, the multiphase alloys also respond to precipitation hardening heat treatment, for which the main alloying elements are copper, manganese, silicon, magnesium, zinc, nickel, and tin (see METALLURGY).

A temper designation system is used for all forms of wrought aluminum and its alloys, and with some restrictions for casting alloys as well. A letter based on the sequences of the basic treatments follows the alloy designation, separated by a dash; subdivisions are indicated by one or more digits following the letter. Designations commencing with the letter H, for example, refer to various forms of strain hardening (cold working) treatments, as follows:

H1X, cold worked. The higher the value of X, the greater the degree of strain hardening.

TABLE 2
Casting Alloys

Code	System
1XX	Al–Cu
2XX	Al–Mg
3XX	Al–Si
4XX	Al–Mn
5XX	Al–Ni
6XX	Al–Zn
7XX	Al–Sn

H2X, cold worked and annealed, with the value of X increasing as above.

H3X, cold worked and stabilized. While X still represents the degree of strain hardening, the stabilization treatment indicates that the alloy has been heated to 30° to 50°C above the maximum service temperature.

Specifications with the letter T refer to alloys that have been subjected to an age-hardening treatment, that is, hardening in air at room temperature. The specific treatments are applied only to the 2XXX, 6XXX, and 7XXX alloy series. Some examples follow:

T1, cooled from the casting process and naturally aged to a substantially stable condition.

T3, solution treated, followed by strain hardening and then natural aging at room temperature; for wrought alloys only.

T6, solution treated followed by artificial (higher temperature) aging; for wrought alloys only.

Building and Construction. The several unique physicochemical and mechanical properties of aluminum and its alloys have resulted in the widespread use of these materials in buildings (Fig. 7) and in associated construction. Aluminum extrusions, frequently anodized to confer both color and protection, are used extensively for windows, door frames, and railings (Fig. 8). Many composite products based on aluminum are now being utilized to construct fire-resistant and noise-reducing panels, and the use of roll-formed products in ROOF SURFACES and siding is expanding rapidly.

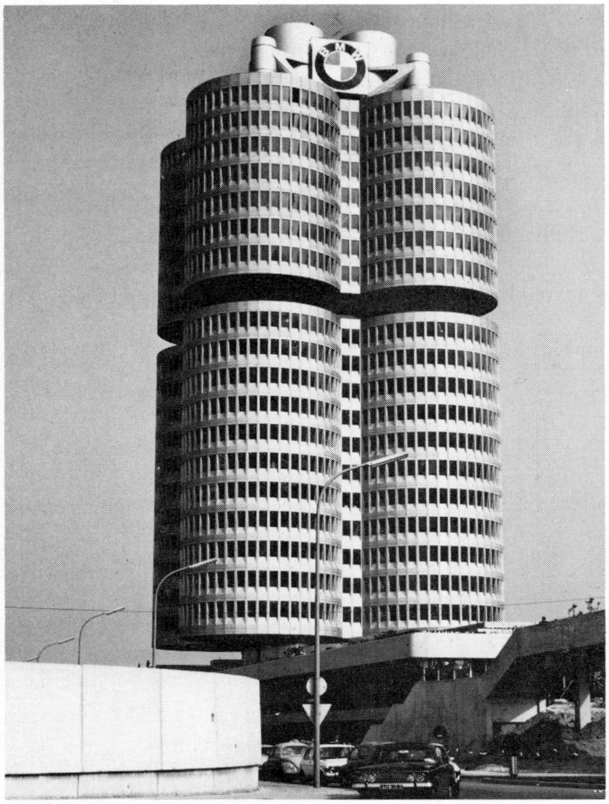

Figure 7 The BMW Building, Munich, West Germany, which has an aluminum curtain wall. (Courtesy of Bayrische Motoren Werke, Munich.)

Figure 8 Anodized aluminum railings. [Reproduced from *Metals Australia*, Vol. 15, No. 5 (1983), by permission of the publishers.]

See also METALLURGY and NONFERROUS METALS.

References

1. Aluminum and Aluminum Alloys. *Metals Handbook*, 8th ed. American Society of Metals, Metals Park, Ohio, 1961.
2. C. A. PAMPILLO et al.: *Aluminum Transformation Technology and Applications*. American Society for Metals, Metals Park, Ohio, 1978.

A.E.J.

ANCIENT ROMAN CONCRETE was the dominant material for monumental buildings in Rome and much of its empire from the 1st to the 4th century A.D. Although the finished product resembles modern CONCRETE, its construction was different. The Latin *caementum* means aggregate (not cement), and the *caementa* were large pieces of rock or debris from demolished buildings placed on top of one another. A mortar of sand and cement was then poured into the interstices.

The formwork, particularly on the outside of the building, generally consisted of brickwork or masonry. Arches were used to strengthen this permanent formwork in multistory construction, and horizontal brick courses were

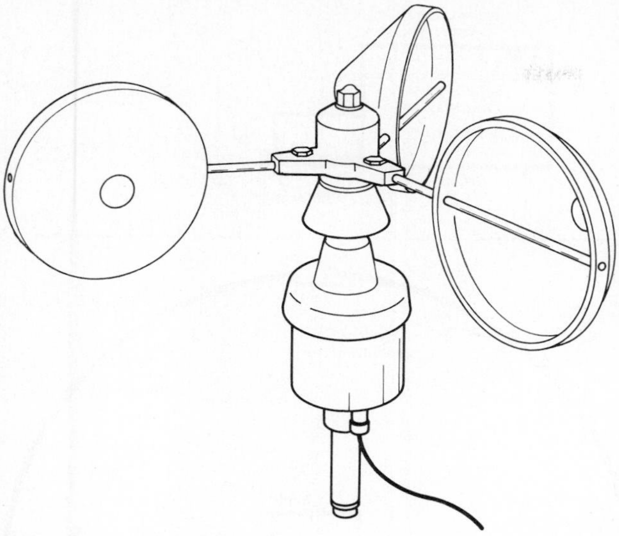

Figure 9 Modern cup anemometer. (Courtesy of Applied Science Publishers.)

sometimes placed in the concrete while it was still wet to act as ties.

For the best Roman concrete, natural cements of volcanic origin were used. Volcanic materials with a chemical composition resembling that of modern PORTLAND CEMENT are found in several parts of Italy. When these were not available, the Romans used a lime–sand mortar to which crushed bricks or tiles were added.

References

1. MARION ELIZABETH BLAKE: *Roman Construction in Italy from Tiberius to the Flavians.* Carnegie Institute Publication 616, Washington, D.C., 1959.
2. M. E. BLAKE: *Italian Construction in Italy from Nerva to the Antonines.* American Philosophical Society, Memoirs, Vol. 96, Philadelphia, 1973.
3. H. J. COWAN: *The Masterbuilders.* Wiley, New York, 1977.

H.J.C.

ANEMOMETERS are instruments for measuring the velocity of a gas flow or, more specifically in architecture, for measuring the speed of the wind or an internal air flow. The measurement is made using a combination of mechanical, fluid, and electrical methods.

Mechanical Anemometers of the cup or propellor type rotate at a speed proportional to wind speed (with some nonlinearity at low speeds), and this rotational speed can be converted into a frequency or amplitude-modulated electrical signal for transmission and recording purposes. A cup anemometer is shown in Fig. 9.

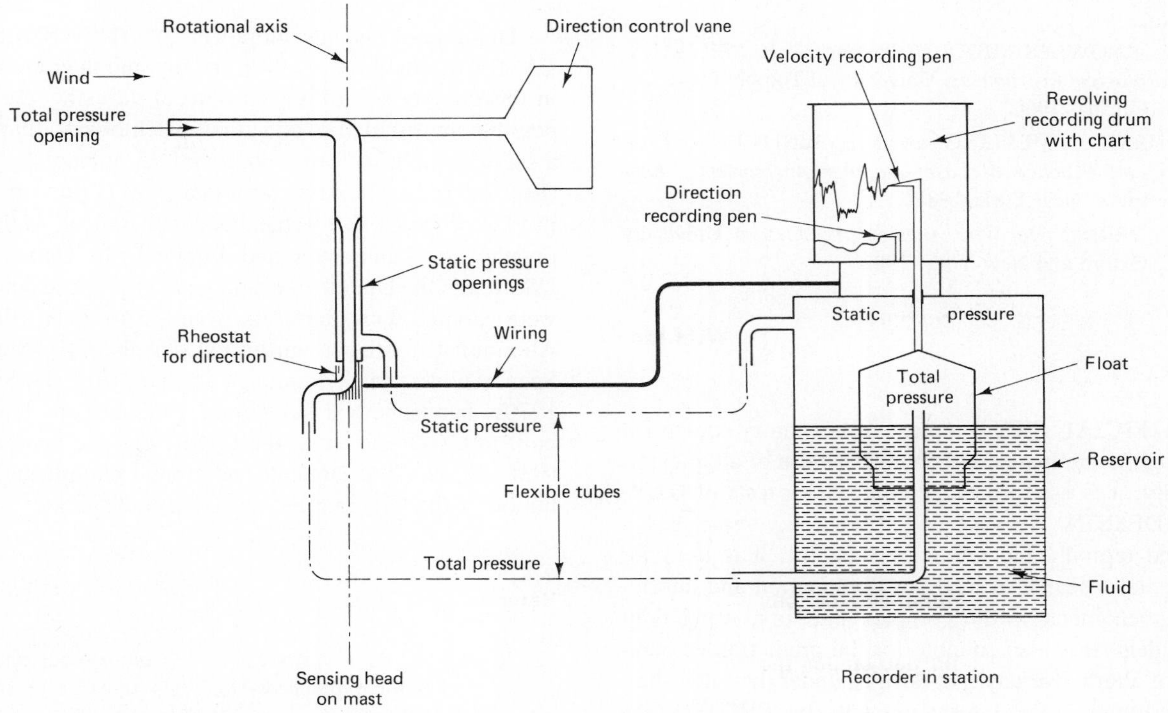

Figure 10 Principal component of Dines pressure tube anemometer. (Courtesy of Applied Science Publishers.)

Fluid Anemometers use the principle of measuring the dynamic pressure that is proportional to velocity squared. The dynamic pressure can be obtained from a pitot-static tube arrangement whereby the dynamic pressure is obtained from the difference between total pressure (from an open tube facing directly into the wind) and static pressure (from a pressure tapping on a streamwise surface). The dynamic pressure can be read directly on a manometer or converted to an electrical signal by a pressure transducer. A Dines anemometer using this principle is shown diagramatically in Fig. 10.

More sophisticated anemometers, with very high frequency response, are based on hot wire or Doppler techniques. *Hot Wire Anemometers* use the dependence of the cooling rate of a heated wire on the velocity of a surrounding fluid to obtain an electrical signal from a bridge circuit that is a function of velocity. Frequency response in excess of 10 kHz can be obtained from a 5-μm-diameter tungsten wire. *Laser–Doppler* and *Sonic Anemometers* are nonintrusive instruments that determine fluid velocity from the Doppler frequency shift of light scattered by solid contaminants in the flow or an acoustic signal, respectively.

See also WIND TUNNELS.

References

1. E. OWER and R. C. PANKHURST: *The Measurement of Air Flow.* Pergamon Press, Oxford and Elmsford, N.Y., 1977.
2. W. E. KNOWLES MIDDLETON and A. F. SPILHAUS: *Meteorological Instruments.* University of Toronto Press, Toronto, Canada, 1965.
3. F. DURST, A. MELLING, and J. H. WHITELAW: *Principles and Practice of Laser–Doppler Anemometry.* Academic Press, New York, 1981.
4. A. E. PERRY: *Hot-Wire Anemometry.* Oxford University Press, Oxford and New York, 1982.

W.H.Me.

An **ARTIFICIAL SKY** is a piece of laboratory equipment simulating the luminance distribution pattern of a typical or design sky; it is used for model analysis or tests of DAYLIGHT DESIGN. Because an artificial sky represents a stable and reproducible reference standard, it is used for experimental research of daylight illumination and interreflection phenomena within complex interiors, using their scale models. It is also suitable for the prediction or comparison of alternative daylight designs under the same standard conditions. Model measurements by PHOTOVOLTAIC CELLS are also used for analyzing the influence of different GLAZING, SUN SHADING devices, or FENESTRATION of complex irregular shapes, for which exact mathematical solutions are not available.

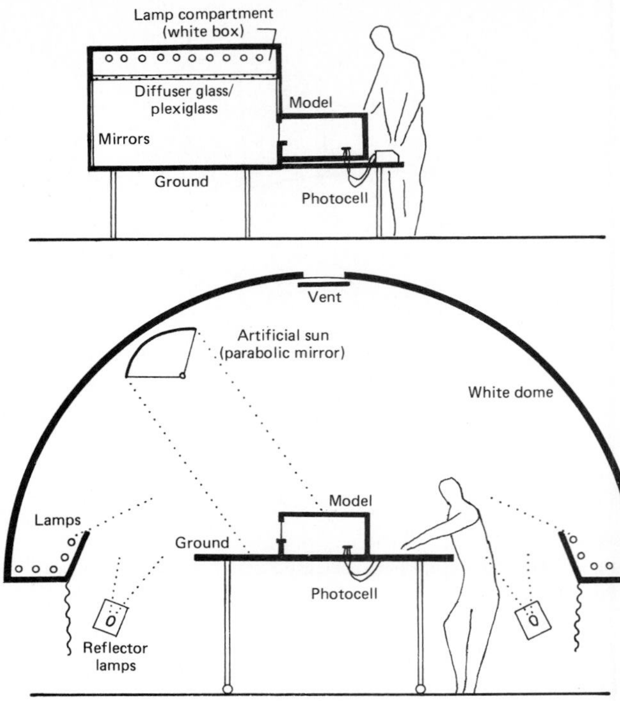

Figure 11 Comparison of the box-type mirror artificial sky with the sophisticated dome type.

History. The development of INCANDESCENT LAMPS during the early 20th century and their encasement in reflective boxes with a translucent diffusing glass made possible the simulation of a small patch of sky illuminating a staircase light well or a courtyard set among skyscrapers (Ref. 1). Larger box-type artificial skies (from 1 m $\times$ 1 m in plan to room-sized chambers with mirror walls) were popular in Scandinavia and England. In Germany, the USSR, and the United States, large multipurpose dome skies were adopted with diameters from 4.5 to 18 m (Fig. 11). All simulated either a uniform sky or the CIE (Moon and Spencer) overcast sky luminance pattern (Ref. 1). Since the 1970s, hemispherical skies with artificial suns have been built in Czechoslovakia, the USSR, and the United States (Refs. 3 and 4) to simulate the complex luminance distribution of the CIE (Kittler and Gusev) clear skies.

References

1. R. KITTLER: A historical review of methods and instrumentation for experimental daylight research by means of models and artificial skies. *Proceedings of the C.I.E. XIVth Session*, Vol. B, pp. 319–334. Publication No. 5, Commission Internationale de l'Éclairage, Paris, 1960.

2. J. LONGMORE: The role of models and artificial skies in daylight design. *Transactions of the Illuminating Engineering Society* (London), Vol. 27 (1962), pp. 121–138.
3. R. KITTLER: A new artificial ''overcast and clear'' sky with an artificial sun for daylight research. *Lighting Research and Technology*, Vol. 6 (1974), pp. 227–229.
4. S. SELKOWITZ: A hemispherical sky simulator for daylighting model studies. *Proceedings of the VI National Passive Solar Conference*, pp. 850–854. American Division of International Solar Energy Society, University of Delaware, Newark, Delaware, 1981.

R.K.

ASBESTOS IN BUILDINGS. Asbestos is a generic term used to describe a number of hydrated crystalline silicate minerals with unique properties such as thin fiber shape, strength, and heat stability. Types with most commercial importance are chrysotile, crocidolite, and amosite, with chrysotile constituting over 90% of production throughout this century.

Health Hazards. All types are considered to be hazardous to health if they become airborne dusts that are inhaled into the lungs. Asbestosis and lung cancer are the major hazards recognized from long-term exposure to high dust concentrations and recommended hygiene standards are set at levels to minimize the risk of such hazards. A further hazard is mesothelioma, a fatal and usually very rare tumor, which can occur many years after short-term exposures insufficient to cause other lung damage. Since crocidolite (blue) asbestos has been found to be significantly the most mesotheliogenic, its use has been banned or severely restricted in many countries and work with existing products containing it is strictly controlled.

Building Products. Asbestos has found application in hundreds of products, including many building materials, particularly asbestos–cement building sheets and pipes, sprayed and preformed thermal insulation, floor tiles, asphalt roof coatings, and plaster patching compounds. Asbestos dust can be emitted during work with these products, although improved work practices (dust extraction, wet handling) can significantly reduce emissions, often to below hygiene standards (see references). Asbestos has now been replaced in many of these products, but in some cases difficulties remain in finding adequate alternatives, for example in some asbestos cement products.

Installation and Removal of asbestos insulation products present the most hazardous operations involving asbestos in building practice, generating extremely high asbestos dust concentrations unless specific precautions are taken. Use of asbestos in insulation products has voluntarily ceased or been legally prohibited in many countries, and replacement products are being used. Asbestos dust concentrations in buildings already insulated with sprayed asbestos are generally very low and present only trivial risk to occupants. Provided the insulation is stable and unlikely to be physically damaged, most authorities recommend it be left alone. Where a health risk is determined to exist the insulation should be sealed with suitable surface coatings or enclosed and should only be removed (by experienced contractors) where such action is not feasible.

See also VERMICULITE.

References

1. U.S. Department of Health, Education and Welfare: *Asbestos: An Information Resource.* R. J. Levine (ed.), DHEW Pub. No. 79-1681, 1978.
2. S. K. Brown: *A Review of Occupational and Environmental Exposure to Asbestos Dust.* CSIRO Division of Building Research Report, Australia, 1981.

S.K.B.

B

BALLOON FRAME. The first balloon frame was erected in Chicago in 1832 by George W. Snow and served as a warehouse. By using machine-made nails to fasten together sawed lumber of small dimensions at a time when neither heavy timber nor the skilled carpenters needed to frame them were available in Chicago, Snow invented a new technique of light frame construction, so light in fact that his invention was soon dubbed ''balloon framing.'' It was a method, however, that proved to be of the greatest significance for the settlement of the treeless regions west of Chicago, because any relatively skilled farmer or mechanic, when supplied by railroad with dimensioned lumber cut in the forests of Wisconsin and Michigan and nails made in the East, could easily erect balloon frames for residential, farm, and commercial buildings. So successful was this new technology that by 1900 the balloon frame had largely superceded the early heavy frames of European origin, now called "braced frames," that relied for their stability on a skeleton of heavy timbers joined together by mortise and tenon joints.

In balloon framing, the walls consist of sawed timbers, generally 2 × 4 inches (50 × 100 mm) in section, called *studs*, which run vertically, and *plates* and *sills*, which finish the wall at its top and bottom, respectively. The studs run continuously from bottom to top.

See also TIMBER JOINTS and TIMBER STRUCTURES.

References

1. G. WOODWARD: *Woodward's Country Homes.* Geo. E. and F. W. Woodward, New York, 1865.
2. C. G. RAMSEY and H. R. SLEEPER: *Architectural Graphic Standards.* Wiley, New York, 1932.
3. P. E. SPRAGUE: The origin of balloon framing. *Journal of the Society of Architectural Historians*, Vol. 40 (1981), pp. 311–319.

P.E.S.

BAMBOO AS A BUILDING MATERIAL has long been in use, particularly in rural buildings in tropical countries. Its easy availability and low cost make it an ideal traditional building material. Some of the densest bamboo forests are found in Southeast Asia.

Species of Bamboo. About a thousand species are known to exist; they differ widely in physical properties. Varying in size from small shrubs, they reach up to 40 m (130 ft) in height. The diameter varies from 0.1 to 3 m (4 in. to 10 ft). The strength varies with species, age, moisture content, and disposition of nodes.

Physical Properties. Bamboos are usually hollow and are divided into culms with rigid cross walls at the nodes, providing extra resistance against bending. Strong, hard tissues of high tensile strength are concentrated near the culm surface. Bamboo has a low specific gravity (0.65 average), a high tensile strength (140 to 280 MPa or 20 to 40 ksi), a high strength to weight ratio, and good resilience. Bamboo can be easily split into strips, both radially and tangentially (Fig. 12) for a multiplicity of uses.

Shortcomings. Bamboo is highly susceptible to attack by termites and powder post beetles and to rot fungi where in contact with moist soil. It can harbor vermin and rodents, and so it must not be used in plague-prone areas. It shows shrinkage cracks on the surface on drying, and it deteriorates quickly in wet conditions. These defects can be minimized by suitable chemical treatment against insects, careful selection and grading of material, kiln-seasoning under controlled conditions, and plugging the open ends of bamboos with wooden plugs to prevent entry of rodents. The Bucherie process provides the most suitable and effective chemical treatment (Fig. 13).

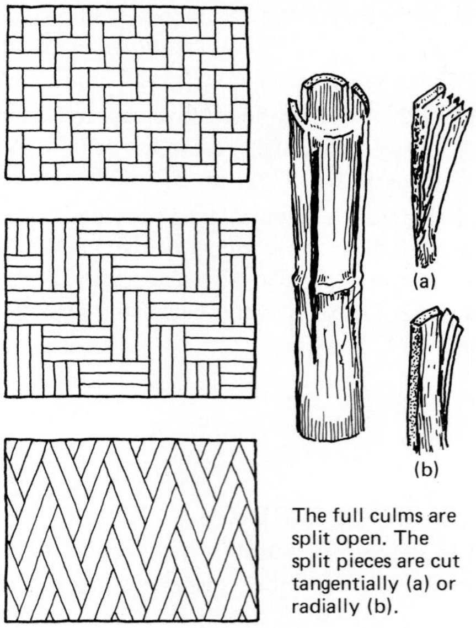

(a)
(b)
The full culms are split open. The split pieces are cut tangentially (a) or radially (b).

Figure 12 Bamboo boards and mats.

Construction. Complete buildings can be constructed with bamboo alone (Fig. 14). However, it is generally used in combination with other materials.

(a) Foundations. Bamboo posts, driven into the ground, provide the foundation for a frame. Although economical, they are susceptible to damage due to dampness, which may reduce their life to only two or three years. However, if the soil as well as the bamboo is treated chemically, it lasts longer. Coating buried ends with hot bitumen or embedding them in concrete or masonry is also useful.

(b) Building Framework. Although steel tubes or hardwood provide a more durable structure, a bamboo frame provides better resistance to earthquakes due to its superior resilience. Its low cost makes it preferable for rural houses.

(c) Walls can be made with solid or split bamboos placed

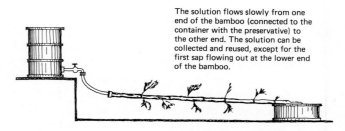

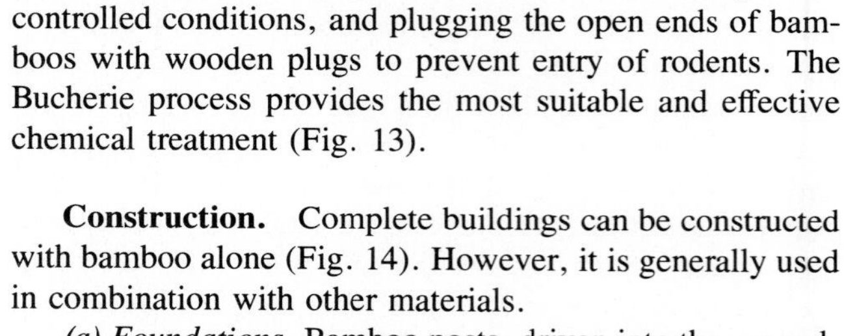

The solution flows slowly from one end of the bamboo (connected to the container with the preservative) to the other end. The solution can be collected and reused, except for the first sap flowing out at the lower end of the bamboo.

Figure 13 Preservation of bamboo by boucherie process.

Figure 14 Experimental structure with walls of bamboo matting, bamboo, and mud at the Forest Institute, Dehra Dun, India.

side by side in a wooden frame; bamboo boards made by opening and flattening out whole culms, fixed to horizontal members fitted in turn with framing poles; plaited bamboo boards made by using open weave with bamboo strips and round bamboos; and wattle walls in a number of variations, like wattle and daub, lath and plaster, and strung-strips construction. In all these, bamboo culms or splints are used as a base for mud plaster. Mud mixed with 5% Aldrine emulsion in water before plastering safeguards against termites. Inner walls can be covered by bamboo mats (Fig. 12), which may or may not be plastered.

(d) Roofing. The entire framework for the roof can be made with thick bamboos for trusses, rafters, and purlins. In a sloping roof the cladding could consist of clay tiles, thatch, reeds, or half-round tiles and shingles of bamboo. Bamboos laid side by side or bamboo boards covered with mud and supported by bamboo purlins provide flat roofs.

Other Uses. Being a strong and resilient material, bamboo poles, lashed together, are used as scaffolding tied to external walls. Tall bamboo scaffolds have withstood typhoons when steel-tubular scaffolding collapsed.

Bamboo can also be used as REINFORCEMENT FOR CONCRETE. The bamboo splints are coated with hot bitumen before using them as reinforcement to avoid shrinkage and loss of bond. Bamboo reinforcement must be safeguarded against termites. Using these precautions, small-span reinforced concrete roofs can be constructed with bamboo reinforcement. However, further research is needed in this field.

Bamboo frames and matting can be used for door and window frames, shutters, ceilings, partitions, decorative panels, furniture, and shelves. Floors can be made with bamboo boards supported on bamboo or timber poles.

Joints. Bamboos have a tendency to split, so they cannot be joined with nails or screws. Joints are therefore made through lashings. Several types of joints have been developed for specific purposes (Fig. 15). Butt joints sheathed with metallic jackets and shear connector pins have been tried at the Central Building Research Institute, India, and at the British Building Research Establishment with encouraging results. However, further research on the subject is needed.

See also TIMBER JOINTS and TIMBER STRUCTURES.

References

1. DINESH MOHAN and D. NARAYAN: *The Use of Bamboo and Reeds in Building Construction.* Department of Economic and Social Affairs, United Nations, New York, 1972.
2. HANNAH SCHRECKENBACH: *Construction Technology for a Tropical Developing Country.* Deutsche Gesellschaft für technische Zusammenarbeit (GTZ), Technische Hochschule, Aachen, German Federal Republic, 1983, for the Department of Architecture, University of Science and Technology, Kumasi, Ghana.
3. G. C. MATHUR: *Durable Bamboo House.* National Building Organization, New Delhi, India, 1972.

R.C.M.

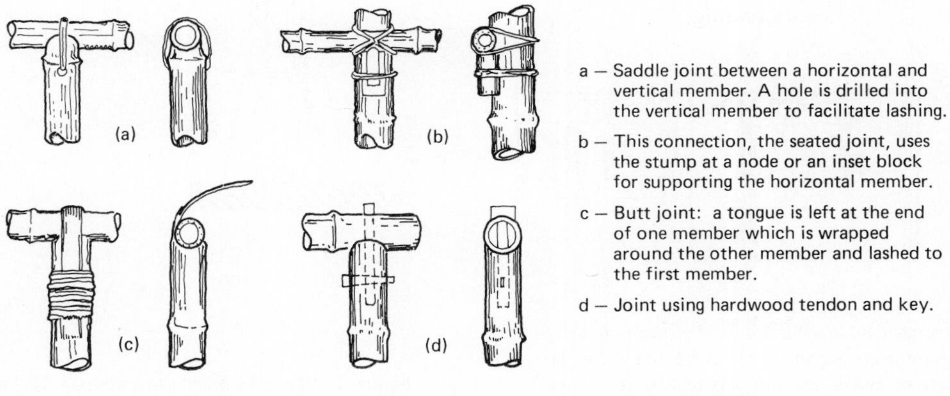

(a)
(b)
(c)
(d)
a — Saddle joint between a horizontal and vertical member. A hole is drilled into the vertical member to facilitate lashing.
b — This connection, the seated joint, uses the stump at a node or an inset block for supporting the horizontal member.
c — Butt joint: a tongue is left at the end of one member which is wrapped around the other member and lashed to the first member.
d — Joint using hardwood tendon and key.

Figure 15 Joints in building with bamboo: butt, saddle, and seated joint.

BARRIER-FREE DESIGN is intended to provide the same opportunities of access and use of buildings by disabled or handicapped people that are available to every citizen. Its scope includes consideration of the site, building entrances, internal circulation, toilet rooms, and communication systems. To provide full access, every piece of equipment and fixture in a building that must be touched during normal building occupancy should be usable by disabled people. A truly accessible building responds not only to the needs of people in wheelchairs, but also to those people who have walking limitations, poor stamina and strength, impairments of reach, manipulation limitations, and sensory impairments.

Specific Issues. Designing accessible barrier-free entrances and circulation spaces means planning dimensional clearances for the wheelchair in motion, accommodating the limited reach inherent in the seated position of wheelchair users, and providing an efficient and obstacle-free wheeling surface.

People with walking limitations and those who use wheelchairs often have severe limitations of stamina and strength and reduced speed of action. Building elements such as ramps, stairs, and self-closing doors must be carefully designed to keep the effort required to use them within the limits of disabled users.

Individuals with low stamina or poor vision, and particularly those who have other disabilities as well, require extended time to react and respond. The timing of automatic elevator doors and warning signals must thus be adjusted to accommodate slowness of action.

People with walking limitations and wheelchair users often have considerable difficulty using conventional toilet room and bathroom facilities (Fig. 16). Aside from the additional space clearances required, grab bars for use in transferring to and from toilets and a higher toilet seat are necessary. Sinks, lavatories, tubs, and showers must accommodate people in a seated position as well as a standing position. Amenities such as mirrors, medicine cabinets, and toilet paper and towel dispensers must be located where clearances allow approach and use of the fixtures with wheelchairs and walking aids.

Many severely disabled people have difficulty manipulating devices with their hands and reaching objects mounted at high or low positions. In general, hardware and equipment controls that require tight grasping and/or turning the wrist should be avoided. The selection of door openers, plumbing fixture controls, window latches, and locks must take these factors into account. For many building products, there is a very limited selection of items that are convenient for disabled people to use. The location of operable devices and storage within range of almost all disabled people can be achieved by using the limits of reach from a wheelchair as criteria. The wheelchair user can usually be accommodated without creating significant inconveniences to ambulatory people.

Sensory impairments particularly affect the design of communication systems in buildings, such as signs, direction-finding aids, warning or emergency signals, hazard markings, and telephone systems. Impairment to hearing or sight may range from reduced thresholds to complete loss of sense reception. The former can be compensated by increasing the level or effectiveness of the *signal.* For examples, signs can be increased in size or made more visible through contrast of figure and background color. Complete loss of hearing or sight must be compenstaed for by providing information to one or more of the other senses. Thus, warning or emergency signals should be both audible and visible or tactile and visible. Figure 17 illustrates the use of a cane to detect tactile information.

Trends. In the early days of barrier-free design, attention was given primarily to accommodating wheelchair users. Today it is recognized that in industrialized countries the rapid increase in the number of older people and the

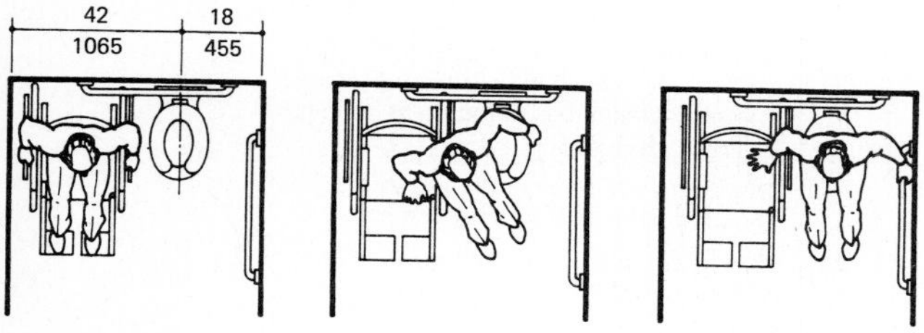

Figure 16 Side approach to transfer from wheelchair to toilet (from Ref. 1). Measurements are in inches and millimeters. (a) Takes transfer position, removes armrest, sets brakes. (b) Transfers. (c) Positions on toilet.

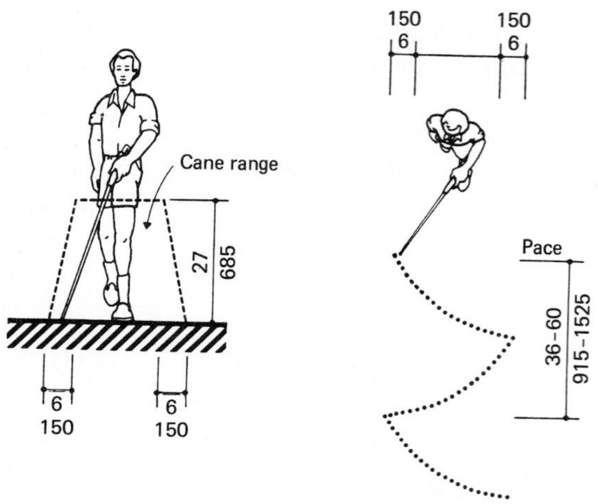

Figure 17 Cane technique for blind person (from Ref. 1). Measurements are in inches and millimeters.

growing independence of severely disabled people through the availability of medical technology is expanding the definition of barrier-free design. The concept of adaptable environments that allow individuals a range of flexibility in workplaces and housing to accommodate their specific needs has become a major design goal; an example is adjustable work surface heights. Finally, it is now evident that providing for the use of buildings by disabled people, if done properly, results in better design for all building users; everyone benefits from the added safety and convenience.

See also ELEVATORS; ESCALATORS; and SANITARY APPLIANCES.

References

1. *American National Standard Specifications for Making Buildings and Facilities Accessible to and Usable by Physically Handicapped People—ANSI A117.1—1980.* American National Standards Institute, New York, 1980.
2. E. S. STEINFELD: Six articles on barrier-free design. *Architectural Record*, Vols. 165–168 (March 1979, May 1979, July 1979, October 1979, March 1980, August 1980).

E.S.

BRICKS AND BLOCKS are manufactured units intended to be joined together, usually with mortar and usually individually by hand, to form an element of a structure. The most common elements are walls, but bricks and blocks are also used to pave roads for vehicles and areas for pedestrians; these units may form part of the structure of the pavement or simply provide a wearing surface on an already adequate structure. Bricks or blocks are made from clay, concrete, or calcium silicate.

In countries where a distinction between bricks and blocks is formalized in standards, a brick is usually a unit with a gross volume of less than 1.2 to 4 $\times$ 10^6 mm^3 (75 to 245 $in.^3$), and a block is a unit of larger volume.

Clay Units were traditionally formed by throwing a clod of soft clay into a sanded mold to form a brick. Mechanized versions of this hand-throwing method have been devised and, in the United States, such units are described as being made by the soft-mud process.

Most clay units are formed either by pressing clay to form rectangular prisms of appropriate size or by extruding a column of clay and wire-cutting the column into separate units. Pressed bricks are made either with an auger that forces moderately wet clay into a die box or with a machine that uses drier clay that flows by gravity into the die box. There are two types of extruded bricks and blocks. One is multiholed and extruded with a fairly high water content on machinery of European origin. The other contains fewer, larger holes; in this case a drier clay is used, and the machinery probably originates in the United States. Having been made and dried, clay bricks must be fired in a kiln that turns the dry clay into an at least partly vitrified ceramic product.

Concrete Bricks and Blocks are made from a mixture of portland cement and a variety of naturally occurring or artificially produced aggregates. Concrete units may be

(a) produced as light-textured units on a press where the mixture of cement, aggregate, and some water flows by gravity into the die boxes;

(b) made on a vibrating machine to produce an open-textured unit, or

(c) formed on a machine that both vibrates and applies pressure to produce a unit with properties intermediate between the other two types.

After being formed, concrete units are not fired, but are steam cured, vapor cured, or autoclaved.

Calcium Silicate (sand lime or silica lime) units are made from a mixture of sand and lime that is pressed in the forming process and then stabilized by autoclaving to produce more tightly grained units than those made of concrete.

Properties. The different ingredients and manufacturing processes produce units with a wide range of properties; few are critical in the nonengineered application where most are used. Perhaps of greatest importance to the user is the fact that all are subject to some long-term changes in their dimensions; if provision is not made to accommodate these changes, cracking can result. Concrete and calcium silicate units tend to shrink as they change from wet to dry. Clay units expand as they take up moisture after firing. These phenomena have caused problems in the past, particularly when growth or shrinkage is high; but the mechanism and magnitude of these changes is now well known, and information is available for the design of appropriate control measures.

In many parts of the world the standard specifications set limits on the shrinkage of concrete and calcium silicate units, and some provide methods for estimating the expansion of clay units.

Most of the current uses of bricks and blocks do not justify their testing in laboratories to determine values for their various properties. However, codes of practice for the engineering design of masonry specify standard methods for the measurement of the relevant properties of the bricks and blocks when they are to be used in demanding situations (Ref. 1). In most countries there are organizations set up by brick and block manufacturers to provide information about their materials (Ref. 2).

References

(1) Standard Specifications or Codes

United States

C7 Specification for Paving Brick
C34 Specification for Structural Clay Load-bearing Wall Tile
C55 Specification for Concrete Building Brick
C56 Specification for Structural Clay Non-load-bearing Tile
C57 Specification for Structural Clay Floor Tile
C62 Specification for Building Block (Solid Masonry Units Made from Clay or Shale)
C67 Sampling and Testing Brick
C73 Specification for Calcium Silicate Face Brick (Sand–lime Brick)
C90 Specification for Hollow Load-bearing Concrete Masonry Units
C112 Sampling and Testing Structural Clay Tile
C129 Specification for Hollow Non-load-bearing Concrete Masonry Units
C140 Sampling and Testing Concrete Masonry Units
C145 Specification for Solid Load-bearing Concrete Masonry Units
C212 Specification for Structural Clay Facing Tile
C410 Specification for Industrial Floor Brick
C426 Test for Drying Shrinkage of Concrete Block
C652 Specification for Hollow Brick (Hollow Masonry Units Made from Clay or Shale)

American Society for Testing and Materials, Philadelphia, Pa.

Great Britain

BS187 Specification for calcium silicate (sand–lime and flintlime) bricks
BS3921 Clay bricks and blocks
BS6073 Part 1. Specification for precast concrete masonry units
Part 2. Method for specifying precast concrete masonry units

British Standards Institution, London.

Australia

AS1225 Australian Standard for Clay Building Bricks
AS1226 Australian Standard Methods for Sampling and Testing Clay Building Bricks
AS1346 Australian Standard for Concrete Building Bricks
AS1500 Australian Standard for Concrete Building Blocks
AS1653 Australian Standard for Calcium Silicate Bricks

Standards Association of Australia, Sydney

(2) Sources of Information about Bricks and Blocks

United States

Brick Institute of America, Reston, Virginia
National Concrete Masonry Association, Herndon, Virginia

Great Britain

Brick Development Association, Winkfield, Windsor, Berkshire
Cement and Concrete Association, Wexham Springs, Slough

Australia

Brick Development Research Institute, Melbourne, Victoria
Concrete Masonry Manufacturers Association of Australia, North Sydney, NSW

T.M.

BRICKWORK AND BLOCKWORK are the assemblages of masonry units (bricks or blocks) with a material, usually mortar, that bonds them together to form a coherent whole. Brickwork and blockwork (termed masonry) can provide not only structural strength, but also the weather protection, thermal and acoustic insulation, fire protection, and space dividing elements needed in a building. Masonry is a relatively inexpensive construction material; when well constructed it is durable and has a pleasing appearance.

Mortar. An easily spreadable mixture of sand, PORTLAND CEMENT, water, and an agent such as lime is required for mortar. Mortar has the dual function of holding the masonry units apart, both in the initial wet state and in the final set state, and of binding the units together. The need for spreadability of the mortar may conflict with the requirement for adequate strength and for crack resistance and watertightness, which may be achieved with good bond, that is, adhesiveness between the units and the set mortar. Good bond can be attained using a mortar with sand containing few very fine particles and retaining water compatible with the suction of the units.

The horizontal bands of mortar between the units are termed *bed joints*, and the vertical bands are termed *perpends*. Bed joints and perpends are normally 10 mm ($\frac{3}{8}$ in.) thick. A bed joint separates adjacent courses of brick as shown in Fig. 18.

Bond Pattern. Masonry units may be positioned relative to each other within brickwork or blockwork in many bond patterns; here the word bond refers to the arrangement of the masonry units, not to adhesiveness. Figure 18 shows two common forms of bond pattern in walls of one unit thickness, termed single-leaf, single-wythe, or single-skin walls. In stack bond, the units are placed one above another to cause the perpends to be in continuous vertical lines, while in stretcher bond the units are staggered, normally by one-half of a unit, in adjacent courses. At perpends the bond between mortar and units tends to be low, so in stack bond the perpends create a line of weakness.

In multiple-leaf walls the bond pattern also describes the relationship of units in the plane of one leaf (stretchers) and units perpendicular to the leaf planes (headers or through-the-wall units). Numerous bond patterns are possible, including Flemish bond, illustrated in Fig. 18, which has alternating headers and stretchers in each course.

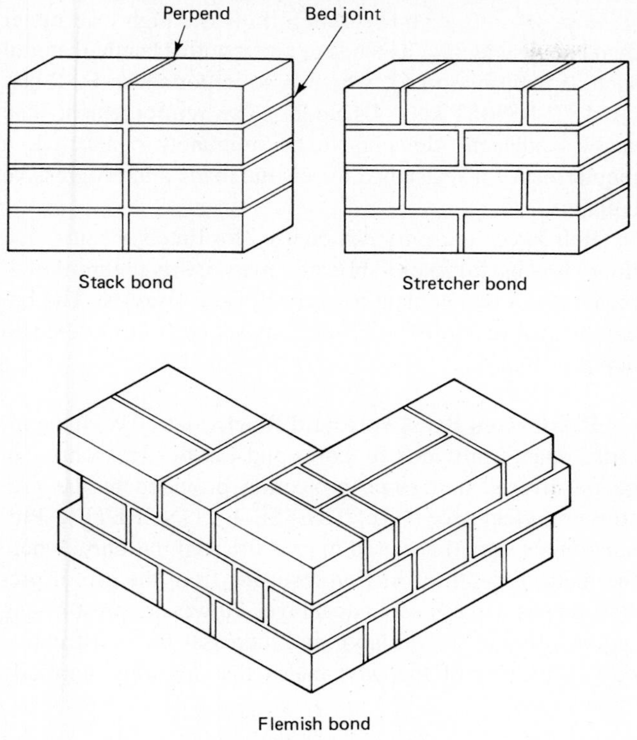

Figure 18 Bond patterns.

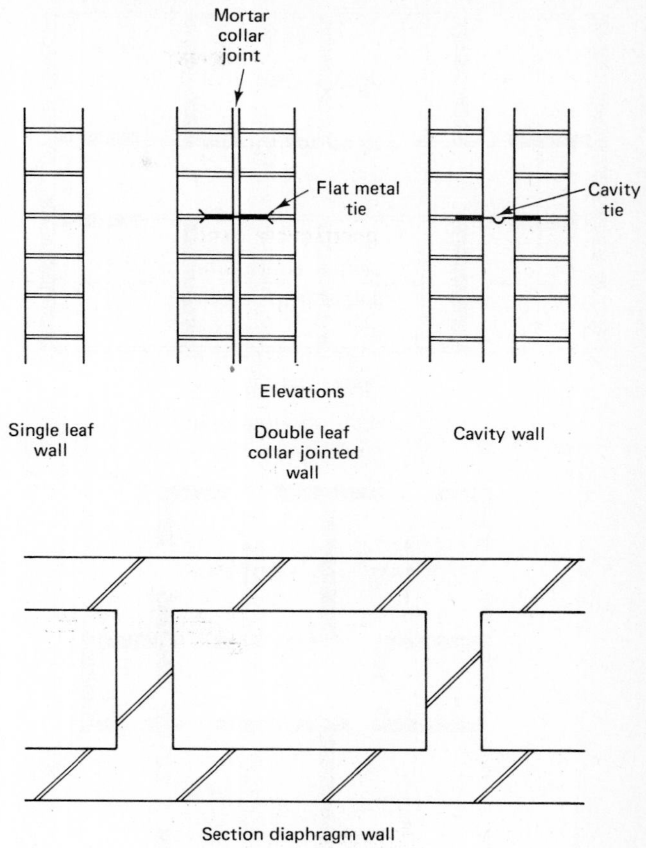

Figure 19 Types of masonry wall.

Masonry Structural Elements. Various wall types are illustrated in Fig. 19. The load-carrying capacity of a wall depends greatly on its thickness, so multiple-leaf walls are stronger than single-leaf walls. In double-leaf collar-jointed walls, two leaves are joined together with a continuous vertical mortar joint, the collar, and with metal ties or mesh reinforcement anchored in some of the bed joints of the two leaves. For additional interlocking between leaves, headers bonded in each leaf may be used.

In a cavity wall the two leaves are separated by a gap of 50 mm (2 in.) or more, and the wall, when well detailed and cleanly executed, can prevent moisture penetration into the internal face. The ties joining the two leaves of a cavity wall need to transfer horizontal forces across the cavity. When the cavity is filled with insulation, or with grout in reinforced masonry construction, the moisture-barrier effect may be destroyed.

A diaphragm wall is essentially a wide cavity wall with masonry cross ribs bonded into each leaf to provide vertical shear resistance. Diaphragm walls are particularly suitable for single-story buildings enclosing large clear plan areas.

Columns, pilasters, and piers are short walls used to carry isolated vertical loads. They are commonly of solid or hollow rectangular section, but hollow circular, cruciform, and I-shaped sections are also used.

Industrial chimneys and brick arches in bridges, cathedrals, and floors were abundant in earlier times. Nowadays their construction is uncommon. Other materials have superseded masonry in these uses because masonry needs to be massive to ensure that no tension exists in the section.

Load-bearing Masonry in Buildings. Due to the low flexural-tension strength of masonry, the floor plans need to be repetitive for economic multi-story construction and consist of small- to medium-sized rooms. Such buildings are usually hotels or apartment blocks.

The possible arrangements of the walls divide roughly into three groups: cross-wall systems, cellular systems, and combinations of these two systems. In a cross-wall system the main bearing walls are at right angles to the longitudinal axis of the building, and the floor slabs span between these cross walls, which also resist lateral wind loads. Some corridor walls resist longitudinal WIND LOADS. The outer longitudinal walls may be non-load bearing. With a cellular system both the internal and external walls are load bearing and the floor slabs span two-way between the walls. The two systems are illustrated in Fig. 20.

Reinforced Brickwork or Blockwork. Where a masonry element is subjected to loads that create significant

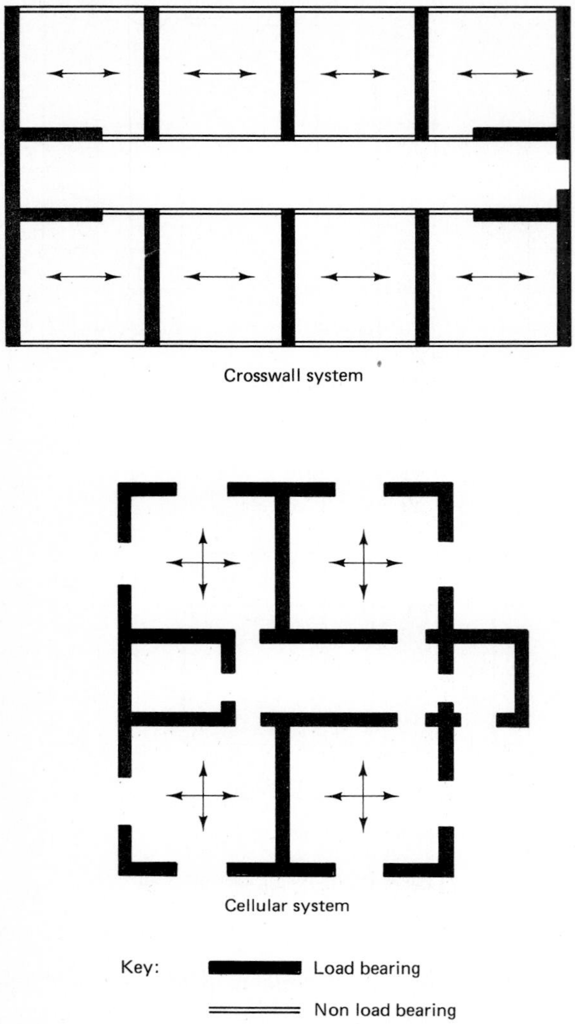

Figure 20 Typical floor plans of wall systems.

tensile stresses, the masonry has to be reinforced. Research has indicated (Ref. 1) that the loaded behavior of reinforced masonry follows the same principles as reinforced concrete. However, with reinforced masonry, due to the restricted geometry of the cores or cavity, it is difficult to confine the longitudinal steel with fitments or stirrups.

The steel reinforcement is incorporated within the hollow cores of the units or in the cavity between the masonry leaves. To ensure composite action, the reinforcement must be bonded to the masonry; this is achieved with grout completely surrounding the reinforcement. Grout is highly workable concrete; so when it is poured from the top of the construction it can flow in confined surroundings, losing water to the masonry units in the process, and yet finally surround all the reinforcement.

Apart from allowing the element to carry tensile stresses, the effect of the reinforcement is generally to increase the load-carrying capacity. Furthermore, it gives the masonry some ductility (the ability to deform under high load rather than break suddenly, as tends to occur with a brittle material such as plain masonry), which is so important in EARTHQUAKE-RESISTANT DESIGN. The reinforcement also enables adjacent elements to be intimately connected, a requirement for resistance to seismic loads and progressive collapse.

Reinforced masonry can be used for lintels, beams, and floor. It is useful in prefabricated masonry as the reinforcement allows the element to carry flexural stresses. The behavior and design of reinforced masonry is considered in depth in Ref. 2.

Prestressed Brickwork and Blockwork. With the internal spaces provided by cores and cavities, masonry can be prestressed and some protection provided to the prestressing steel (see PRESTRESSED CONCRETE). Prestressing is sometimes used in prefabricated masonry, where the reduced depth of section resulting from the use of prestressing is a particular advantage. However, prestressing requires highly skilled labor and increases the construction cost; thus two of the advantages that masonry normally enjoys are lost.

Structural Design of Brickwork and Blockwork. The design of masonry buildings based on structural engineering principles is relatively new. Up to the 1950s loadbearing masonry walls were designed by empirical rules that led to heavy and uneconomical structures, not competitive with reinforced concrete and steel framed structures. In the 1960s, a number of countries introduced building regulations in which a rational basis for structural calculations of masonry was required. With increasing research and with the experience of buildings designed more rationally (Refs. 3 and 4), knowledge of masonry behavior has increased dramatically within the last 20 years, leading to refined masonry codes and competitive structures.

See also ADOBE and REINFORCEMENT FOR CONCRETE.

References

1. J. C. SCRIVENER: Reinforced masonry—seismic behaviour and design. *Bulletin of New Zealand Society for Earthquake Engineering,* Vol. 5 (1972), pp. 143–155.
2. R. R. SCHNEIDER and W. L. DICKEY: *Reinforced Masonry Design.* Prentice-Hall, Englewood Cliffs, N.J., 1980.
3. A. W. HENDRY: *Structural Brickwork.* Macmillan, London, 1981.
4. S. SAHLIN: *Structural Masonry.* Prentice-Hall, Englewood Cliffs, N.J., 1970.

J.C.S.

BUCKLING occurs in compression members that are slender, that is, members with a high ratio of length to thickness. Due to very small imperfections in the straightness of the member or the centering of the load, which are in practice unavoidable, slender compression members deflect sideways and buckle when a critical load is reached. This *buckling load* depends on the elastic modulus of the material, but not on its strength. Thus ALUMINUM alloys are more likely to buckle than STEEL, which has an elastic modulus three times as high. Furthermore, the buckling load of high-tensile steel is the same as that of lower-strength structural steel, because their elastic modulus is the same.

Buckling is a particular problem in steel, and even more in aluminum structures, whose members are relatively slender because of the high strength and high cost of the materials. Compression members in timber and concrete structures are generally less slender and therefore less susceptible to buckling failure. Buckling is prevented in slender compression members by suitable detailing, which uses wide-flange I-sections, tubes, or latticed construction to reduce the slenderness ratio.

See also STEEL STRUCTURES.

References

1. LAMBERT TALL (ed.): *Structural Steel Design*, 2nd ed. Ronald Press, New York, 1974. Chapters 9–11 on Compression Members.
2. R. J. ROARK: *Formulas for Stress and Strain.* McGraw-Hill, New York, 1943. Chapter 11, Columns and Other Compression Members.

H.J.C.

C

CABLE AND SUSPENSION STRUCTURES, THEORY OF. Suspended cable structures resist applied load by a combination of an alteration in the state of stress in the main tensile elements and an alteration in the profile geometry. Unless special measures are taken, cable structures can be quite flexible under those live loadings that differ significantly in spatial extent from the dead load supported.

The mechanics of cable structures is a subject of some antiquity, dating at least from Stevin, who established the triangle of forces in 1586. The discovery of the CATENARY, the shape of a cable hanging under its own weight, is generally attributed jointly to Leibniz, Huygens, and James and John Bernoulli about 100 years later. Cable theory was understood earlier than beam theory. In beams, in general, applied load is resisted by changes in the internal state of stress; changes in profile geometry are not necessary for the task.

The solution for a uniformly distributed load is STATICALLY DETERMINATE. However, the theory for concentrated loads, which occur in cable networks due to the interaction of the cables, is complex and statically indeterminate. Exact solutions, where they exist, are almost invariably rather cumbersome and approximations are required. These can be achieved for suspended cables of relatively flat profile, which are generally those of most structural advantage in any event.

Irvine (Ref. 1) has developed a comprehensive theory of cable structures for both static and dynamic loads. Figure 21 shows a vertical point load P acting a distance x_1 from the left-hand support of a flat-sag cable, the self-weight profile of which is shown dotted.

The profile z of the unloaded cable is given by the parabola

$$z = \frac{mgl^2}{2H}\frac{x}{l}\left(1 - \frac{x}{l}\right)$$

where mg is its weight per unit length, l the span, and H the horizontal component of cable tension ($H = mgl^2/8d$, in which d is the sag). After the point load is applied, statements of vertical equilibrium can be written for either side of the point of application. We need another equation,

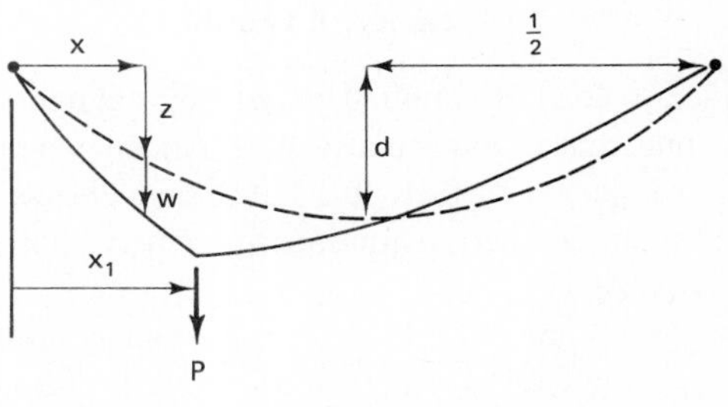

Figure 21 Deformation of cable under the action of a point load. The dotted line shows the unloaded cable, and the full line shows the profile of the cable under the action of the point P, acting at a distance x_1 from the left-hand support. d = sag at mid-span; z = sag at a distance x from the support; w = deformation of cable at x due to load P.

and this comes from Hooke's law and the strain–displacement compatibility relationship. In this way we are able to link additional tension directly with a nonlinear expression for the additional displacement. The result (Ref. 1, pp. 43–58) is the following cubic, which permits the complete solution to be established. It is expressed in dimensionless form:

$$h^3 + \frac{2 + \lambda^2}{24} h_*^2 + \frac{1 + \lambda^2}{12} h_* - \frac{\lambda^2 x_{1_*}(1 - x_{1_*})P_* \ (1 + P_*)}{2} = 0$$

using the dimensionless ratios $h_* = h/H$, $x_{1_*} = x_1/l$, $P_* = P/mgl$, and $\lambda^2 \simeq (mgl/H)^2(EA/H)$, where E is the modulus of elasticity, and A is the cross-sectional area of the cable.

The independent parameter λ^2 is of fundamental importance in the static and dynamic response of suspended cables. The parameter accounts for geometric and elastic effects. It resembles the stiffness of two springs in series. In a sagging cable, in which the ends are being stretched apart, some of the resistance supplied is geometric because the sag is being reduced: this stiffness is given by the expression $12\ H/[l(mgl/H)^2]$. The remaining stiffness is axial and quantified by a term like EA/l.

The preceding cubic has only one positive real root. When the point load is small, we obtain a linearized solution.

$$\frac{1 + \lambda^2}{12} h_* = \frac{\lambda^2 x_*(1 - x_*)\ P_*}{2}$$

When λ^2 is small, we obtain response akin to that of a taut string: very little additional tension is generated (unless the load is large), because resistance to applied load is found to be supplied almost solely by changes in geometry.

On the other hand, when λ^2 is larger, as in a suspension bridge cable, where the profile is moderately deep, the solution is independent of λ^2, being found from the quadratic.

$$h_*^2 + 2h_* - 12x_{1_*}(1 - x_{1_*})P_*(1 + P_*) = 0$$

Resistance to load is supplied mainly by changes in cable tension, and this is particularly pronounced when the applied load is large. In fact, this latter case allows a useful description of geometric nonlinearity when applied loads are appreciable.

A sagging cable to which a point load is applied will deflect until the load is such that the cable profile is almost triangular. After that, further increases in applied load engender almost no increase in deflection. This nonlinear stiffening effect characterizes all suspended cable behavior when the loads are appreciable. Nevertheless, it is an elastic nonlinearity; removal of the load causes the cable to recover its original profile.

Values of h_*/P_* when the load is placed at midspan, which is what generates the largest additional tensions, are given in Table 2.1 of Ref. 1 (p. 58).

See also TENSILE ROOFS.

Reference

1. H. M. IRVINE: *Cable Structures*, M.I.T. Press, Cambridge, Mass., 1981.

H.M.I.

The **CATENARY** is the curve assumed by a uniform chain or CABLE freely hanging from two points. If these points are at the same height, the shape of the catenary is

$$y = c \cosh \frac{x}{c} = \tfrac{1}{2}c(e^{x/c} + e^{-x/c})$$

where x and y = horizontal and vertical coordinates
e = base of the natural logarithm
c = constant.

If the sag of the cable is small, the catenary is very similar to a parabola,

$$y = cx^2$$

which is the curve assumed by a chain whose weight is uniformly distributed in plan (not along its length).

The chain and the cable are flexible structures that cannot resist bending and are consequently in pure tension. If the shape is reversed, a structure in pure compression, free from bending, is produced. Thus a catenary-shaped arch is in pure compression, and a parabolic arch is almost free from bending. This principle was discovered in the 17th century, probably by Robert Hooke (Ref. 1), and used in the design of the DOME of St. Paul's Cathedral. It is still used today in LOAD BALANCING in prestressed concrete.

Arches whose shape resembles a catenary have been used in the Middle East at least since the 6th century A.D. There is, however, no evidence that the catenary principle was discovered at that early time. It is more likely that these are masonry copies of wooden arches and vaults, produced by bending thin pieces of wood, bamboo, or reeds to arch shape, a form of construction already described by Vitruvius (Ref. 2). Since this shape is similar to a catenary, the arches and vaults would be almost free from bending and thus particularly useful in regions, such as Mesopotamia, where both natural stone and fuel for burning bricks were in short supply.

The pointed Gothic arch, formed by two circular arcs, was imported by the crusaders from the Middle East, where it originated. It is much easier to set out than the near-catenary arch, and its structural advantages are similar.

References

1. H. J. COWAN: Domes—ancient and modern. *Journal of the Royal Society of Arts*, Vol. 131 (1983), pp. 181–198.
2. H. J. COWAN: Some observations on the structural design of masonry arches and domes before the age of structural mechanics. *Architectural Science Review*, Vol. 24 (1981), pp. 94–102.

H.J.C.

CEILING, the inner roof of a room, probably derives from the Latin *Caelum*, the vault of heaven, and the French *ciel*. The concept of ceilings proper began in the 14th century when what had hitherto been merely the underside of the room above became the ornamental feature of the room below.

The decoration of ceilings reached its peak in France and Italy in the 16th century, when plaster molding, gilding, and painting contributed to the overall splendor of the room design. In England, the late 18th century ceilings of the Adam brothers (Fig. 22) added elegance of form and light and shade, which contrasted with the highly ornate baroque of the previous period. Ceilings have evolved from these showpieces of lofty compartments to the generally functional and comparatively plain linings that make up the "inner roofs" of present-day domestic and commercial buildings. The functional purpose of these linings may be:

Figure 22 Centerpiece of Adam ceiling from *The Works in Architecture of Robert and James Adam*, London, 1773.

To conceal the overlying structure.

To provide a space for, and a covering to, a complex network of services.

To adjust space proportions.

To provide a control of acoustic and heat transmission and reflection.

Ceilings may be divided into three main categories: (1) flat linings fixed directly to the underside of the structures, (2) flat linings suspended below the structure, and (3) profiled linings suspended below the structure.

Flat Linings Fixed Directly to the Structure. Ceilings in this category include the following:

1. Plaster, applied directly to the structure.
2. Sheet plasterboard and a skim coat of plaster, fixed to the underside of the floor members or to battens fixed to the underside of the structure.
3. Thick spray coatings based on mineral fibers.
4. Thin spray coatings based on cement or resins.
5. Fiberboard or foamed plastic boards fixed to joists or battens.
6. Acoustic tiles fixed to joists or battens or by adhesive to the underside of the structure.

Flat Linings Suspended below the Structure

Grid Systems. The most common form of suspended ceiling consists of a light structural grid supporting panels or sheets. As panel and sheet materials are usually lightweight, grid can be of light aluminum or galvanized steel construction. They may be concealed (Fig. 23) or, where exposed (Fig. 24), they are stove-enameled, plastic-coated, or anodized.

The grid may consist of main runners and cross members or of primary channels that support through clips a system of main runners and cross members.

In ceilings formed from panels, the main runners are usually Z- or T-sections, and the cross members either T-sections or flat splines. They are joined with fixing clips and suspensed from the structure with wires or rods. Module sizes are multiples of 300 mm (12 in.). A 1200 mm × 600 mm (4 ft × 2 ft) module is usually the most economical, particularly if the system has main runners at 1200 mm (4 ft) centers and cross members at 600 mm (2 ft) centers, both exposed.

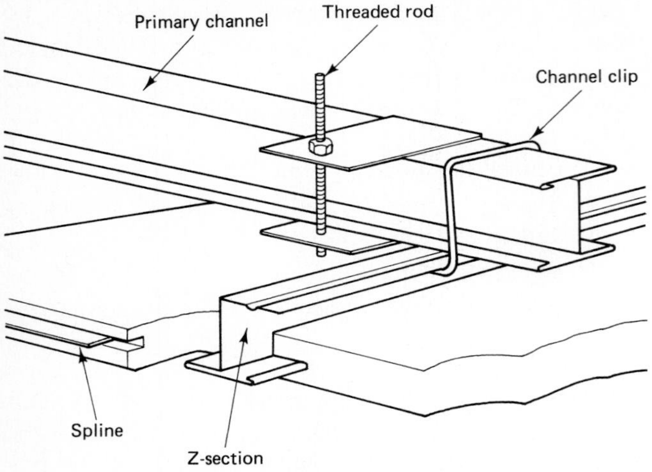

Figure 23 Concealed-grid ceiling system from technical literature of Alphacoustic Ceiling Industries, Sydney, Australia.

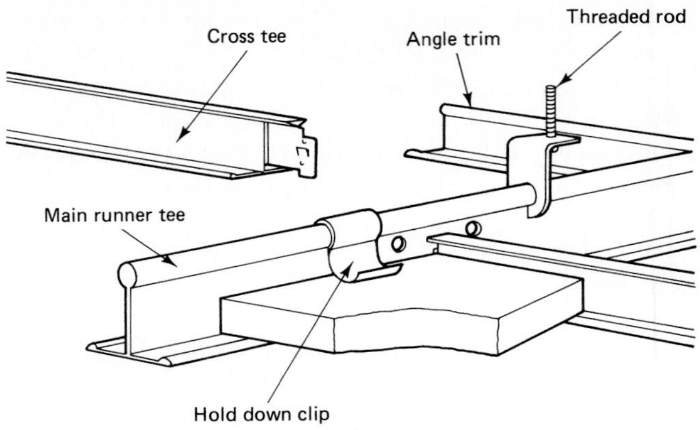

Figure 24 Exposed grid ceiling system.

For sheeted ceilings, the main runners are clip-fixed to furring channels. The sheet materials, usually plasterboard, are screw-fixed to these channels. Joints and screw holes are filled with plaster to give a smooth, seamless surface.

Tiles and Panels may be made from ordinary fiber board, mineral-based boards, or plastics. Nearly all are obtained either plain or with various finishes.

Integrated Ceilings are used when lighting and ventilation services and sound and thermal insulation are incorporated into the ceiling. An illuminated translucent ceiling lit from above is an integrated ceiling (Fig. 25).

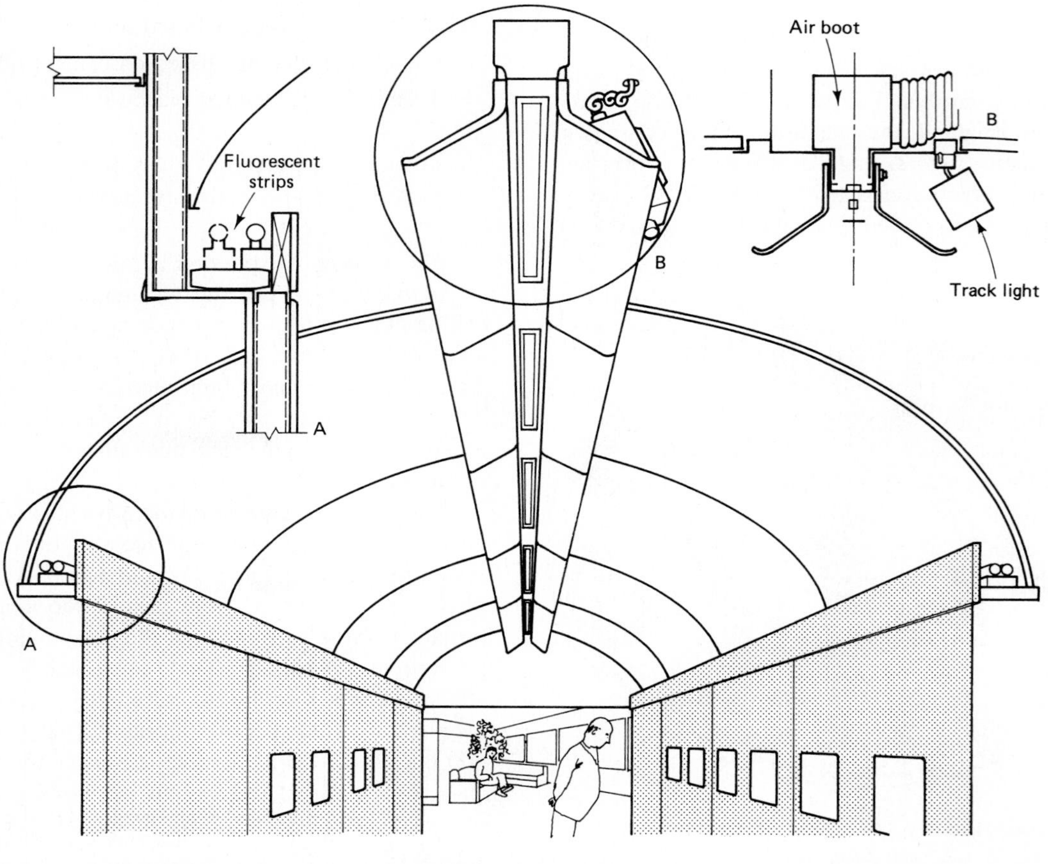

Figure 25 Integrated ceiling. (From Peter George, A kaleidoscope of ceilings, *Architectural Record*, Vol. 170, No. 11 (Sept. 1982), p. 136.)

Open Grid Ceilings. Early examples of such ceilings consisted of a suspended grid of timber or metal ribs intersecting vertically and spaced comparatively far apart. The depth of the ribs, their spacing, and their height above eye level determines the visual cutoff angle. Proprietary systems are now available, which use a variety of materials.

Profiled Suspended Linings. Ceilings of this type are used mainly in buildings such as auditoriums and in special areas of commercial buildings, such as foyers and lobbies. They are designed to produce a decorative effect while satisfying functional requirements. There are three main types:

1. Molded or sculptured tiles (usually of plaster) used in a normal grid.
2. Translucent panels utilizing prismatic or grid effects, usually combined with illumination from above.
3. Louvered ceilings using metal fins whose bases are cut to achieve a decorative effect.

References

1. A. J. ELDER: *Ceilings*, Section 9 in *AJ Handbook of Building Enclosure* (eds. A. J. ELDER and M. VANDENBERG). Architectural Press, London, 1975, pp. 337–364.
2. B. HOGAN: Overheads. *Interior Design, Westbourne Journals, England*, March 1979, p. 50.

R.G.C.O.

CELLULAR CONCRETE is an aerated cementitious material with or without sand, having a unit mass between 150 and 1600 kg/m^3 (9 and 100 lb/ft^3). Aeration is achieved by one of the following processes:

1. *Mechanical.* A preformed foam is injected into a mix until the desired unit mass is obtained. Alternatively, a foaming agent is added to the basic mix and the whole whisked to achieve aeration.
2. *Chemical.* An agent that reacts with cement, giving off bubbles of gas, is added to the basic mix. The bubbles are trapped within the cement matrix. Flakey aluminum powder is one such agent.
3. *Micropore.* Hydrated lime and siliceous material are mixed into a paste and placed in sealed steel molds. The material is then autoclaved. Excess water is turned to steam to develop a micropore structure.

Advantages of cellular concrete lie in its lower unit mass and improved insulation properties, but these are offset by low strengths and increased moisture movement. Dimensional stability is improved by high-pressure steam treatment.

See also LIGHTWEIGHT CONCRETE.

Reference

W. H. TAYLOR: *Concrete Technology and Practice*. McGraw-Hill, Sydney, 1977.

F.A.B.

CERAMIC FLOOR AND WALL TILES. Ceramic tiles are used in the building industry as a cladding material for walls, partitions, and floors in locations where cleanliness is required (Fig. 26). They are also used for cladding internal and external surfaces of structures.

Ceramic tiles evolved from earthenware composed of clay and silica. These were superseded by clay-calcareous bodies with opacified tin oxide glazes and decorated with overglaze colors and by clay-feldspathic products, which sometimes have transparent glazes. At present the bulk of tile production consists of these two types of ceramic materials made in a variety of sizes, glazed or unglazed, and in a variety of colors and decorative finishes. Tiles are made in highly mechanized and automated plants. The production cycle, which used to be counted in weeks, is now reduced to less than one day.

Figure 26 A1 nonglazed stoneware tiles on the roof of the Sydney Opera House, made from a special clay to minimize expansion. They were placed in the curved molds in an inverted position, and the concrete slabs were then poured over them.

TABLE 1
Classification of Tiles by the Comité Européen de Normalisation (Ref. 1).

Method of Shaping	Water Absorption			
	Below 3%	3%–6%	6%–10%	Above 10%
Extruded	AI	AIIa	AIIb	AIII
Pressed dust	BI	BIIa	BIIb	BIII

Classification of Tiles. Because of the multitude of ceramic tiles available it is sometimes difficult to choose the proper product for a given application. To remedy the growing number of complaints about failures experienced by the building industry, the *Comité Européen de Normalisation* (CEN) in 1982 approved a classification and new standards for ceramic tiles. The tiles are classified by CEN in eight groups, based on the method of shaping and the value of the water absorption or porosity of the finished product (Table 1).

Products commercially known under the names of klinker, split tiles, and quarry tiles are classified as extruded (A). Products commercially known as fully vitrified stoneware, faience (earthenware), red quarry tiles, monocottura, steingut, and majolica are classified as dust-pressed (B).

Each group must satisfy the appropriate standard performance of the tiles under normal conditions of use. The mechanical, physical, and chemical properties of the tiles are determined by measuring their modulus of rupture, water absorption, abrasion resistance, and crazing resistance.

Selection of Tiles. Ceramic tiles are brittle and should preferably be used under compressive stress to prevent failure by cracking (Ref. 2). Tensile stresses should be avoided; for example, ceramic tiles should not be placed on a flexible underlayer, such as a wooden floor. In the selection of tiles the performance under service conditions should be given priority over esthetic considerations; for instance, wall tiles do not require the same mechanical strength as floor tiles for an air terminal. Standards that document minimum performance values protect the user against low-quality tiles and failure in service. The standards and test methods vary from country to country (Ref. 3).

References

1. *Ci News* (*Ceramics International, Faenza, Italy*) Vol. 2, No. 4 (Dec. 1982), p. 6.
2. D. W. BUDWORTH: *An Introduction to Ceramic Science*. Pergamon Press, Oxford and Elmsford, N.Y., 1970, pp. 206–242.
3. *Specification for Clay Tiles for Flooring, BS 1286–1974*. British Standards Institution, London, July 1974; *Recommended Standard Specifications for Ceramic Tiles*. Tile Council of America, Princeton, N.J., 1976; *Feinkeramische Fliesen, DIN-18155*. Deutscher Normenausschuss, Berlin, January 1973.

E.T.

CLEANING OF BRICKWORK AND MASONRY. Exposed brickwork and masonry need to be cleaned because certain contaminants give them an unsightly appearance. In general, cleaning methods include (1) water-jetting spray, steam or high pressure, (2) mechanical methods, such as sandblasting, and (3) chemical methods.

Contaminants, such as dirt and dust and mortar, bitumen, tar, and soot stains that may occur during construction or on exposure to atmospheric pollutants, may be removed as follows. Smears and dirt can be partially removed with a scraper, chisel, or wire brush and the remainder with muriatic acid (1 : 9 acid : water). The surface should then be rinsed and washed thoroughly with water. Sandblasting is also effective for removing such stains but should only be used if the brick or mortar is strong enough to be unaffected. Bitumen and tar stains can be removed by first scraping them and then applying a poultice containing solvents such as xylene, toluene, trichloroethylene, or mineral spirits.

Efflorescence occurring on the surface should be washed with water or scrubbed and then rinsed with water. If this method is not satisfactory, the surface can be washed with muriatic acid followed by a water rinse. Efflorescence due to vanadium salts can be removed by a solution of sodium hydroxide.

Stains from metals are treated as follows. Iron stains can be removed with an oxalic acid (1 : 10 acid : water) solution containing ammonium bifluoride (50 g/L; $\frac{1}{2}$ lb/gal); if the stain persists, a heated solution or a poultice of sodium citrate, glycerine, and water (1 : 7 : 6) can be used. Copper and bronze stains can be removed with a poultice containing 1 part NH_4Cl and 4 parts talc or diatomaceous earth to which ammonia is added.

Paints on building surfaces can be removed with commercial paint stripper or with a propane or blow torch.

Microorganisms, such as algae, lichens and moss, may be destroyed by applying ammonium sulfamate or a 1 : 40 solution of zinc or magnesium silicofluoride.

When a stain has been identified, the cleaning method should first be tried on an inconspicuous area before complete application. Also, certain precautions are warranted in using all these methods, which are described in the references listed below.

See also BRICKWORK and NATURAL BUILDING STONE.

References

1. T. RITCHIE: Cleaning of Brickwork, *Canadian Building Digest*, No. 194, Division of Building Research, National Research Council Canada (1978), 4 p.
2. V. S. RAMACHANDRAN and J. J. BEAUDOIN: Removal of stains from Concrete Surfaces, *Canadian Building Digest*, No. 153, Division of Building Research, National Research Council Canada (1973), 4 p.
3. B. M. FEILDEN: *Conservation of Historic Buildings*. Butterworth, London, 1982.
4. H. J. ELDRIDGE: *Common Defects in Buildings*, Her Majesty's Stationery Office, London, 1974.

V.S.R. & J.J.B.

CLIMATE AND BUILDING. One main purpose of a building is to create a comfortable interior microclimate different from that outdoors. The difficulty of doing this depends on the nature of the outside climate and should be assessed as a preliminary to planning land use and building design. Often the climate is vaguely described as either temperate, hot-dry, hot-wet, or cold or else in terms such as those of the Koeppen classification of climate, which is based on botanical considerations. It is better to consider the January and July conditions quantitatively and separately as regards sunshine, temperature, humidity, rainfall, wind strength, and wind direction. In some places it is necessary also to consider tornadoes, snowfall, or tropical cyclones.

Data on particular places can be obtained from the national meteorological bureau or the nearest airport. Otherwise, measurements can be made locally using standard equipment and procedures or values can be obtained by estimation (Ref. 1).

Radiation (see also DAYLIGHT DESIGN, FENESTRATION, PASSIVE SOLAR DESIGN, SOLAR ENERGY, and SUN). Solar radiation (Q_s) is measured in watts per square meter (W/m^2); it varies with latitude, season, time of day, cloudiness, and air pollution. The amount received on a wall depends on its orientation (Fig. 27). In the absence of measurements, one can roughly estimate the monthly average radiation intensities as follows (Ref. 1):

$$\text{Coldest month: } Q_s = 165 + 0.13A - 0.059A^2 + 0.014h \quad \text{W/m}^2$$

$$\text{Warmest month: } Q_s = 207 + 6.45A - 0.108A^2 + 0.014h \quad \text{W/m}^2$$

where A = latitude (degrees)
h = altitude (meters)

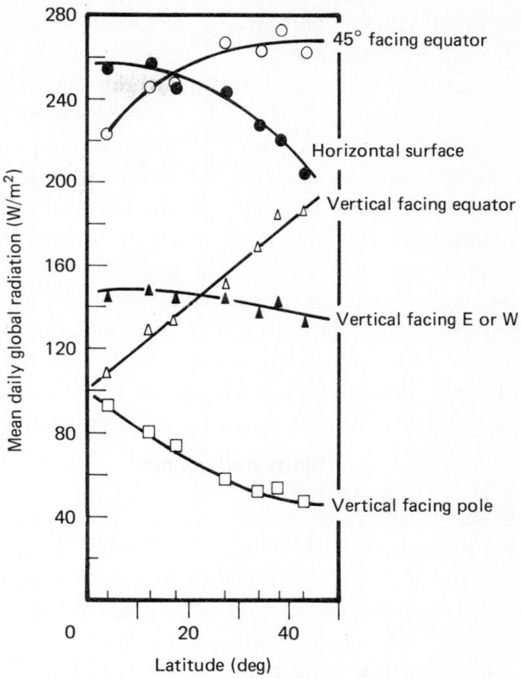

Figure 27 Daily mean solar radiation on two surfaces with various orientations. The top curve refers to a surface tilted at 45° to face the equator. (Reproduced from K. G. Martin, Australian Climatic Data and Weathering. *Australian Oil and Color Chemists' Association, Proceedings and News, No. 13.* Canberra, 1976, by courtesy of the Association.)

If values are available on the duration (n) of bright sunshine (over 200 W/m^2) in daylight of N hours, then Q_s can be estimated from Prescott's formula:

$$Q_s = Q_A \left(\frac{0.25 + 0.5\,n}{N} \right) \quad \text{W/m}^2$$

where Q_A is the radiation intensity on to a horizontal surface above the atmosphere. Both N and Q_A are given in tables (Ref. 2, p. 241) for any month or latitude. The maximum instantaneous value of Q_s is about 1000 W/m^2 with the sun overhead in a clear, clean sky, but total cloud would reduce that to about 330 W/m^2.

Cloudiness governs the complementary proportions of diffuse and direct radiation. In a clear, clean sky, about 20% of the solar radiation is diffuse, whereas it is all diffuse when the sky is overcast. About a third of all solar radiation energy is visible, that is, light.

Temperature (see also MEASUREMENT OF TEMPERATURE AND HUMIDITY and THERMAL INSULATION). The air temperature outdoors depends on the latitude, season, and time of day (which determines the radiation intensity), direction of the incoming wind, distance from the sea, and altitude. Temperatures fall by about 6°C for each 1000-m ascent.

The highest temperatures are limited by the escape of hot air in convection upward, so they vary less over a wide range of latitudes and locations than the minimum temperatures. Particularly low temperatures are found in sheltered valley bottoms, where relatively heavy cold air collects, and on mountain tops, which are cooler because of the height; an intermediate zone is warmest. The daily range is greater in dry climates inland and at higher latitudes.

The daily mean temperature is taken as the average of the daily minimum (at dawn) and the daily maximum (in the early afternoon). The monthly mean is the average of the daily means. The difference between the daily mean and some threshold value is a measure of the amount of heating or cooling required that day. If the mean is below 18.3°C (65°F) in North America [15.5°C (60°F) in Britain], the difference is the number of *heating degree-days* that day, and the total for all days in a typical year is a measure of climatic rigor. Similarly, when the mean exceeds 18.3°C (65°F) in North America [24°C (75°F) in Australia], there is a corresponding number of cooling degree-days, which determines air-conditioning needs.

Special attention must be paid in building design to the frequency of extremely hot or cold days, those with maximum temperatures over 33°C or minimums below 0°C, and the likelihood of runs of days with such extreme conditions.

Humidity (see also CONDENSATION and MEASUREMENT OF TEMPERATURE AND HUMIDITY). All the surface atmosphere contains water vapor, even in the desert, and especially at low latitudes. Traditionally, the yardstick of measurement has been the *relative humidity* (RH), but this has the disadvantage of depending on the temperature, not only on the air's moisture content, so it fluctuates widely during the day. Also, it is inaccurate to take averages of RH values. A better measure is the *dew-point temperature*, which varies little during the day and gives directly the temperature at which condensation will occur if the air is cooled. If the dew point is close to the air temperature, this is equivalent to a high relative humidity.

Rain (see also HYDROLOGY OF ROOF WATER COLLECTION). Rainfall results from clouds, which are caused by local uplift of the atmosphere by either hills (orographic clouds), convection (due to heating of the surface by the sun), or fronts (the zones of collision between air masses of different temperature and humidity). The three kinds of clouds cause characteristic sorts of rainfall. Orographic rain and frontal rain tend to be steady and prolonged; convective rain is intense and brief, as in equatorial regions. Areas affected by monsoonal winds (with a switch of the prevailing direction by at least 120° between seasons) or tropical cyclones receive particularly high rainfalls in certain months of the year. (For places between the tropics

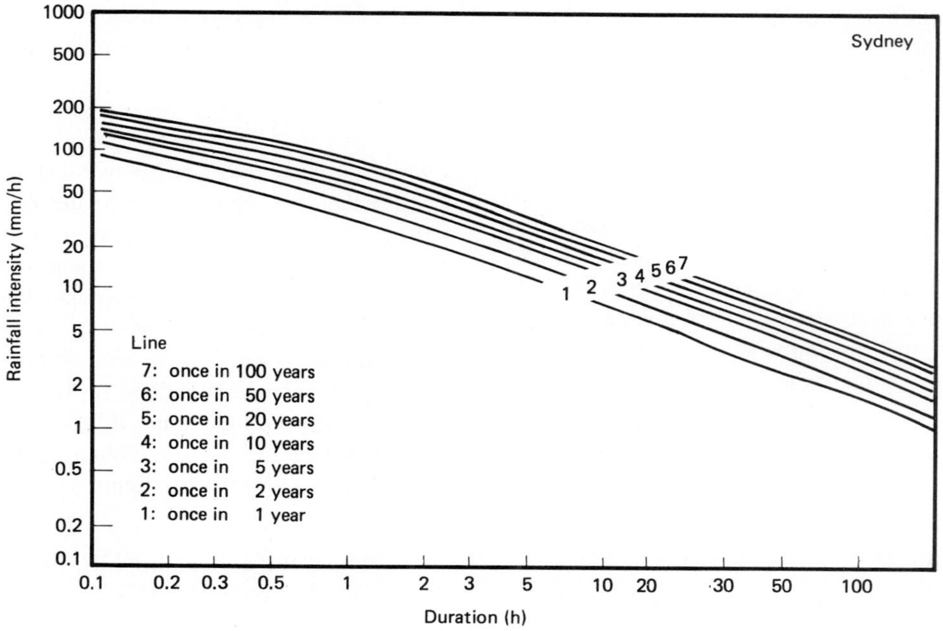

Figure 28 Diagram of the frequency–duration–intensity of rainfall at Sydney. (Reproduced from *Australian Rainfall and Runoff.* Institution of Engineers Australia, Canberra, 1977, by courtesy of the Institution.)

one talks of the wet and dry seasons, instead of summer and winter.)

Charts are available for many cities that show the frequency of a particular intensity of rainfall. For example, Fig. 28 shows that a rainfall of at least 35 mm/h will last for an hour about once a year at Sydney.

Rainfall is variable over small distances, especially in hilly districts. Two suburbs in Sydney that are only 20 km apart, receive respectively 800 and 1300 mm per year, in the long term. Differences on a monthly or daily basis are even greater.

Snow affects building design, because of its weight on the roof and the obstruction of windows and doorways. It occurs at either high latitude or high altitude and is associated with daily mean temperatures below 0°C (32°F).

Wind (see also NATURAL VENTILATION). A major factor in building design is wind direction, especially at times when it is raining, when the wind is extremely strong, or when the wind is cool or extremely hot in a hot climate, and conversely in a cold climate. The need for protection from driving rain, strong gusts of wind, and cold winds in winter or hot in summer from particular directions should be assessed in terms of the frequencies in January and July, along with the opportunities for welcome cooling by sea breezes near the coast in a hot climate.

The frequency of calms is an index of poor ventilation and high air pollution conditions from nearby chimneys and traffic.

The especially strong winds associated with tornadoes (Ref. 2, p. 66) need particular attention in regions such as the central United States.

Tropical Cyclones (or hurricanes or typhoons) affect building on the east and equatorward coasts of the continents and nearby islands such as Japan and Jamaica. They occur when sea-surface temperatures rise above 26°C and start only within a 5 to 15° latitude, subsequently drifting westward, and then poleward. A tropical cyclone represents a huge convection, creating winds up to 60 m/s, intense rainfalls for a day or two, and a surge wave several meters high, which can engulf the coast.

References

1. E. T. LINACRE: *Climatic Data and Resources*, Australian Professional Publications, Sydney, Australia, 1987.
2. E. T. LINACRE and J. E. HOBBS: *The Australian Climatic Environment*. Jacaranda Wiley, Brisbane, Australia, 1977.

E.T.L.

COLORED CONCRETE. To avoid the dull natural gray color and add esthetic value and warmth, architectural concrete may be colored by methods other than opaque coatings. They are color admixtures, white and colored cements, mineral oxide pigments, dry-shake hardeners, and chemical stains.

Color Admixtures are homogeneous blends of carefully proportioned compatible ingredients that reduce water demand, increase strength, and disperse pigments and cement thoroughly for ease of placement and good color uniformity. Colors are generally pastel and pleasing for mass concrete structures.

White and Colored Cements are available. White PORTLAND CEMENT is manufactured with minimum quantities of iron and manganese oxide, reducing or eliminating the natural gray color. Pigments and generally white portland cement are blended by the manufacturers into pigmented cements. Special color cements are manufactured in shades of tan to beige.

Mineral Oxide Pigments are used mostly in precast concrete pavers or products where some variation in color shade is acceptable. Pigments must be carefully selected and proportioned, as many are not lime proof or fade resistant, and these should not be used in concrete.

Dry-Shake Colored Hardeners are streak-free intergrinds of pigments, surface-conditioning and dispersing agents, and portland cement, combined with specially graded aggregates and resulting in a ready-to-use product. They are applied monolithically to the surface of fresh floated concrete. Colors are more intense than by other methods. They are used in floors and walkways and with stamped, patterned concrete.

Chemical Stains are a method of coloring the matrix of concrete after the concrete is cured. They are not pigmented coatings, but water-solutions of metallic salts that penetrate and react with alkaline particles to produce insoluble, abrasion-resistant color deposits in the pores of the concrete. These stains enhance decorative textures in walls and horizontal surfaces, providing interesting and unusual effects.

A number of sealers and coatings, both clear and pigmented, can be used to enhance and maintain the surface appearance.

Reference

C. M. DABNEY: Five ways to color concrete. *Concrete Construction*, Vol. 27, No. 1 (Jan. 1982), p. 19.

C.M.D.

COLOR MEASUREMENT SYSTEMS. Various systems have been developed that allow the specification or description of color without the need to physically match samples. The CIE (Commission International de l'Éclairage) has specified a number of *illuminants* and *standard light sources* for color measurement and comparison to represent *incandescent lighting*, *direct noon sunlight*, *overcast sky daylight*, and *overcast sky daylight with a component of ultraviolet radiation* (Ref. 1).

Colored Surfaces need three dimensions for their specification: *hue*, *saturation*, and *lightness*. There are three methods for specification: color names, material color standards, and spectrally based colorimetric systems. Sometimes color names are sufficient; for example, "white" may adequately specify the color required for painted road lines. But, in general, names are too imprecise for color specification, as there is no internationally agreed system for color names. However, the Inter-Society Color Council and the U.S. National Bureau of Standards have proposed a *Universal Color Language* (Ref. 2).

Color can also be expressed in terms of *material color standards*. One of the best known atlases of colored chips is the *Munsell Book of Color* (Ref. 3). The Munsell system uses three subjective attributes that describe a three-dimensional *color solid* of all possible surface colors (Fig. 29). *Munsell hue* defines red, green, yellow, blue, purple, or intermediate colors between some adjacent pairs. There are 10 hues and each hue has 10 subdivisions; for example, 5BG is a blue-green in the center of the BG hue range. *Munsell value* defines lightness related to surface reflectance from 0/ (black) to 10/ (white). A Munsell value of 5/ is a gray midway between black and white. *Munsell chroma* defines saturation (vividness) on a scale extending from /0 (no color), by steps of approximately equal visual importance, to /20 for the most saturated specimens producible. The *Munsell notation* for a color is hue value/chroma; for example, 2.5 YR 5/10 indicates a hue of 2.5 YR (yellow-red), a value of 5, and a chroma of 10.

A spectrally based colorimetric system is used where precise color specification is essential. The CIE system (Ref. 1) allows the finest subdivision of color, and it has been estimated (Ref. 2) that it provides 5 million identifiable colors. It uses two chromatic variables (x, y) to specify the color of the sample (Fig. 30) and a third variable (Y) to specify its *luminous reflectance* or *transmittance*. Tables allow a given Munsell notation to be converted to the CIE system.

Light Sources can be described by their *correlated color temperature*, which is the temperature of a black-body radiator whose perceived color most closely resembles that of the light source, measured in kelvins (K).

The *CIE General Color Rendering Index* (Ref. 4) is a method of measuring the ability of a light source to render colors compared with that of a full radiator at a temperature close to the correlated color temperature of the source. It is calculated from the shift in the chromaticity coordinates of eight Munsell color chips under both sources.

In many commercial applications, *brightening agents* are used to improve or alter the appearance of surfaces by adding color or fluorescent material. The latter make the measurement and specification of color more difficult because these *nonoptical brighteners* convert ultraviolet radiation to visible light.

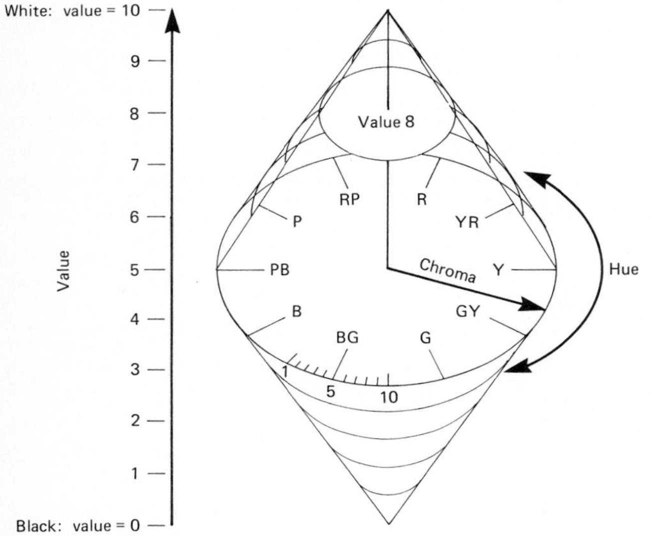

Figure 29 This idealized color solid shows how colored surfaces are represented in the Munsell nomenclature.

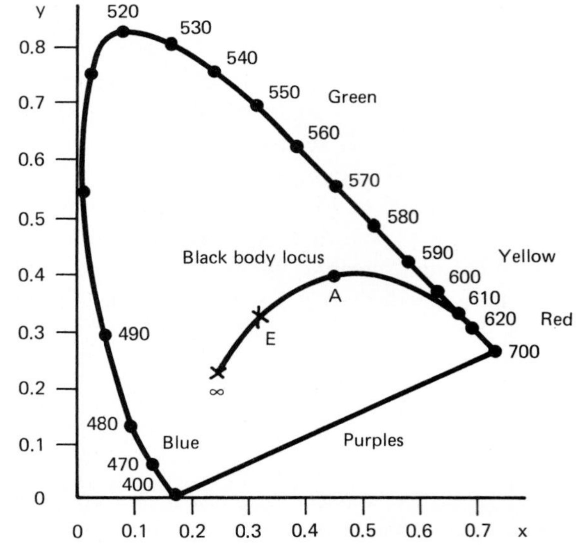

Figure 30 CIE chromaticity diagram. The spectral colors form the boundary, with the nonspectral purples along the base. The black-body locus is shown extending from red into the white center of the diagram.

References

1. *Colorimetry*, CIE No. 15. Commission Internationale de l'Éclairage, Paris, 1971.
2. K. KELLY and D. B. JUDD: *Color. Universal Language and Dictionary of Names*. National Bureau of Standards, Publication 440, Washington, D.C., 1976.
3. *Munsell Book of Color*, Munsell Color Company, Baltimore, Md., 1983.
4. *Method of Measuring and Specifying Color Rendering Properties of Light Sources*. CIE No. 13.2, Commission Internationale de l'Éclairage, Paris, 1974.

W.G.J.

COMPOSITE CONSTRUCTION is an improvement of the commonly used concrete slab resting on supporting beams and girders. Composite beams may consist of either steel, PRECAST CONCRETE, or timber beams, with either a lightweight or regular concrete slab. Instead of merely being supported by the beams, the concrete slab is made a part of the beam and serves the dual purpose of enhancing the properties of the supporting member and serving as a floor slab. The composite concept is also used with slabs and columns. Since composite construction came into use in the 1930s, a variety of proprietary systems has been patented.

The design of composite members may be by the elastic theory or by ultimate strength. In the United States, the design of steel or timber beams with a composite slab normally follows an elastic procedure in which the concrete area is transformed into an equivalent area of the beam material. The design of a member with a precast concrete beam and concrete slab is based on ultimate strength.

Composite Slab. The usual type of composite slab consists of a formed metal decking with bonding deformations for developing the shear transfer between the metal decking and the concrete slab. The metal decking serves as the slab reinforcement and as a stay-in-place form (permanent FORM WORK). The design information for composite slabs is provided by the manufacturers of the metal decking. This type of composite slab is shown in Fig. 31. Another type of composite slab uses flat steel plates or checker plates with shear connectors. These may be used as floor construction for very heavy loads or as hatch covers.

Steel–Concrete Composite Beams are very common. They consist of steel beams with shear connectors welded to the top flange, which project into the concrete slab. Headed studs are the most frequently employed shear connectors, but other types of welded or bolted devices may be used. The steel beam may also be encased in concrete and the shear connection provided by bond without mechanical shear connectors.

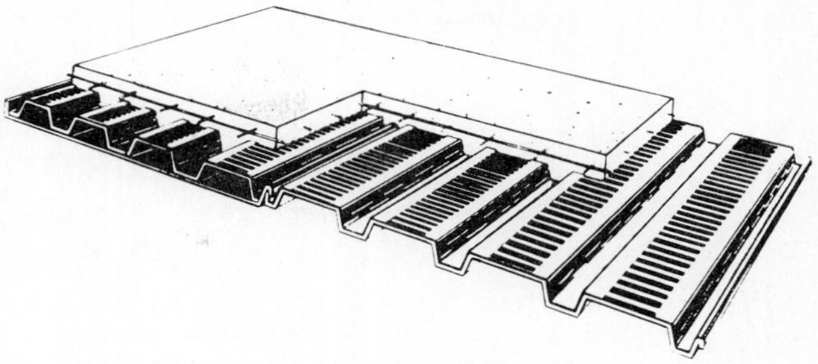

Figure 31 Composite slab with metal decking as reinforcement.

The concrete slab may be formed by plywood; alternatively, metal decking may be used as permanent FORM WORK. When composite metal decking is used, the resulting construction consists of both composite slabs and composite beams, as shown in Fig. 32. Shear connectors welded through the decking develop shear transfer for the composite beam and provide a very ductile type of member, as shown in Fig. 33.

Concrete–Concrete Composite Beams consist of precast reinforced or PRESTRESSED CONCRETE joists or beams, cast composite with floor or roof slabs. The shear connection for this type of construction consists of a rough top surface on the precast member and of reinforcing steel that projects from the top of the precast member into the concrete slab.

Composite Columns. Several types of composite columns have been used. They consist of a steel section combined with concrete, and sometimes also additional reinforcing steel. Ultimate strength design procedures are used

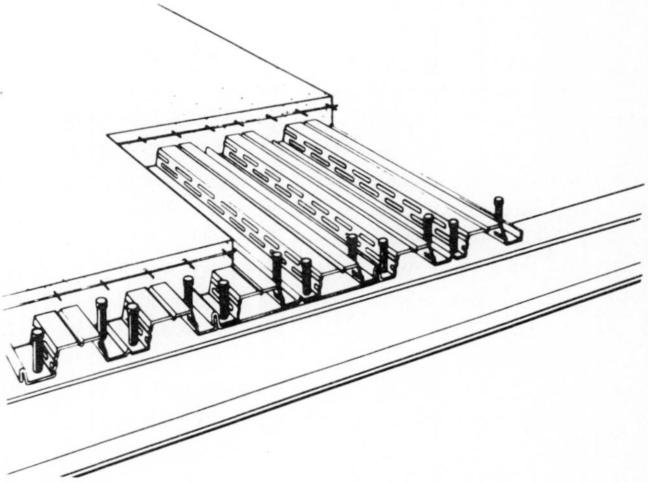

Figure 32 Composite beam showing stud shear connectors.

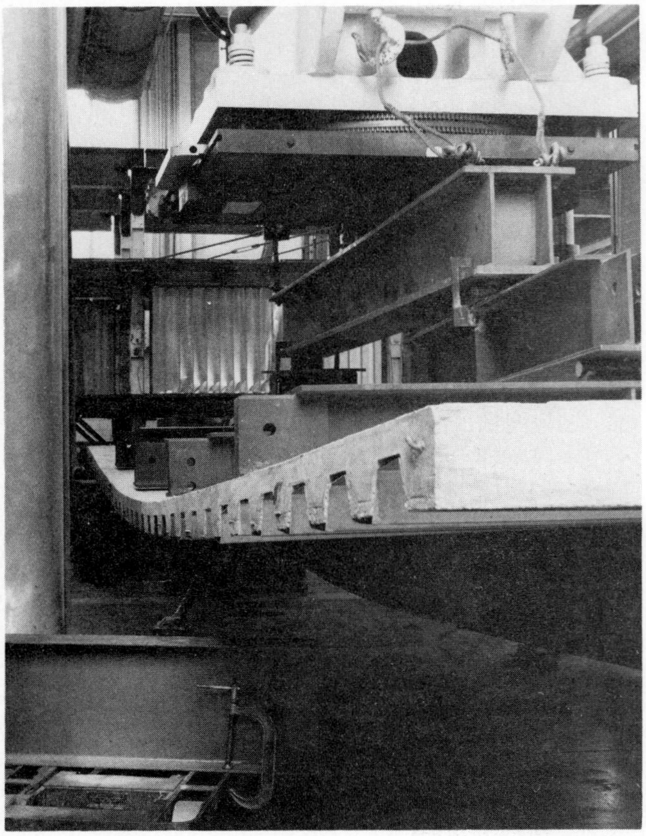

Figure 33 Testing composite beam with loading at four points along the span.

for the proportioning of these members in accordance with the ACI code (Ref. 4).

See also REINFORCED CONCRETE DESIGN and STEEL STRUCTURES.

References

1. R. G. SLUTTER and G. C. DRISCOLL: Flexural strength of steel–concrete composite beams. *Journal of the Structural Division, ASCE*, Vol. 91, No. ST2 (1965), Procedure Paper 4294.
2. Survey of the sub-committee on the state of the art: composite steel concrete construction. *Journal of the Structural Division of ASCE*, Vol. 100, No. ST5 (May 1974).
3. J. A. GRANT, J. W. FISHER, and R. G. SLUTTER: Shear Connector Behavior in Composite Beams with Metal Decking. *ASCE Convention, Las Vegas, Nevada, April 26–30, 1982*, Reprint.
4. *Building Code Requirements for Reinforced Concrete, ACI 318-83*. American Concrete Institute, Detroit, Mich., 1983.

R.G.S.

COMPUTER-AIDED ARCHITECTURAL DESIGN, taken in the broadest possible sense, might refer to any application of DIGITAL COMPUTER technology in the building feasibility study, programming, design, documentation, construction, management, maintenance, renovation and reuse, and demolition cycle. Since the mid 1970s, though, it has increasingly been understood in a narrower sense, as the use of comprehensive *computer-aided design systems* (Fig. 34) to support an architectural design process from beginning to end.

A computer-aided design system has four basic functions (Fig. 35). First, it provides facilities for *input and editing of building descriptions* (much as a word processor provides facilities for input and editing of text). Second, it

Figure 34 A CAD workstation. (Photograph by courtesy of the Calcomp Corporation.)

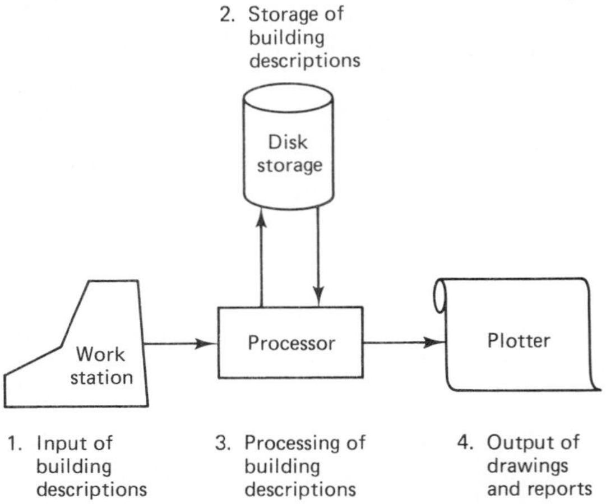

Figure 35 The four basic functions of a computer-aided design system.

provides capacity for *storage* of building descriptions in digital format, just as a word processor allows storage of text on disk. Third, it has software for *processing* building descriptions in various ways, for example, to execute searches (just as a word processor can search for specified character strings), to take off counts, areas, and volumes, and to perform engineering and cost analyses. Finally, it provides for *output* of drawings (plans, elevations, sections, details, perspectives) and printed reports (such as equipment schedules), just as a word processor provides for output of final printed copy.

The heart of a computer-aided architectural design system is its software for maintaining building descriptions. The characteristics of this software will determine the kinds of input and editing operations that the system can provide, the kinds of processing that it can support, and the kinds of output that it can produce.

The simplest computer-aided design systems (Fig. 36) represent buildings as sets of stored *two-dimensional* drawings (plans, elevations, etc.). These have very limited functions and are mostly used to enhance productivity in routine drafting tasks.

Slightly more sophisticated systems recognize structures such as closed polygons and networks in two-dimensional drawings (Fig. 37). Those can, to a limited degree, support automated production of schedules and bills of materials from drawings, automated layout and sizing of ductwork, and other such operations.

Increasingly, now, computer-aided design systems incorporate *three-dimensional* modeling capabilities. There are several levels of sophistication (Fig. 38). *Wire-frame* modeling systems represent objects by their edge lines; these can do little more than produce simple perspectives (without hidden lines removed) and allow modeling of structural frames. *Surface* modeling systems represent the surface properties of solid objects and so support the production of realistic shaded and colored perspectives and the simulation of lighting and acoustic effects (Fig. 39). *Solid* modeling systems are geometrically complete and allow the representation of material, volume, and mass properties of solid objects. These can support such applications as automated spatial clash checking and finite-element analysis.

It is clear that three-dimensional modeling systems can do more than two-dimensional systems and that solid mod-

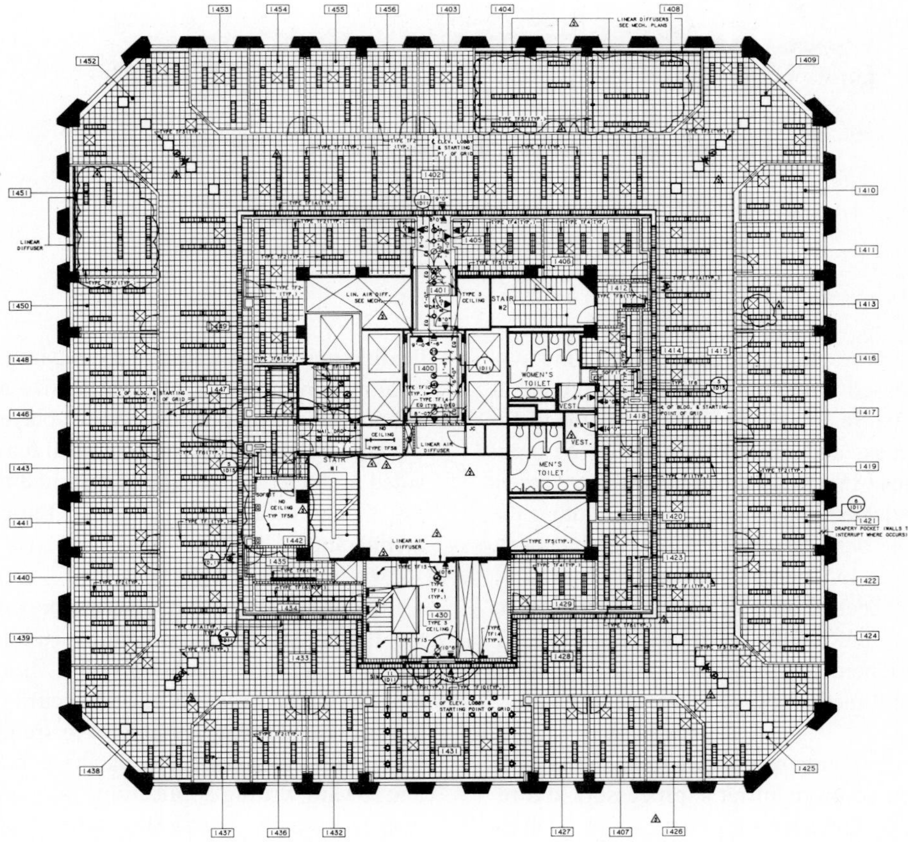

Figure 36 Plans generated by a drafting system.

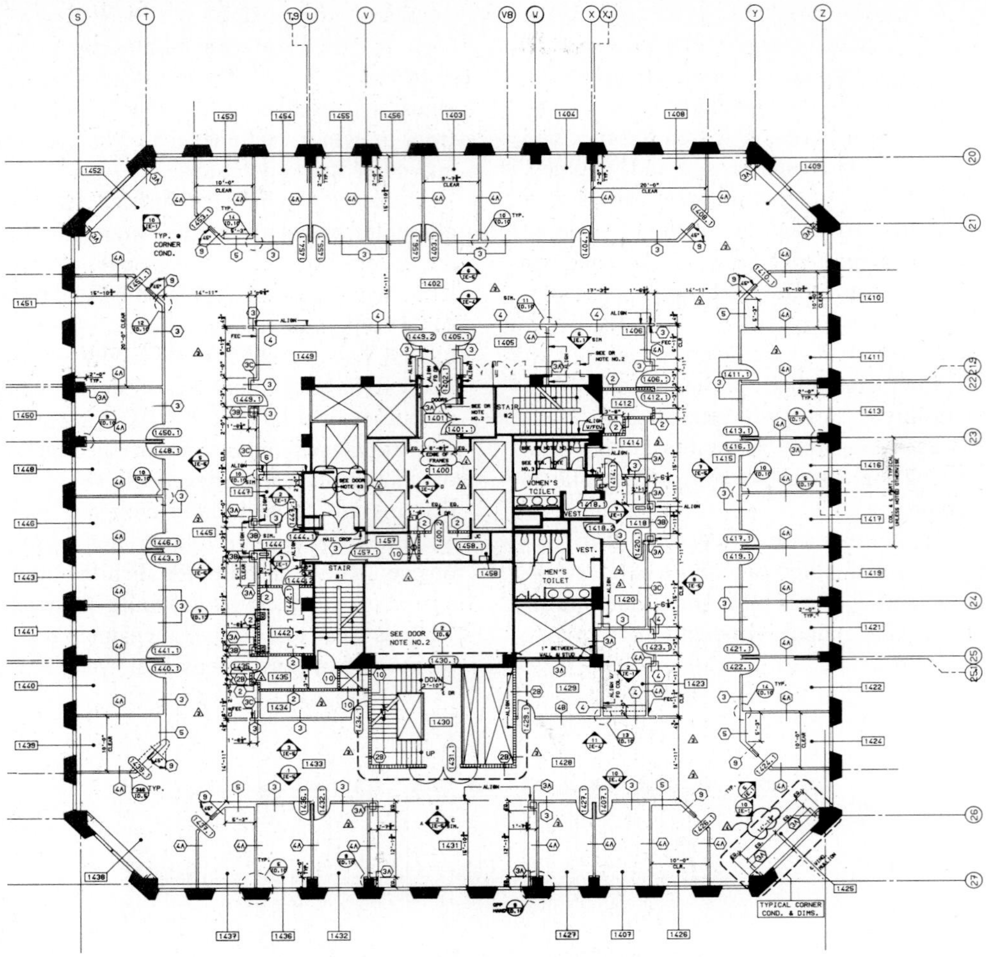

Figure 36 *(cont.)*

eling systems can do more than surface or wire-frame systems. However, increased functionality is always bought at a price of greater software complexity, greater demand for computational resources, and greater complexity of input and editing operations. In practice, one does not always want the most sophisticated possible system, but rather the one that represents the best balance of functionality, cost, and operating efficiency for the task at hand.

So far there have been four generations of computer-aided design systems, and a fifth is on the horizon. Each generation has represented a substantial advance in both functionality and cost-effectiveness over its predecessor.

First-Generation Systems began to appear in the early 1960s. These were based on mainframe processors and refreshed vector displays. Software for graphics and database management was, by today's standards, primitive and inflexible and very limited in its capacity to handle architectural design problems of realistic size and complexity. So, although these pioneering systems played the important role of demonstrating the basic technical feasibility of computer-aided architectural design and excited a good deal of interest, they had negligible effect on the everyday practice of architecture.

Second-Generation Systems prevailed through most of the 1970s. These were driven by 16-bit minicomputers and used storage tube graphic displays. Issues of speed, capacity, flexibility, reliability, and security were seriously addressed by software. There were trends toward providing increasingly sophisticated database management software and toward writing applications programs in high-level languages (usually FORTRAN) rather than in assembly language. This generation of technology precipitated the initial

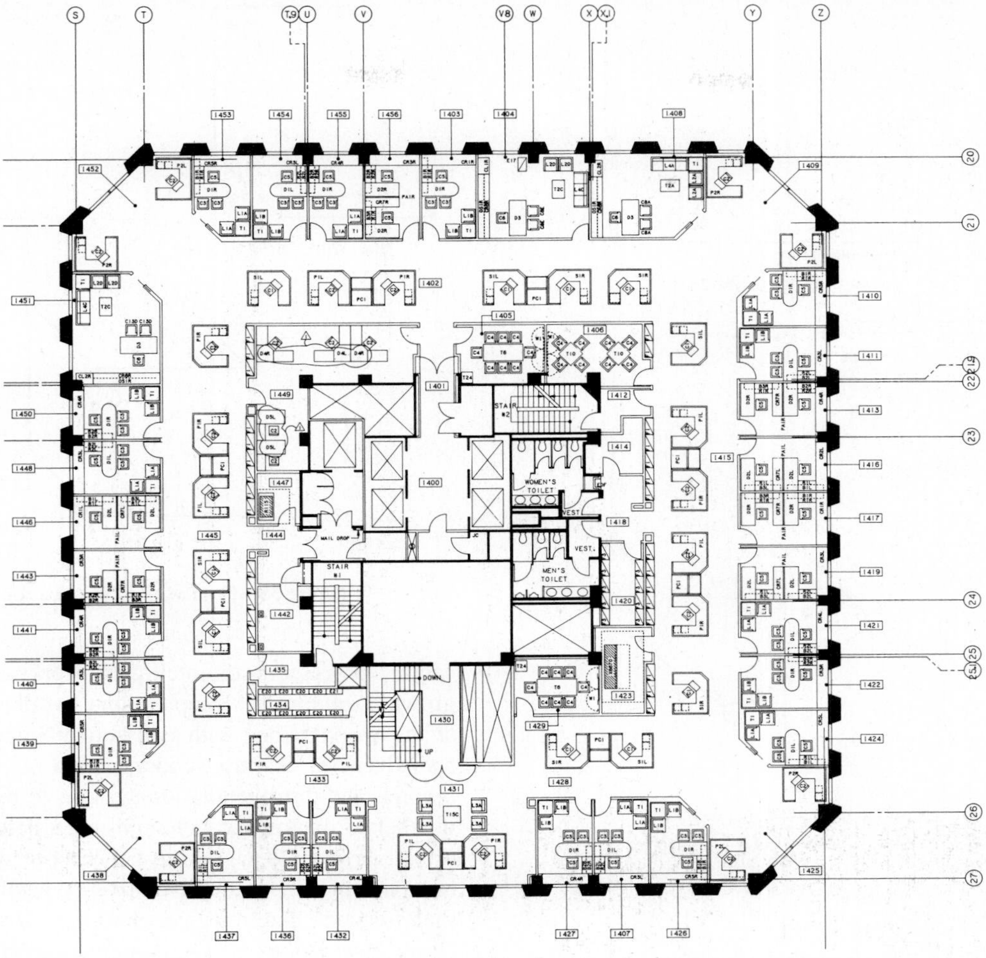

Figure 36 *(cont.)*

growth of what has now become the very large turnkey CAD/CAM (computer-aided design/computer-aided manufacture) system industry. During the 1970s, many large architectural offices began to experiment seriously with second-generation computer-aided design systems, mostly in working drawing production applications.

Third-Generation Systems began to appear in the late 1970s and became very popular in the early 1980s. These are based on 32-bit superminicomputers, such as the VAX. Increasingly, raster graphics displays have replaced storage tubes, and the use of color displays has become commonplace. Electrostatic plotting of graphic output has become increasingly competitive with pen plotting, which prevailed in earlier generations. Software, generally, has become much more sophisticated. The availability of sophisticated virtual memory operating systems and of powerful software engineering tools has facilitated the development of increasingly ambitious software. More application-specific intelligence has been built into software, and increasing attention has been paid to the design of user interfaces and issues of user training and assistance. This generation of technology ushered in a period of explosive growth for the CAD/CAM industry; by 1984, in the United States, there were several dozen active vendors, with combined yearly sales in the billions of dollars. About half of the sales were in the mechanical engineering field and another quarter in electronics. Architecture and construction accounted for 10% to 20%, but was the fastest growing segment of the market.

In 1984, in the United States, an installation of a third-generation CAD system typically cost several hundred thousand dollars. The yearly cost per work station came out to about $50,000, approximately the yearly cost (with over-

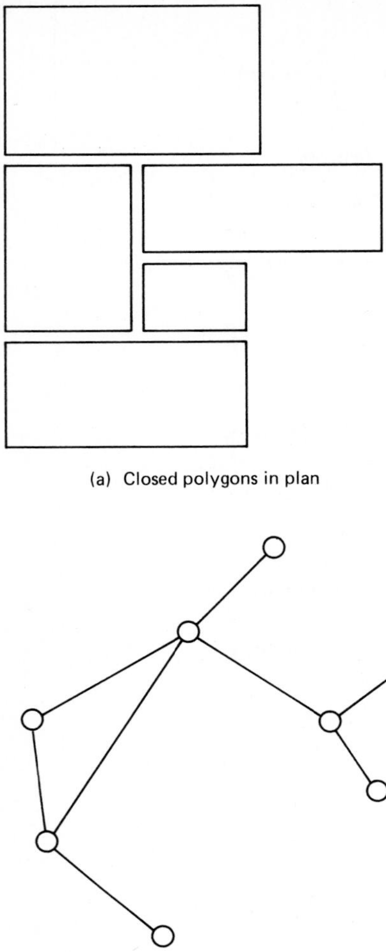

Figure 37 Closed polygons and networks.

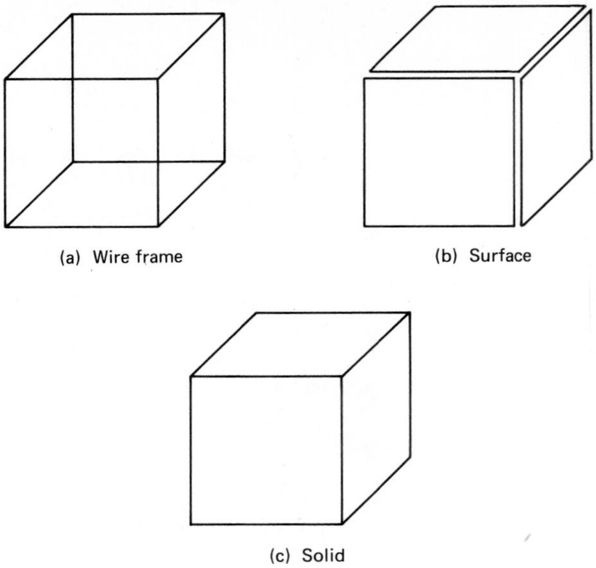

Figure 38 Three ways to model a solid object.

head) of a single technical employee. This meant that, if drafting productivity was assumed to be the principal benefit of a CAD system, achievement of a productivity ratio of about 2 : 1 was the threshold for cost effectiveness. Within this economic framework, many large architectural offices, with high, steady flows of relatively repetitive work, were able to make CAD cost effective. Smaller firms, however, found that the high level of initial capital investment and the higher levels of work flow needed for cost effectiveness still constituted formidable barriers to getting into CAD.

Fourth-Generation Systems began, in 1983, to challenge established third-generation systems. The emergence of very cheap, powerful processor and memory chips had, by then, generated a strong trend toward the development of distributed intelligence computer networks supporting distributed databases, and the new CAD generation began to take advantage of this. Established CAD system vendors began, increasingly, to move functions away from large central processors to intelligent workstations, so that a growing amount of computational work was done locally at the workstation, with consequent improvements in responsiveness. Startup vendors, with no installed base of second- and third-generation systems to protect, were even quicker to exploit the potentials of a networked computational environment. Fourth-generation systems generally have a much more fluid, natural style of user interaction than second- and third-generation systems. Whereas the earlier generations of CAD systems mostly focused on providing the general graphic, geometric modeling, and database functions needed in a variety of design fields, third-generation systems have become increasingly highly specialized to specific design disciplines. In other words, architects have begun to make use of *architectural* CAD systems, which incorporate a substantial amount of professional

Figure 39 Shaded perspective displayed by a CAD workstation.

knowledge in their software and databases and so can automatically perform relatively high level architectural analysis and synthesis tasks.

Fourth-generation systems have also begun to change the conditions of the financial feasibility of CAD. Whereas earlier generations had required a large initial investment to acquire a central unit before any workstations could be operated, fourth-generation installations can begin with a single, relatively inexpensive, stand-alone workstation and can grow station by station to a large size. Thus it becomes more feasible for small- and medium-sized architectural firms to become involved in CAD.

Fifth-Generation Systems will no doubt emerge as a result of ongoing developments in nontraditional computer architectures, special-purpose processors for efficiently handling basic graphics and database functions, and knowledge engineering. Some very clear trends can be observed in the development of computer-aided architectural design and seem likely to characterize the field in the foreseeable future. Basic hardware costs will continue to drop. Large-scale networks of powerful, 32-bit workstations with high-quality raster graphics will become the standard CAD environment. Such networks will provide on-line access to databases of building products information, construction cost data, standard element and detail libraries, and so on. Specialized software modules and databases of various kinds will support a widening range of architectural functions and will do so with increasing sophistication. As a result, the major part of the investment that an architectural firm makes in CAD will be in software, databases, and user education. Systems will be increasingly highly integrated around comprehensive building description databases and unified user interfaces. Data transfer standards will emerge, and digital communication of construction information will begin to supplant transmission via drawings and text on paper. Digital building descriptions will be used not only in design and construction, but also for ongoing facility management.

The traditional architectural office is organized on a 19th-century pattern; it is labor intensive, processes large quantities of paper by hand, relies very heavily on the knowledge and experience carried in the heads of its members, and operates within a legal and institutional definition of professionalism that was established in the 19th century. The emerging electronic architectural office is capital intensive. There is a powerful CAD workstation on every desk, backed up by digital communications and extensive software and databases. Much of the expertise that the firm offers is carried in software and databases and so remains intact as staff come and go, and can accumulate over time. Consequently, traditional definitions of the professional architect's role and of the education and experience required for it are rapidly becoming obsolete.

See also COMPUTER GRAPHICS and DRAFTING.

References

1. WILLIAM J. MITCHELL: *Computer-aided Architectural Design*. Van Nostrand Reinhold, New York, 1977.
2. WILLIAM J. MEWMAN and ROBERT F. SPROULL: *Principles of Interactive Computer Graphics*, 2nd ed. McGraw-Hill, New York, 1979.
3. CHARLES M. FOUNDYLLER: *Turnkey CAD/CAM Computer Graphics*. Daratech Associates, Cambridge, Mass., 1985.

W.J.M.

COMPUTER GRAPHICS. The development of computer graphics (the use of computers to generate graphics) can be traced directly to work of Ivan Sutherland at MIT in the early 1960s. Computers generate graphics through the use of display screens where the computer controls the drawing of lines or vectors on the screen. There are many different screen technologies; three make use of an electron gun that fires electrons at a screen coated with phosphor. When the electrons hit the phosphor, it glows briefly causing the viewer to see an image. Since this glow persists for only a short time, some mechanism must be introduced to make the image appear constantly on the display.

History. The first displays, called *vector* displays, made use of a graphics memory. This was used to constantly redraw the image on the screen, a process called *refreshing*. Changing the contents of the graphics memory changes the displayed image. Since the optic nerve samples at approximately 25 times per second, the image needs to be refreshed at least that often to prevent the perception of flicker. This technology was very expensive and was quickly replaced for most applications by the *storage tube* display, a name synonymous with its manufacturer, Tektronix. This contained a fine wire grid placed next to the phosphor-coated surface of the screen. The electrons fired at the screen hit this grid. It stored the charge and continued to emit electrons that hit the screen, thus storing the image. This has the advantage over vector-refresh displays in that no graphics memory is required. Its main disadvantage is the static nature of the image; it is impossible to alter a single line on the screen without first erasing the entire screen and then redrawing the modified image. Storage displays were the standard graphics displays for a decade, commencing in the early 1970s. They continue to be used today, although less frequently.

Resolution. The large-scale production of television screens resulted in substantial cost reductions and increasing reliability. Coupled with a substantial reduction in the cost of memory, this produced the *raster* display. This still fires electrons at phosphor on the screen, but does so in a serial raster form. That is, the screen is divided into rows of picture elements, called *pixels*, and the electron gun targets each pixel in turn, row by row. If a particular pixel is part of a line, the gun fires an electron; otherwise, it moves on to the next pixel. Thus a line is simply a series of discrete pixels. Raster displays are very cheap, compared to both vector-refreshed and storage screens. They came to popularity with their use in personal computers. Raster displays, however, have a significant functional advantage over the other technologies; they can exhibit color by using a mix of red, green, and blue. Whereas raster displays initially had a low resolution, the displays in typical computer-aided design systems now have a resolution of 1000 by 1000 pixels.

Interaction. Building designers can interact graphically with a computer by using a variety of pointing devices, which control a cursor on the screen. These range from joysticks, thumbwheels, and lightpens to mice. A lightpen is used to point directly at the screen. A mouse, so called because of its appearance, has a wire connecting it to the computer; this looks like a tail. Because the device sits on a platen or surface under the hand of the operator, it is ergonomically sounder than the other methods. Graphical interaction between the user and the system is important for effective use. This is carried out by an interface.

Fundamental to the use of computer graphics in building and architecture is the notion of a computer model of the object to be drawn. When humans produce drawings, they carry a model of what is to be drawn in their heads. Similarly, for computers to be effective, a computer-readable (digital) representation is essential. This is called *geometric modeling*. Once a model of a building has been constructed inside the computer, the production of drawings can be carried out automatically, using computer programs designed for this purpose. Computer-aided drafting systems fall into two categories, those without and those with modeling capabilities. Those without models can only be used as highly sophisticated drawing machines; those with modeling extend the productive range of computers (Figs. 40–42). Typical drawings of buildings produced by computers are shown under the DRAFTING entry.

See also COMPUTER-AIDED ARCHITECTURAL DESIGN and DIGITAL COMPUTERS.

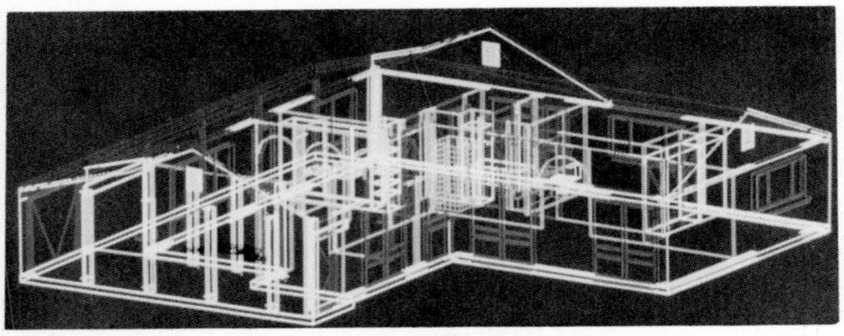

Figure 40 Wire-frame (or outline) perspective view of a three-bedroom house drawn by computer from a geometric model stored in the computer. (By courtesy of Huxley Homes, Sydney, Australia, using the EASINET CAD System.)

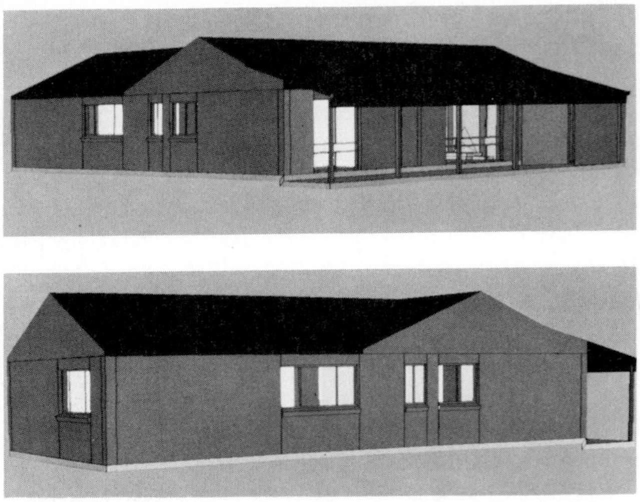

Figure 41 Solid perspective with hidden surfaces removed drawn from the same geometric model used to draw Fig. 40. The two views demonstrate that the perspective can be drawn from any location. (By courtesy of Huxley Homes, Sydney, Australia, using the EASINET CAD System.)

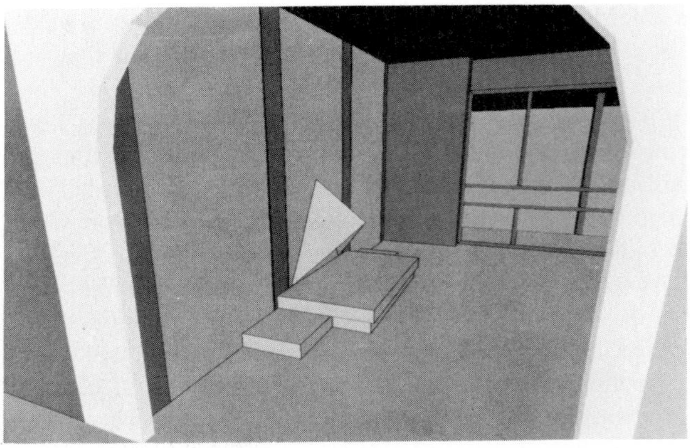

Figure 42 With a full geometric model it is possible to generate internal views of buildings. This is a view of the living room of the same house shown in Figs. 40 and 41. (By courtesy of Huxley Homes, Sydney, Australia, using the EASINET CAD System.)

References

1. J. D. FOLEY and A. VAN DAM: *Fundamentals of Interactive Computer Graphics*. Addison-Wesley, Reading, Mass., 1982.
2. W. M. NEWMAN and R. F. SPROULL: *Principles of In-*

teractive Computer Graphics, 2nd ed. McGraw-Hill, New York, 1979.

J. S. G.

COMPUTER METHODS IN STRUCTURAL ANALYSIS are used for the analysis of beams, rigid frames, and trusses to predict their response to load. Prior to the arrival of the computer in the early 1950s, methods were developed that minimized the amount of arithmetical work to be done. The type of structural form used was determined often by its suitability for stress analysis and the tail wagged the dog. In general, STATICALLY DETERMINATE frames were preferred to indeterminate ones, since effective graphical techniques were available for solving quite large systems on the drawing board. For rigidly connected concrete frames, where a high degree of redundancy had to be accepted, moment distribution was the predominant method from the mid 1930s until recently.

Structural analysis is an operation for converting the mathematical model into a tractable numerical procedure capable of producing solutions of relevance without exorbitant effort. Modern computer methods are more general forms of the earlier *slope-deflection method*, which becomes impracticable as a manual method when the number of displacement unknowns exceeds five or six. *Stiffness methods* (explained later) generally are avoided in manual solutions, not necessarily because of the large number of simultaneous equations involved, but because the nature of the equations does not permit of an effective uncoupling with determinate frames to solve joint equilibrium equations a few at a time to find the values of stress resultants; the same approach rarely worked if one was solving for *joint displacements*. The stiffness method gives rise to very large sets of simultaneous equations with displacements as the unknowns. The first real task of the computer is then the solution of simultaneous equations.

First-generation computers were not well suited to solving equations because of deficient memory capacities (Ref. 1). With the provision of larger memories, compiler languages became possible, giving faster execution using arrays where more and more of the elements could be in the high-speed memory. The microcomputer of the early 1980s has memory enough to accommodate problems of frame analysis that required the mainframes of the 1960s and the minicomputers of the 1970s.

Linear-Elastic Analysis. The automation of the stiffness method of frame analysis was achieved by the end of the 1950s, driven by the urgent need to analyze airframes of unusual shape to achieve supersonic flight. The computer model of a framework involves the presentation to a computer of coordinate and connectivity data to define numerically the frame's topology, plus member proportions and loading data. The computer models the structure by generating and solving quite large sets of simultaneous equations that describe the physical conditions of equilibrium at each joint in the structure.

The main operation in assembling the equations of joint equilibrium is the generation of the stiffness matrices of each element in turn and their movement to form the complete stiffness matrix of the frame. With the successful solution of the equation sets, further computation produces the final stresses and reactions. The theory of the stiffness method of analysis for elastic frames (Ref. 2) involves only the fundamental considerations of equilibrium, elasticity, and compatibility expressed in the symbols of matrix algebra.

Matrix Formulation. The stiffness method of linear-elastic analysis of structures sometimes goes by the name of the displacement or equilibrium method. The reason for choosing various names is that we concentrate on element and frame stiffness (rather than flexibilities); the equations are solved for displacements or deformations as the unknowns (rather than for redundant forces or actions); finally, the equations themselves express equilibrium of forces and actions (rather than compatibility of deformations in members and displacements of the joints).

We express the fact that the forces and moments acting on every joint in a frame's structure are in a state of equilibrium by the matrix expression

$$\{W\} = [A] \cdot \{SR\} \tag{1}$$

In this equation the vector or list of joint loads is represented by $\{W\}$ and the vector of stress resultants by the term $\{SR\}$. A stress resultant can be a force, bending moment, or twisting moment, depending on the type of structure undergoing analysis. The term $[A]$ represents the array or matrix of coefficients of the joint equilibrium equations and is called the statics matrix. Such an array will be square only for a STATICALLY DETERMINATE frame, but is rectangular for any redundant structure. Rectangular arrays cannot be inverted, which is another way of saying that the ELASTIC ANALYSIS OF A STATICALLY INDETERMINATE frame that takes into account only the condition of equilibrium is not possible.

It is a simple matter to calculate the deformations at the ends of a straight bar or strut caused by moments and forces applied at the ends. Hooke's law allows us to calculate the extension of a bar carrying an axial force. The simple theory of bending produces the end slopes of a uniform beam as functions of the applied end moments. For all the members of a frame, we can express the simple relationships between member stress resultants and the corresponding and deformations by a group of flexibility equations represented by

$$\{x\} = [f] \cdot \{SR\} \tag{2}$$

Since there must be as many member deformations as there are stress resultants, the member flexibility matrix (f) is square and may be inverted to provide a more useful member stiffness relationship:

$$\{SR\} = [S] \cdot \{x\} \tag{3}$$

Equations (1) and (3) may be combined to relate the externally applied loads $\{W\}$ to the member deformations $\{x\}$:

$$\{W\} = [A] \cdot [S] \cdot x \tag{4}$$

We relate the member of deformations $\{x\}$ to the joint deformations, which will be a vector $\{X\}$, to get a set of equations relating the joint loads to the joint displacements. There must be a set of relationships connecting the joint to the member deformations; if the members do not deform, the joints will not move, and vice versa. We displace each joint with one movement at a time and work out from compatibility or kinematic considerations the corresponding member deformations. The result is a set of compatibility equations:

$$\{x\} = [C] \cdot \{X\} \tag{5}$$

The array $[C]$ is simply related to the earlier array $[A]$, or the statics matrix. In fact, it is the transposed form of the array $[A]$:

$$[C] = [A]^T \tag{6}$$

Equation (4) expresses the joint equilibrium equations in the form

$$[W] = [A] \cdot [S] \cdot [A]^T \cdot [X] \tag{7}$$

The triple matrix product in Eq. (7) generates the frame stiffness matrix or the array of coefficients of the joint equilibrium equations; if we denote this by $[K]$, the set of equations can be written as

$$[K] \cdot [x] = [W] \tag{8}$$

Once these equations are solved for the joint deformations $[X]$, Eqs. (3) and (4) may be used directly to evaluate the member stress resultants:

$$\begin{aligned}\{SR\} &= [S] \cdot \{x\} \\ &= [S] \cdot [A]^T \cdot \{X\} \qquad (9)\end{aligned}$$

Finally, we may premultiply the vector of the stress resultants by the statics matrix. This produces all the initial joint loads (so providing a check on the accuracy of the arithmetic), including the external reactions, which are usually required in any event.

$$\begin{aligned}[A] \cdot \{SR\} &= [A] \cdot [S] \cdot [A]^T \cdot \{X\} \\ &= [K] \cdot \{X\} \\ &= [K] \cdot [K]^{-1} \cdot \{W\} \\ &= \{W\} \qquad (10)\end{aligned}$$

Equations (1) to (10) represent the entire theory of the matrix stiffness method of frame analysis or, for that matter, the finite element method in general. Admittedly, we have glossed over a few tedious details, such as the development of the member stiffness matrix for various frame elements. Additionally, the process chosen to solve the stiffness equations or to invert the frame stiffness matrix will clearly call for some consideration.

Application to Practical Structures. The preceding matrix theory applies to any stiff structure. Furthermore, it applies to a complete structure composed of many members, but it also applies to the simplest structure, one made of a single member. In early applications to complete structures, it was found that some very large and sparse matrices were involved, which quickly filled the available computer memories. Much more efficient use of memory capacity can be made when we apply the theory to each member in turn and then assemble the overall frame stiffness matrix by inserting each member's contribution, one at a time.

There is much work to be done in implementing the theory to produce computer software to analyze some real frames. In a plane truss with moment-free connections, there is only one stress resultant per member, the axial force or tension. Furthermore, only two static equations apply at each joint. The statics matrix $[A]$ for a single member will be a 4×1 column matrix or vector, and the member stiffness matrix will be a single scalar quantity. Hence the frame stiffness matrix $[K]$ will be an assembly of the 16 elements of the 4×4 member stiffness matrices associated with each member. The situation is more complex for a rigidly connected space structure. There will be six stress resultants per member and six degrees of freedom at each joint, so the statics matrix $[A]$ will be a 12×6 array and the $[S]$ matrix will be 6×6. Consequently, the result of the triple matrix produced, $[A] \cdot [S] \cdot [A]^T$, will be a 12×12 array. Each member then can contribute 144 different elements to the frame stiffness matrix in a methodical, swift, and error-free manner for each of the frame members in turn. The manual assembly of such an array would be an impossible task.

With the stiffness matrix assembled, steps have to be taken to account for the constraints at supports, for otherwise the array is certain to be declared "singular" in the equation solution stage. The right-hand sides of the stiffness equations have to be assembled from the loading information and the set of simultaneous equations processed by the chosen equation-solving routine.

Solution of the Stiffness Equations. The solution of simultaneous equations is perhaps the most common operation in any modeling of physical systems. Structural engineers have to deal usually with either equilibrium or compatibility equations, depending on the mathematical model chosen to represent the behavior of the structure under consideration. For the present purpose, we are concerned with the solution of equilibrium equations with joint deformations as the unknowns.

The best understood method of solution of linear equations is Gaussian elimination, which is taught to school children. Unless there is some special characteristic of the equations to be solved, the school-taught method of elimination is not only the most popular but also perhaps the most efficient. Gaussian elimination involves the removal of the first unknown from all but the first equation by a series of multiplications and subtractions. Eventually, the coefficients are reduced to a triangular form so that the last unknown can be determined directly from the last equation. Thereafter, successive back substitution will evaluate in turn each of the other unknowns. The algorithms for triangulation of the coefficients and back substitution have been set out in Chapter 5 of Ref. 2, where simple FORTRAN lists are also included.

If the coefficients of the stiffness matrix generated by any linear elastic frame analysis program are examined, they will be found to be symmetrical about the leading diagonal. There is no need to either generate or store in memory something less than half the total number of coefficients in the stiffness equations. Furthermore, it will be found that the array of coefficients is often strongly banded in the sense that only zeros occur from the edge of the central band of numbers to the outermost coefficients. This banding is in fact influenced by the node numbering system that is adopted by the engineer, and the band width will be minimized if the engineer keeps the difference between member-connected node numbers as small as possible. Two important benefits follow the appreciation of the banded nature of a computer-generated frame stiffness matrix: (1) there is no need to store the outermost zero coefficients, and (2) there is no need to include such numbers in the equation-solution process.

In taking advantage of the banded, symmetrical nature of stiffness matrices as generated by programmed linear-elastic methods, it is necessary that the extent of the banding be preserved during the elimination procedures. The basis of the ''compact elimination'' methods is the possibility that the original array may be ''decomposed'' into the product of a lower and an upper triangular array. A two-stage back-substitution process may then be used to complete the solution of the equations. An advantage over the direct Gaussian elimination method is that the right-hand sides of the equations are not involved in the decomposition process but only in the back substitutions. For large sets of simultaneous equations arising from frame analysis by the stiffness method, the use of banded, symmetric equations solving systems such as the Cholesky method or the trimatrix decomposition method (Ref. 2) leads to the very effective use of current-model mini- and microcomputers for the stress analysis of framework by structural engineers.

References

1. H. H. GOLDSTINE: *The Computer from Pascal to von Neumann*. Princeton University Press, Princeton, N.J., 1972.
2. H. B. HARRISON: *Structural Analysis and Design: Some Minicomputer Methods*. Pergamon Press, Oxford and Elmsford, N.Y., 1980.

H. H.

CONCRETE is an artificial stone made from an inert aggregate, usually broken pieces of NATURAL BUILDING STONE, and a binding agent, usually PORTLAND CEMENT. In this encyclopedia it is discussed under the following headings: ANCIENT ROMAN CONCRETE; CELLULAR CONCRETE; COLORED CONCRETE; CONCRETE ADMIXTURES; CONCRETE MIX PROPORTIONING; CONCRETE STRUCTURES; DETAILING OF REINFORCED CONCRETE STRUCTURES; FOLDED PLATES; FORMWORK; LIFT-SLAB CONSTRUCTION; LIGHTWEIGHT CONCRETE; LOAD BALANCING IN PRESTRESSED CONCRETE; PLACING OF CONCRETE; PRECAST CONCRETE; PRESTRESSED CONCRETE; REINFORCED CONCRETE DESIGN; REINFORCED CONCRETE SLABS; REINFORCEMENT FOR CONCRETE; SHELL STRUCTURES; SHRINKAGE AND CREEP; and TILT-UP CONSTRUCTION.

CONCRETE ADMIXTURES are materials other than water, aggregates, and hydraulic cement that are used as ingredients of concrete or mortar and are added to the batch immediately before or during its mixing. Admixtures are no substitute for good concrete practice, suitable materials, and appropriate mix design and construction. They may be used, however, to achieve economy (reduce cement content, allow early form removal, provide ease of placing, and save energy), modify the properties of fresh concrete (increase workability with reduced water content, pumpability, and time of setting), improve the quality of the hardened concrete or mortar (rate of strength development, high strength, durability, permeability, etc.), and, in some cases, may be the only means of achieving a desired result (e.g., air entraining for frost resistance). Generally, admixtures work by influencing the hydration reaction of cement, creating a new phase (e.g., entrained air), or forming new compounds with cementitious properties (e.g., the reaction of pozzolan with the calcium hydroxide from cement hydration). Various materials are known to have been used as admixtures since ancient times, such as hog's lard, curdled milk, buttermilk, and blood (probably the first air-entraining admixture). During the last 50 years, reliable concrete admixtures have developed greatly, with recent major steps being the development and use of superplasticizers. At present, the great majority of concrete produced in North America uses some form of concrete admixture. Admixtures can be classified according to their major function into five groups.

Accelerating Admixtures shorten the time of setting and increase the early strength of concrete. This enables earlier FORMWORK removal, surface finishing, and completion of construction. Many soluble inorganic salts can be used, the most widely employed being calcium chloride, up to a maximum of 2% by weight of cement. Soluble organic compounds include triethanolamine, and quick-setting admixtures, for example, for shotcreting operations, include ferric salts, sodium fluoride, and aluminum chloride. Care should be exercised in the use of calcium chloride since it may increase creep and drying shrinkage and reduce resistance of the concrete to sulfate attack. Because of possible corrosion hazards, no calcium chloride admixture should be used in prestressed concrete.

Air-Entraining Admixtures enable the concrete to resist the action of freezing and thawing and the determined effects of deicing agents. Watertightness is improved and, in the fresh state, the concrete is more workable and cohesive and exhibits reduced segregation and bleeding. This is achieved by the creation of entrained microscopic air bubbles uniformly dispersed throughout the material. The active ingredients in these admixtures are alkylbenzene sulfonates, detergents, or salts of fatty acids. Excessive air entrainment should be avoided since it can cause greatly reduced strength.

Water-Reducing and Set-Controlling Admixtures reduce the water requirements for a given consistency (slump), thus increasing the strength for a certain cement content, or increase the slump without additional water. They may also modify the setting properties. Retarding admixtures slow down the rate of setting, in order to offset the accelerating effect of hot weather, or, under difficult placement conditions, to allow for longer construction times. Examples of these admixtures are lignosulfonic and hydroxylated carboxylic acids and their salts. Some water-reducing admixtures may show undesirable increases in drying shrinkage. A significant recent advance has been the development of the high-range water-reducing admixtures (*superplasticizers*). They allow the use of "flowing concrete" of great workability and pumpability with water content, or "high-strength concrete" with water reductions of 20% to 33% while maintaining normal workability. The latter material, with its high early strength, leads to economic advantages of early finishing and form removal. Superplasticizers are having a great impact on the competitive position of reinforced concrete and its use in high-rise construction, high-strength columns, and slab foundations. Typical materials are sulfonated melamine- or naphthalene-formaldehyde condensates or modified lignosulfonates, which change cement particle surface charges and have a dispersing and lubricating effect. Caution is necessary to avoid undesirable side effects and possible rapid slump loss.

Finely Divided Mineral Admixtures contribute to better workability and finishing, increased sulfate resistance, reduced permeability and heat generation, and added strength development. Reductions in cement content, possible with these admixtures, result in cost and energy savings. Prominent mineral admixtures include granulated blast-furnace slags, natural pozzolans, and fly ash, which is a finely divided residue from the combustion of ground or powdered coal used in about one-third of the ready-mixed concrete produced in the United States.

Miscellaneous Admixtures are available for special purposes. Gas-forming admixtures, for example, aluminum powder, generate bubbles in the fresh mixture to reduce density (see CELLULAR CONCRETE), and counteract settlement and bleeding. Grouting admixtures impart special properties in a grout; expansion-producing aids minimize the effects of drying shrinkage, pumping aids improve concrete pumpability, and coloring admixtures, using various pigments, provide desired color. Fungicidal and germicidal additives inhibit the growth of bacteria and fungus, and dampproofing admixtures help prevent leakage. Others aim to reduce alkali-aggregate expansion and inhibit the corrosion of embedded reinforcement.

The successful use of concrete admixtures requires care in selection, testing, and batching. Various specifications for admixtures are available. Among the most widely used are ASTM C494 (Chemical Admixtures), C260 (Air-entraining), and C618 (Fly Ash and Natural Pozzolan). Proper selection is paramount. The effects of admixtures are sometimes multiple, and include possible undesirable side effects. Combining the admixtures is sometimes advisable, but avoiding those that may not be compatible. Testing of admixtures is desirable with job materials under the anticipated ambient conditions and construction procedures. Batching could be by weight or by volume in powder or liquid forms. Accepted procedures should be followed as to accurate dispensing at appropriate times in the mixing schedule. Based on past history and trends and on potential benefits, the use of admixtures in the future will probably increase. See also COLORED CONCRETE and CONCRETE MIX PROPORTIONING.

References

1. *Admixtures for Concrete*. ACI Committee 212 (ACI 212.1 R-81). American Concrete Institute, Detroit, 1981.
2. *Guide for Use of Admixtures in Concrete*. ACI Committee 212 (ACI 212.2 R-81). American Concrete Institute, Detroit, 1981.
3. *Superplasticizers in Concrete*. Publication SP-62. American Concrete Institute, Detroit, 1979.

S. A.

CONCRETE MIX PROPORTIONING is the process whereby suitable local ingredients of concrete are selected and their relative quantities determined with the objective of producing as economically as possible concrete of certain properties, such as consistency, workability, strength, and durability. The concrete mix design depends on the properties specified for the hardened concrete, and the properties required of the fresh concrete as dictated by the techniques to be used for transporting, placing, and finishing the concrete.

Concrete can be simply specified by the proportions of volume of cement, fine aggregate, and coarse aggregate, such as 1 : 2 : 4 or $1 : 1\frac{1}{2} : 3$. Current practice is to proportion the mix more precisely to give the performance required, for example compressive strength. The mix can then be designed, given the compressive strength and the properties of the coarse and fine aggregate and the cement to be used. However, the proportions must be adjusted if chemical CONCRETE ADMIXTURES, air-entraining agents, or fly ash are used.

There are several proven methods of mix design. Those of the American Concrete Institute use either an estimated unit weight or the absolute volume; in both cases the water content is derived from the workability required for the maximum size of aggregate. Alternatively, the techniques used in the United Kingdom may be adopted. Water : cement and aggregate : cement ratios are selected, and the mix is then designed by the use of tables and graphs.

Having completed the mix design, the individual weights of the materials are calculated for the required batch size of concrete. It is most important to carry out a small-scale trial mix, using the constituents in the proportions calculated from the mix design, to ensure that the properties specified have been achieved.

References

1. AMERICAN CONCRETE INSTITUTE: *ACI Manual of Concrete Practice Part 1*. Materials and General Properties of Concrete. American Concrete Institute, Detroit, 1983.
2. J. D. McINTOSH. *Concrete Mix Design* and *Concrete Mix Design Data*. Cement and Concrete Association, London, 1966.

W.G.J.R.

CONCRETE STRUCTURES have become an accepted part of the environment, be they bridges, cooling towers, industrial structures, lamp standards, or park benches. Concrete structures are also the backbones of buildings (Fig. 43), clad on the outside with brick, stone, or glass and on the inside by plaster or masked by false ceilings. Occasionally, they are exposed and form part of the weathertight skin of the building.

But What Is Concrete? Concrete is an agglomerate of stone, gravel, and sand formed into a mass by a binder

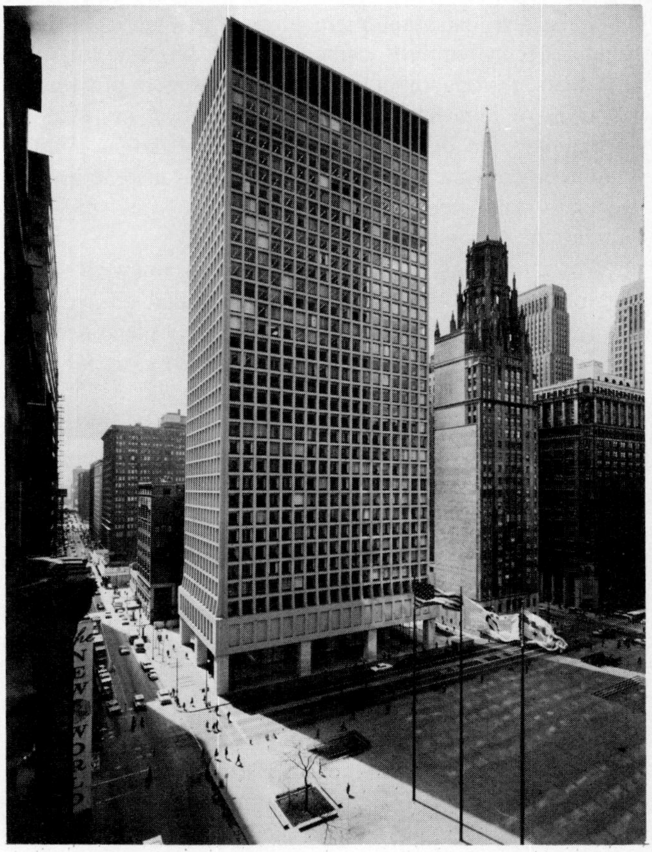

Figure 43 Brunswick Building, Chicago (architects and engineers: Skidmore, Owings and Merrill).

such as lime, cement, or resin. It is fluid when mixed and hardens in the shape in which it is cast. We can therefore shape a concrete structure to our needs.

As concrete has a low tensile strength, early structures were essentially compression structures, such as walls, arches, and DOMES. The Romans used their natural pozzolana cements in the construction of some outstanding structures, the dome of the Pantheon in Rome being one of the most adventurous. Later, builders strove to find a cement that would harden under water, and by the mid-19th century artificial cements had been invented. Foundations for bridges and harbor walls were the first uses of mass concrete, while concrete blocks or artificial stone proved an economic alternative to natural stone and brick.

Reinforcing Concrete to take tension, as in the bending of a beam, was successfully attempted from about 1850. In England, Wilkinson took out a patent in 1854, but the first important patent was by Monier in 1877, to be followed in 1890 by Coignet. Many alternative systems of reinforcing concrete were patented in the 1890s, and some, like that of Cottançin, led to innovative structures that were peculiar to concrete. There was little structural theory, and following the tradition of the great railroad engineers, concrete engineers test-loaded their structures. It was left to a French builder, Hennebique, to carry out extensive tests for some five years, introducing a complete system with proven calculations of reinforced concrete construction in 1892. As an example, he built his own house in Paris in 1900 in reinforced concrete. In the United States, Ransome was the leader in reinforced concrete construction from 1888. His first complete concrete frame came in 1903.

It is interesting that the introduction of a new structural material, such as cast iron, laminated timber, or steel, has always been developed in the field of civil engineering, particularly of bridges. Such was also the case for reinforced concrete.

The Swiss engineer Maillart, having left the university in 1894, first worked on the design of a mass concrete arched bridge in 1899. A year later, on his second arched bridge, he turned to a much lighter reinforced concrete construction of similar form. His investigation of the cracks that appeared in the bridge led him to appreciate the distribution of loads through the structure and to produce later some of the finest reinforced concrete bridges of minimum structural content (Fig. 44).

In the field of building, architects such as Perret expressed the basic characteristics of poured concrete in houses and factories, and his church at Raincy is a landmark in PRECAST CONCRETE construction.

Prestressed Concrete. For engineers, the most significant contribution to concrete construction was made by Freyssinet. Intrigued by the low strength of concrete in tension, he sought a way to prevent tension stresses occurring in a bending structure. From observation of the properties of concrete under load over a period of years, he successfully invented a system of PRESTRESSED CONCRETE with highly tensioned wires, whereby the concrete was initially compressed so that it could effectively take tension. This was more of a state of mind than a technique, for it led to a new esthetic of concrete structures, in particular for bridges. In a similar manner to the early days of reinforced concrete, many prestressing systems evolved.

Figure 44 Bridge at Saginatobel in Switzerland, completed in 1930 (engineer: Robert Maillart).

Three-Dimensional Structures. In his early career, Freyssinet also built concrete shells, or three-dimensional structures, such as his airship hangars at Orly. The theory for such structures had been developed in the 1930s, but it was not until 20 years later that the SHELL roof came into its own when minimum-material structures were required.

Nervi in Italy is probably the engineer–architect best known for his true understanding of the technical properties and construction possibilities of reinforced concrete. His hangar near Orvieto, completed in 1939, is a masterpiece. He later molded concrete slabs into stiff forms to span large distances. His arched roofs for the Exhibition Hall in Turin in 1948 and his domes for the two Olympic Stadia in Rome in 1953 are outstanding. Because he was also a contractor there was no divorce between theory and practice.

Factory-Produced Precast Concrete units for houses and multistory blocks of apartments became the answer to the problem of rebuilding in Europe after World War II. Different exposed concrete finishes appeared, and story-height walls were cast with window or door openings. Systems were developed, in France in particular, and they spread to England and the rest of Europe. Problems of damp penetration, weathering, and, in one case, failure have today led to their disappearance from the building field.

Felix Samuely in England designed his first and best structure of site-cast precast and prestressed concrete for a factory in Bristol (Fig. 45) in 1949. It was his first use of these techniques and, as has often occurred in history, their essential qualities were immediately clearly appreciated and utilized. He also developed FOLDED PLATE structures instead of curved shells, for he believed them to offer architects more freedom of form.

The Art of Reinforced Concrete Design. It must be remembered that the purpose of structure is to transfer to the ground all self-weight and applied loads with safety and economy without undue deformation or vibration. There are no laws or rules as to how structural members should be positioned in space. Therein lies the art of the engineer.

Concrete structures have their own characteristics, which engineers have exploited in their conception of structural forms. To those engineers already mentioned should be added the names of Torroja, Candela, Morandi, Arup, Leonhardt, and Menn, if a list were to be compiled of those who have furthered the development of concrete structures with their art.

Historically, the development of concrete has also been influenced by the ever-increasing quality and strength of cement. Today, very high strengths of PORTLAND CEMENT are possible under laboratory conditions, and practical uses for it have yet to be found. Yet portland cement has some unsatisfactory properties, one being that it is subject to SHRINKAGE. This leads in many cases to the CRACKING of concrete and in time can result in exposed conditions to rusting of reinforcement and spalling of concrete. In fact, concrete weathers very badly and needs maintenance. Portland cement, however, remains the most economic system of binding sand and stone to form concrete. On the other hand, the science of glues is advancing, and in time a new concrete with new binders and methods of reinforcing will appear, bringing with it new forms and extended uses.

See also ANCIENT ROMAN CONCRETE; REINFORCED CONCRETE DESIGN; REINFORCED CONCRETE SLABS; REINFORCEMENT FOR CONCRETE; SPAN; and TALL BUILDINGS.

Figure 45 Factory in Bristol, completed in 1949 (engineer: F. J. Samuely).

References

1. BOAGA BONI: *The Concrete Architecture of Riccardo Morandi*, Tiranti, London, 1965.
2. C. F. MARSH: *Reinforced Concrete*. Constable, London, 1904.
3. P. COLLINS: *Concrete—The Vision of a New Architecture*. Faber, London, 1959.
4. D. P. BILLINGTON: *Robert Maillart's Bridges: The Art of Engineering*. Princeton University Press, Princeton, N.J., 1979
5. P. L. NERVI: *Aesthetics and Technology in Building*. Harvard University Press, Cambridge, Mass., 1966.
6. E. TORROJA: *The Structures of Eduardo Torroja*. Dodge, New York, 1958.

F.N.

CONDENSATION, FORMATION OF. The amount of water vapor that can be held by a given quantity of air is determined by the temperature of the mixture; warm air can hold more water vapor than cold air (see PSYCHROMETRIC CHART). As the air/vapor mixture is cooled, a temperature called the dew-point temperature will be reached, at which the water vapor in the mixture saturates the air. If the mixture is cooled to below this temperature, some of the vapor will condense either as water if the temperature is above 0°C (32°F) or as hoarfrost below 0°C.

In buildings the cooling of the air/vapor mixture that leads to condensation usually occurs when the mixture moves from a warm location to a cooler one where it comes in contact with a cold surface on which the condensation can form. Windows in heated buildings are frequently colder than the dew-point temperature of the air and often become covered with visible condensation. Such condensation is not only annoying but can damage the window sill and wall below if it is not mopped up. A more insidious form of condensation may take place in concealed locations in the building envelope as warm air leaks out of the building, carrying the water vapor to cooler locations where it may condense. If such condensation is as hoarfrost, a considerable quantity may accumulate before it melts with a rise in temperature. If the interior of the building is cooler than the outside environment, an inward movement of air may produce condensation. In all cases the air/vapor mixture must move from a warmer location to a colder one.

Condensation may also be produced, but usually to a lesser extent, by the diffusion of the water vapor alone through materials of the building envelope under the action of a difference in vapor pressure. If the diffusion of water vapor from a warmer location to a cooler one is faster than the diffusion of water vapor away from that cooler location, then the excess vapor will condense. The quantity of condensation produced in this manner is seldom serious in buildings unless it continues over a long period without an opportunity to evaporate.

See also VAPOR BARRIERS.

References

1. A. G. WILSON and G. K. GARDEN: *Moisture Accumulation in Walls Due to Air Leakage*; RILEM/CIB Symposium, Helsinki, 1965, on "Moisture Problems in Buildings"; reprinted as DBR Technical Paper No. 227, NRCC 9131, Division of Building Research, National Research Council Canada, 1966.
2. J. K. LATTA and R. K. BEACH: Vapour Diffusion and Condensation. *Canadian Building Digest*, No. 57, Division of Building Research, National Research Council Canada, 1964.
3. J. K. LATTA: Vapour Barriers: What Are They? Are They Effective? *Canadian Building Digest*, No. 175, Division of Building Research, National Research Council Canada, 1976.

J.K.L.

CONIC SECTIONS are formed when a right circular cone is cut by a plane. Horizontal cuts produce *circles* of varying diameter. A cut parallel to a line on the surface of the cone that passes through its vertex generates a *parabola*. Cuts at a shallower angle produce *ellipses*, which are closed curves. Cuts at steeper angles produce *hyperbolas*, which, like the parabola, are open-ended curves.

Conic sections have been discussed by Greek mathematicians since the 4th century BC, but only the circle and ellipse were used in structures prior to the 18th century, particularly in arches, DOMES, and vaults. The circle is easily generated by a radius of fixed length; for example, if one end of a string is tied to a fixed peg, a pencil attached to the other end of the string describes a circle. An ellipse can also be generated with a string by placing its two ends at the foci of the ellipse; if the string is pulled tight with a pencil, it draws an ellipse as it moves along the string.

Today, the parabola is used for arches, because parabolic arches are free of bending moment when they carry a uniformly distributed load (see CATENARY). Conic sections are used for SHELL ROOFS to form hyperbolic paraboloids, elliptical paraboloids, and parabolic or elliptical conoids. However, the circle remains the most important conic section because of the ease with which circular shapes can be constructed, even though other curves may produce lighter structures (see CATENARY).

References

1. D. HILBERT and S. COHN-VOSSEN: *Geometry and the Imagination*. Chelsea Publishing, New York, 1932.
2. L. M. BLUMENTHAL and K. MENGER: *Studies in Geometry*. Freeman, San Francisco, 1970.

H.J.C.

CONSERVATION OF ENERGY IN BUILDINGS. Conservation of energy implies the saving of energy. It should be considered as the avoidance of unnecessary energy use. The more efficient use of energy must be expanded.

After 1970, there was a sharp escalation in petroleum prices, soon followed by proportionately large increases in the cost of electricity, gas, and coal. It immediately became economically imperative for consumers at all levels, homeowners and corporations, to reduce energy purchases as much as they could. Certain standards were reexamined and found to be higher than was necessary to provide acceptable comfort. The effort to reduce energy use, particularly petroleum use, became a concern of governments. The states in the United States developed energy codes, calling for better performing building components, as a supplement to existing building codes. Nontraditional building methods that would result in lower energy use were permitted if computations indicated that they would match or exceed the performance achievable using the component standards in the new codes. Generally, these were based on a model code developed by ASHRAE (American Society of Heating, Refrigeration, and Air-conditioning Engineers).

Building Types. The existing stock of buildings will constitute the bulk of the built environment for decades to come. Conservation, or energy reduction, is applicable both to the existing and the new. The potential, to have a stock of buildings in the next two decades that operate at 50% of present levels (expressed in energy use per unit area), is made possible by the fundamental inefficiency of our entire collection of buildings, new and old. The old buildings were generally designed to take advantage of natural light and ventilation, but are virtually devoid of insulation, suffer from massive infiltration through cracks in and around doors and windows, and have relatively inefficient mechanical equipment and controls. Most new buildings are sited with complete disregard for orientation; they have large interior areas that must be lighted, ventilated, and temperature controlled at all times, and have inelegant mechanical systems, replete with conflicting delivery criteria and gross control systems. Against this background it is not too difficult to project and achieve 25% to 35% energy reductions by modifying existing buildings and 50% to 60% reductions in the design of new buildings (using the standard of the early 1970s as a benchmark).

In the recent ten-year period since the question of energy use in buildings was first looked into, energy-saving strategies were introduced into a sufficiently large number of buildings to allow evaluation. This has resulted in a general concentration on a few strategies and the downgrading of other approaches that had been considered promising. It is now widely recognized that different building types have significantly different energy-use patterns. For example, residences are most affected by thermal gains and losses through the skin. Large office towers with extensive internal areas remote from the perimeter produce excessive heat during most of the year through occupants, lights, and equipment. Museums must control internal temperature and humidity conditions in display areas, within small tolerances, on a 24 hour a day, 365 day a year basis. Local climatic conditions make a solution that would be excellent in one location inappropriate in another.

The changes that have taken place are discussed next.

Greater Insulation. The amount and quality of insulation have increased. Where a 100-mm (4-in.) stud wall with 50 mm (2 in.) of fiber-glass insulation would have been considered good practice in residential construction, 150-mm (6-in.) studs with 140 mm ($5\frac{1}{2}$ in.) of insulation are increasingly common. Position and integrity of VAPOR BARRIERS has become important. Double glazing is standard. Triple glazing is becoming more common. Movable insulation at windows is used increasingly although not extensively. Rigid foamed insulation boards are used more widely in walls of commercial and institutional buildings and also in roofs. Placement of insulation to reduce leaks around openings and at corners is watched more closely. Cellar slab and foundation insulation is provided.

Tighter Buildings. Air leaks at windows and doors have been reduced by better construction, closer tolerances, and better weather-stripping. Caulking where window and door assemblies meet walls and at other penetrations has been improved, both in care of placement and in the quality and longevity of the caulking materials. Where previously a well-constructed building would have one air change an hour through infiltration, in some new buildings infiltration has been virtually eliminated, raising new questions about the quality of the indoor air, the presence of some damaging chemicals, particles and gases in the space, and the necessity to introduce air changes through mechanical means that use energy in the process.

Better Use of Daylighting. The lighting of commercial buildings has been the largest single energy user in this category. Moreover, the use is of electricity, energy in its highest quality and most expensive form. Through more effective use of daylight at the perimeter and through skylights and light-directing devices, artificial lighting loads may be cut down by 60% and more. Studies have been carried out that have developed methods for introducing natural light without glare.

More Sophisticated Use of Passive Solar Gain. While all buildings respond to the impact of the sun, some beneficially, some uncomfortably, there has been a period of extensive investigation into ways to control and utilize this occurrence through the use of selective shading devices, films, and thermal storage in heavy masonry materials. Where possible, openings are concentrated on the south side and minimized on the north (in the northern hemisphere). Through the use of double and triple glazing and movable insulation, a net heat gain can be realized.

More Precise Delivery Systems. By allowing systems to operate when they are needed and where they are needed, they operate more efficiently. Examples are task lighting delivering required levels of light only on the work plane, heating and air-conditioning zones that permit different conditions to be provided at different times according to varying needs, and ventilating systems that can be modulated according to occupancy or use.

Reclamation of Waste Heat. Waste heat is produced by many processes that take place within buildings: use of lights, motorized equipment, refrigeration, electric generation, cooking, sterilizing, laboratory work, some industrial processes, and human metabolism. The heat can be utilized in various ways, by direct transfer to an area where it is needed, by being extracted from an air stream to preheat cold outside air, and through various heat exchangers. Cogeneration, the use of the high-grade waste heat from electric generation, is effective in improving the thermal rate for the generating process, providing there is a nearby use for the waste heat that is related to its quantity and availability.

Revision of Criteria and Standards. Some mandatory requirements that have dictated high energy use are being studied. These include air change requirements, lighting levels, and temperature and humidity standards. When these studies are completed, there may be significant savings in certain building types, such as museums and hospitals. Recommended lighting levels for schools and commercial buildings have already been lowered.

Improved Equipment Performance. There has been some improvement in the performance of building components. There are more choices in high-intensity discharge (HID) lamps. The coefficient of performance of heat pumps is better. The part load efficiency of boilers has been improved. These and other equipment improvements contribute to the overall gain in energy efficiency. Controls have been extensively improved, allowing much more specific control of systems in complex buildings from central control points. The use of micro-computers has developed rapidly to permit more sophisticated and particularized controls.

Elimination of Counterproductive Systems. The terminal reheat system for air-conditioning delivery has been largely replaced with variable air volume systems. Under the former, air was chilled below the temperature required, and then brought up to the desired temperature by adding heat at the point of delivery, resulting in unnecessary heating and cooling. Dual duct systems that mix hot and cold air at mixing boxes have a similar problem. The variable air volume system, now in extensive use, as its name implies, delivers varying amounts of cool air as required to maintain predetermined temperatures.

Reduced Use of Electric Resistance Heating. Electric resistance heating has a thermal efficiency of about 36%; it takes about 10 MJ (10,000 Btu) at the generating plant to produce 1 kWh of electricity, which gives off 3.5 MJ (3500 Btu) of heat through a resistance heater. A heat source burning fossil fuel at the building operates at 60% to 85% efficiency. Offsetting the simplicity and economics of installing electric resistance systems is the cost of operating them, which has become so high that their use has been sharply curtailed.

Energy-Determined Building Configuration. Natural lighting, access to sunlight, and open window ventilation have combined to make the shape of buildings more responsive to their performance with natural energy sources. Although there are examples of this approach to building design, most buildings are still shaped in response to lot boundaries and real estate considerations.

Use of Solar Hot Water Heating. The great interest in active solar collection systems, particularly those using flat plate collectors, that was evident through the 1970s has diminished. There is, however, a growing use of collectors for domestic hot water, particularly in single-family residences. Since hot water can represent 20% of the entire energy load of a house, a 50% reduction through the use of collectors is significant.

User Involvement. Once the systems are in place, the critical factor is the degree to which the user takes advantage of the installation. In 1973 and 1974 there was a widespread concern about energy use. Individuals, as well as corporate and institutional users, exercised discipline in operating systems. The zeal has abated somewhat and discretionary use has gone up recently. There are also numerous examples of systems that have greater complexity than the capabilities of operating personnel can cope with. Many buildings built in the last decade have failed to perform at the level of the predicted performance.

As the performance of buildings improves, the factor of embodied energy becomes more important. The energy required to produce the materials and construct a building may now be the equivalent of 20-year operation of the building. Improvements in manufacturing efficiency, more advanced construction methods, less waste on the job and in the plant, greater use of regional materials to cut transportation energy, and planning for better space utilization can all contribute to significant conservation in the preuse phase of a building's life. In addition, the longer the life of the building, the lower the embodied energy will be on a prorated basis.

See also AIR CONDITIONING AND HEATING; DAYLIGHT DESIGN; FENESTRATION; LIGHTING LEVELS APPROPRIATE FOR VARIOUS VISUAL TASKS; NATURAL VENTILATION; PASSIVE SOLAR DESIGN; SOLAR COLLECTORS FOR SOLAR THERMAL SYSTEMS; SUN SHADING AND SUN-CONTROL DEVICES; and THERMAL INSULATION.

References

1. DONALD WATSON (ed.): *Energy Conservation through Building Design*. McGraw-Hill, New York, 1979.
2. *Statistical Abstract of the United States 1982*. U.S. Bureau of the Census. Washington, D.C., 1982.
3. R. A. FAZZOLARE and C. B. SMITH (eds.): *Energy Use Management, Proceedings of the International Conference*. Volumes I, II, III. Pergamon Press, Elmsford, New York, 1977.
4. R. G. STEIN, C. STEIN, M. BUCKLEY, and M. GREEN: *Handbook for Energy Use in Building Construction*. National Technical Information Service (NTIS), Springfield, Va., 1981.

R.G.St.

CORROSION OF METALS is caused by chemical or electrochemical reaction with their environment (see CORROSION PROTECTION).

Steel Corrosion is caused by the formation of minute positive and negative areas on the surface of the metal. Iron goes into solution at the positively charged areas. The oxygen in the solution can combine with water and electrons (*e*) as a negative reaction to form hydroxyl ions (OH^-):

$$O_2 + 2H_2O + 4e^- \rightarrow 4OH^-$$

The hydroxyl ions combine with ferrous ions to form ferrous hydroxide. This is oxidized by oxygen to form ferric hydroxide, which precipitates as rust.

Weathering STEEL has a lower maintenance cost because its corrosion resistance enables it to be used without coating. Exposure causes the surface to slowly weather to a dark color due to the formation of a dense protective oxide film. However, adequate drainage must be provided to prevent rust from staining concrete at lower levels.

Alloying steel with nickel, chromium, and sometimes molybdenum produces STAINLESS STEEL.

Aluminum is corrosion resistant because a thin, adherent oxide film forms instantaneously on its surface when exposed to the atmosphere. Serious corrosion only occurs when the film is damaged and prevented from re-forming. Anodizing thickens the natural oxide film and thus increases its corrosion resistance.

When other metals are joined to aluminum, care must be taken to avoid galvanic corrosion (see CORROSION PROTECTION). Because zinc, cadmium, and lead are close to ALUMINUM in the galvanic series, cadmium-plated or galvanized steel may generally be used in contact with it. Lead washers are frequently used with galvanized steel nails to fix aluminum roofing. Contact with copper or brass should be avoided, especially in a marine environment.

Aluminum sheet and bitumen-coated strip are frequently used for damp-proof courses and flashings. Since aluminum is attacked by alkalies present in wet lime and cement mortars, it must be coated with bitumen for use in walls.

Crevice corrosion in aluminum building products may occur close to BRICKS or CONCRETE because of entrapped water or CONDENSATION. Pitting corrosion of aluminum may occur when the protective film is damaged and cannot repair itself. Adequate metal thickness should be provided in such cases.

Copper reacts with air to produce a brown oxide surface that gradually changes to a green patina (basic copper carbonate; this may take up to ten years). In industrial atmospheres, weathering may proceed more rapidly to produce basic copper sulfate, which forms a protective film and inhibits further corrosion. Copper chloride may also be present in a marine environment.

Lead-coated copper (either hot-dipped or electrodeposited) may be used where water runoff from uncoated copper could discolor concrete, stone, or other building materials by leaving a green stain.

Copper tubing is frequently used for water pipes although low-zinc brass is more resistant to water containing free carbon dioxide. Such water may dissolve traces of copper, which could cause blue or green stains on plumbing fixtures, for example, by combining with soap. Hard water may protect the copper by depositing a lime scale on its surface.

Brasses comprise a large range of alloys of copper with up to 50% zinc, with or without the addition of small quantities of other elements. Brasses containing more than approximately 37% zinc frequently suffer dezincification, resulting in loss of strength and ductility. It can be inhibited by the addition of small quantities of arsenic.

Lead has long been used for flashings and damp-proof courses. When freshly cut and exposed to the atmosphere, lead forms a surface film of oxide and then a patina of lead carbonate by reaction with atmospheric carbon dioxide. This is strongly adherent and insoluble in normal atmospheric moisture. In industrial atmospheres, the lead patina may also contain a sulfate, but its protective effect is not impaired. Lead can normally be used in contact with metals such as copper, zinc, aluminum, or steel without galvanic action.

Fresh cement mortar or concrete may encourage corrosion of lead if carbonation and drying are delayed; but an underlay provides protection. When lead is fixed into mortar joints, the free lime is quickly carbonated and rendered harmless.

References

1. U. R. EVANS: *The Corrosion of Metals*, 3 volumes. Edward Arnold, London, 1960–1976.
2. J. C. SCULLY: *Fundamentals of Corrosion*. Pergamon, London and Elmsford, N.Y., 1966.

S.M.R.

CORROSION PROTECTION. When an electrically neutral piece of metal is immersed in an electrolyte, the metal tends to dissolve and send positive ions into solution, leaving the metal negatively charged. This results in an electrical potential known as the *electrode potential* between the metal and solution.

The electrode potential of a metal in contact with a solution of its ions of unit activity is known as the *standard* or *normal* electrode potential. Metals can then be arranged in a table in the order of their normal electrode potentials at, say, 25°C.

However, the table does not necessarily indicate the behavior of a metal in service because of possible reactions between the metal and its environment. Because of this and the phenomenon of passivity, an alternative series based on observed behavior and known as the galvanic series has been prepared (Table 1).

A galvanic couple results when two dissimilar metals are joined and immersed in an electrolyte. The corroded metal is the anode and the protected one the cathode.

The galvanic series only indicates the tendency for galvanic corrosion. Corrosion can only proceed if there is a flow of electric current and this is affected by the following:

1. The *electrical conductivity of the circuit*, which is related to the effectiveness of the contact between the metals and the electrical resistance of the solution.

TABLE 1
Galvanic Series of Metals and Alloys in Seawater

Noble and cathodic end	Platinum
	Gold
	Titanium
	Silver
	18 Cr–8 Ni–Mo stainless steel (passive)
	18 Cr–8 Ni stainless steel (passive)
	11%–30% Cr stainless steel (passive)
	Copper
	Brasses
	Nickel
	Tin
	Lead
	Steel
	Cadmium
	Aluminum
	Zinc
Active and anodic end	Magnesium

2. The *relative position of the two metals in the galvanic series*. The farther apart in the series, the greater the tendency for galvanic corrosion.
3. *Polarization*. The effect of a potential difference between the anode and cathode may be reduced by polarization at either the anode through accumulation of corrosion products or at the cathode by deposition of hydrogen. Dissolved oxygen is the usual depolarizer and unites with hydrogen to form water.
4. *The relative areas of cathode and anode*. Although the ratio of the anode to the cathode does not affect the initial potential difference between them, the amount of current generated is largely dependent on this.

Metallic Coatings are produced by electroplating or hot dipping. There are basically three types of electroplating:

1. *Cathodic metals* are cathodic to the underlying metal, for example, nickel on steel; they are protective because of their superior corrosion resistance.
2. *Anodic metals* are anodic to the underlying metal, for example, zinc on steel; they are protective because the latter receives cathodic protection.
3. *Protective film metals*, such as chromium on steel, develop a protective film.

For hot dipping, the basis metal is cleaned and then dipped in a bath of molten coating metal covered by a flux to prevent atmospheric oxidation. Galvanizing is an example where a protective zinc coating is formed on steel.

See also CORROSION OF METALS.

References

1. F. N. SPELLER: *Corrosion—Causes and Prevention*, 3rd ed. McGraw-Hill, New York, 1951.
2. INSTITUTION OF METALLURGISTS: *Corrosion and Protection of Metals*. Iliffe, London, 1965.

S.M.R.

CRACKS IN BRITTLE MATERIALS appear either as surface defects and notches produced during fabrication (e.g., machining) or as internal voids and structural discontinuities introduced during manufacture (e.g., sintering). Brittle materials such as glasses and ceramics are sensitive to the presence of cracks. It is found that the tensile strength of these materials varies inversely with the square root of the size of the crack. Crack propagation in brittle materials is usually sudden and unstable. There is little plastic deformation associated with the fracture.

Whether or not the cracks will grow catastrophically when subjected to given service loadings is dealt with by a new science called *fracture mechanics*. The ability of a material to control fracture propagation is determined by its *fracture toughness*. This is a new material property, which is measured by loading to fracture specimens containing artificial sharp cracks of known lengths according to test standards that have been recently established.

See also TESTING OF MATERIALS FOR STRENGTH AND HARDNESS.

References

1. B. R. LAWN and T. R. WILSHAW: *Fracture of Brittle Solids*. Cambridge University Press, Cambridge, 1975.
2. J. F. KNOTT: *Fundamentals of Fracture Mechanics*. Butterworths, London, 1973.
3. Y. W. MAI and A. G. ATKINS: Crack stability in fracture toughness testing. *Journal of Strain Analysis*, Vol. 15 (1980), pp. 63–74.

Y.W.M.

CURTAIN WALLS. The Oxford English Dictionary describes "curtain" as "a piece of plain wall not supporting a roof" and "wall" as a "vertical and solid structure, narrow in proportion to length and height, serving a house."

The curtain wall, or exterior wall of a structure, may be of many materials: steel, aluminum, bronze, stainless steel, precast concrete, stone, marble, granite, ceramic tiles, glass, and so on.

A curtain wall is analogous to the human skin. It is conceived as an active filter, rather than a passive barrier. The wall must be flexible and in constant motion. It must control the transfer of heat and cold, control the passage of light and sound, and prevent the passage of moisture, dirt, vermin, and, of course, people.

The most common, practical, and functional curtain wall is the aluminum and glass wall developed since the late 1930s. The first glass curtain wall on a multifloored structure was on the Willis Polk's Hallidie Building in San Francisco in 1917.

Today's metal curtain walls are simpler and more sophisticated (Fig. 46). Although the basic principles of good curtain wall design have not changed, the greatest development has been from the original reliance on sealants to prevent entry of water and air to the application of the pressure equalization or rain-screen principle of design.

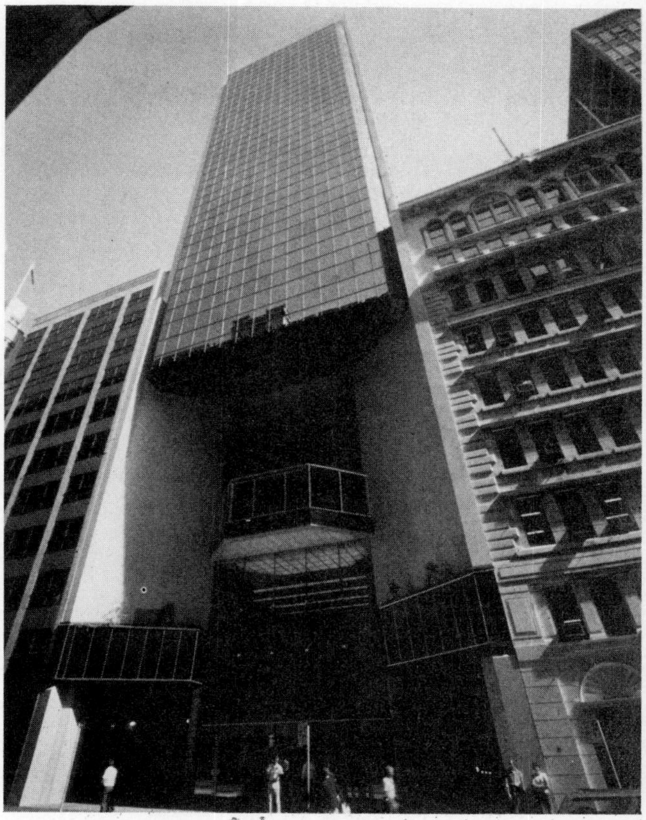

Figure 46 Curtain wall at 15 Castlereagh Street, Sydney. (Photo by courtesy of Peddle Thorp and Walker.)

Types of Curtain Walls. Because of the wide and increasing variety of aluminum curtain wall designs, it is difficult to identify each design precisely as representing one or another of the basic types. There are five generally recognized systems of curtain wall.

1. In the *stick wall system*, the wall is installed piece by piece, first the mullion units, followed by the horizontal members. Current designs offer a wide variety of esthetic concepts (Fig. 47).
2. In the *unit system*, the wall is composed entirely of large framed units, one or two floors high and one module wide. The units are preassembled, and glazing and spandrel panels can be installed in the factory.
3. The *unit and mullion system* is a combination of types 1 and 2. The mullions are installed first. Preassembled framed units are then placed between them. The framed sections may be of full floor height or divided into vision and spandrel panels.
4. The *panel system* is similar in concept to the unit system. However, instead of preassembled framed units, homogeneous panels formed of sheet metal or cast metal, precast concrete, or natural stone are used. Jointing is thus reduced to a minimum. The panel system is particularly suitable for molded facades.

Figure 47 AMP Place, Brisbane. (Photo by courtesy of Peddle Thorp and Walker.)

5. The *column cover and spandrel system* consists of column cover sections, long spandrel units spanning between columns, and infill glazing units that may be preassembled (or separate framing members glazed on the building site) (Fig. 48).

Figure 48 Dalgety House, Melbourne. (Photo by courtesy of Peddle Thorp and Walker.)

The **Design of Curtain Walls** requires careful attention to the natural forces and their effect on the materials of the wall, the structure on which the wall is to be installed, and the esthetics and function of the curtain walls. Walls should have esthetic characteristics that do not negate the requirements of a functional curtain wall, such as structural integrity, provision for movement of both the wall itself and the building structure, allowance for tolerances, water tightness, air tightness, moisture control, thermal insulation and energy conservation, sound transmission, and provision for cleaning of the wall.

Structural Integrity is one of the primary concerns of curtain wall design. The requirement of stiffness, rather than strength, usually governs. Vertical loads in the wall system are relatively light, but structural design must provide proper resistance to lateral wind forces. Maximum wind velocities vary with geographic location, height above ground, nature of the surrounding buildings, and the shape of the building itself.

Sufficient attention must be given to negative wind loadings, or suction forces, acting on the curtain wall. Together with internal building pressures due to air conditioning, these are often critical for design, and wind damage to a high-rise building is more frequently a "blow out" rather than a "blow in."

Cleaning gantries may be attached to the mullions or to the face of the wall by the suction or "hovercraft" principle. They produce pressure loads.

Provision for Movement is a vital factor in designing a curtain wall. Failure to allow room for movement results in failure of the wall. The wall components expand and contract; they move relative to one another and relative to the structural frame. These movements are caused by temperature changes, gravity forces, and wind acting on the curtain wall and by the deformation of the building frame due to dead and live loads, and due to creep if the structural frame is of concrete.

Lateral movement occurs in buildings due to earthquake and wind loads. Its extent depends on the shape of the building; unsymmetrical shapes add torsional movement.

Temperature changes are particularly important for the movement of aluminum walls because of aluminum's high coefficient of expansion. In some regions the temperature range on a metal surface could be as high as 93°C (200°F). This causes a thermal movement of up to 8 mm ($\frac{5}{16}$ in.) in a 3-m (10-ft) length of aluminum. A sheet of glass has only about one-third of this movement. However, the color finish to the aluminum and the area exposed relative to the total area of the section affect the temperature differential.

Compression and creep of building frames vary for structural steel and reinforced concrete. Deflection of perimeter beams or slabs causes additional movement allowance requirements, especially where cantilevers occur.

Tolerances should always be in addition to the necessary allowance for clearance, and the two should not be confused. Allow for those of the building structural frame.

Material production tolerances are required for extruded aluminum for the dimensions of the cross sections, as well as for bow and twist. A tolerance must be allowed for the thickness of glass. Separate tolerances are needed for fabrication and assembly. Structural tolerances for steel-framed buildings differ from those for reinforced concrete.

Watertightness. Since curtain walls are in constant, if minor, movement due to structure flexing, thermal variation, and wind loads, jointing and assembly techniques must accommodate this movement, yet avoid water and air penetration.

High winds cause rainwater to flow in all directions over the face of the wall and much of it collects at the joints, the major point of vulnerability. The early curtain walls were designed to provide a full water seal at the outer surface of the wall, but many failures occurred due to the continual movement and reliance on sealants. Two methods of water seal have been developed since then. The *internal-drainage* system acknowledges that water enters the wall. The *pressure-equalization* system keeps the water out of the building by balancing pressures within the wall.

Air Tightness. Curtain walls that perform well under air infiltration tests are usually excellent in their resistance to water penetration, although the converse is not necessarily the case. Infiltration of air can result in discomfort, impose an undue load on the air-conditioning and heating systems, and allow penetration of noise and dust. It can also produce sound effects in windy conditions. Continuity and durability of gaskets and sealants are especially important for the airtightness of curtain walls.

Moisture Control. Condensation occurs because the metal and glass of a curtain wall are impermeable to moisture and have low heat retention capacity. This condensation must be controlled.

Thermal Insulation and Energy Conservation. Although a curtain wall is lightweight and may be predominantly of glass, by careful selection of glass it can be a more efficient thermal insulator than a masonry facade. Metal sheeting must be backed with an insulating material.

Sound Transmission. The intrusion of outdoor noise into buildings in built-up areas must be considered in the design of curtain walls.

Cleaning. Most multistory buildings have fixed windows so that external washing and maintenance machines are required. They may be manually operated or fully automatic. The curtain wall must be designed to allow for the attachment of the cleaning gantry to the mullions or to specially designed attachment devices, or retained to the glass or infill panels.

Testing of Curtain Walls. The thermal efficiency, watertightness, airtightness, and structural design criteria for wind and other superimposed loads must be verified before proceeding with the manufacture and installation of a curtain wall. Government research stations and systems manufacturers have developed extensive test procedures, and many curtain wall manufacturers and installers are prepared to guarantee their product for periods of up to 10 years.

Test Methods. The dynamic test employs a wind generator, usually an aircraft motor or a large propeller, to simulate wind, while water is fed into the air stream and thus onto the wall. The static test utilizes an airtight chamber, with the curtain wall test specimen forming one of the sides of the box. The static test is more accurate than the dynamic test and higher pressures can be applied.

Sealants. Caulking and sealant materials were originally developed for the aircraft industry, and these materials were subsequently adopted by the construction industry and modified for buildings. An effective sealant must permit the necessary expansion and contraction, while maintaining the seal. The materials that do this most satisfactorily are silicone, polysulfide, polyurethane, acrylic, and butyl compounds. Resilient neoprene rubber and extruded silicone gaskets are used in conjunction with, or as alternatives to, sealant materials.

See also ALUMINUM; CONDENSATION; CONSERVATION OF ENERGY IN BUILDINGS; NATURAL BUILDING STONE; SEALANTS AND JOINTING; SOUND INSULATION AND ISOLATION; STAINLESS STEEL; THERMAL INSULATION; and TOLERANCES.

References

1. *Aluminum Curtain Wall Design Guide Manual*. Architectural Aluminum Manufacturers Association, Chicago, 1979.
2. GORDON H. SMITH: Design and testing of metal/glass curtain walls. *Architectural Record*, Vol. 170, No. 4 (March 1982), pp. 71–83.
3. CHARLES H. THORNTON: Avoiding wall problems by understanding structural movement. *Architectural Record*, Vol. 169, No. 16 (Dec. 1981), pp. 108–111.
4. *Aluminium in Construction*, No. 1–11. Aluminium Development Council of Australia, Sydney, 1984–85.

A.B.K.

D

DAMP-PROOF COURSES are physical or chemical barriers built into masonry structures to inhibit the diffusion of water, in the vertical and horizontal directions, through porous media. Physical barriers have proved to be the most effective in inhibiting moisture transfer. Water penetrates masonry because of hydraulic pressure near foundations, gravity forces on water intercepted by walls, capillary flow due to surface tension forces, or molecular diffusion.

Physical and cosmetic deterioration of masonry may be caused by swelling and shrinkage stresses due to uptake and loss of water and included salts, the influence of moisture changes on surface textures, or the growth of mold and mildew at optimum conditions.

In some early structures, the processes of moisture transfer and its subsequent damage were not always realized and necessary precautions were often omitted. In others, moisture barriers were included, in that impervious materials, such as basalt or fired clay products, were used to construct foundations to above the ground line. In later structures, damp-proof courses were constructed in walls using strips of durable timber, sheet metal, slate, or glass to act as physical barriers or admixtures of mortar, consisting of sand, lime, or portland cement, and coal tar or chimney soot to act as chemical barriers.

Modern masonry structures are protected against moisture diffusion by physical barriers embedded in mortar courses, consisting of continuous or well-lapped strips of coated metals, bituminous felts, or plastic materials to present an impervious barrier to an approaching moisture front under moderate hydraulic pressure. Alternatively or additionally, chemical barriers of silicones, stearates, latex, bitumen, or other water repellants may be used, incorporated as admixtures in composition mortars to reduce surface tension forces or even block voids, to interrupt capillary flow systems. Furthermore, physical and chemical barriers are often applied to existing structures to remedy dampness problems.

It is important to ensure that damp-proof courses in masonry are never bridged by porous media such as surface renders, soil, or paving that may conduct water; that proper drainage is provided to remove water; and that adequate ventilation is installed to exhaust water vapor.

See also BRICKWORK AND BLOCKWORK; DRAINAGE, DAMP-PROOFING, AND WATERPROOFING OF BASEMENTS; and VAPOR BARRIERS.

References

1. L. D. ARMSTRONG: Inhibiting dampness in buildings. *Architectural Science Review*, Vol. 23, (1980), pp. 4–6.
2. H. J. ELDRIDGE: *Common Defects in Buildings*. Property Services Agency, Department of Environment, London, 1978.

L.D.A.

DAYLIGHT DESIGN is a general term describing the deliberate utilization of daylight in architectural spaces with the aim of distributing natural light and its reflected components to those places where visual tasks are performed and visibility problems are expected.

Although the close link between daylighting and the architectural design process has always been recognized, it is only since the 1950s that daylight design has been expressed in logical and quantitative terms with scientifically based criteria relating human, hygienic, ergonomic, and psychological requirements to LIGHTING LEVELS APPROPRIATE FOR VARIOUS VISUAL TASKS and to qualitatively important luminance patterns. Daylight design aims to determine the position and size of windows and other apertures in buildings, that is, the dimensions of window heads and sills, the height of the ceiling, the placing of roof lights, adoption of SUN SHADING AND SUN-CONTROL DEVICES, internal light wells or courtyards, according to the depth of the rooms, the arrangement and location of the working places, and the circulation space. Hence daylight design governs the efficiency of utilization of the available site area with respect to daylight, climate, and energy received from the SUN. Furthermore, the overall architectural composition and the building shape are influenced by the daylight design and the FENESTRATION system chosen. Thus the principles of daylight design are concerned as much with the interior comfort and environmental satisfaction of the users as with the level and quality of the illumination and the resulting energy conservation.

History. Although scientific methods for daylight design have been devised in comparatively recent times, the Law of Light and legal cases with respect to the infringement of skylight are quite old, as documented by easements in the ancient Roman Law of Light (Ref. 1). In the 18th century, P. Bouguer and H. Lambert used basic photometric principles to study the effect of the size and shape of a window aperture on the resulting sky illuminance and considered the problems of calculation and measurement. Methods for solid angle projection based on the principle of the unit sphere were developed in the late 19th century in Europe; these led to the concepts of *sky factor* and *daylight factor*, thus establishing the relative quantitative criteria of daylight design in the 1920s and 1930s. While the first half of the 20th century has been devoted mainly to practical calculation methods of sky factors and sky components, model studies, especially under ARTIFICIAL SKIES, during the 1950s and 1960s were concentrated on the reflected components. More recently, computer-based daylight design and evaluation methods have been developed.

Principal Methods Employed. Daylight design as a part of the architectural design process is performed by applying either practical experience or calculation and graphical methods:

1. Windows or skylights can be designed according to routine rules, such as the following:
   a. Prescribing the fraction of the floor area required for glazing, such as $\frac{1}{8}$ to $\frac{1}{10}$ for offices, shops, and factories, $\frac{1}{7}$ to $\frac{1}{10}$ for dwellings and public buildings, and $\frac{1}{5}$ to $\frac{1}{7}$ for schoolrooms.
   b. Defining the minimum or required ratio of window area to facade area for rooms with daylighting alone (35% to 50% is sufficient). Thirty percent is required for rooms with only perimeter daylighting and for deep offices with permanent supplementary artificial lighting of interiors (PSALI). A further reduction to 10% to 20% is possible in rooms with permanent artificial lighting (PAL) when using windows only as vision strips.
2. Window or daylight aperture size can be chosen to achieve the standard illuminance level by applying light-flux methods (Ref. 2) or the simplified lumen prediction methods (Ref. 3).
3. *Point-by-point* daylight factor values required in standardized critical places, usually on the working plane 0.75 to 0.85 m above floor level, can be determined for specified visual tasks (Fig. 49). The evaluation procedure for the daylight factor is based on one of the following means:
   a. Precalculated design tables, for example, the BRS (British Building Research Station) daylight tables (Ref. 4 or 5).
   b. Graphical methods, for example the Waldram, BRS, or Kittler protractors (Refs. 5 and 6), the Danilyuk (Refs. 6 and 7), or the pepper pot diagrams (Ref.

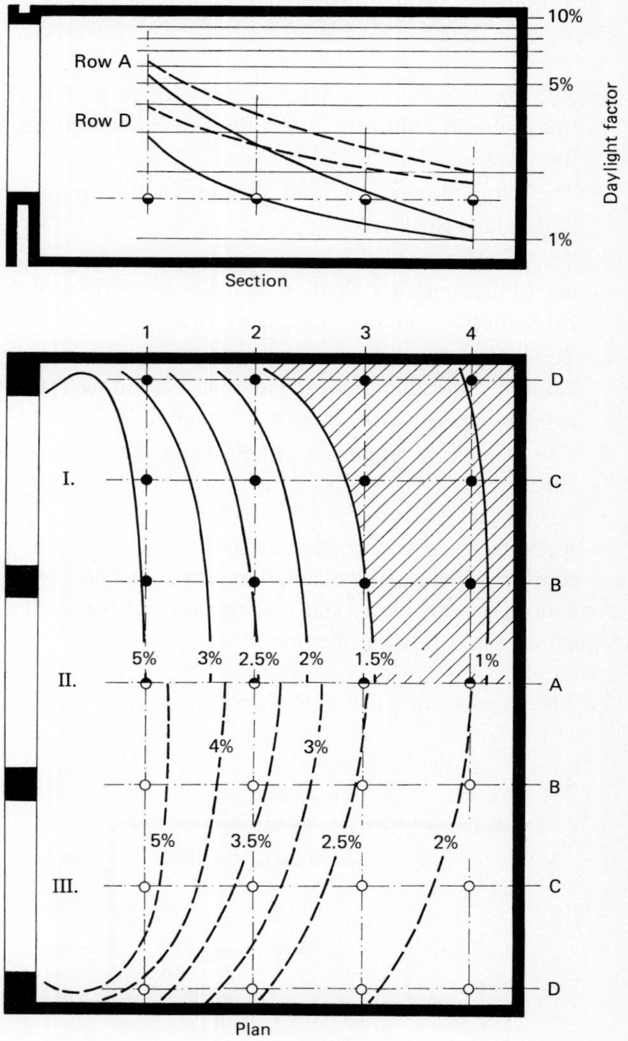

Figure 49 Daylight factor contours representing the daylight distribution on the plan and section of a classroom under overcast sky conditions. The full lines show the conditions for normal dark ground. The dashed lines show, for comparison, the influence of snow-covered ground.

8) for the sky component; and BRS nomograms or Kittler's diagrams for the reflected component of the daylight factor (Ref. 9).

c. Numerical calculation methods using formulas or computer programs, for example, Quicklite, Superlite, Microlite, or Solite (Ref. 7).

d. Model measurements outdoors or under an artificial sky.

At the earliest stages in the design of a building, the amount of fenestration, the shading, and the internal reflectances are determined approximately, using single-stage methods. This is later supplemented by a more precise daylight design method (Fig. 50) utilizing more elaborate calculations or model simulation. All these methods are concerned mainly with the quantitative aspects of the daylight design:

1. The size of operative GLAZING with respect to the transmission, diffusion, and reflection characteristics of the glass.
2. The room dimensions and location within rooms of particular visual tasks.
3. The location and spacing of apertures and the possible use of elements for shading or the redirection of light, for example, louvers, blinds, or awnings.
4. The effect of external and/or internal obstructions, including opposite facades, balconies, window frames, accumulated dirt, or interior equipment.
5. The reflecting properties of the exterior and interior surfaces.

In addition, certain qualitative aspects affect efficiency and comfort. These are associated mainly with glare, visual distraction, loss of view, and loss of visual privacy. They depend on the following factors:

1. The freedom of users to change their visual orientation with respect to GLARE sources, such as the sun, windows, mirrors, or reflective parts of machinery.
2. The position of daylight apertures in the field of view and their location in one or more walls and in the ceiling.
3. The amount of sky visible, its luminance pattern, and the construction details of the window wall, such as glazing bars, reveals, and prismatic glass or blocks, which provide graded brightness steps between the wall and the sky.
4. The utilization of different adaptation levels with veiling glare to achieve visual privacy, for example, by mesh curtains or mirrored reflective glazing.
5. The interior decoration, lightness, color patterns, and contrast, whose emotional and esthetic color harmony may attract the viewer in the desired visual orientation.
6. The purposeful flow of light, using nonuniform illumination with modeling by shadow effects when it is desirable to reveal form and texture.
7. The reduction of gloominess in a room by a balanced distribution of daylight by using high-reflectance finishes on deep rear walls or by using supplementary lighting.

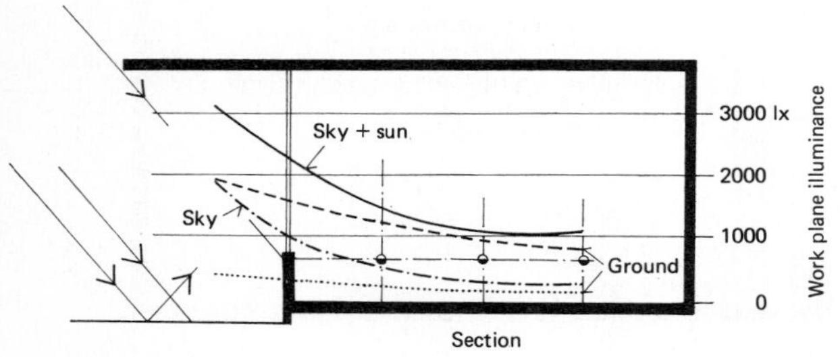

Figure 50 Daylight illuminances in a room showing the effect of a horizontal sunbreak. Separate illuminance curves are shown for the overcast and the clear sky conditions. In the hot-arid tropics the light reflected by the ground is important, as it forms the main source of the interior illuminance levels.

REFERENCES

1. J. SWARBRICK: *Easements of Light*. Batsford, London, 1933.
2. B. V. EVANS: *Daylighting for Architecture*. Architectural Record Books, New York, 1981.
3. J. W. GRIFFITH, W. J. ARNER, and E. W. CONOVER: A modified lumen method of daylight design. *Illumination Engineering*, Vol. 50 (1955), pp. 103–112.
4. J. W. T. WALSH: *The Science of Daylight*. Macdonald, London, 1961.
5. R. G. HOPKINSON, P. PETHERBRIDGE, and J. LONGMORE: *Daylighting*. Heinemann, London, 1966.
6. R. KITTLER and L. KITTLEROVÁ: *The Design and Evaluation of Daylighting*. Alfa, Bratislava, Czechoslovakia, 1975.
7. *1983 International Daylighting Conference*. General Proceedings. Phoenix, Ariz., 1983.
8. J. A. LYNES: *Principles of Natural Lighting*. Elsevier, London, 1968.
9. *Daylight*. Publication No. 16. Commission International de l'Éclairage (CIE), Paris, 1970.

R.K.

DETAILING OF CONCRETE STRUCTURES deals with the proper arrangement of the REINFORCEMENT within the CONCRETE. The following must be observed to ensure correct detailing.

Cover, the minimum distance between the outside of the reinforcement and the surface of the concrete (Fig. 51), must ensure that the bars are surrounded by sufficient concrete to establish adequate bond between the concrete and the steel, to protect the steel from FIRE, and to protect the steel against rusting. Protection against rusting depends on the degree of the exposure of the concrete to the natural environment.

Bar Spacing refers to the distance between reinforcing bars (Fig. 51). A minimum spacing is specified by the concrete code (Ref. 3) to ensure that the concrete can penetrate the space between the bars and be compacted so that voids are absent. A maximum spacing is also specified so that local failure under concentrated load will not occur. Maximum spacing is also specified to ensure an even distribution of CRACKS.

Shrinkage and Temperature Reinforcement is the minimum amount of reinforcement that must be provided in slabs and walls, or concrete members with large surfaces, to ensure that the shrinkage and temperature cracks remain small and evenly distributed. A maximum spacing of the bars for this reinforcement is also specified. These cracks are caused by restraints to contraction of the concrete due to SHRINKAGE and also due to the alternate contraction and expansion caused by fluctuations in temperature. A minimum amount of reinforcement is also specified for bar beams so that sudden collapse after cracking is avoided.

Ties must be provided in beams and columns to restrain longitudinal compressive bars from BUCKLING (Figs. 51 and 52).

Termination of Bars. The force in a terminated bar must be transferred through the surface bond between the steel and the concrete before the bar can be cut off (Fig. 52). The *development length* required depends on the stress in the bar and the bar diameter. If the development length is too long for a straight bar, a cog or hook is used to reduce the development length.

Splicing of reinforcing bars is necessary when the required lengths are in excess of those commercially available. The cheapest method is to transfer the stress through the concrete by bond from one bar to another parallel bar for a splice length that depends on the development length. If there is insufficient space, bars may be spliced at a higher cost by welding or by using mechanical connections.

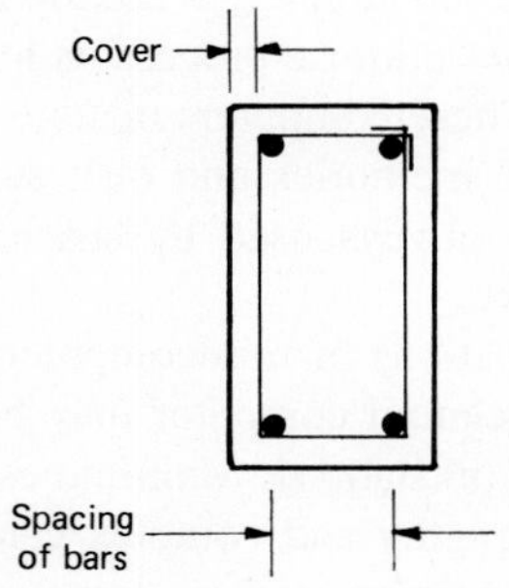

Figure 51 Cover, spacing of bars, and ties.

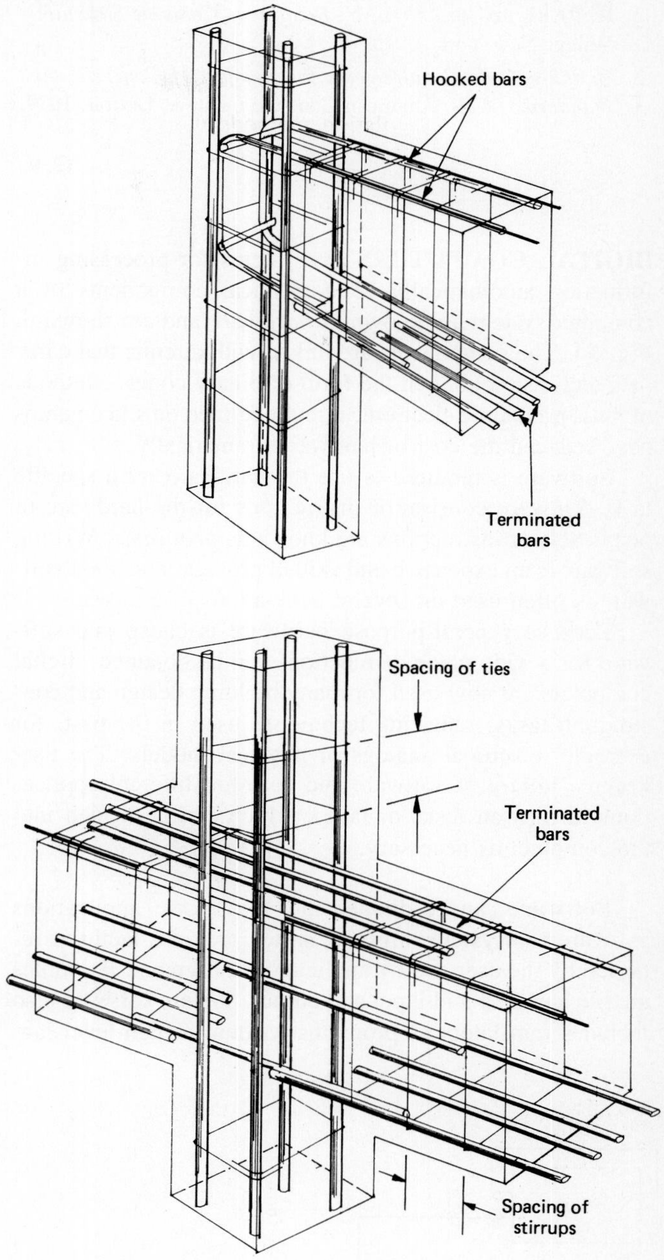

Figure 52 Spacing of ties and termination of bars.

References

1. H. J. COWAN: *Design of Reinforced Concrete Structures*. Prentice-Hall, Englewood Cliffs, N.J., 1982.

2. R. PARK and T. PAULAY: *Reinforced Concrete Structures*. Wiley, New York, 1975.

3. *Building Code Requirements for Reinforced Concrete. ACI Standard 318-79*. American Concrete Institute, Detroit, 1979.

G.V.

DIGITAL COMPUTERS are systems for processing information automatically. The physical components of a computer system are known as *hardware* and are shown in Fig. 53. The components are linked with circuits that carry data and instructions in the form of binary codes. Methods of mass producing electronic circuits formed on silicon chips have reduced the cost of hardware dramatically.

Software is required to use the hardware for a specific task. *Software* consists of instructions for the hardware to obey. Sets of instructions are known as *programs*. Writing software is an expensive and skilled process, and good software is often used on several generations of hardware.

Because general-purpose hardware is cheap and software for a wide range of functions can be obtained, digital computers are now used for many building design and construction tasks, replacing techniques used in the past, for example, electrical analogs or physical models. The user keeps a library of software and data for different applications in files on disks or tapes, which can be loaded into the computer as necessary.

Software can be divided into two types: applications software and system software, which provides facilities required by the user and by applications programs. Examples are file handling and input and output. System software also includes translators of programs written in high-level languages, such as BASIC, FORTRAN and Pascal, which programmers can understand, to languages the machine understands. System software, which is often purchased with the hardware, determines which applications can be used, because programs, even those in high-level languages, contain references to it.

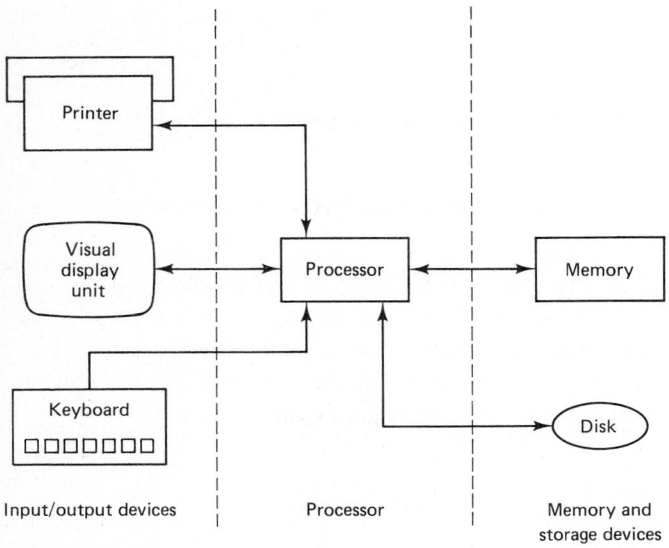

Figure 53 Physical components of a digital computer.

Small **microcomputers** have evolved to standard configurations based on a few types of processor chips and costing about $1500 in 1986. This enables standard system software to be used, for which there are many utility application programs available such as word processors and spreadsheet calculators. Small technical applications such as lighting design and static thermal analysis are also available for microcomputers with standard systems.

The memory and processing of the smaller microcomputers operate using a group of 8 binary digits (or bits), known as a *byte*, which is sufficient to contain a binary code representing a character or an address in memory. Numbers are stored by concatenating bytes. Thus 8-bit microcomputers can process text well, but have limited memory and are slow at calculations involving numbers such as energy analysis or drawing data. Larger, faster computers are required for these applications.

Newer 16-bit micros use a 16-bit processor chip. This makes the processing of numbers faster, and the processor can use more memory. In the simpler 16-bit computers, all other data transfers are in 8 bits. This type of computer costs 50% more then the 8-bit micro but is also a single-user system with the same range of applications.

Larger 16-bit micros have 16-bit data transfers and can use extended memories. These computers can support more powerful system software and four to eight simultaneous users, but they cost four or more times as much as the 8-bit micros. They are suitable for applications such as energy analysis and drawing systems. Like the smaller microcomputers, standard system software and application programs for them are increasing. The larger of these systems have the same performance as smaller minicomputers, but at a lower cost.

Minicomputers and Mainframes are larger, more complex computers containing, in addition to the main processor, many special-purpose processors to speed data flows and processing. These computers operate in 32 or more bit words, with large memories and disk stores. These computers are almost always used by sharing their resources among many users.

Typical applications of minicomputers are for drawing systems where a central computer may be shared between several graphics workstations. Mainframes are used for large data banks, for lengthy and complex calculations, or as a central source of processing for a large organization.

The system software of mainframes and minicomputers

is usually unique to the hardware manufacturer, although this is changing under the influence of microcomputer standards. This often limits the choice of application software to that developed for the particular computer.

The power of microcomputers is rising rapidly, and large computers are now often replaced by groups of smaller computers with communications between them. This type of configuration suits many tasks that are better not centralized. In many cases this also enables micros with standard system software to be used, enabling the user to choose from a wide range of hardware and applications software.

References

1. S. BENSASSON: *Micros in Construction*. Design Office Consortium (now Construction Industry Computing Association), Guildhall Place, Cambridge, England, 1980.
2. J. CHALMERS: Guide to hardware. *Architects Journal*, Vol. 175, (12 May 1982), pp. 71–83.
3. J. CHALMERS: Guide to software and systems. *Architects Journal*, Vol. 175 (19 May 1982), pp. 65–74.

J.C.

DIMENSIONAL COORDINATION. The concept of coordinating the dimensions of buildings and their parts with the dimensions of manufactured components or assemblies was pioneered in the United States in the 1920s and 1930s. The terms *modular coordination* and *dimensional coordination* were adopted for the application of a basic building module and preferred dimensions to building design, product manufacture, and construction.

While the American approach to dimensional coordination used a basic module of 4 in. (101.4 mm), the European nations developed metric systems of coordination after World War II based on a 100-mm module and, in Germany, an octametric module of 125 mm. Despite the differences in the basic unit of size, the principles of coordination were remarkably similar. In the 1960s, the development of international standards for modular coordination commenced under the auspices of the International Organization for Standardization (ISO), based on the 100-mm module (Refs. 1 and 2). A range of ISO standards has been published. To simplify description, the symbol M may be used for the international building module; thus 12M would represent 1200 mm.

Principles of Dimensional Coordination. The coordination of dimensions of buildings and rectilinear building parts involves several related concepts:

1. Use of the basic module and selected multiples (multimodules and submodules) for building dimensions and component sizes.
2. Use of a controlling reference system of lines, planes, or grids to provide reference locations in space for the positioning and coordination of building elements and the components that form them.
3. Controlling dimensions in the horizontal and vertical plane to control the location and size of major building parts or elements.
4. Coordinating dimensions for building components to reduce the variety of sizes and promote better fit within the controlling reference system.
5. Rules for the sizing of building components, including their tolerances (permissible deviations), and the design of joints for fit and functional performance.
6. Conventions and symbols for the representation of dimensionally coordinated designs on drawings.

Multimodules are whole multiples of the basic module, which are used for structural and planning grids and for determining ranges of sizes for building elements and components. Different multiples may be used for horizontal and vertical dimensions. Preferred horizontal multimodules are 300, 600, 1200, 3000, and 6000 mm (1, 2, 4, 10, and 20 ft); 900 or 1500 mm (3 or 5 ft) may be used for special applications. Preferred vertical multimodules are 200, 300, and 600 mm (8 in., 1 ft, and 2 ft). Typical submodules are 25, 50, and 75 mm (1, 2, and 3 in.).

The **Dimensional Reference System** is a rectilinear system used for dimensionally coordinated design and detailing of building parts. It may be one of the following:

1. A standard modular grid in two of three dimensions.
2. A multimodular grid in two of three dimensions, which can employ different modules in different directions.
3. A selected set of reference planes or lines spaced at modular intervals.

Controlling Dimensions are dominant dimensions that determine the principal geometric relationships and, therefore, influence the sizes of many building elements, assemblies, or components. Vertical controlling dimensions (Fig. 54) determine the height of building parts, elements, or assemblies; horizontal controlling dimensions (Fig. 55) determine the length and width of building parts, elements, and assemblies. Three are two types of controlling dimension:

1. Axial controlling dimensions, taken from centerline to centerline of building elements or parts.
2. Boundary controlling dimensions, taken from the controlling face of one building part to another.

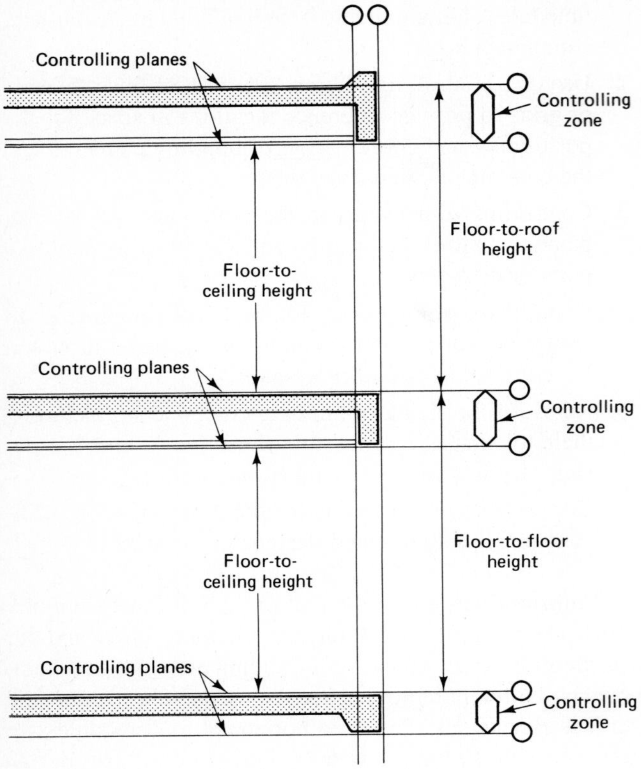

Figure 54 Vertical controlling dimensions.

The modular space between boundary controlling lines that is filled by walls, floor systems, or roofs is called a *controlling zone*. A nonmodular zone is called a *neutral zone*.

Standardization. Dimensional/modular coordination standards have been issued by all major countries. A study by the Center for Building Technology at the U.S. National Bureau of Standards in 1980 (Ref. 3) lists international standards dealing with dimensional or modular coordination, as well as national standards issued by 53 countries.

The reference system of modular grid lines constitutes an ideal basis for COMPUTER-AIDED dimensionally coordinated architectural design. Numerous modular details of components and joints can be stored and reused in a wide variety of designs, and documentation is speeded up by computer-aided DRAFTING.

See also TOLERANCES.

References

1. *Modular Coordination—Basic Module—ISO1006-1973*. International Organization for Standardization, Geneva, 1973.
2. *Modular Coordination—Principles and Rules—ISO 2848-1974*. International Organization for Standardization, Geneva, 1974.

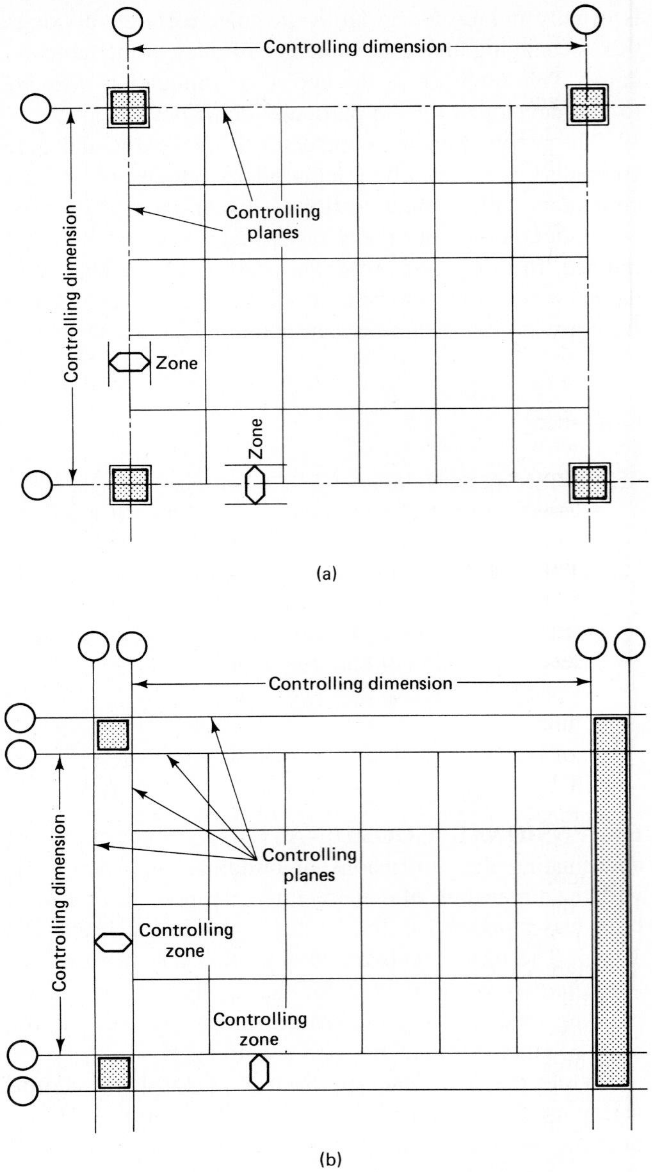

Figure 55 Horizontal controlling dimensions. (a) Axial controlling planes. (b) Boundary controlling planes.

3. *International and National Standards on Dimensional Coordination, Modular Coordination, Tolerances and Joints in Building*. Special Publication 535. U.S. National Bureau of Standards, Washington, D.C., 1980.

H.J.M.

DISCHARGE LAMPS produce light by excitation of an ionized gas or vapor between two electrodes in a discharge tube (electrical discharge), sometimes in combination with the luminescence of phosphors that are excited by the ultraviolet radiation of the discharge. Discharge lamps have a higher, sometimes much higher, efficacy (lumens emitted per watt of power consumed) than that of incandescent lamps, and their life is considerably longer.

Discharge lamps are operated in conjunction with a current-limiting device, or ballast, connected in the lamp circuit. The ballast is normally a combination of chokes and capacitors. A starter or an ignitor is employed to initiate the discharge. Alone or together with the ballast, it delivers voltage pulses that ionize the discharge path and so bring about ignition. Ignition is followed by lamp run-up, which can last for up to several minutes, according to lamp type.

The discharge creates a multiline spectrum whose composition is determined by the gas or vapor filling and the pressure within the discharge tube. Apart from certain types of fluorescent lamps, color rendering is inferior to that given by incandescent lamps.

Fluorescent Lamps are low-pressure mercury discharge lamps in which the light is produced predominantly by fluorescent powders, or phosphors, activated by the ultraviolet energy of the discharge (Fig. 56). The composition of the phosphor coating on the inner surface of the bulb determines the quantity, color, and color-rendering properties of the light emitted. The fluorescent lamp is characterized by its relatively low luminance, variety of color appearance (from warm white to daylight), and the choice of the color-rendering properties offered.

Fluorescent lamps have a combination of line spectra (from the discharge) and continuous spectra (from the phosphors). The application of special fluorescent powders emitting light in three well-defined wavelength bands makes it possible to develop fluorescent lamps with a high efficacy combined with good color rendering.

Three types of fluorescent lamps are available; they are defined by the means of lamp starting: preheat (switch-start) lamps, instant start lamps, and rapid start lamps. Recently, electronic ballasts for high-frequency operation of fluorescent lamps were introduced, resulting in both increased efficacy of the lamp and lower energy losses in the ballast. These can be fitted with dimming devices at only slight extra cost.

Tubular fluorescent lamps vary in diameter from approximately 16 mm ($\frac{5}{8}$ in.) to 54 mm ($2\frac{1}{8}$ in.) and in overall length from a nominal 150 to 2440 mm (6 to 96 in.). Shapes other than straight tubes include circular lamps (circline), U-shaped lamps, and compact single-ended fluorescent lamps. An integrated solution of the latter consists of a miniature fluorescent lamp, complete with its control gear, contained inside a glass bulb and fitted with a conventional incandescent lamp cap (Fig. 57). It can be employed as a direct replacement for an incandescent lamp and offers great economic advantages in shops, hotels, and the main lights in the home.

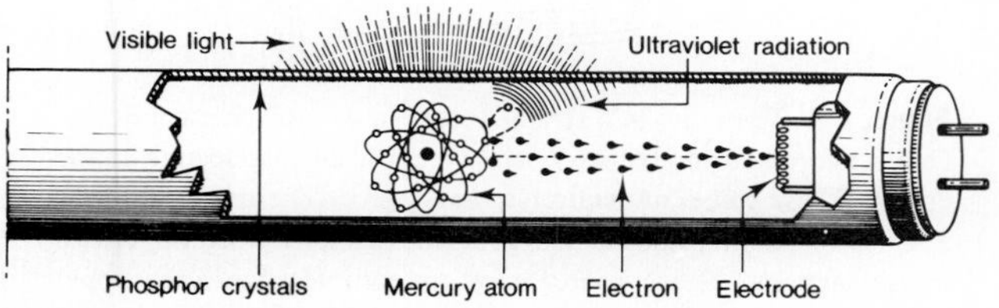

Figure 56 Part section of a tubular fluorescent lamp, illustrating the mechanism of discharge.

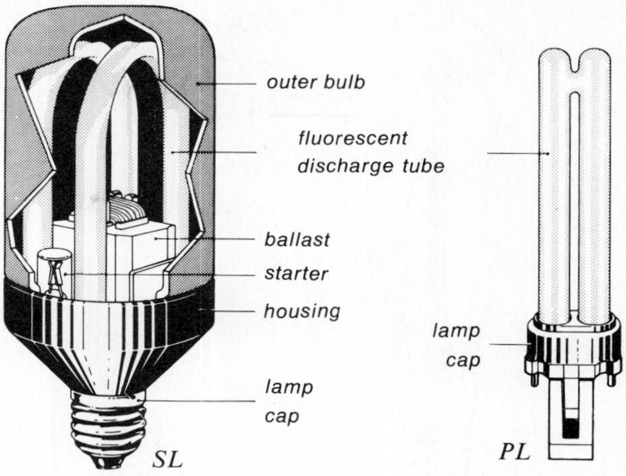

Figure 57 Principal parts of compact fluorescent lamps. The SL lamp consists of a miniature fluorescent lamp, complete with its control gear, contained inside a glass bulb and fitted with a conventional incandescent lamp cap. The PL lamp operates with a separate ballast and is not interchangeable with normal incandescent lamps.

High-Intensity Discharge Lamps (HID lamps) are high-pressure discharge lamps including the groups of lamps commonly known as high-pressure mercury, metal halide, and high-pressure sodium. The compact discharge tubes of these lamps are enclosed in outer bulbs or tubes.

High-Pressure Mercury Lamps with clear bulb appear bluish-white due to the mercury line spectrum. By use of a phosphor coating on the inside of the outer envelope, the ultraviolet energy from the discharge can be made to introduce a red component, thus improving both color rendering and color appearance. The efficacy, but also the price, of high-pressure mercury lamps is lower than that of other high-pressure discharge lamps.

Metal Halide Lamps are similar in construction to high-pressure mercury lamps. The main difference is that the metal halide arc tube contains various metal halides in addition to mercury in order to give a substantial improvement in efficacy and color rendering. By improving the color without the need for phosphor coating, the lamp approaches a point source needed for good optical control.

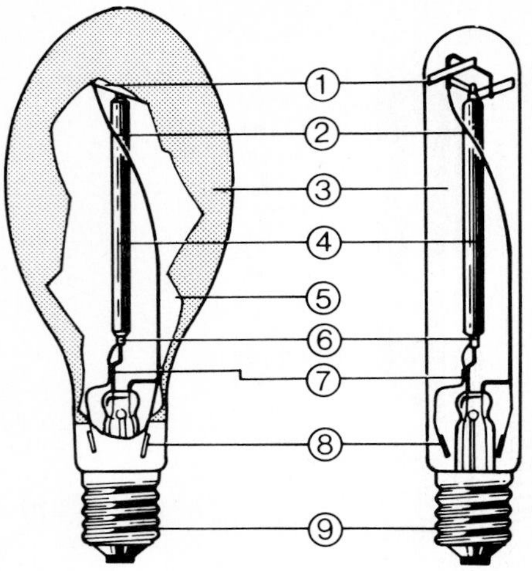

Figure 58 High-pressure sodium lamps are available in clear and diffuse versions. 1, support springs to maintain discharge tube alignment; 2, lead-in wire; 3, hard-glass outer bulb; 4, translucent aluminum oxide discharge tube; 5, inner phosphor coating; 6, end cap of discharge tube; 7, lead-in wire/support; 8, getter rings for maintaining high vacuum; 9, screw base.

High-Pressure Sodium Lamps (Fig. 58) have about twice the efficacy of high-pressure mercury lamps. The light emitted by high-pressure sodium lamps is concentrated in the yellow-orange part of the spectrum and is consequently golden-white in color. Increasing sodium pressure increases the percentage of red radiation and thus improves color rendering. This occurs, however, at the expense of life and efficacy.

Low-Pressure Sodium Lamps have the highest luminous efficacy of all existing lamps, but they are monochromatic yellow. Therefore, all colors appear as different shades of yellow. While this matters little in case of, say, interior security lighting, it effectively eliminates the lamp for consideration in connection with interior lighting in general.

See also INCANDESCENT LAMPS.

References

1. *IES Lighting Handbook*. Reference Volume. Illuminating Engineering Society of North America, New York, 1981.
2. J. F. WAYMOUTH: *Electric Discharge Lamps*. M.I.T. Press, Cambridge, Mass., 1971.
3. W. ELENBAAS: *Fluorescent Lamps*. MacMillan, London, 1971.
4. W. ELENBAAS: *Light Sources*. MacMillan, London, and Crane, Russak & Co., New York, 1972.
5. M. H. van de WEIJER: Light generation in the eighties. *International Lighting Review*, Vol. 32 (1981), pp. 64–69.

D.F.

DOMES have held the record for the interior span in a building from the days of Ancient Rome until the present time, except for a short break in the mid-19th century, after the development of iron trusses and before that of iron domes (Table 1, SPAN). This is due to their structural and, to some extent, their esthetic qualities.

Masonry Domes. Before IRON became available as a major structural material in the 18th century, the designer had a choice between materials, such as timber and reeds, that possess good tensile strength, but are liable to be destroyed by fire, insects, or fungi, and masonry materials (NATURAL BUILDING STONE, or BRICK, or ADOBE) that lacked tensile strength. Durable structures therefore had to be designed to minimize tension; masonry vaults and domes do this most effectively. Their structural behavior is discussed in the articles on MASONRY ARCHES, VAULTS, AND DOMES and CATENARY.

A masonry dome, however, can span much further than an arch or a vault. Furthermore, an arch exerts an outward thrust, and this complicates the construction of arches and vaults. The problems encountered by the Gothic master builders in bringing the thrust from their vaults to the ground are well known. In a dome the arches are arranged on a vertical axis through the crown, tied together by horizontal hoop forces (Fig. 59). These are compressive near the crown of the dome, but they become tensile in the lower portion of a hemispherical dome. This hoop tension presents the major structural problem in the design of masonry domes (Refs. 1 and 2); but it also ties the vertical arches together so that a hemispherical dome, unlike an arch, exerts no outward thrust. It is consequently possible to construct it without formwork, which is not possible for an arch. Furthermore, a masonry dome can be left with an opening, an "eye," at the top, because due to the closed hoops no keystone is required.

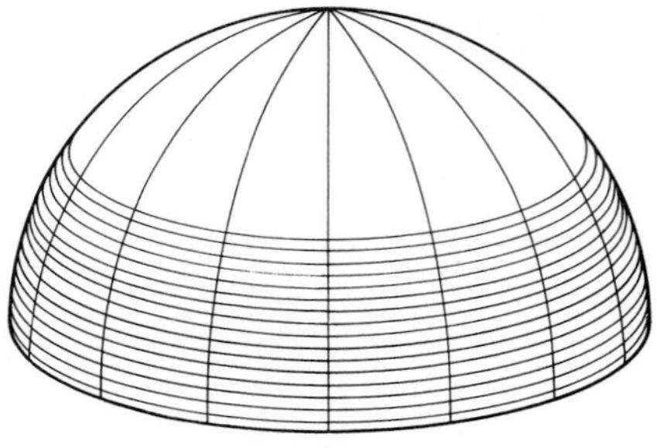

Figure 59 There are two sets of forces in a hemispherical dome. One set forms a number of vertical arches, all intersecting at the crown. These forces are entirely compressive. The other set forms a series of parallel hoops, increasing in diameter from the crown to the supports. These hoop forces are compressive in the upper part of the dome, and they gradually reduce to zero at a horizontal circle that makes an angle of 52° with the crown. Below 52° they become tensile and reach maximum tension at the supports (Drawing by Forrest Wilson, from Ref 1.)

Modern Domes. Iron was used for the structure of a dome for the first time in 1842, in St. Isaac's Cathedral in Leningrad. The economic advantages were immediately apparent, and a number of masonry domes were redesigned in iron, including that of the Congress Building in Washington, D.C. Reinforced concrete domes were introduced in the first decade of the 20th century. Their behavior is not unlike that of masonry domes, but, since the tensile hoop forces can be absorbed by reinforcement, they can be built much thinner.

Reinforced concrete domes and steel domes are briefly discussed in the articles on SHELLS and SPACE FRAMES.

Structural Economics. Using steel and reinforced concrete, the dome has retained its position as the structure with the longest interior span (see Table 1, SPAN), because it curves continuously toward its supports. The loads are therefore gradually transmitted to this lowest level, which, in modern domes, is generally at or near ground level. Because of these efficient load paths, only a minimum amount of material is required to transmit the loads to the ground.

A good measure of structural economy is the ratio of thickness to span. It is not possible to compare iron or steel and masonry domes; but reinforced concrete and masonry are similar materials. In Table 1, thicknesses are averaged over the dome, and the combined thickness is used for domes with an inner and outer shell.

TABLE 1
Reduction in the Amount of Material Used in Concrete and Masonry Domes

Year of Completion, A.D.	Name and Construction of Dome	Ratio of Span to Thickness of Dome
123	Pantheon, Rome. A solid concrete dome, faced with masonry.	11
1434	S. Maria del Fiore, Florence. A double dome of brick and stone.	21
1710	St. Paul's Cathedral, London. A brick dome, surmounted by a brick cone; the lightest of the great classical domes.	36
—	Hen's egg	Approx. 100
1927	Planetarium, Jena, East Germany. First reinforced concrete dome designed by the membrane theory.	420
1958	C.N.I.T. Exhibition Hall, Paris, A double-shelled reinforced concrete dome; longest interior span in concrete.	1700

The egg shell occupies an interesting position in this hierarchy, as all reinforced concrete shell domes are relatively thinner.

References

1. H. J. COWAN: Domes: ancient and modern. *Journal of the Royal Society of Arts*, Vol. 131 (1983), pp. 181–198.
2. H. J. COWAN: A review of the structural dimensions of masonry and concrete domes in Antiquity, in the Renaissance, and at the present time. *Atti dell'Accademia Pontaniana, Nuove Serie*, Vol. 30 (1982), pp. 349–386.

H.J.C.

DOORS are wooden or metal structures, hinged or sliding, giving access to a building, room, passage, or cupboard.

Door Types are named according to their construction: ledged; ledged and braced; framed ledged; framed ledged and braced; glazed; paneled; bead and butt; or bead and flush.

1. A *ledged and braced door* consists of a number of vertical boards nailed to horizontal rails or ledges. Timber braces run diagonally between the ledges. This is the most common form of external door, used mainly for outhouses, sheds, and temporary structures (Fig. 60).
2. A *paneled door* relates to the number of vertical or lay (horizontal) panels used. It may be solid or self-molded, insertion-molded, or bolection molded (Fig. 61).
3. A *flush door* can be constructed as a solid core or as a honeycomb core with a ply covering. The edges are covered with a margin strip to cover the end grain.

Hinges. Doors can be hinged to the door frame by butt hinges, rising-butt hinges, pin hinges, black flaps, counter flaps, strap hinges, or hook-and-eye hinges.

Locks and Fasteners. Doors can be locked with mortice lock, mortice latches (A), tubular latches (B), tubular locking latches (C), night latches (D1), cylinder rim dead latches (D2), rim locks (E), or padlocks (F), as shown in Fig. 62.

References

1. GRAHAM BLACKBURN: *Illustrated Interior Carpentry*. Evans, London, 1979.
2. N. H. ROBERTS: *How to Build Your Own Wood-Framed House from Scratch*. TAB Books, Blue Ridge Summit, Pa., 1981.
3. JOHN FRASER: *How to Build Frame Houses and Summer Houses*. Coles Publishing Co., Toronto, 1976.

R.J.S.

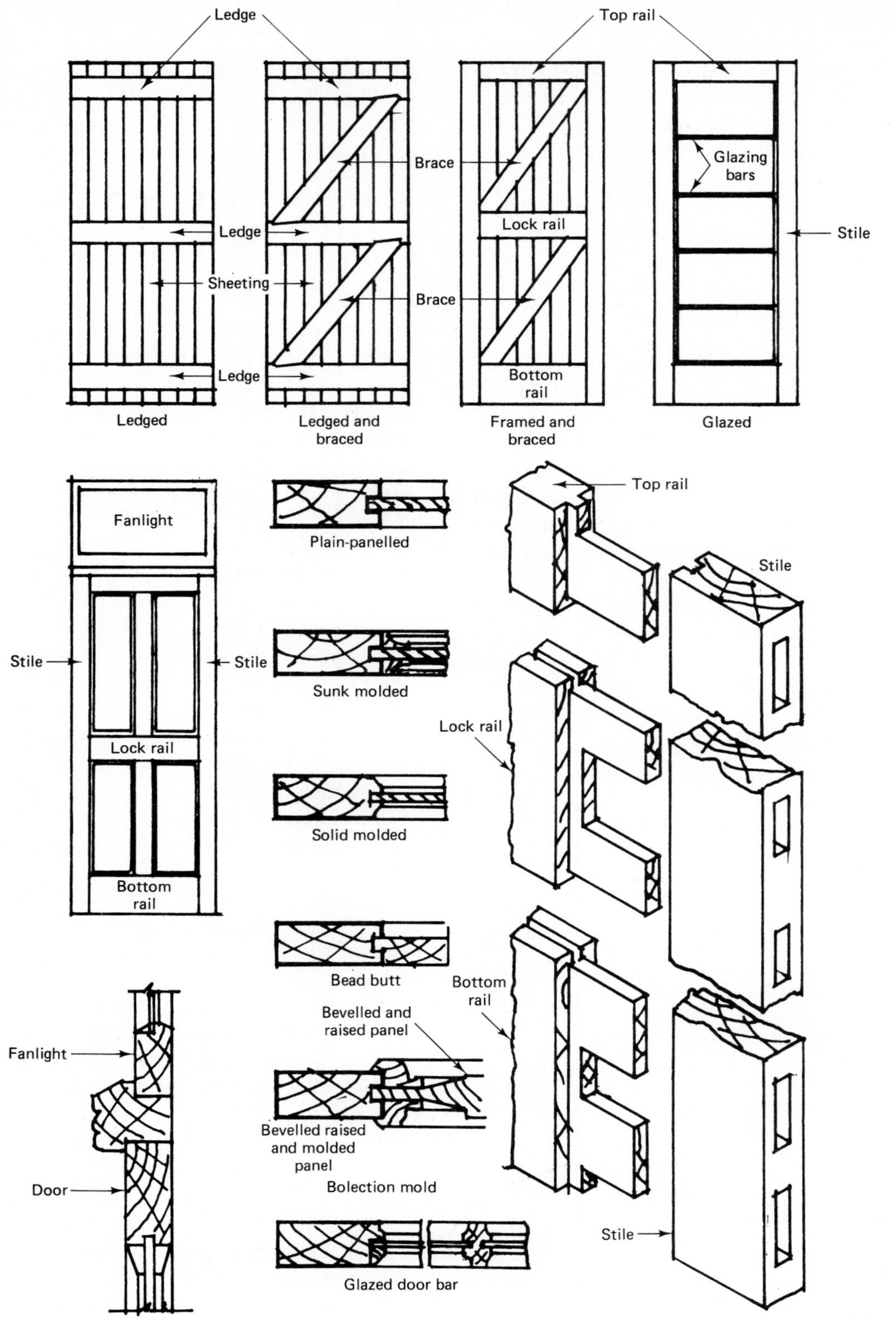

Figure 60 Ledged, braced framed, and glazed doors.

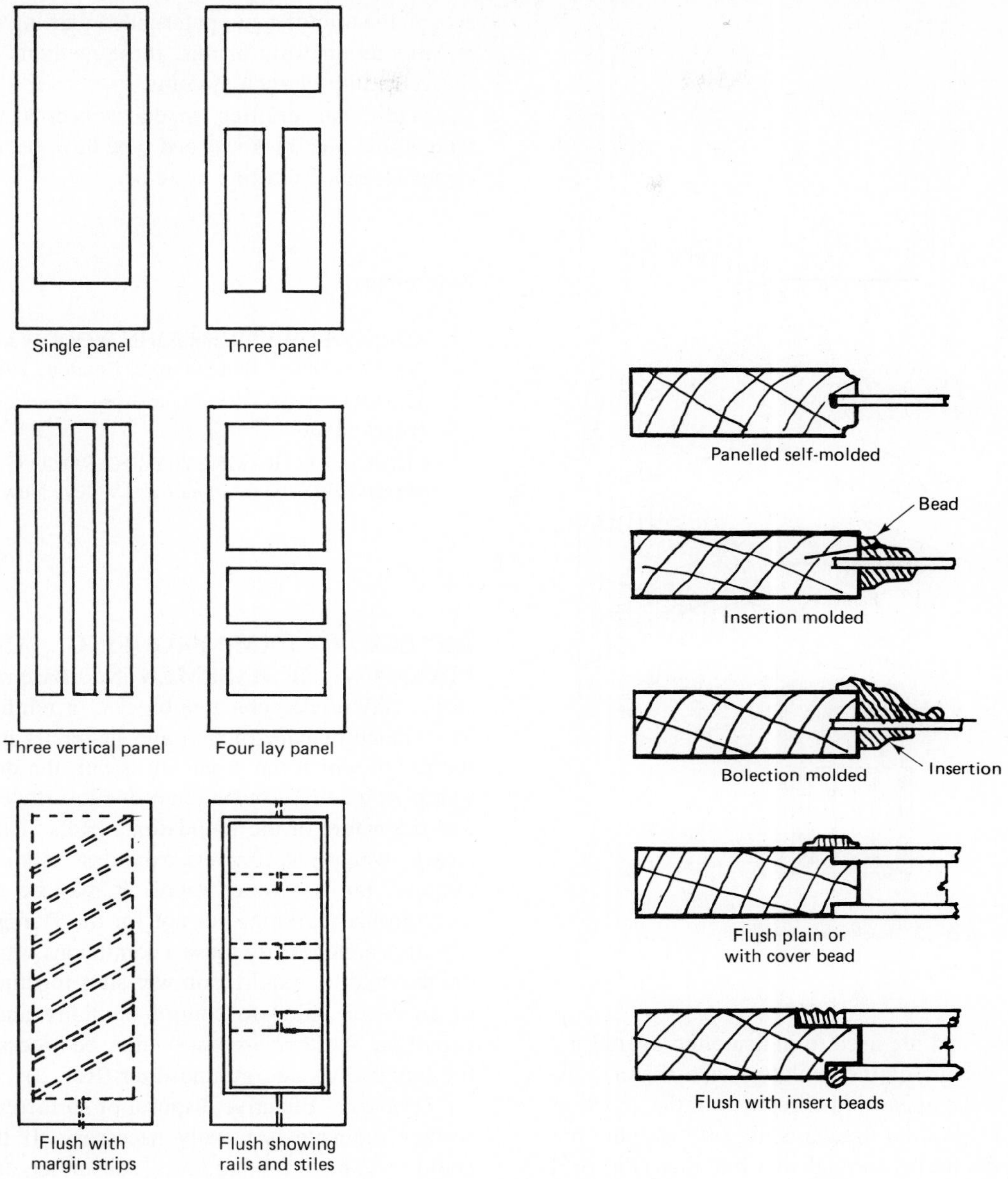

Figure 61 Paneled doors.

DRAFTING. Drawings are the primary communication medium between all persons associated with a building. The architect uses drawings to pass on design ideas to his or her client and to the consultants and provides drawings to the local authorities to obtain approval for the design. The drawings, along with written specifications and sometimes a bill of quantities, are the definition of the building to be constructed; these documents are generally referred to as the *contract documents*. Drafting is the process of creating these all-important drawings.

Drafting involves the creation of a graphical representation of a building or some part of it. The function of a drawing determines what sort of graphical representation will be used. To represent three-dimensional objects on two-dimensional paper, it is necessary to use PROJECTIONS. *Presentation* drawings, which aim to provide an overall visual impression, generally make use of the *perspective* projection. The figures in the COMPUTER GRAPHICS entry are all perspective projections.

Plan and *elevation* drawings make use of the *ortho-*

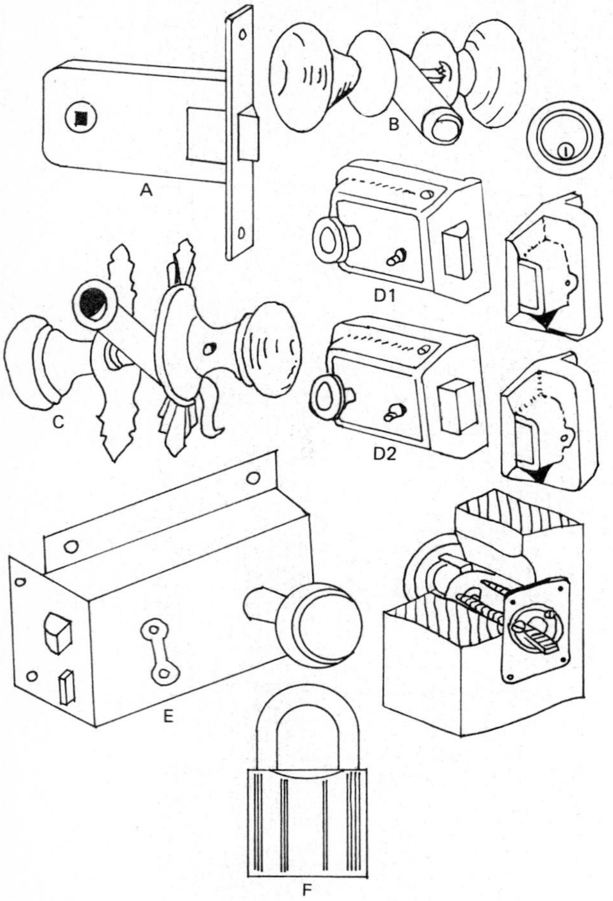

Figure 62 Locks and fasteners.

graphic projection and are used to indicate the overall layout and facade design of the building in a manner that allows lengths to be measured directly off the drawings (Figs. 63 and 64). *Section* drawings are orthographic presentations of vertical cuts through the building (Fig. 65). *Detail* drawings also make use of the orthographic projection and represent a small number of building elements in some detail (Fig. 66).

Drafting involves the use of conventions and conventional symbols. These conventions are often formalized in standards at the office level, national level, and even the international level. Until recently, all drawings were generated manually, but now there is an increasing trend to utilize computer-aided drafting systems in the production of drawings. All the drawings shown here were computer produced. *Computer-aided drafting* is beginning slowly to replace manual drafting because of productivity benefits and the versatility of the computer. New mathematical methods for the production of the projections have unified all the disparate projections in both two and three dimensions. Although the user of a computer-aided drafting system is likely to be quite unaware of this, these methods have increased the versatility of such systems.

Traditional drafting media concerns such as paper weight and line thickness and type have been transferred to computer-aided drafting systems.

References

1. ADRIAN BOWYER and JOHN WOODWARK: *A Programmer's Geometry*. Butterworths, London, 1983.
2. ALBERT BOUNDY: *Engineering Drawing*. McGraw-Hill, Sydney, 1980.
3. CHARLES G. RAMSAY and HAROLD G. SLEEPER: *Architectural Graphic Standards*. Wiley, New York, 1970.

J.S.G.

DRAINAGE, DAMP-PROOFING, AND WATERPROOFING OF BASEMENTS. Basements can be of stone, clay bricks, concrete blocks, or reinforced concrete. The choice of construction and materials is influenced by the use to which the basement is put, the degree of watertightness required, the amount of subsurface water present, and the nature of the foundations (Refs. 1 and 2). For example, watertight concrete excludes visible moisture penetration, but not water vapor. It may be satisfactory for underground parking, but not for the storage of moisture-sensitive archives; for these a continuous, impervious membrane could be required on wet sites together with heating or air-conditioning to control humidity and prevent condensation. A site investigation may be necessary to examine the foundations and groundwater (Ref. 3).

Drainage. Effective disposal of surface runoff and subsurface drainage is usually necessary. If the water table could rise above footing level, subsurface drains should be installed. On waterlogged sites, pumping may be required during construction, as membranes may require dry application.

Membranes. Solid, liquid, and sheet membranes are available for tanking basements. Mastic asphalt, pure polyurethane, and tar-modified polyurethanes are applied by troweling; rubber-latex-bitumen or bitumen emulsions by brush or spray. Single-layer sheet membranes include butyl rubber and glass-fiber-reinforced polyvinyl chloride (PVC). Composite multilayer membranes made from asphalt with organic or inorganic reinforcement are applied with flood coats of hot asphalt. In another composite membrane, swelling of wetted montmorillonite clay creates the moisture barrier. Factors that influence membrane choice include strength, extensibility, durability, and resistance to

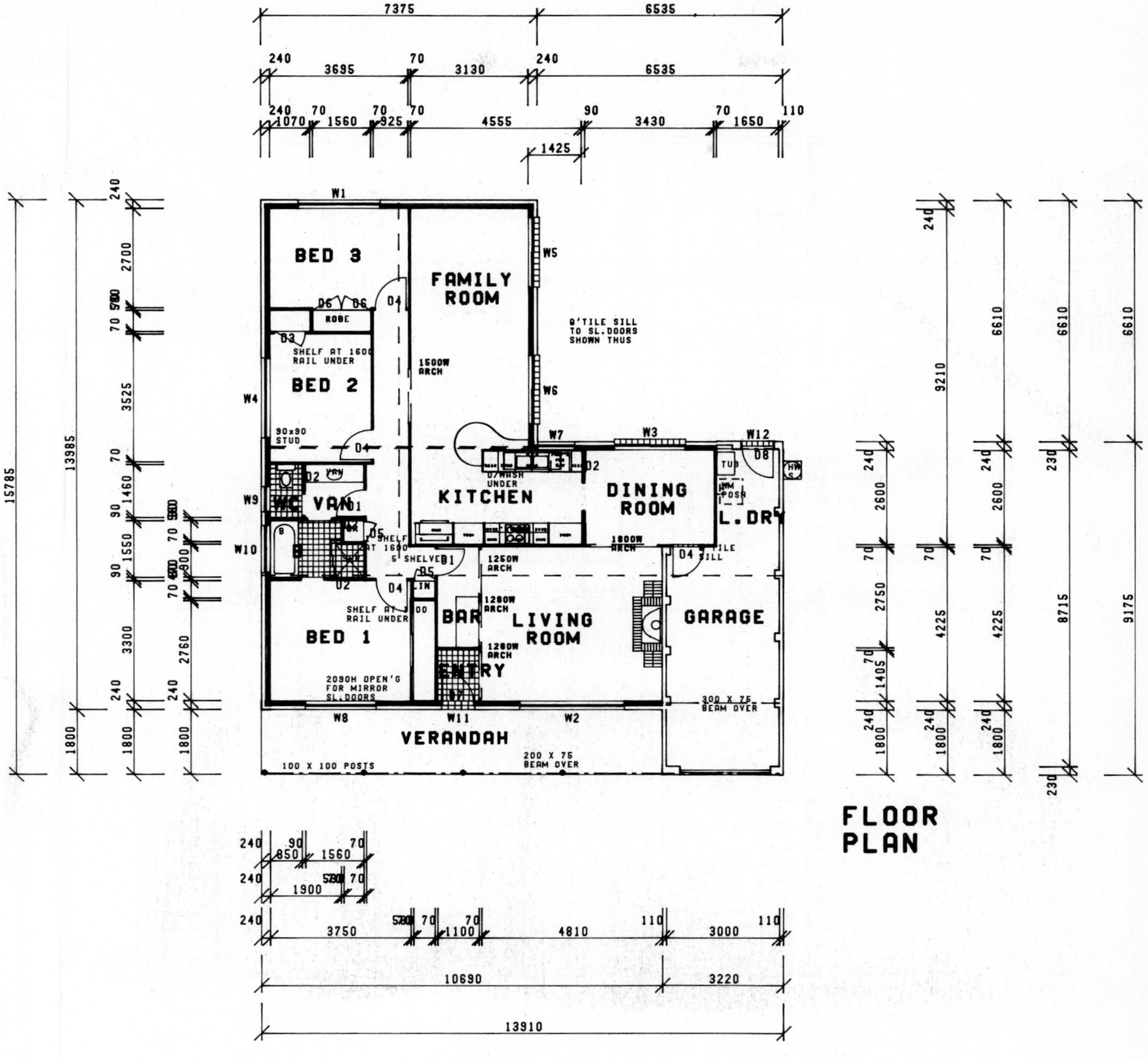

Figure 63 Plan of a three-bedroom house, drafted by a computer using an orthographic projection. (By courtesy of Huxley Homes, Sydney, Australia, using the EASINET CAD System.)

damage during installation (Ref. 3). Membranes are applied internally or externally. External application protects structural concrete exposed to aggressive groundwater. Restricted external access may require internal application.

Insulation. In cold climates, insulation may be required (Ref. 4).

Repairs. Pressure grouting and watertight renders are used to repair leaks (Ref. 5).

References

1. *Guide to the Design of Waterproof Basements.* Construction Industry Research Association, London, 1978.
2. *Code of Practice for Protection of Building against Water from the Ground, CP102.* British Standards Institution, London, 1973.
3. *Waterproofing Basements,* Notes on the Science of Building,

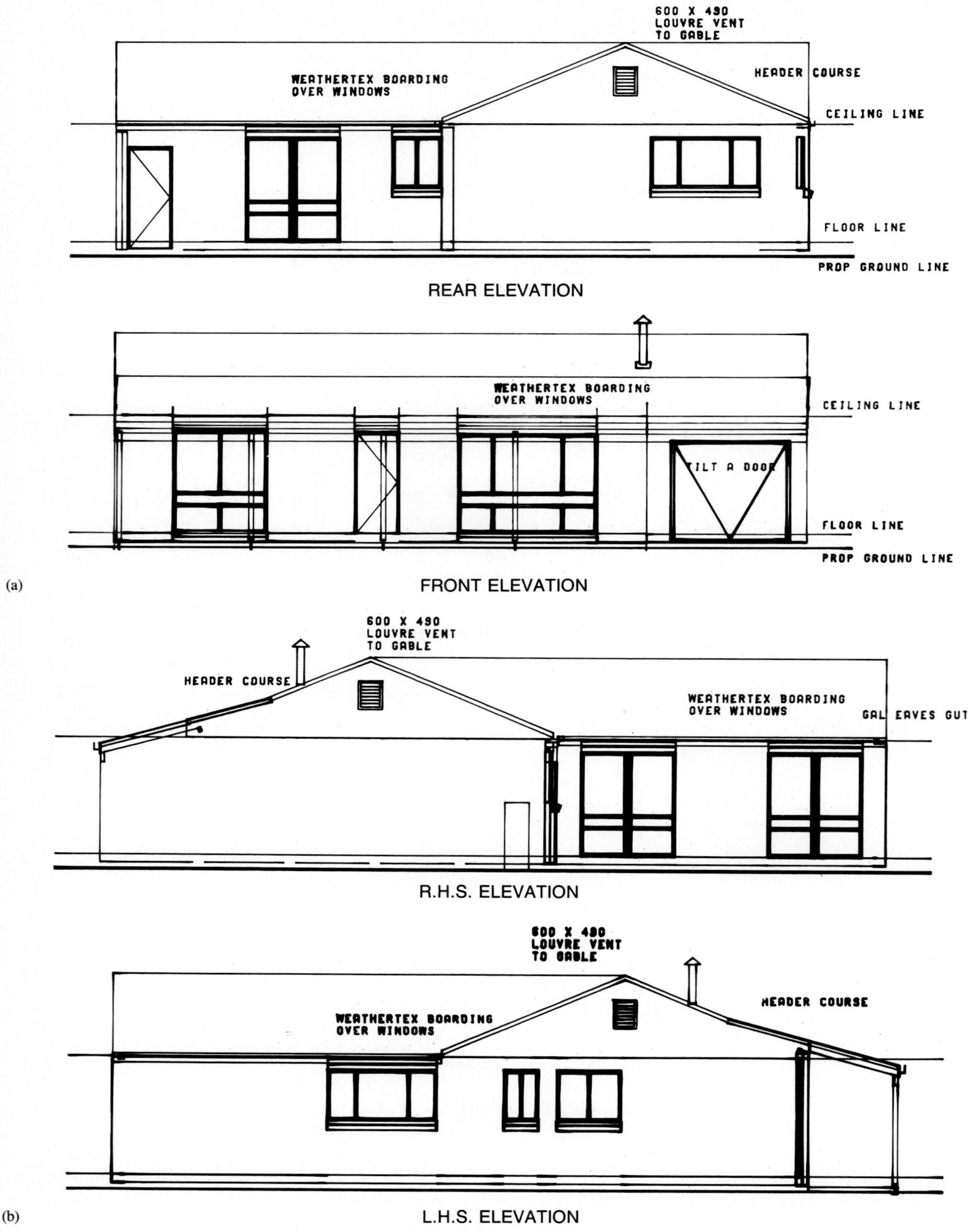

Figure 64 (a) Front and rear elevation of the house in Fig. 63. (b) Side elevations of the house in Fig. 63. (By courtesy of Huxley Homes, Sydney, Australia, using the EASINET CAD System.)

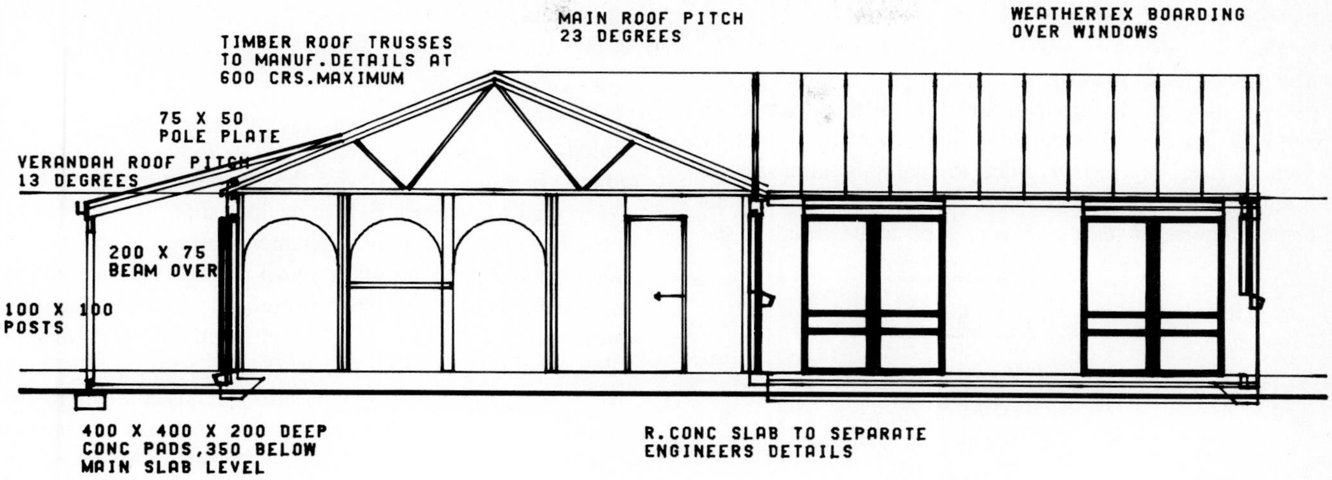

Figure 65 Section of the house in Fig. 63. (By courtesy of Huxley Homes, Sydney, Australia, using the EASINET CAD System.)

No. 160. Australian Government Publishing Service, Canberra, 1984.

4. A. ELMROTH and I. HOGLUND: *New Basement Wall Designs for Below-Grade Living Space, CNRTT-1801*. Division of Building Research, National Research Council Canada, Ottawa, 1975.

5. *Watertight Basements, Parts 1 and 2*, Advisory Leaflets 51 and 52. Her Majesty's Stationery Office, London, 1971.

J.L.H.

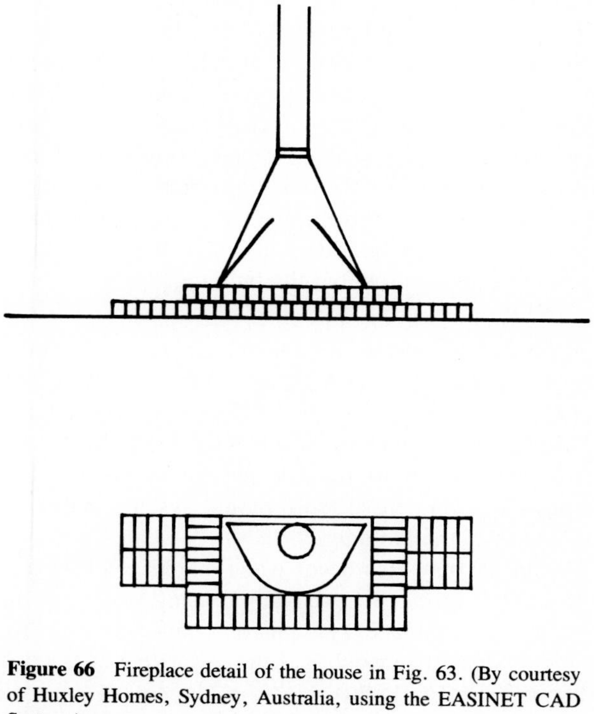

Figure 66 Fireplace detail of the house in Fig. 63. (By courtesy of Huxley Homes, Sydney, Australia, using the EASINET CAD System.)

DRAINAGE INSTALLATIONS. Almost every building intended for human habitation, occupancy, or use should be provided with installations to convey sewage from sanitary fixtures or appliances (such as baths, basins, kitchen sinks, water closets, and clothes and dishwashers) to a disposal system. Whether discharge is to the public sewer or to private disposal, the building drainage installation is essentially the same, consisting of a system of pipework connecting the sanitary fixtures to the underground drains that terminate at the sewer, septic tank, or other disposal system.

Water Supply. All equipment or fixtures that discharge into the drainage installation must be finished with a water supply so that solids will be flushed through the pipelines by water carriage; to achieve this it is necessary to install the pipes and fittings at suitable grades to ensure self-cleaning rates of flow.

Traps. Water-seal traps are furnished at all inlets to drainage installations near the outlets of the sanitary fixture to preclude the escape into the building from the interior of the drainage and sewerage system of odors, gases, vermin, and noise from discharging water.

Pressure and Vacuum Effects. Flow within the drainage pipework tends to remove part or all of the water seals by pressure or partial vacuum produced if the pipes run full bore. Variations in pressure are kept within close limits by not installing pipework that is too small in diameter, or too long, or laid at such flat grades that normal discharge will completely fill the pipes in any part of the system for anything but brief intervals. Where it is thought possible that trap seals might be lost either wholly or partly, good practice requires ventilation pipes (vents) to be fitted near the outlet of water-seal traps, terminating in open air, normally above the roof of the building.

Vent pipes are required in any case at the upper ends of vertical pipes that run between floor levels of multistory buildings (i.e., stacks) in order to maintain atmospheric pressure within the main pipelines of the system; most authorities require at least one vent pipe to be fitted to the main drains of any system where there are no vertical drains in the installation (Fig. 67).

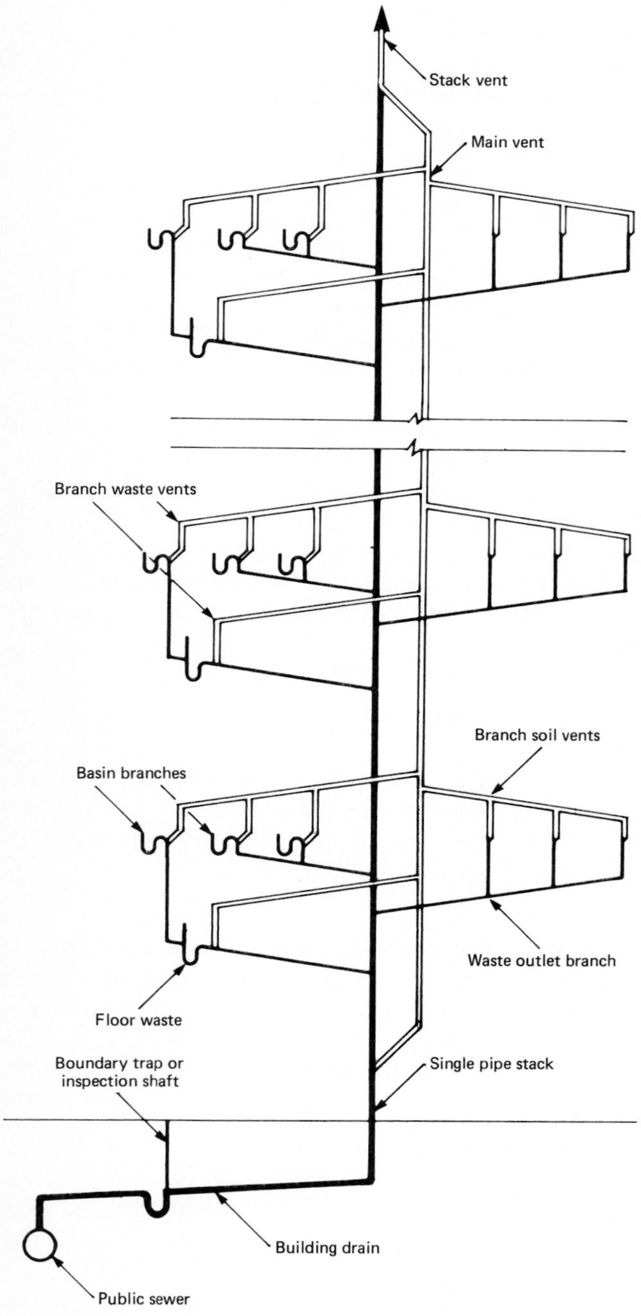

Figure 67 Principal features of a building drainage installation for a tall building.

Single- and Two-Pipe Systems. Modern building drainage is of the single-pipe type, in which one system of pipework is used to convey the waterborne wastes from every kind of sanitary fixture or appliance. In some countries, however, one may encounter drainage installed as a two-pipe system, where one set of pipework above ground is used for conveying those waterborne wastes containing human excrement (i.e., soil), and a separate one for handling all other permitted wastes (i.e., waste). Oddly, these two separate pipework systems subsequently join where they enter the underground drains, so the one conduit is used ultimately to drain everything into the sewer or septic tank.

Combined Stormwater and Sewage. Although it is usual for stormwater drainage to be separated from sanitary sewage, some older sewerage systems handle combined flows. In such cases it may be permissible to convey both discharges to the combined sewer in a combined drainage installation sized to carry the larger flows. Some situations may also justify the conveyance in the sanitary sewers of small flows originating from subsurface drains from building foundation areas.

Drainage Installations below Sewer Level. The public sewerage system is not always laid at sufficient depths that all parts of a building or all buildings on a large property will be capable of draining their fixtures by gravitation. Extensive industrial complexes and multistory buildings with several basement levels below the street may therefore need to be provided with means of collecting and elevating the sewage from fixtures on the lower levels or at the most distant areas so that it may then gravitate to the sewer. Sumps with a pumping plant or receiving tanks forming part of air ejector installations are normal means adopted for dealing with this problem; the latter are more commonly employed with city buildings in which it is necessary to house the equipment within the building.

Preventing Sewage Backflow. Pump or ejector discharge lines and pipelines that drain by gravity at levels that are below ground, but not so low as to require raising by ejector or pump, need to be fitted with check valves (backwater valves) to prevent sewage from high levels flowing back into these parts of the system, possibly to flood areas of the building.

Backflow may also occur when undersized sewers run full, usually during wet weather, causing sewage to flow back up the drainage systems of connected properties. Where this risk may exist, it is necessary to install a check valve on the main drain.

Plumbing Codes. Most municipalities, cities, states, and countries have produced or adopted codes that specify requirements in regard to the building drainage installations that shall be used in their areas. Codes cover such aspects of building drainage as the standards to be complied with; the materials to be used; allowable loadings in the plumbing and drainage systems; the sizes, gradients, lengths, and arrangement of pipework; and venting, trapping, securing, and access for maintenance. Many codes also provide that the installation of these services shall be executed by licensed or authorized personnel.

References

1. L. S. NIELSEN: *Standard Plumbing Engineering Design*. McGraw-Hill, New York, 1963.
2. H. E. BABBITT: *Plumbing*, 3rd ed. McGraw-Hill, New York, 1960.
3. L. BLENDERMANN: *Design of Plumbing and Drainage Systems*, 2nd ed. Industrial Press, New York, 1963.
4. BRITISH STANDARDS INSTITUTION: *Code of Practice for Sanitary Pipework B.S. 5572: 1978*. British Standards Institution, London, 1978.

W.L.H.

DURABILITY is a qualitative term describing the property of a material (or a product) to be long-lasting, despite the service conditions to which it is subjected. With increased emphasis on the need to conserve material resources, there are widespread efforts to increase the durability of building materials and to develop methods of predicting durability quantitatively in terms of service lifetimes of product.

Degradation Mechanisms. Materials of poor durability rapidly degrade in important properties due to environmental factors. The degradation mechanism is the process by which a property declines. These processes are usually very complex. They may be chemical and/or physical and/or biological. Each process is usually specific to the material and in some cases to a particular product. Some of the more important processes are corrosion of metals, stress fatigue of metals, wet rot of timbers, erosion of stone, photo-oxidation of paints and plastics, moisture expansion of fired clay bodies, moisture detachment of adhesive bonds, and freeze–thaw disruption of concrete.

Environmental Factors. Degradation is caused by factors in the service environment to which the material is susceptible. Weathering is the degradation experienced by all building materials due to climatic factors, such as solar irradiance, ultraviolet radiation, relative humidity, surface wetness, wind velocity, airborne particulates, and air pollutants. Rates of degradation may be influenced by cyclic changes, as well as by the level of these factors.

Durability Prediction. The prediction of service life requires an understanding of the decline of the critical property that results in failure or loss of acceptable product function. The failure criterion is the minimum level of that property for acceptable product function. Predictions of durability are based on the duration of exposure to the relevant aggressive environmental factors that is required for the product to fail. The most reliable predictions are those based on field simulations at known very hazardous sites without reducing the time scale, for example, outdoor weathering exposure of commercial products, fixed as in practice. To get quicker predictions of durability, however, a shorter time scale is usually desired in weathering studies. This is achieved by exposure under controlled laboratory conditions and measurement of changes in a particular property by a method that relates to the function of the product (see ACCELERATED WEATHERING).

References

1. AMERICAN SOCIETY FOR TESTING AND MATERIALS: *Durability of Building Materials and Components*. Proceedings of the First International Conference. Special Technical Publication 691, ASTM, Philadelphia, 1980.
2. P. J. SEREDA (ed.-in-chief): *Durability of Building Materials*. Elsevier Scientific Publishing Company International Journal. Vol. 1 was published in 1981.
3. STANDARDS ASSOCIATION OF AUSTRALIA: *Outdoor weathering of plastics in the Australian environment—Guide for design purposes*. AS 1745 Part 2, S.A.A., Sydney, 1975.
4. AMERICAN SOCIETY FOR TESTING AND MATERIALS: *Standard Recommended Practices for Increasing Durability of Building Construction against Water Damage*. E241-1979, ASTM, Philadelphia, 1974.

K.G.M.

E

EARTH-COVERED ARCHITECTURE. Earth-covered buildings have earth-coupled floors with a high proportion of roofs and walls covered with soil of sufficient depth to support vegetation and modify air temperature. Openings, skylights, and shafts are left uncovered. Earth-covered buildings (also referred to as underground buildings) (Ref. 1) may have all or a proportion of the building envelope earth-integrated, that is, "made whole" with the earth. The infinite thermal mass of the enclosing soil then becomes a heat sink and/or heat source. If the interior of the building is partially or wholly isolated thermally from the soil, for example, by insulation or by a building envelope of low thermal conductivity, the building is earth-integrated only where the building envelope is in direct contact with the soil; it is then described as earth-covered or earth-sheltered with a specified degree of earth-integration (Ref. 2).

Earth-sheltered buildings have no earth cover to the roof. They may be physically in contact with or only sheltered behind earth berms or mounds.

Advantages. The advantages of earth-integrated or earth-covered buildings depend on the correct incorporation of passive-solar principles, correct detailing, a high degree of earth-coupling of the building to the soil, adequate insulation if required, vegetation for earth temperature modification, and appropriate user management of interior thermal conditions. Earth-covered architecture facilitates multiple land use, and the natural visual qualities of a site are maintained. It is energy efficient because the ground temperature cycle has a much smaller annual range than that of the ambient air for any given site (Fig. 68). Day-to-day running and maintenance costs are low. Long-term environmental impact is minimized and airborne noise is reduced. Dust can be excluded. The building is protected from the destructive effects of high winds, such as cyclones, hurricanes, and tornadoes; bushfire and nuclear blast protection is possible (Ref. 2).

Disadvantages. Earth-covered buildings are not suitable for flood-prone land. On flat land it may be necessary to pump subfloor seepage from drains. For such buildings to function correctly, solar access must be guaranteed to ensure continuous passive-solar heat gain.

Subterranean Thermal Environment. The thermal environment of the ground for a building in Sydney, Australia, is shown in Fig. 68. Curve A gives the annual vari-

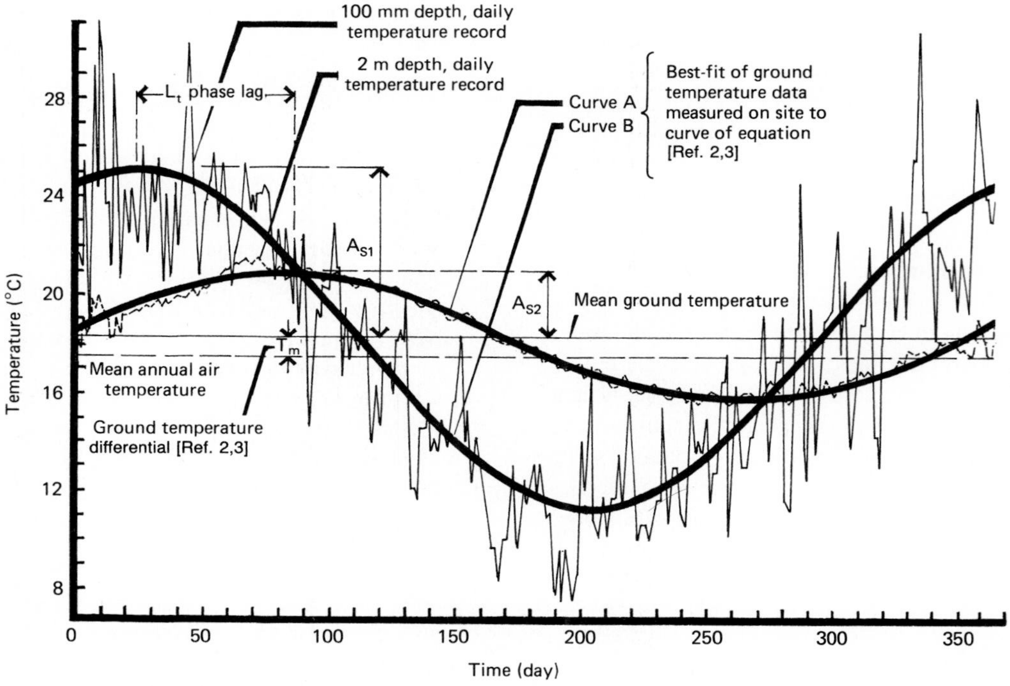

Figure 68 Typical variation of annual ground temperature 100 mm and 2 m below the surface (recorded in Sydney, Australia, 1978–1979).

ation of ground temperature at the shallow depth of 100 mm (4 in.), and curve B does so at the much greater depth of 2 m (6 ft 9 in.). The measured data can be fitted to the sinusoidal curves shown in Fig. 68, using an equation for predicting periodic ground temperature (Ref. 3). As Fig. 68 shows, the temperature amplitude, that is, the difference between the maximum and the mean ground temperature, decreases from A_{s1} to A_{s2}. This damping in amplitude continues until A_s becomes negligible at depths varying from 8 to 18 m (26 to 59 ft), depending on the average thermal diffusivity of the soil (Fig. 69). When variations occur in the temperature above ground, heat is gained or lost from the earth's surface. These variations are delayed in their effect on the soil because of the phase lag; they are also decreased in magnitude. Both effects are utilized in the design of earth-covered buildings to improve their energy efficiency.

Evergreen vegetation on the roof and surrounds of an earth-covered building further decreases temperature amplitude, although a small elevation of annual average temperature may occur beneath extensive ground cover (Ref. 4). The layer of still air between foliage and ground surface warms the soil during winter when the earth-covered building benefits most; it cools it in summer. Thus vegetation, even beneath snow, provides additional insulation.

See also DRAINAGE, DAMP-PROOFING AND WATERPROOFING OF BASEMENTS and PASSIVE SOLAR DESIGN.

References

1. S. CAMPBELL. *The Underground House Book.* Garden Way Publishing Co., Charlotte, Vt., 1980.
2. S. A. BAGGS, JOAN C. BAGGS, and D. W. BAGGS. *Solar Earth-covered Building: Answers to Your Questions about Earth-covered Architecture in Australia.* New South Wales University Press, Kensington, 1984.
3. S. A. BAGGS: Remote prediction of ground temperature in Australian soils and mapping its distribution. *Solar Energy,* Vol. 30, No. 4 (1983), pp. 351–366.
4. S. A. BAGGS: Effects of vegetation upon earth cooling potential. In L. BOYER (ed.): *Proceedings, Earth Shelter Performance and Evaluation, Second National Technical Conference,* Tulsa, Okla. (1981), pp. 81–90.

S.A.B.

EARTHQUAKE can mean either a rock fracture within the earth that causes elastic waves or the waves themselves that cause damage or that we record with a seismograph. The fracture is usually a shear fracture as, for example, when the western edge of California is forced northwest relative to the rest of California along the San Andreas fault systems or the eastern edge of the Pacific Ocean is forced down in Peru and Chile under the Andes mountains.

The waves that cause damage or that we record are P

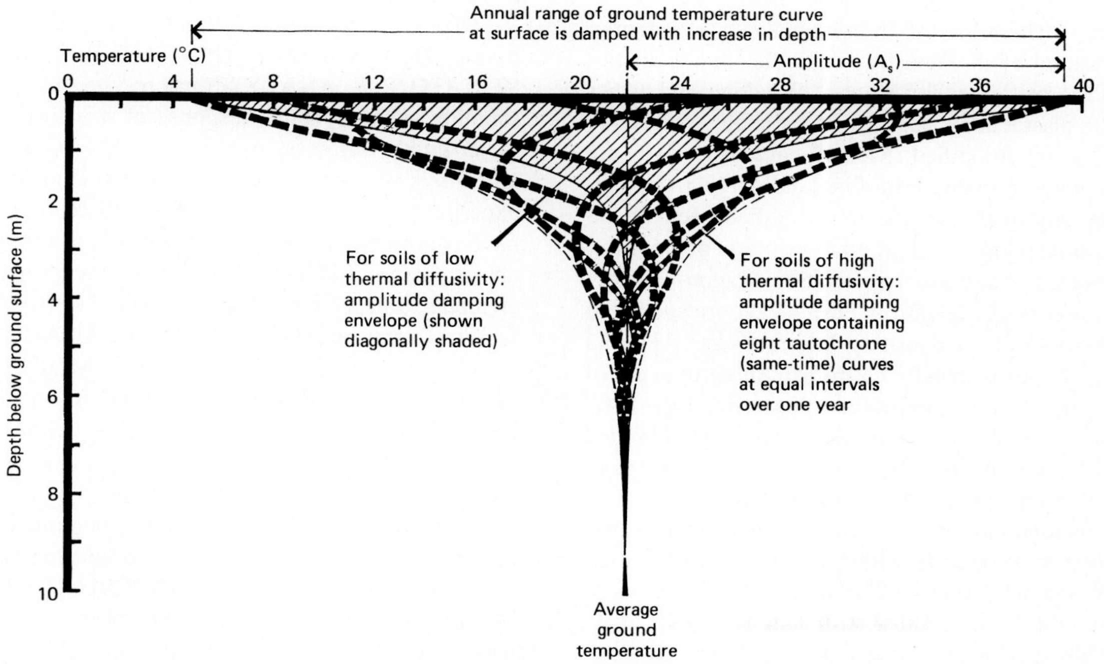

Figure 69 Variation of ground temperature with depth at eight equal intervals throughout the year in a hot-arid climate. (White Cliffs, New South Wales, Australia).

waves, S waves, and surface waves. P or primary waves are compressional dilatational sound waves that arrive first from an earthquake. S or secondary waves are transverse shear waves that travel at about two-thirds the velocity of P waves (depending on the value of Poisson's ratio of the material). Finally, surface waves spread around the surface of the earth and do not attenuate with distance from an earthquake as rapidly as do P and S waves (which propagate in all directions in the interior of the earth).

The **Magnitude** of an earthquake is the logarithm of the half-amplitude (in micrometers) of the motion it would cause on a Wood–Anderson seismograph 100 km from the *epicenter*, the point on the earth's surface above the initial fracture. Thus an earthquake of magnitude 5 would cause a motion of 10 cm half-amplitude on a Wood–Anderson seismograph 100 km from its epicenter. C. F. Richter in 1935 constructed a table for estimating magnitudes. Nowdays, magnitudes are estimated from P, S, and surface waves (Refs. 1 and 2).

The relation between the wave energy E released by an earthquake and its surface wave magnitude M is

$$\log E = 4.8 + 1.5M$$

Thus an earthquake of magnitude 8 on the Richter scale releases 27,000 times as much wave energy (6.3×10^{16} J) as an earthquake of magnitude 5.

Intensity. The effect of an earthquake at a particular place is measured by its intensity. Thus, the characteristics of modified Mercalli intensity VI are "Windows and dishes broken; weak plaster and masonry of poor quality cracked." Characteristics of modified Mercalli intensity VII are "Weak chimneys broken at roofline; fall of plaster; masonry of ordinary quality cracked." Modified Mercalli intensity VI corresponds to a ground velocity of approximately 50 mm/s and a ground acceleration of approximately 600 mm/s^2. Modified Mercalli intensity VII corresponds to a ground velocity of approximately 100 mm/s and a ground acceleration of approximately 1.2 m/s^2. An earthquake of magnitude 5 to 5.5 can be expected to have modified Mercalli intensity VII in its epicentral region. The modified Mercalli scale goes up to XII. The Rossi–Forel intensity scale is similar, except that it goes up to X.

The word *magnitude* refers to an earthquake in the sense of the original rock fracture within the earth and the total wave energy radiated from it. The word *intensity* refers to the effects of these waves in a particular place. Intensity decreases with increasing distance from the original rock fracture.

References

1. B. A. BOLT: *Earthquakes, A Primer*. W. H. Freeman, San Francisco, 1978.
2. H. KANAMORI: Quantification of earthquakes. *Nature*, Vol. 271 (1978), pp. 411–414.

L.A.D.

EARTHQUAKE-RESISTANT DESIGN. Observations of performance of contemporary steel and reinforced concrete structures in recent EARTHQUAKES have demonstrated that, if designed by modern codes, these structures can have the capacity to resist earthquakes of major intensity. Early attempts to provide earthquake resistance in buildings in the 1920s were handicapped by the lack of proper analytical approaches and reliable earthquake records. As time progressed, the prescribed lateral loads and their distribution along the height of a building were upgraded, and details for ductility improved.

Important Developments. Earthquake engineering is gradually adding scientific components to what was previously considered an art:

The Earthquake Phenomenon. Seismological observations and accumulation of records have yielded insight into the mechanisms of earthquakes and have provided data on the distribution and intensity of ground accelerations. These may possibly also lead to techniques for prediction of future earthquakes.

Dynamic Response of Structures to Earthquake Excitation. Development of COMPUTER METHODS OF STRUCTURAL ANALYSIS has resulted in greater understanding of the dynamic response of structures.

Strength, Stiffness, and Ductility of Structural Elements and Their Assemblies. In the last two decades, extensive experimental and analytical studies of member properties have contributed to the development of an ultimate strength approach (see LIMIT DESIGN) for the proportioning of elements; this is more realistic than the previously used working stress approach or ELASTIC ANALYSIS. Experimental studies (see STRUCTURAL MODEL ANALYSIS) have also generated background information on the postelastic (hinging) behavior of beams, columns, and their assemblies.

Ductility, which is the term for toughness as opposed to brittleness, is an extremely important factor in seismic resistance (see TESTING OF MATERIALS and CRACKS IN BRITTLE MATERIALS). Ductility of structural elements, which is their postelastic deformation capacity, is attained when the elements are proportioned to be governed by flexural strength, and not to fail in shear. Brittle failures, such as those caused by shear and compression, are undesirable in earthquake-resistant structures. Only ductile behavior, associated with yielding, produces the redistribution of stresses and dissipation of energy needed to resist large seismic deformations. Details of beams, columns, and their connections needed to achieve shear resistance in potential hinging regions have been developed and adopted by the seismic codes.

Earthquake Forces are not externally applied loads, but are internal forces generated in response to motions of the ground upon which the structure is founded. These inertia forces generated by a particular ground motion depend on the weight and the stiffness of the structure. The inertia forces increase with increasing weight and are smaller for lighter structures. In addition, the stiffness of the structure significantly affects the inertia forces. Stiff structures (with a short period of VIBRATION) inherently deform less and, as a result, must move back and forth as a whole, following the ground motion; the generated inertia forces are thus a high portion of their weight. In a very flexible structure (with a long period of vibration), only the bottom part may move back and forth when the ground moves quickly, while the rest of the structure remains in place; the generated inertia forces are a reasonably small portion of the weight. In both cases, the structure must have the ability to resist the corresponding structural deformations.

Structural Response. When structures deform under lateral loads, the sections that are stressed to capacity start yielding. The yielding sections act like fuses limiting the inertia forces to the level at which yielding occurs.

The yield strength of the structure sets the level of maximum lateral forces to which the structure may be subjected; it also determines the amount of inelastic (postyield) deformations the structure will undergo. The designer may elect either a low yield and accept large inelastic deformations or provide a higher yield level with correspondingly lower inelastic deformations.

This balance between strength and ductility is the basis for earthquake design of inelastic structures. All modern codes assume inelastic response of structures. Only nuclear power plants are designed to respond elastically to earthquakes. Therefore, earthquake resistance is the only area of structural engineering in which loads and corresponding details for ductility must be considered concurrently at all times. In addition to prescribing earthquake forces, all seismic codes require details for ductility. The prescribed force levels are adequate and safe only when the corresponding detailing requirements for ductility are met.

The force level specified in seismic codes is about one-quarter to one-sixth of the inertia forces that would occur if the structure were to respond elastically (i.e., with no yielding) to an EI Centro (1940) type earthquake. It is assumed that a structure designed to the code forces can develop a maximum deformation four to six times as much as the deformation at which the structure starts yielding. The code-specified force levels have validity only if used in structures with an available overall ductility in the preceding range. Structures made of brittle materials without ductility must be designed to considerably higher force levels than those specified in the codes.

Recent Developments include the following:

1. Inelastic dynamic approaches, by which a structure is analyzed in a step-by-step procedure for earthquake records to determine the degree of inelastic behavior of the individual member.
2. Isolation techniques by which sandwiches of elastomeric material are placed between foundation and structure to reduce significantly the amount of shaking transmitted during an earthquake.

References

1. J. A. BLUME, N. M. NEWMARK, and L. H. CORNING: *Design of Multistory Reinforced Concrete Buildings for Earthquake Motions*, Publication EB032.01D, Portland Cement Association, Skokie, Ill., 1961.
2. V. V. BERTERO and E. P. POPOV: Seismic behavior of ductile moment-resisting reinforced concrete frames. *Reinforced Concrete Structures in Seismic Zones*, American Concrete Institute, Special Publication SP-53, Detroit, 1977, pp. 247–291.
3. M. FINTEL and S. K. GOSH: Structural systems for earthquake resistance concrete buildings. *Proceedings of Workshop on Earthquake-resistant Reinforced Concrete Building Construction*, University of California, Berkeley, 1977, Vol. 2, pp. 707–741.
4. R. PARK: Accomplishments and research and development needs in New Zealand. *Proceedings of Workshop on Earthquake-resistant Reinforced Concrete Building Construction*, University of California, Berkeley, 1977, Vol. 2, pp. 255–295.

M.F.

ELASTIC DESIGN OF STATICALLY INDETERMINATE STRUCTURES. Statical indeterminacy is due to elements, rigid joints, or supports that are supernumerary to the requirements of statics for equilibrium of the structure under load. The four-legged stool is a simple example: it requires only three legs to fulfill its function, but for reasons of esthetics and general convenience, four legs are common. When all four share the load, the distribution of the load depends on their elasticities. The usual materials of construction (steel, reinforced concrete and timber) exhibit elasticity under load, and the strain, extension, or compression is directly proportional to that load.

In contrast with a STATICALLY DETERMINATE structure, the distribution of load between the various parts or elements of a statically indeterminate structure depends not only on the laws of equilibrium, but also on the elasticities of those elements. The design problem is therefore complicated.

The elastic design problem requires the solution of simultaneous equations derived from the equations of statics and those of geometry for the strains of the elements of the structure. This laborious process is avoided by STRUCTURAL MODEL ANALYSIS, and more recently COMPUTER METHODS IN STRUCTURAL ANALYSIS have eliminated the tedium of design calculations entirely. Thus, the computer will deal with the problem if provided with the essential data for loads and permissible stresses for tension and compression; the latter are often determined by BUCKLING criteria. There is no unique solution for the sizes of the elements for specified loads and permissible stresses. OPTIMIZATION for the least weight of structural material is not readily obtained without the aid of a computer.

Structures are usually chosen to be statically indeterminate for reasons of economy, although esthetics may be a consideration. The former include weight saving (especially for beam structures), ease of manufacture and erection on the site, and replacement of elements during maintenance. Although rigid JOINTS may introduce indeterminacy, they are much less expensive to provide. Indeed, they were often the cause of (unidentified) statical indeterminacy in early timber and iron structures until the middle of the last century. The sustained stimulus of the railroad era was an important factor in the development of precise design methods for indeterminate structures, notably by Mohr in Germany, after 1868. Navier in France and Maxwell in Britain had independently solved the essential problem previously, in 1819 and 1864, respectively.

Among the familiar statically indeterminate structures are the propped cantilever, the encastré (fixed-ended or built-in) beam, and the two-pin arch (Fig. 70). It was well-known in carpentry before 1800 that a fixed-ended beam of uniform rectangular cross section need be only one-half of the depth of the corresponding beam simply supported. The fixing of the ends and consequent indeterminacy causes the material to be used more efficiently. A two-pin arch (Fig. 70c) joined to rock or concrete foundations can be made much lighter than a similar arch with one foot free to move on a horizontal roller bearing, because it is statically indeterminate due to the horizontal restraint of the feet of the arch by the foundation; the horizontal thrust increases the strength of the structure.

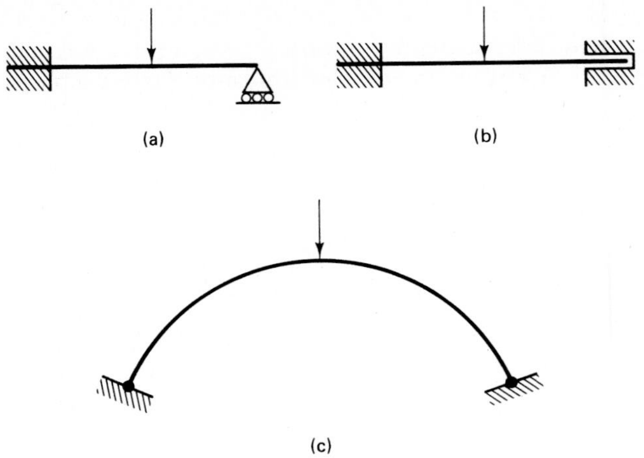

Figure 70 Simple examples of statistically indeterminate structures. (a) Propped cantilever. (b) Encastre (fixed-ended or built-in) beam. (c) Two-pinned arch.

Statically indeterminate structures often cannot be identified immediately. For example, a simple triangular truss is STATICALLY DETERMINATE if the joints are pinned or very flexible, but is indeterminate if the joints are made rigid by welding. In fact, if the bars are slender, this source of indeterminacy may be ignored for design purposes.

One of the earliest statically indeterminate structures to be designed elastically anywhere in the world using correct principles was the spectacular Britannia tubular railway bridge over the Menai Straits; it was completed by Robert Stephenson in 1850. It utilized a pair of massive tubular iron-plate beams, resting unbroken on five supports, one at each end and three intermediate; the total length is 461 m (1511 ft). Navier's principles were used for the design by H. Moseley, author of the first book in English to deal with statical indeterminacy of beams and arches (Ref. 1).

See also LIMIT DESIGN OF STEEL STRUCTURES and LIMIT STATES DESIGN.

References

1. H. MOSELEY: *The Mechanical Principles of Engineering and Architecture.* Longmans, London, 1843.
2. R. C. COATES, M. G. COUTIES, and F. K. KONG: *Structural Analysis*, 2nd ed. Nelson, London, 1980.

T.M.C.

ELECTRICAL MEASUREMENTS indicate and record the electrical quantities needed for the safe and efficient operation of the electrical services in buildings. They can be used directly to meter or record the energy demands of a building or indirectly to operate control and protection equipment, such as circuit breakers, pumps, and emergency generators. The most commonly measured quantities are voltage, current, energy, power, and power factor.

Supply Voltage. Most electrical systems require that the supply voltage to the equipment be maintained under all load conditions to within 10% of its design voltage. In alternating-current systems, voltage is measured in volts (rms), usually called simply volts, and the symbol V indicates the rms value of an alternating voltage (rms is the abbreviation for root mean square, which is the mathematical process involved in determining the effective voltage of an alternating source). Meters can be designed to perform the rms process electromechanically or electronically, and either indicates effective voltage.

Meters can have additional indicators to show the maximum and minimum voltages over any recording period. Control and protection circuits sense voltage and disconnect equipment, for example, electric motors in the event of under- or overvoltage. Furthermore, the level of fuel or water in a tank can be determined by measuring the voltage across a mechanically variable resistance.

Current is the time rate of flow of electric charge. It is measured in *amperes*, usually abbreviated *amps* (A), and the same mathematical averaging is used to obtain the effective current from an alternating current. Thus a 15-A circuit breaker opens a circuit when the current exceeds 15 A (rms). Current meters can also be fitted with additional indicators for maximum and minimum current. An important application of the current meter is the *maximum demand indicator*, which indicates the maximum current used by an electrical installation. This quantity is often used in conjunction with energy measurements to determine the charges made to large electrical consumers. The current that flows through electrical circuits must not exceed the safety limits of the cables used for its conduction. *Overcurrent* sensing is used to protect an electrical installation and the equipment connected to it, chiefly through *circuit breakers* or *fuses*. Current sensing can also be employed to measure temperature, and some fire-detection systems are based on changes of current with changes of ambient temperature.

Electrical Energy is the basis of charging for electricity consumption; however, for most large users the charge also depends on the maximum current demand. It is usual to measure electrical energy in *kilowatt hours* (kWh) rather than joules, and the instruments are called kWh meters. These measure the real energy used by an electrical system, that is, the vector product of voltage, current, and time; this may be less than the arithmetic product of those quantities. Electrical utility companies insist that the consumer maintain a high *power factor*, which measures the efficiency of the utilization of an alternating-current system. Some large customers with many electric motors also monitor their power factor using a suitable meter.

The **power** used in an electrical system is measured in *watts* (or more commonly in kilowatts, kW, or megawatts, MW). Power is the vector product of voltage and current; it can also be expressed as the arithmetic product of voltage, current, and power factor. Thus the power, in watts, can be less than the apparent power measured in *volt-amps*. Meters are available to measure watts (W), volt-amps (VA), and the difference between them, volt-amps-reactive (VAR). VA and VAR are carefully monitored in large installations, because penalty charges are usually incurred for high demands and inefficient utilization.

References

1. A. E. FITZGERALD, DAVID E. HIGGINBOTHAM, and ARVIN GRABEL: *Basic Electrical Engineering: Circuits, Electronics, Machines Control*, 4th ed. McGraw-Hill, New York, 1975.
2. EDWARD HUGHES: *Electrical Technology*, 5th ed. in SI units. Longman, London, 1977.
3. RALPH J. SMITH: *Circuits, Devices and Systems: A First Course in Electrical Engineering*, 3rd ed. Wiley, New York, 1976.

W.G.J.

ELECTRICAL SERVICES. Electric power is generated by utility companies at high voltage levels to minimize transmission line losses; these are largely a function of the square of the current. For a given power requirement, the current decreases as the voltage increases. As the power is transported farther from the point of generation toward centers of load, the voltage is reduced to levels consistent with the load and area to be served. Electric service, therefore, is the final extension of the transmission system, whereby generated electric power is introduced to the using facility.

Transmission voltages commonly encountered in the United States are 69, 138, and 345 kV, while 4160 V, 15 kV, and 34.5 kV are used for local distribution to the using facility, where further transformation by either the utility company or the user reduces the voltage to utilization levels. In the United States these are 208 or 480 V, three phase, three wire or four wire, 60 Hz, and elsewhere, in general, 380 or 415 V, 50 Hz.

Service Voltage Selection. While influenced by the load, and the utility company's rules, regulations, and rate schedules, the user has flexibility in the selection of the voltage level at which it purchases the electric service. If the electric power is purchased and metered at utilization levels, that is, 208 to 480 V, all transformation down to that level from the utility company's distribution level is furnished and installed by the utility. Power purchased and metered at medium distribution voltage levels, that is, 4160 V to 34.5 kV, is generally reduced to utilization voltage levels by transformers furnished and installed by the user. Utility companies can, therefore, offer rate reductions in their billing schedules to customers purchasing at the higher voltage levels commensurate with the cost of transformer equipment and associated transformer losses, which are not sensed by the billing meter, since it is connected ahead of the transformer.

The service voltage selection process requires a life-cycle cost analysis that compares capital expenditure by the user for transformer and necessary auxiliary equipment, maintenance (including equipment servicing, repair and/or replacement), and the cost of borrowed money versus the reduction in the cost of electricity. An additional factor to be considered in the equation is the potential economy of having a higher voltage available in the using facility for internal distribution. Depending on the geometry of the building, the reduction in cable sizes that results from decreased currents as a function of higher voltages can more than offset the cost of transformation to the final voltage levels.

Rate Schedule. Most utility company rate schedules consist of three separate categories of charges: demand, energy, and fuel adjustment.

The demand load is the maximum load imposed on the system integrated over a given time period, generally 15 to 30 minutes. The demand charge is the means whereby the utility company recovers its capital investment in equipment for generating capacity, transmission lines, and substations (including transformers and other materials that can be influenced by the demand load on the system) that serve the user's needs. The spinning generator capacity, transmission lines, and substations are all sized to satisfy the total projected demand load of the geographic area served by the particular utility company, including sufficient spare or redundant capacity to allow for equipment being inoperative for servicing and the requirements for repairs and replacement.

The energy portion of the rate schedule is based on the operating cost of the utility company; it is a function of the time rate of load usage. Whereas the demand relates to capacity and is expressed in kilowatts (kW), energy relates to the hours of use of that capacity and is measured in kilowatt-hours (kWh). Energy consumption, therefore, influences maintenance, repairs, replacement, and fuel consumption.

Fuel adjustment charges did not exist prior to 1974. They were instituted as a result of the rapid escalation of fuel costs in the subsequent years to enable the utility company to recover the incremental cost of fuel above a base price; this replaced the periodic revision of the energy-charge portion of the rate schedule to keep pace with the frequent price changes. Each utility company, in accordance with its unique formula and influenced by its mix of diverse generating plant fuel requirements, determines a surcharge to compensate it for the increase in oil and gas costs. The utilities whose generating plants are fired predominantly by gas and oil impose the highest fuel adjustment charges, and those using coal and nuclear power incorporate the lowest adjustments in their rate schedules. This charge is made per kilowatt-hour, since it is directly related to energy consumption.

Thus, rate reductions offered for the use of higher electric service voltages are almost proportional to the demand load of the using facility, because the utility company is spared the cost of the substation equipment that would be required to reduce the distribution voltage to utilization levels.

Reliability. Probably the most important aspect of an electric service is its reliability. It must provide continuity of service during operating hours, in spite of servicing requirements and equipment failure. A high reliability is achieved by including varying degrees of redundancy into the electrical services. This ensures that routine servicing or failure of one element of the system does not compromise the entire system. Reliability is influenced by the following questions:

1. Are the transmission and distribution lines overhead or underground?
2. Is the site served by a single or by multiple redundant feeders? If multiple, are they from separate substations, traveling along separate routes?
3. Can full service be maintained if one service transformer fails?
4. Is an alternate source of supply available for emergency loads in the event of total utility failure?

Electric Service Arrangement. Considerable flexibility can be exercised in the design of a system for the reception and distribution of the incoming electric service in order to select a scheme that optimizes cost in relation to reliability, while considering the limitations or opportunities offered by the type of service available from the utility company. The system arrangements that accomplish these objectives fall into the following four basic categories, each of which is capable of being modified to suit the requirements of the specific installation:

1. Radial
2. Primary selective
3. Secondary selective
4. Secondary network

The **Radial Arrangement** is illustrated in Fig. 71. There is a single incoming feeder from the utility company and a single service entrance protective device from which individual feeders radiate to substations or other load groupings. This is the least expensive of the systems, but its reliability is adversely affected by the multiplicity of single points of failure. These are the single incoming feeder, the single feeder to each substation, the single transformer serving a large load group; any one can cause the entire system, or large portions of the system, to shut down. There is no method whereby the service can be restored until the cause of failure is determined and the necessary repairs and/or replacements are made.

The **Primary Selective System** rectifies one of the shortcomings of the radial arrangement, that is, the single feeder to each substation. As illustrated in Fig. 72a, two primary feeders, each sized for the full load, are available at each substation. The loads are distributed by each feeder as evenly as possible. Upon failure of either feeder, the loads can be manually switched to the one remaining in operation, thereby reestablishing full service.

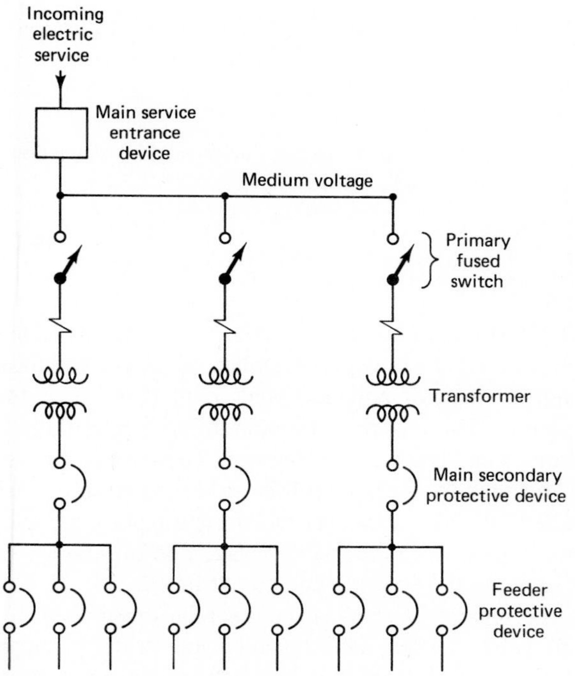

Figure 71 Radial arrangement.

Reliability can be further enhanced by a second incoming utility line sized for the full load of the facility. This arrangement (Fig. 72b) eliminates two of the three single failure points inherent in the radial system. As in all the following arrangements, the increased reliability must be balanced against the incremental cost for additional equipment and material, and a judgment must be made regarding its cost effectiveness for the particular facility involved.

In the **Secondary Selective System,** each primary feeder and transformer (Fig. 73) is sized to provide full redundancy. All single points of system failure contained in the radial arrangement are eliminated by this design, if two full-capacity incoming lines are available from the utility company. In the event of a feeder failure, the respective secondary main protective devices are manually opened and the secondary tie devices closed. This enables the entire substation to be fed through the remaining energized transformer. The same procedure is followed on failure of a transformer. This arrangement can be marginally improved by incorporating the transfer arrangement described under the primary selective system. This transfer arrangement is placed ahead of each end of the substation to enable each transformer to be transferred to the remaining feeder should one feeder fail. This procedure eliminates the necessity of operating any transformer at full load or overload for an extensive period of time.

By interlocking the two main secondary protective devices with the tie device, the paralleling of the two transformers is prevented. This eliminates a condition that could increase the available secondary fault current to a level beyond the capability of the secondary protective devices to interrupt. The arrangement can be significantly enhanced by adding a second incoming line arranged on the medium voltage side, as illustrated in Fig. 72b.

Secondary Network. In the preceding systems, the failure of any major element, such as an incoming utility line, a feeder, or a transformer, would cause all, or an appreciable block of, load to be de-energized. Redundancy enables service to be manually restored without having to wait for repair or replacement. Nevertheless, an interruption of service will occur, causing inconvenience, wasted time, and possible financial loss. A secondary network system eliminates single points of failure. In addition, as illustrated in Fig. 74, it is connected, in parallel, on the load side of the network protectors, with at least one transformer more than is necessary for the full facility load, and preferably with an incoming line for each transformer. Loss of an incoming line or a transformer does not cause a disruption to any of the facility systems, and operation in this mode can be maintained until repairs or replacements are completed. Certain utility companies design their networks on a double-contingency basis to provide continuity of full

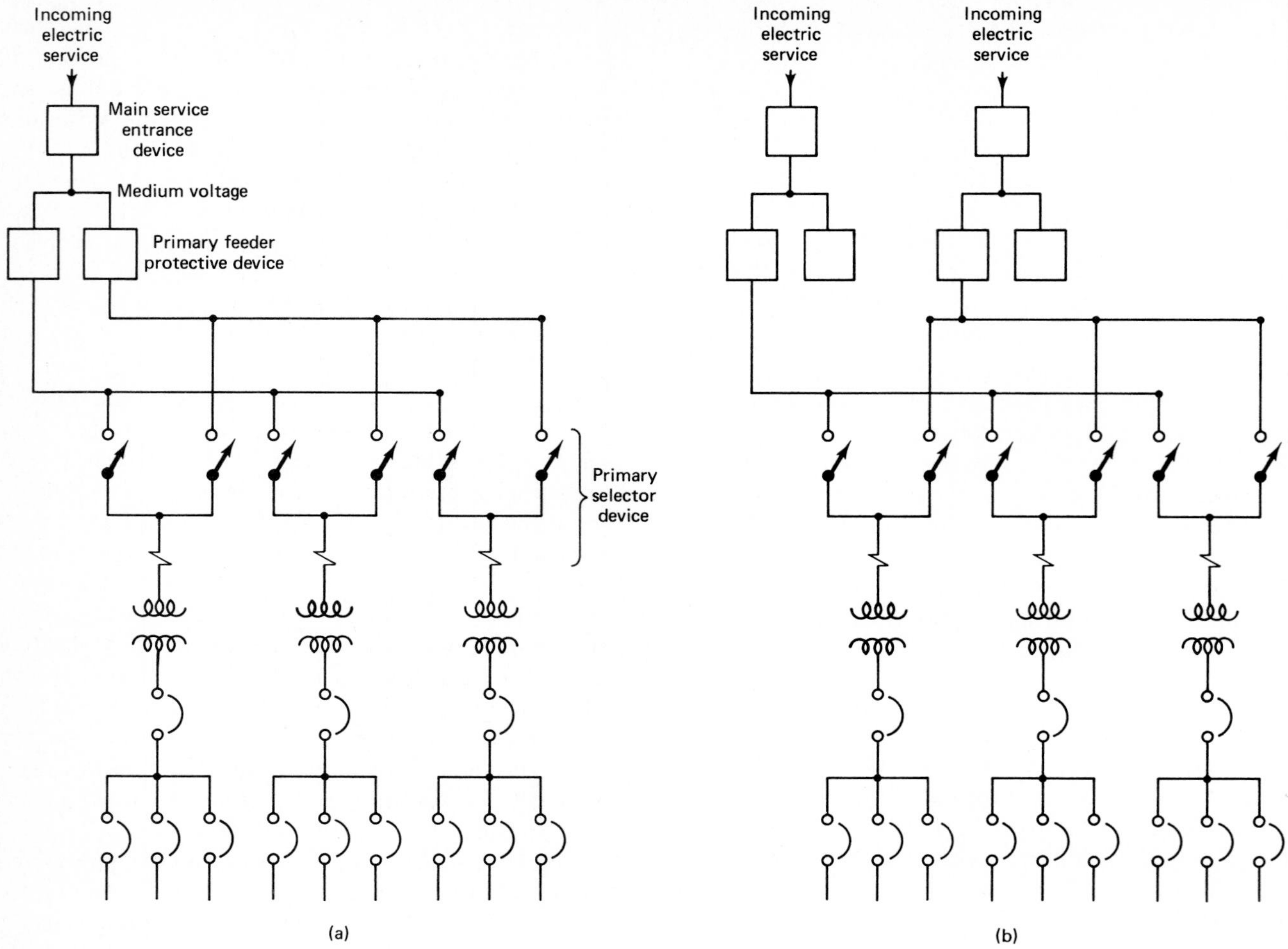

Figure 72 (a) Primary selective arrangement with single incoming utility line. (b) Primary selective arrangement with two incoming utility lines.

service when loss of an incoming line and loss of a transformer fed from another line occur simultaneously.

Generally, the secondary network equipment is furnished and installed by the utility company, and the user who purchases power is metered at a low voltage.

Paralleling transformers has the inherent disadvantage that the impedance of the system is reduced. This causes higher available fault currents to be encountered downstream of the secondary network bus. This will probably adversely affect the cost of the user's secondary distribution equipment, which must have the capability of interrupting the higher fault currents.

References

1. *IEEE Recommended Practice for Electric Power Systems in Commercial Buildings*. Institute of Electrical and Electronics Engineers, New York, 1983.
2. *National Electrical Code® 70*. National Fire Protection Association, Quincy, Mass., 1984.

H.O.A.

ELEVATORS (LIFTS). Vertical conveyances have existed since early Roman times, when pulleys or windlasses, driven by men or animals, were used to raise or lower containers. These were replaced in the early part of the 19th century with steam-driven engines. These early elevators, put together by local foundries or machine shops, often resulted in major accidents when the supporting ropes or chains broke, since they had no way to defeat gravity when the supporting or driving equipment failed.

The elevator became a commercial concept in 1853, when Elisha Graves Otis invented the "safety" elevator that would retain its integrity when the driving or supporting means failed. He utilized wooden guide rails, common at

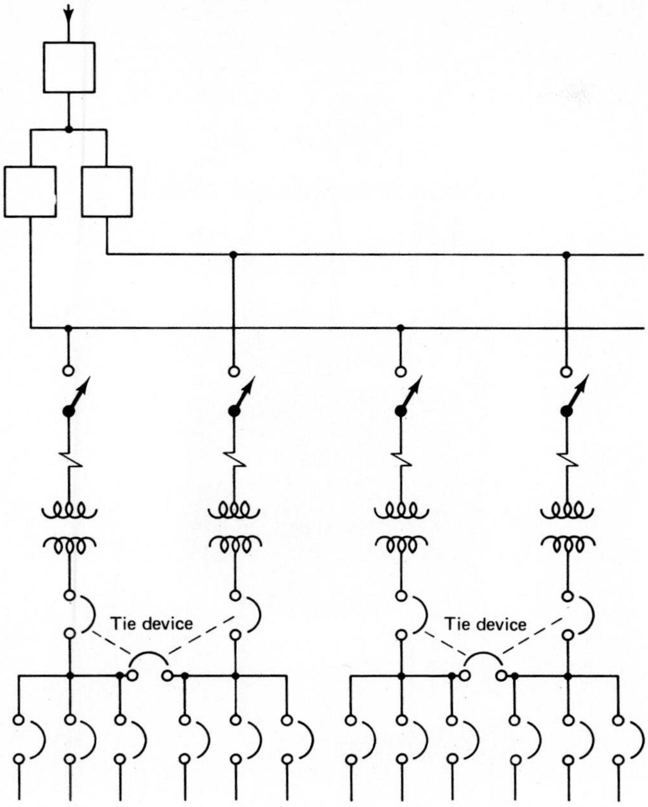

Figure 73 Secondary selective arrangement.

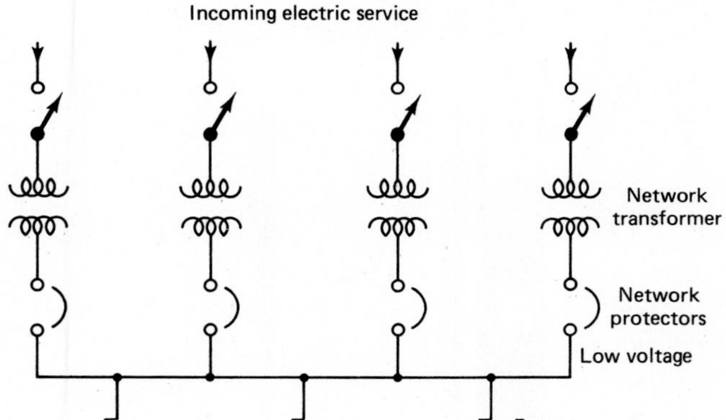

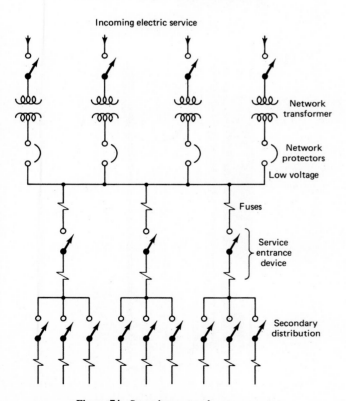

Figure 74 Secondary network arrangement.

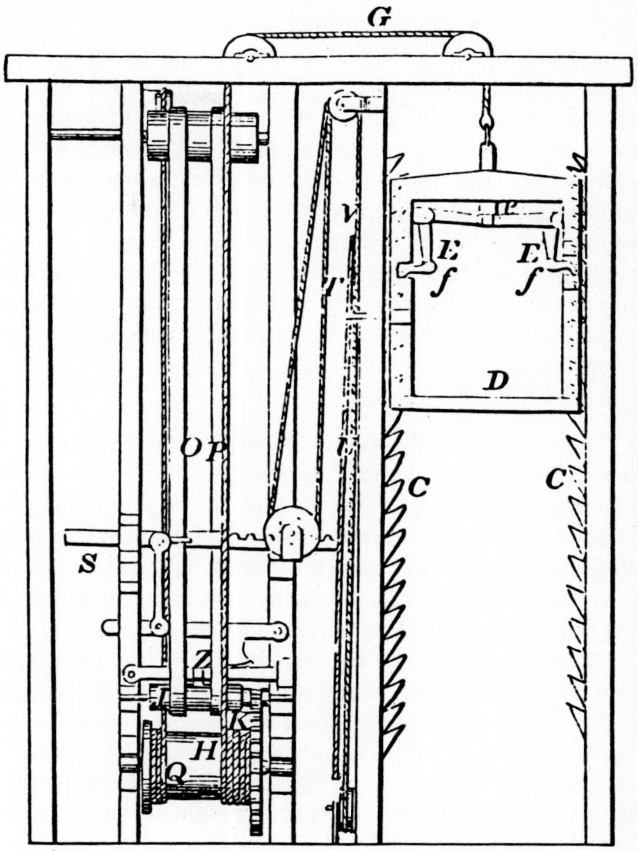

Figure 75 Elevator safety device, patented by Elisha Otis. (By courtesy of the Otis Elevator Company.)

that time, to stop the platform when the supporting ropes parted by engaging cogs in the rails with a wagon-spring pawl (Fig. 75). The first passenger elevator was commissioned in 1857 in the E. V. Haughwout & Company Building at Broadway and Broome Streets, New York. It was steam powered and served the five stories in less than 60 seconds.

For the next 40 years, elevators were built with winding drums, although electric motors were substituted for steam engines. The traction elevator was invented just prior to the 20th century; it used a counterweight at one end of the hoist ropes and the car with its load at the other as a viable means of propulsion. This removed the rise limitations of the winding drum drive. The traction concept made possible higher rises and also increased the safety, because the bottoming of the counterweight in the elevator pit prevented the elevator car from being drawn into the overhead construction even though the drive machine continued to turn. The elevator now became a mature and safe technology whose rise was limited only by the height of the building.

In the early 1900s, the introduction of automatic leveling and door operation permitted higher speeds with safe passenger transfer. In the second half of the 20th century, attention has been directed to eliminating the human operator and replacing manual operation with automatic operation; this provides prompt and efficient elevator service among a multiplicity of floors with a minimum of confusion to the passenger and serves the needs of the resident population with the minimum number of elevators. Today's passenger elevator can travel at 10 m/s (2000 ft/min) and utilizes sophisticated feedback speed and leveling control techniques and high-speed computer logic operation to optimize landing call assignments that efficiently serve commercial and institutional structures of widely varying uses and occupancies.

Elevator Classification. Elevators are classified by the A17.1 Safety Code (Ref. 1) as either passenger or freight elevators (Figs. 76 and 77). A third unrecognized classification is the service elevator, which has passenger classification but is primarily used for moving materials and other "freight." It is used predominantly in buildings where people other than employees may ride on the elevator.

Passenger elevators are "batch" conveyors of people that must have horizontally sliding doors for passenger safety in transfer. Freight elevators are also "batch" conveyors, but they may have either horizontally or vertically sliding doors, depending on the size and weight of the materials to be transported. The use of freight elevators by the general public is prohibited except in emergencies. The shape of a passenger car platform is wide and shallow for the ease of passenger transfer at multiple landings, whereas the size of the platform for a freight elevator is usually narrow and deep, relating to the prime function of moving equipment or materials on skids or dollies.

Elevator Types relate primarily to the driving machine. *Hydraulic elevators* are used in low-rise, low-speed elevator installations; *geared traction elevators* in medium-speed, medium-rise applications; and *gearless traction elevators* in high-speed, high-rise applications. Since the geared elevator concept overlaps the upper range of the hydraulic elevator and the gearless elevator overlaps the upper range of the geared elevator, performance, capital and operating costs, and architectural or structural considerations should all be evaluated before a final selection is made.

The *Hydraulic Elevator* is the least expensive, and it has the slowest speed, utilizing a direct-acting hydraulic piston with oil pumped by a hydraulic system (Fig. 78). It travels downward by gravity and returns the oil through a controlled valve to the storage tank. It has the advantage of low overhead height and shallow pit depth, as well as the opportunity to locate the machine room remotely from the hoistway. The hydraulic elevator is usually installed without a counterweight and safety devices. It can operate under special circumstances at up speeds of 1.0 m/s (200

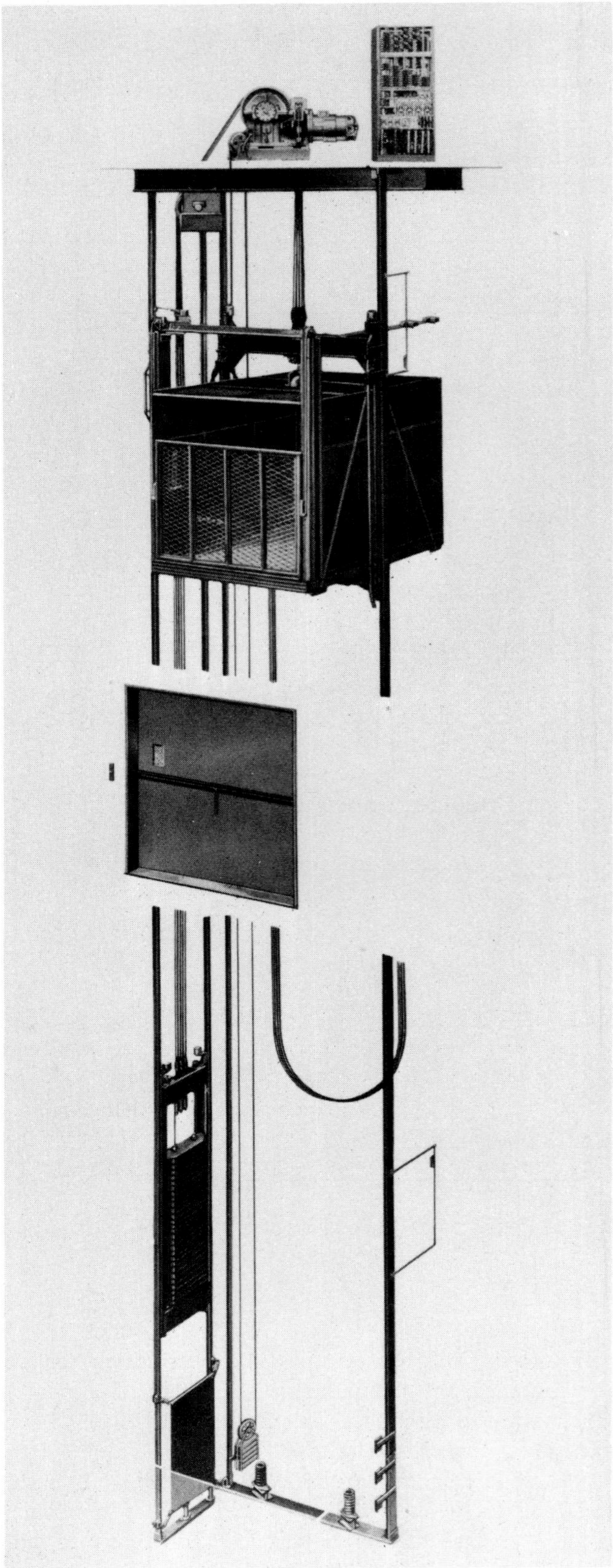

Figure 76 Electric geared freight elevator. (By courtesy of the Otis Elevator Company.)

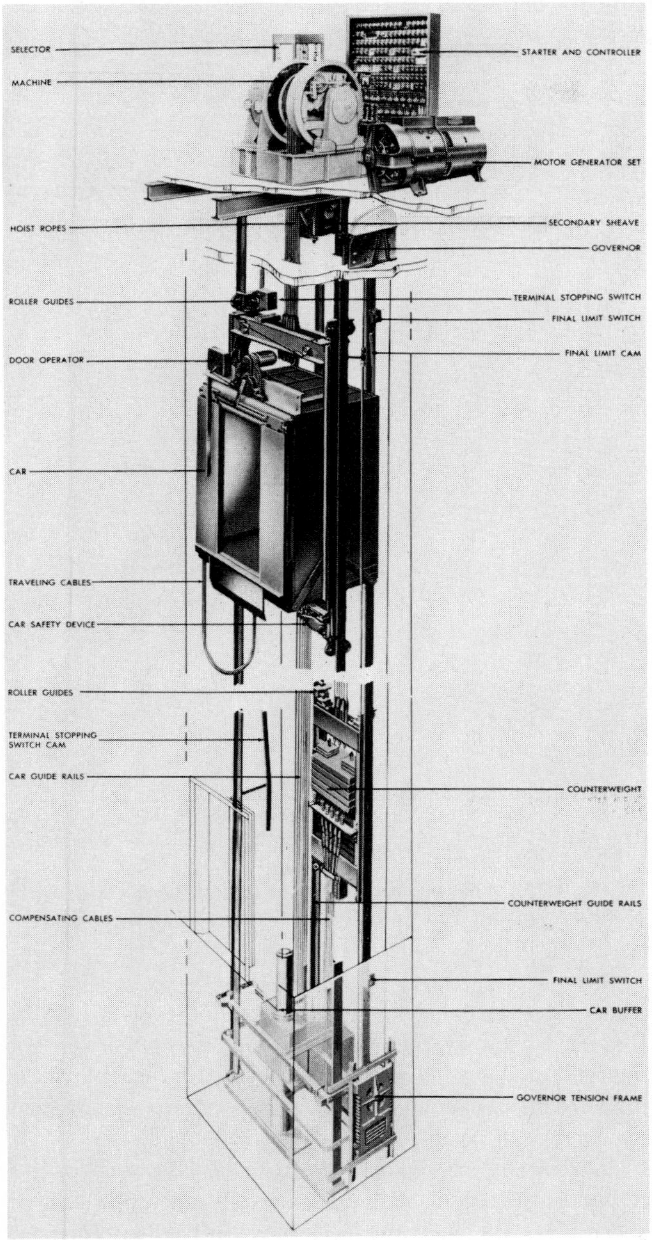

Figure 77 Electric gearless passenger elevator. (By courtesy of the Otis Elevator Company.)

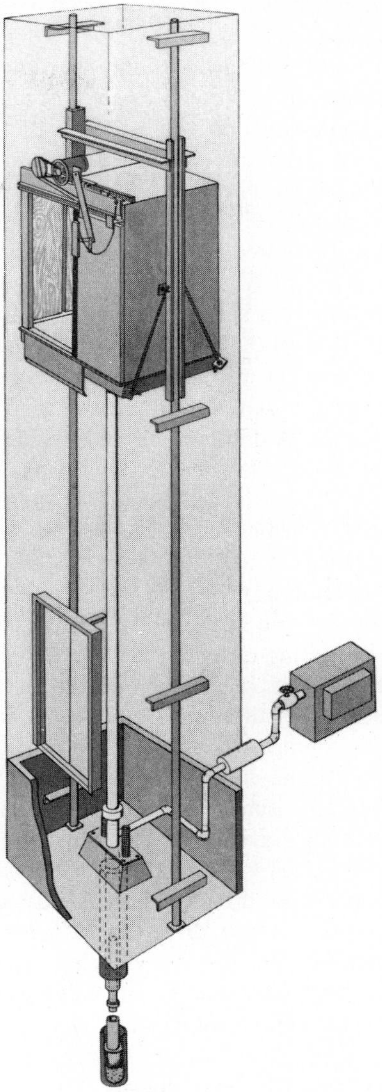

Figure 78 Hydraulic passenger elevator. (By courtesy of the Otis Elevator Company.)

ft/min), but generally operates at 0.6 m/s (125 ft/min). Its maximum vertical rise is limited to approximately 20 m (70 ft).

The *Geared Traction Elevator* is used for intermediate speed ranges and intermediate travel distances. It has an upper speed of approximately 2.3 m/s (450 ft/min), with a maximum travel of approximately 20 stories, depending on the size of the elevator. It uses a high-speed, low-torque electric motor, with a worm-to-worm gear reduction between the motor shaft and the driving sheave of the hoist ropes. It is roped 1 : 1 between the car and the counterweight in the higher speed ranges and 2 : 1 in the lower speed ranges.

The *Gearless Traction Elevator* has a low-speed, high-torque electric motor with a drive sheave mounted directly on the motor armature shaft (Fig. 77). Its speed ranges from 2.5 to 10 m/s (500 to 2000 ft/min). Its vertical rise is virtually unlimited. It is most effective with a roping of 2 : 1 between the car and the counterweight in the lower speed and lower travel ranges and 1 : 1 in the higher speed and higher travel ranges.

Special-Purpose Elevators are recognized by the codes to permit unusual applications, restricted uses, and temporary uses, provided certain safety aspects are accommodated.

Observation Elevators are installed on building interior and exterior walls.

Inclined Elevators have recently become popular in unique residential applications, as they accommodate the handicapped in conjunction with a 30° or 35° escalator in special terminal installations. Currently, most inclined elevators for the handicapped are in Europe. It is anticipated that their acceptance will rapidly spread where a retrofit of transit stations is required to accommodate the handicapped that already have ESCALATORS.

A *Dumbwaiter* is a special form of elevator; its height is limited to 1.2 m (4.0 ft) and its platform area to 0.8 m^2 (9.0 ft^2). It is used in medium-rise commercial or institutional buildings for mail handling and cart, package, or drawing deliveries. A dumbwaiter has a winding drum machine for the lower rises and a traction machine for the higher rises. Its speed is limited to 5 m/s (1000 ft/min). Its maximum dimensions prevent passengers from attempting to ride on them.

A *Material Lift* is a special form of dumbwaiter or small-sized elevator with an automated transfer device for the loading and unloading of tote boxes or carts. It is used for special mail handling, package handling, or cart handling in commercial and institutional buildings. It is specifically designed to discourage passengers from riding and is limited in speed, like the dumbwaiter.

A *Stage Lift* is usually hydraulic, with special equipment to equalize the travel of the multiple hydraulic pistons.

A *Temporary Personnel and Material Hoist* is a special kind of elevator whose drive machine is located on the car. Multiple hydraulic motors drive pinion gears engaging a full-height rack over the distance traveled. The hoist is installed to assist the construction of buildings prior to the availability of the permanent elevator.

Elevator Capacities and Speeds. The minimum capacity of a passenger elevator is a function of its net inside area (Fig. 79). The capacity of a freight elevator is related to its loading category and the severity of its full load of fork-lift trucks.

The speed of an elevator system should be selected to provide the anticipated throughput requirements for either passenger or freight at the anticipated time of the peak traffice for up, down, or two-way traffic.

Passenger Elevator Arrangements. The maximum number of elevators that can be programmed to operate in a single group is eight cars, that is, four cars facing four cars with a common lobby. The maximum number of elevators in a single bank should not exceed four. If there are more than four elevators in a bank, the outermost elevators must have longer landing call dwell times to permit passengers to reach these cars. This limitation holds true for all elevators.

The *Single-deck Elevator* is the most commonly used. It has a single passenger compartment with doors at the

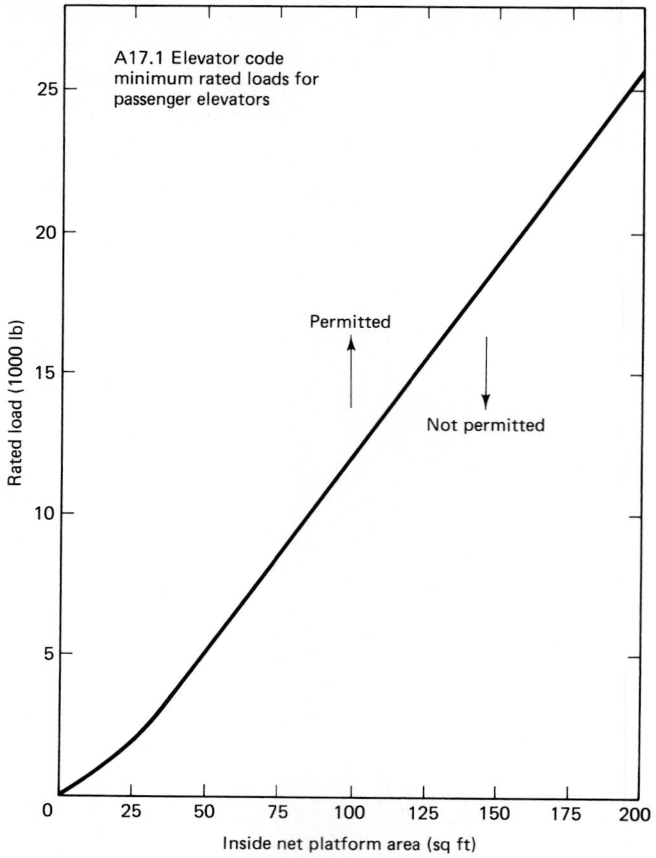

Figure 79 Minimum rated load for passenger elevators in accordance with the A17.1 Code (Ref 1). (By courtesy of Jaros, Baum and Bolles.)

front of the car and serves a multiplicity of levels (Fig. 80). It is used for low-, medium-, and high-rise buildings, but multiple groups of elevators can be used to accommodate different interval and handling capacity criteria, determined by the type of occupancy.

Double-deck Elevators have recently been designed to respond to the demand for a more efficient utilization of space (Fig. 81). A double-deck elevator has two separate, vertically connected, passenger compartments that stop simultaneously at adjacent floors. This reduces by approximately 25% the number of elevator hoistways required and allows a commensurate increase in the net usable floor area. The floors served by double-deck elevators must be constructed with a closely controlled floor-to-floor tolerance so that the simultaneous stopping can be accommodated without generating a tripping hazard. Double-deck elevators must also be carefully designed to minimize passenger confusion at the entrance lobbies. Peak traffic must be accommodated even though an elevator may be out of service.

Building Arrangements fall into two categories, with variations according to the combination of single-deck or double-deck local elevators. The conventional arrangement has all elevators originating at the main or street lobby. The

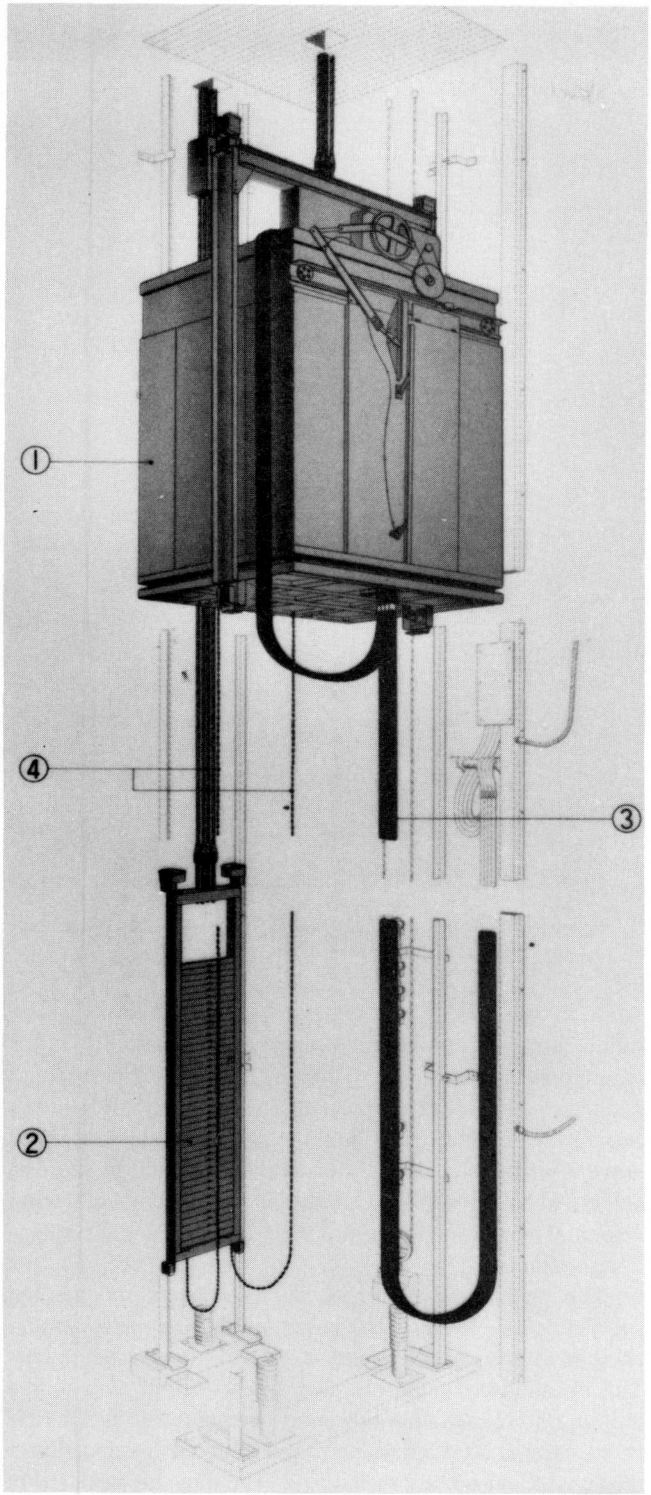

Figure 80 Single-deck passenger elevator.

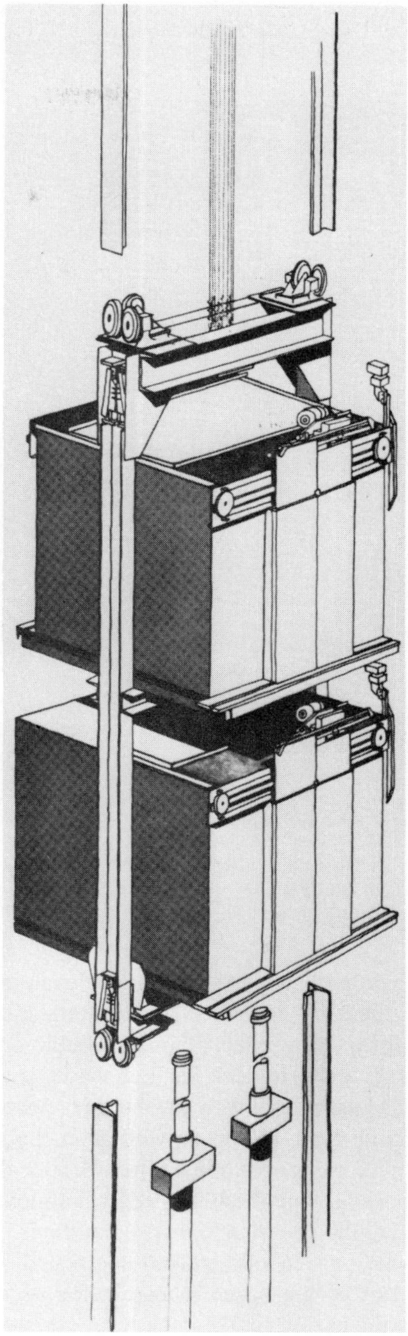

Figure 81 Double-deck passenger elevator.

sky lobby arrangement has shuttle elevators, usually one-half or two-thirds, traveling between the street floor and a sky lobby. From there local elevators travel to stories above and below the sky lobby.

Of the elevator arrangements in large high-rise buildings, the conventional arrangement with single-deck local elevators requires the most space dedicated to the elevator system. This can be seen from Fig. 82, which indicates the usual as well as the alternate transfer floors, depending on the occupancy of the building. Conventional double-deck elevators for a similar building are shown in Fig. 83, which indicates a reduction in the number of elevator hoistways.

The sky lobby concept for single-deck local elevators,

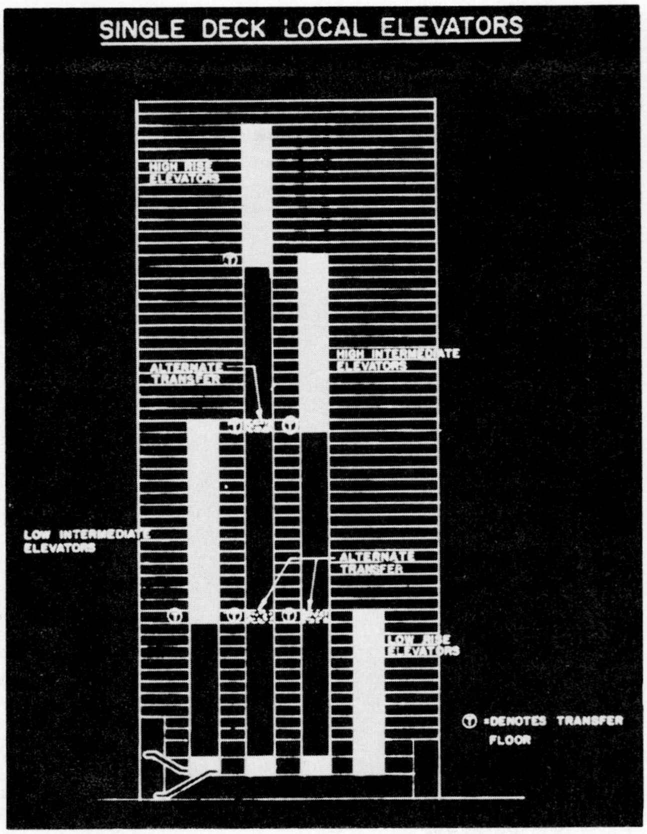

Figure 82 Single-deck local elevator arrangement. (By courtesy of Jaros, Baum, and Bolles.)

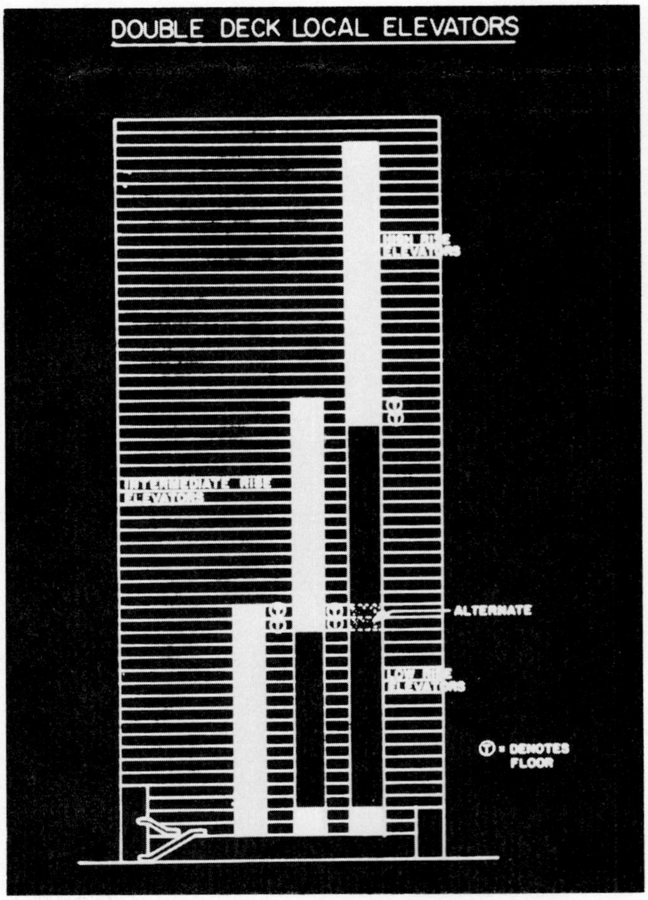

Figure 83 Double-deck local elevator arrangement. (By courtesy of Jaros, Baum, and Bolles.)

served by a group of single-deck shuttle elevators, is shown in Fig. 84; the area dedicated to elevators is further reduced. The sky-lobby concept using double-deck shuttle elevators to serve single-deck local elevators is also shown in Fig. 85. The use of double-deck shuttle elevators requires escalators at the main lobby, as well as at the sky lobby, to accommodate passenger transfer between local groups of elevators under all conditions of pedestrian movement.

When studying complex high-rise buildings, it is essential that all the concepts be studied and tested for efficacy against the net usable space available for occupancy. In many cases, the use of double-deck local elevators requires more space than a sky lobby arrangement, and it may be substantially more expensive with respect to both capital and operating costs.

Building Interfaces. Elevator systems interface with a building in its architectural, structural, electrical, and mechanical design.

Architectural Considerations must take into account the horizontal pedestrian circulation, with respect to the width of the elevator lobbies, and the arrangement of the hoistway doors at the main, as well as at the typical, floors. Special consideration must be given to elevators that travel in a single blind hoistway, since emergency access doors are required every 11 m (36 ft) for the removal of passengers trapped in the blind portion of the hoistway. Revolving doors may be required in both the main and sky lobbies to prevent air transfer deleterious to the operation of the elevators, to avoid a stack effect, and to prevent high wind velocities or the incorrect operation of mechanical air-handling systems.

The *Structural Requirements* must accommodate the support of the elevator equipment in the *machine room* and *pit*, allowing for a fully loaded car and the counterweight. The installation of the elevator and its connection to the building structure must take into consideration the anticipated sway and compression of the building, as this affects the subsequent quality of the ride. The elevator installation must be in accordance with the seismic risk zone appropriate to its location. Vibration isolation measures should be taken with respect to the machine room and the pits. If the elevator runs in hoistways adjacent to occupied residential or office space, the guide-rail brackets may require vibration isolation.

The *Electrical Power Requirements* under normal op-

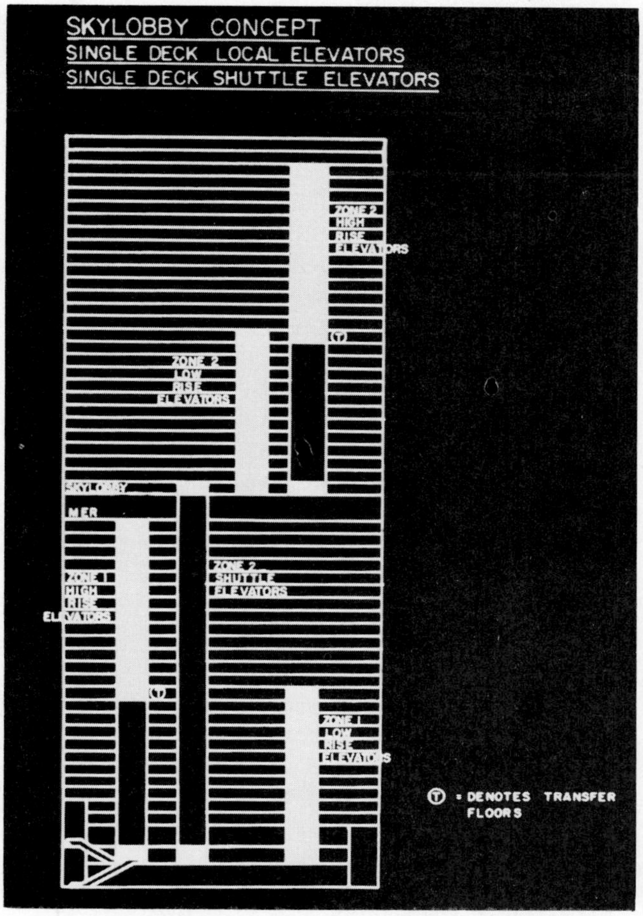

Figure 84 Sky-lobby arrangement with single-deck local and single-deck shuttle elevators. (By courtesy of Jaros, Baum, and Bolles.)

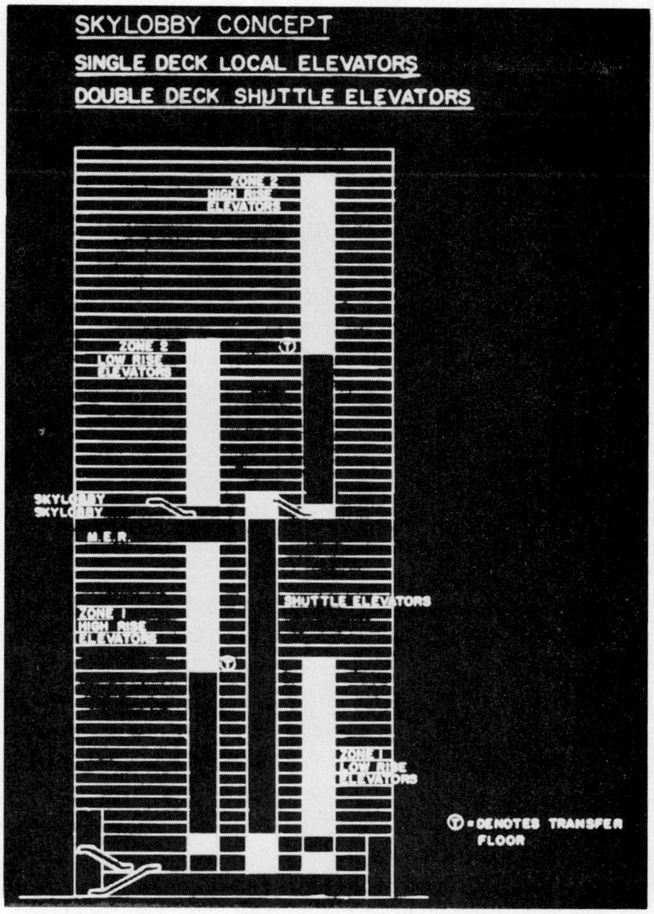

Figure 85 Sky-lobby arrangement with single-deck local and double-deck shuttle elevators. (By courtesy of Jaros, Baum, and Bolles.)

eration are consistent throughout the industry, but special consideration must be given to the operation of the building on standby power. The life safety aspects of the elevator system must conform to local code requirements for manual and automatic phase I recall, as well as for subsequent phase II operation of the elevators by the local fire department.

The *Mechanical System* must provide sufficient air to dissipate the heat released in the elevator machine rooms to maintain an ambient air temperature not less than 15°C (60°F) nor more than 35°C (95°F). The maximum temperature has become particularly important since the introduction of microprocessor and minicomputer logic systems, which require lower temperatures than the previous relay logic systems.

References

1. *American National Standard Safety Code for Elevators, Dumbwaiters, Escalators and Moving Walks—ANSI/ASME A17.1.* American Society of Mechanical Engineers, New York, issued annually.
2. F. A. ANSETT: *Elevators: Electric and Electrohydraulic Elevators, Escalators, Moving Sidewalks and Ramps*, 3rd ed. McGraw-Hill, New York, 1960.
3. *Safety Code for Elevators, Dumbwaiters, Escalators and Moving Walks—B44.* Canadian Standards Association, Ottawa, issued annually.
4. JOHN J. FRUIN: *Pedestrian Planning and Design.* Metropolitan Association of Urban Designers and Environmental Planners, New York, 1970.
5. COUNCIL ON TALL BUILDINGS AND URBAN HABITAT: *Monograph on Planning and Design of Tall Buildings*, Volume SC, Tall Building Systems and Concepts. Chapter SC4, *Vertical and Horizontal Transportation.* American Society of Civil Engineers, New York, 1980.
6. R. STRAKOSCH: *Vertical Transportation: Elevators and Escalators*, 2nd ed. Wiley, New York, 1983.
7. *Elevator World* (monthly magazine). Various issues.

W.S.S.

ERGONOMICS is the study of people's performance and behavior in their working environments. It is derived from the Greek *ergos*, work, and *nomos*, science. In the United States the terms human factors, human factors engineering, or just human engineering are commonly used. For human operators, the workability of tools, machines, and environments is circumscribed by physiological, psychological, and other behavioral parameters. Ergonomists study these factors and search for principles that can be applied to improve performance, reduce fatigue, increase safety, and bring tasks within the capacities of more people.

Public awareness of ergonomics as an area of scientific study has not been high, although, since the march of computer-linked visual display units (VDUs) into workplaces, media references to ergonomics are on the increase. For example, "prolonged VDU work can lead to symptoms of fatigue which can be largely prevented by the use of human factors methodology or ergonomics as it is more fashionably called" (Ref. 1).

For buildings, the most relevant information is codified in general handbooks, (Refs. 2 and 3) or in books dedicated to more specialized functions (Ref. 4). Journals reporting research findings in ergonomics include *Applied Ergonomics*, *Ergonomics Abstracts*, *Ergonomics*, and *Human Factors*. Various national organizations, for example, the Ergonomics Research Society in Britain and the Human Factors Society in the United States, hold regular conferences, and the International Ergonomics Association provides an international forum. Most ergonomics research is carried out at the level of the workstation. *Anthropometric data* giving figures for average height, reach, and forces that can be exerted for various movements, and other relevant measures, are included in most ergonomics textbooks (Ref. 5). To help designers make the built world physically fit the people working in it, it is usual to cite the values that mark the cutoff point for the top and bottom 5% of the population, in addition to the averages for men and women. Some of these data are incorporated in specific standards; for example, British Standard 3044 (Ref. 6), which deals with furniture, points out that the common desk and chair heights of 750 and 450 mm (2 ft 6 in. and 1 ft 6 in.), respectively, are too high for most people.

The effects of *ambient conditions* on performance are generally well understood (see LIGHTING LEVELS APPROPRIATE TO VARIOUS VISUAL TASKS, NOISE, and THERMAL COMFORT CRITERIA). Others, notably the color, texture, or esthetic quality of the environment, are recognized to be just as important, but the body of empirical evidence concerning these is smaller and less reliable.

Ergonomic considerations in buildings that are on a larger scale than the workstation include *traffic through and around buildings*, many *safety* features, the use of auditory and visual (even tactual in the case of blind people) symbols for *sign posting* and *information communication* and an enormous range of *control devices* like push buttons, switches, levers and handles, all of which are necessary for successful interaction with the modern built environment.

The amount of useful ergonomic data, which has been amassed over the years, is now such that when society dictates a change in norms, like the relatively recent improved BARRIER-FREE DESIGN for the disabled in many countries, the information needed to put that change into practice is already at hand. Based on this information, firm guidelines that lobbies and passageways should not be less than 1200 mm (3 ft 11 in.) in width for wheelchair users or that elevators (lifts) should not be less than 1400 mm (4 ft 5 in.) in depth and 1100 mm (3 ft 7 in.) in width, with door openings at least 800 mm (2 ft 7½ in.) across, can be offered to those bearing the responsibility for making the necessary changes in an effective way.

References

1. *The Times*. June 28, 1983.
2. S. V. SZOKOLAY: *Environmental Science Handbook for Architects and Builders*. Halsted Press, New York, 1980.
3. W. E. WOODSON: *Human Factors Design Handbook*. McGraw-Hill, New York, 1981.
4. J. BREBNER: *Environmental Technology in Building Design*. Applied Science, London, 1982.
5. E. J. McCORMICK: *Human Factors in Engineering and Design*, 4th ed. McGraw-Hill, New York, 1976.
6. *Ergonomic Aspects of Furniture Design, BS 3044*. British Standards Institution, London, 1958.

J.B.

ESCALATORS AND MOVING WALKS are continuous-flow people movers, as opposed to ELEVATORS or other conveyances that assemble people in a batch and then proceed to transport them to their destination. Moving walks and escalators emerged at the turn of the 20th century in parallel developments. The Paris Exposition of 1900 was the most successful demonstration of a continuous series of pallets as a moving-walk system. The successful installation of an escalator by the Otis Elevator Company was demonstrated in their Yonkers factory in 1899, with subsequent commercial installation in 1902.

The Escalator of today is a motor-driven gear-reduction drive of a continuous series of steps attached to two parallel sets of drive chains that are track-guided to provide a horizontal boarding and exit area; this gradually forms steps along a 30° or 35° incline (Fig. 86). The 30° incline is standard in all applications; the 35° incline is permitted

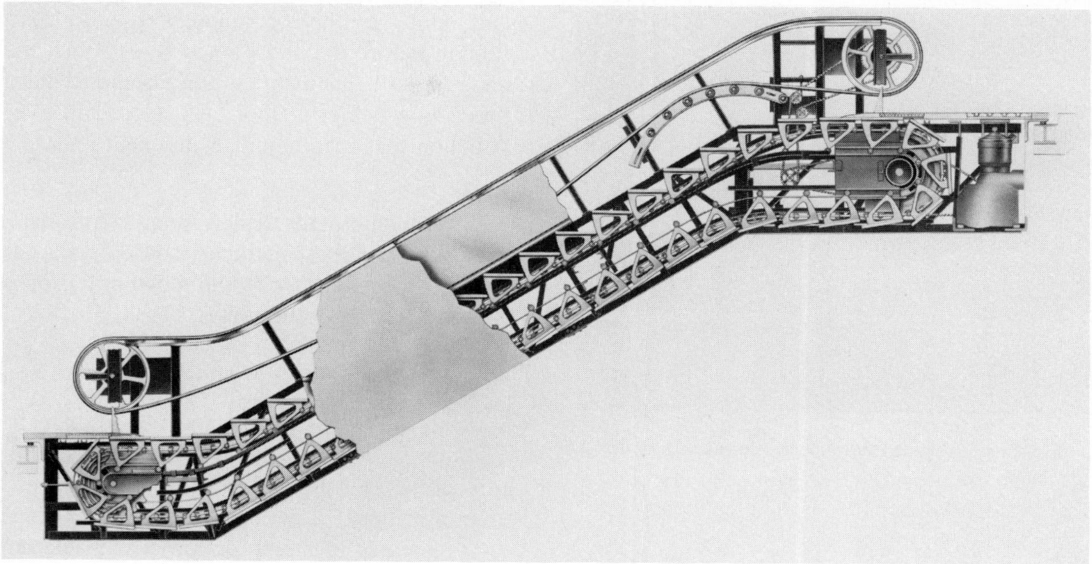

Figure 86 Cutaway section of a typical escalator. (By courtesy of the Otis Elevator Company.)

in some jurisdictions under the European CEN Code, with speed and rise restrictions.

The current design of **Moving Walks,** which perform a similar function without steps, is limited to inclines not exceeding 15°. The moving-walk treadway takes two forms in current design: one has a pallet or flat-step design, the other a structural-belt design. While the latter design provides secure footing on the greater inclines, the most successful moving-walk designs currently are applications using pallets on the lower inclines.

Moving walks provide an escalator-type synchronized handrail at each side of the treadway to provide passenger support. Extended newels permit pedestrian engagement with the handrail prior to boarding and after exiting for maximum stability and balance at these points.

Escalator Width, Speeds, Capacities. The standard 30° escalator has a nominal speed of 0.45 m/s (90 fpm) or 0.6 m/s (120 fpm), with a tread width of 600 mm (24 in.) or 1.0 m (40 in.). Emerging from the European manufacturers under the CEN Code is an intermediate standard tread width escalator with a nominal width of 800 mm (32 in.).

The nominal capacity of a 600 mm (24 in.) tread width escalator is based on one person for every two steps and the nominal capacity of a 1.0 m (40 in.) tread width on one person on every step. These capacities deliver approximately 2000 people per hour and 2700 people per hour on the 600-mm (24-in.) escalator operating at 0.45 m/s (90 fpm) or 0.6 m/s (120 fpm), respectively. The 1.0-m (40-in.) escalator has a capacity of approximately 4100 people per hour and 5400 per hour at 0.45 and 0.6 m/s (90 and 120 fpm) speeds, respectively.

Moving-walk Widths, Speeds, and Capacities. The nominal capacity of moving walks based on the usual tread width of 1.0 m (40 in.) varies with the speed. Speeds range from 0.9 m/s (180 fpm) on an incline up to 3° to 0.63 m/s (125 fpm) on a maximum slope of from 12° to 15°. Additional speed is permitted on the higher inclines if there is a level boarding and exiting area. The nominal capacity varies from 7200 people per hour for a level walk to 5000 people per hour for one with a 15° incline. For practical purposes, even though moving walks have been built in narrower widths, the 1.0-m (40-in.) tread width is the accepted standard, which permits a walking passenger to pass a standing passenger on the treadway.

Building Interfaces. Escalators and moving walks each have specific requirements with respect to the interface with the building and safe pedestrian movements.

The primary **Architectural Consideration** is to provide an alternative and parallel means of pedestrian movement whenever an escalator is not available due to preventive or corrective maintenance conditions. This usually takes the form of an adjacent stairway with two escalators, one up and one down, or three escalators, which will, under all circumstances, provide at least one in each direction in that escalator core.

In heavily traveled buildings, it is extremely difficult to provide equivalent elevator service to that provided by a pair of escalators. The standard arrangements are parallel

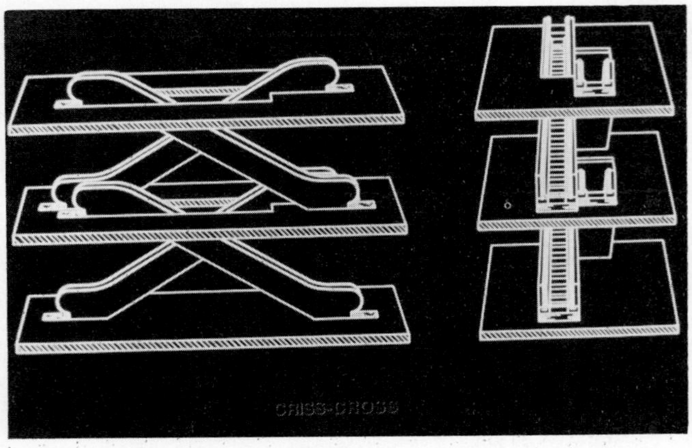

Figure 87 Criss-cross escalator arrangement. (By courtesy of the Otis Elevator Company.)

(Fig. 87) or criss-cross (Fig. 88) to accommodate the desired pedestrian flow at each level. The parallel arrangement provides the shortest on-floor travel distance, while the criss-cross arrangement may provide the same travel distance or a longer travel distance, depending on how the escalators are operated with respect to the up and down directions. The latter is most often used in merchandising to provide the maximum exposure of displays in the vicinity of the escalator core.

It is important to provide ample queuing space between the escalator or moving-walk newel and the nearest obstruction or wall. Nominally, this should not be less than 3 m (10 ft); in an optimum configuration it should be 3.7 m (12 ft). While the design of an escalator requires that there be at least 1.3 to 1.5 flat step configurations at both the top and the bottom, it has been ascertained that a configuration of 2.5 to 3.0 flat steps at both top and bottom provides the optimum in safe boarding and exiting; since the flat steps exceed the maximum stride length, the horizontal boarding

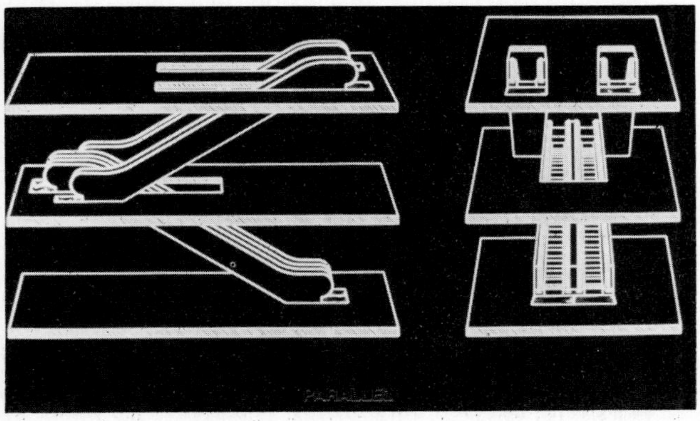

Figure 88 Parallel escalator arrangement. (By courtesy of the Otis Elevator Company.)

and exiting can be accommodated with a flat tread configuration before the steps begin to establish the riser. Most codes require that the truss and associated machinery spaces protruding below a floor be enclosed in a fire-rated construction that accommodates the heat release.

The **Structural Design** should consider the potential vibration of the escalator or moving walk carrying its anticipated pedestrian load and make provision for the damping of structure-borne noise.

The **Electrical Energy** consumed by an escalator or moving walk is minimal, since the passenger load is a minor addition to the almost constant friction load of the system. As can be seen from Fig. 89, the normal escalator, when running empty, draws approximately 2 kW. The consumption increases in the up direction with the number of passengers, but it decreases in the down direction with the number of passengers to the point that energy is regenerated to the building's electrical system if it is loaded consistently with more than 16 passengers.

The design of escalators and moving walks should include careful consideration of the *safety of passengers*, since both are unforgiving conveyors once boarded and can be exited only at the end of the treadway. An optional feature of a *controlled stop* can be obtained from some manufacturers; this retards the escalator speed more slowly and safely than the emergency stop. A controlled stop also permits an escalator to be stopped remotely by a smoke detector (see FIRE AND SMOKE DETECTORS).

See also ELEVATORS (LIFTS).

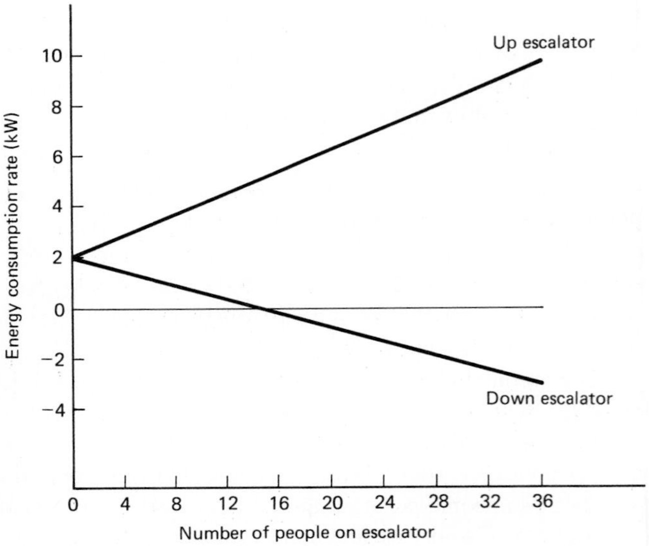

Figure 89 Escalator energy consumption versus loading. (Reproduced by courtesy of *Elevator World*.)

References

1. *American National Standard Safety Code for Elevators, Dumbwaiters, Escalators and Moving Walks—ANSI/ASME A17.1*, American Society for Mechanical Engineers, New York. Issued annually.
2. F. A. ANNETT: *Elevators: Electric and Electrohydraulic Elevators, Escalators, Moving Sidewalks and Ramps*, 3rd ed., McGraw-Hill, New York, 1960.
3. *Safety Code for Elevators, Dumbwaiters, Escalators and Moving Walks—B44*. Canadian Standards Association, Ottawa. Issued annually.
4. JOHN J. FRUIN: *Pedestrian Planning and Design*. Metropolitan Association of Urban Designers and Environmental Planners, New York, 1970.
5. *Vertical and Horizontal Transportation*. Chapter SC-4 in Volume SC of *Monograph on Planning and Design of Tall Buildings*. American Society of Civil Engineers, New York, 1980.
6. G. R. STRAKOSCH: *Vertical Transportation: Elevators and Escalators*, 2nd ed. Wiley, New York, 1983.
7. *Elevator World*. Monthly magazine, various issues.

W.S.L.

EVACUATION OF BUILDINGS IN CASE OF FIRE. Strategies for minimizing the effects of fire have generally focused on the physical facilities in buildings, for example, flame and smoke control measures and exit design. Evacuation procedures, however, are often based on tradition and some questionable assumptions about human behavior in emergencies.

Evacuation maxims are drawn from a variety of factors, including evacuation-time predictions, occupant density, flow rates down stairs, and exit geometry. Current standards appear to have been derived from experiments using a nonrepresentative sample of people who achieved high densities and flow rates on stairs by disregarding conventions of personal space and safety (Ref. 1). This approach, called the "hydraulic" or "carrying-capacity" model because people are supposed to evacuate in the same way water flows through a pipe, has been criticized for placing undue emphasis on the size of the "safe end" rather than on the "threat end," where most fatalities occur (Ref. 2). Moreover, it reflects the erroneous assumption that behavior in fire is a simple process controlled by alarm and exit facilities.

Human response to fire is neither simple nor characterized by irrational, antisocial acts. Recent investigations reveal largely adaptive behavior patterns as people seek more information, warn or help others, make decisions, or otherwise cope. Assumptions to the contrary have led to such ill-advised practices as denying people needed information for fear of panic. Nevertheless, at variance with planning expectations, well-executed evacuations are uncommon. Exits that are not customarily used will carry significantly fewer people in emergencies since building occupants often choose familiar exits in preference to closer but less familiar ones (Ref. 3). Thus, as studies of selective and total evacuations have shown, evacuation-time predictions should be considered minimums, not maximums (Ref. 4).

A life safety plan, if it is to ensure sufficient time for evacuation, must incorporate not only quantitative factors, but also qualitative ones that include normal occupancy conditions and the actual behavior of people in emergencies.

See also HUMAN BEHAVIOR IN FIRES and STAIRS.

References

1. J. L. PAULS: Building evacuation: research findings and recommendations. *Fires and Human Behaviour*, D. Canter (ed.). Wiley, New York, 1980, pp. 251–275.
2. F. I. STAHL and J. ARCHEA: *An Assessment of the Technical Literature on Emergency Egress from Buildings*. National Bureau of Standards Report No. NBSIR 77-1313, Washington, D.C., October 1977.
3. J. L. PAULS and B. K. JONES: Research in human behaviour. *Fire Journal*, Vol. 74, No. 3 (May 1980), pp. 31–41.
4. J. L. PAULS and B. K. JONES: Building evacuation: research methods and case studies. *Fires and Human Behaviour*, D. Canter (ed.). Wiley, New York, 1980, pp. 227–249.

B.K.J.

EXPANSION AND CONTRACTION JOINTS are used to control the accumulation of linear movements in structures. In metal structures, the average effects of temperature change of structural elements cause the elements to expand and contract in their own plane. In concrete, timber, and masonry structures, the average effects of moisture change, as well as temperature change, cause linear structural movements. Gradients of temperature and moisture through the depth of structural elements cause the elements to flex by bending and warping. Neither expansion nor contraction joints in a structure control flexural movements.

In concrete structures, portland cement without expanding additives is normally used. The drying SHRINKAGE movement of such structures is the initial and dominant linear movement. Subsequent expansion movements, due to rise in temperature or moisture, often do not exceed the initial shrinkage movement. Control joints in concrete structures thus normally require provision for one-directional movement and are called *contraction joints* (Fig. 90). In all other structures, control joints must provide for struc-

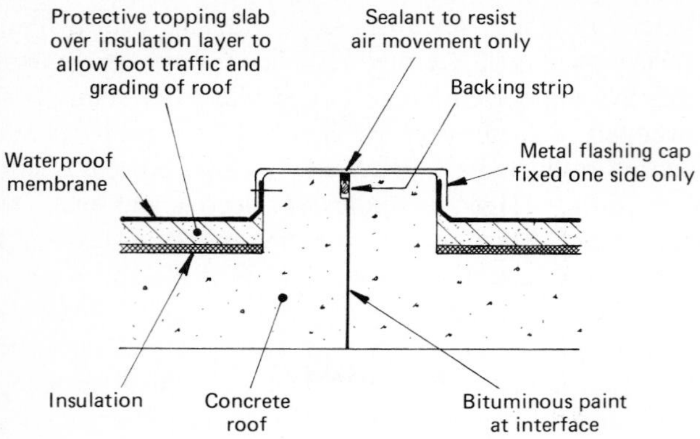

Figure 90 Concrete roof contraction joint where watertightness and ease of maintenance are the prime objectives.

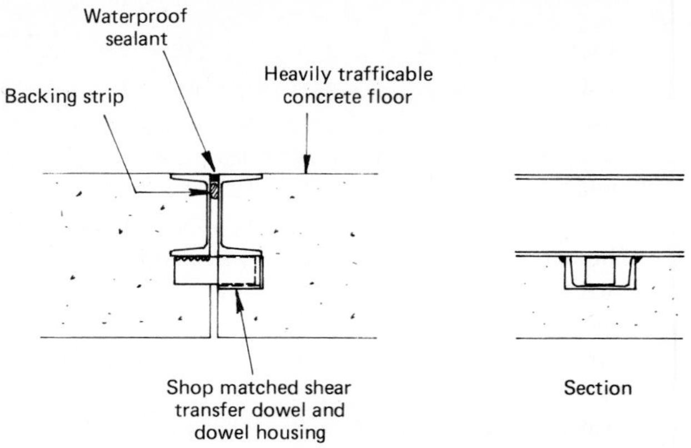

Figure 91 Floor control joint where shear transfer capability and edge wear resistance are the dominant objectives.

tural element expansion or contraction in a nonpredetermined manner. Two-directional joint movement provision is required. Such control joints are called *expansion joints* (Fig. 91).

Before a particular control joint is designed in detail it is first necessary to assess the likely amount of movement and define the design objectives. The objectives are likely to be one or more of the following:

Shear transfer capability
Water tightness
Edge wear resistance
Ease of maintenance
Airtightness
Fire resistance

Reference

H. J. COWAN and T. JUMIKIS: Chapter 24, Environmental Design, in *Handbook of Structural Concrete*. Pitman, London, 1983.

T.J.

F

FENESTRATION is normally defined as any opening that admits light and air into a building. Windows, doors, and skylights are all part of the fenestration system of a building, and this includes glazing, framing, openable or fixed sashes, doors, skylights, external and internal shading devices, and protective devices.

History. Fenestration began as simple openings, as found in Greek temples, or in the form of pierced or latticed windows as found in the buildings of the Indian subcontinent. The Chinese built ''window-walls'' to frame external views and covered them with rice paper. The Romans succeeded in the building of domes with central openings such as the ''eye'' of the Pantheon and used split alabaster for glazing. Windows were sometimes covered with thin marble sheets during the Romanesque period. The early basilicas had clerestory lighting as well as aisle windows, while the Byzantine builders constructed small windows as seen around the dome of St. Sophia, Istanbul. Elaborately sculptured orders were placed around windows and inside rose windows.

By the 13th century, clear GLASS was available for Westminster Abbey. In the Renaissance, windows dominated the facade, and glass houses were introduced in the 18th century in England by Sir Robert Mansell and others. During the 19th century, fenestration was further improved by the technique of glazing with cast iron frames, which made possible the construction of the Crystal Palace.

Large windows became important building components in commercial buildings during the 20th century. From the St. Louis river front, the Chicago School, the work of Walter Gropius, Le Corbusier, Mies van de Rohe, to the present era, windows have been used either as infills between structural elements or as cladding outside the structural frames.

Optical and Thermal Properties. Ordinary glass is diathermanous, as it is transparent to the spectrum of solar radiation, but opaque to long-wave infrared radiation normally emitted by heated bodies. This greenhouse effect is useful when heat is to be retained in the interior of a building. However, if solar radiation is to be excluded, ordinary glass should be protected by SUN SHADING DEVICES or replaced by glass of different optical properties. ''Anti-sun'' glass may be made heat absorbent by retaining the solar heat, which is reradiated indoors and outdoors after a time lag. Reflective glass is more effective in redirecting solar heat away from the building, although it adds it to adjacent buildings. The extent to which ''anti-sun'' glass may exclude solar heat depends on its *shading coefficient*, which is defined as the ratio of the solar heat gain of *any* glass-and-shading combination to the solar heat gain through 3-mm ($\frac{1}{8}$-in.) *unshaded* clear glass.

Shading of glass may be by external and internal shading devices (Figs. 92 and 93). In general, external shading such as awnings, fins, and louvers is more efficient in excluding solar heat from the building than the internal devices. However, a certain amount of diffuse radiation is unavoidable. Adjustable devices, including automated blinds controlled by photocells, are able to exclude direct solar heat gain while admitting DAYLIGHT into the interior.

Ordinary glass has little thermal insulation, and its thermal conductance depends on the thin layers of air on the internal and external surfaces. That of the external air film varies with the wind speed over the fenestration. To improve the thermal resistance of glass, double or triple glazing is used. The air cavity between the two panes of glass may induce convection and reduce the thermal resistance of the fenestration, if the cavity is more than 100 mm (4 in.) in width.

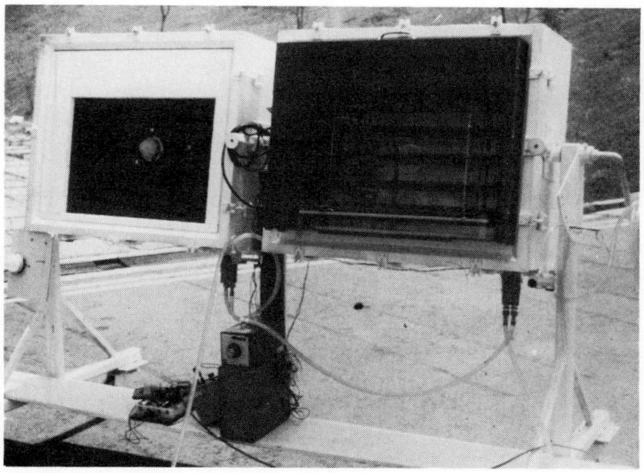

Figure 92 Solar calorimeter for the evaluation of shading coefficients of fenestration. (Department of Building Science, National University of Singapore.)

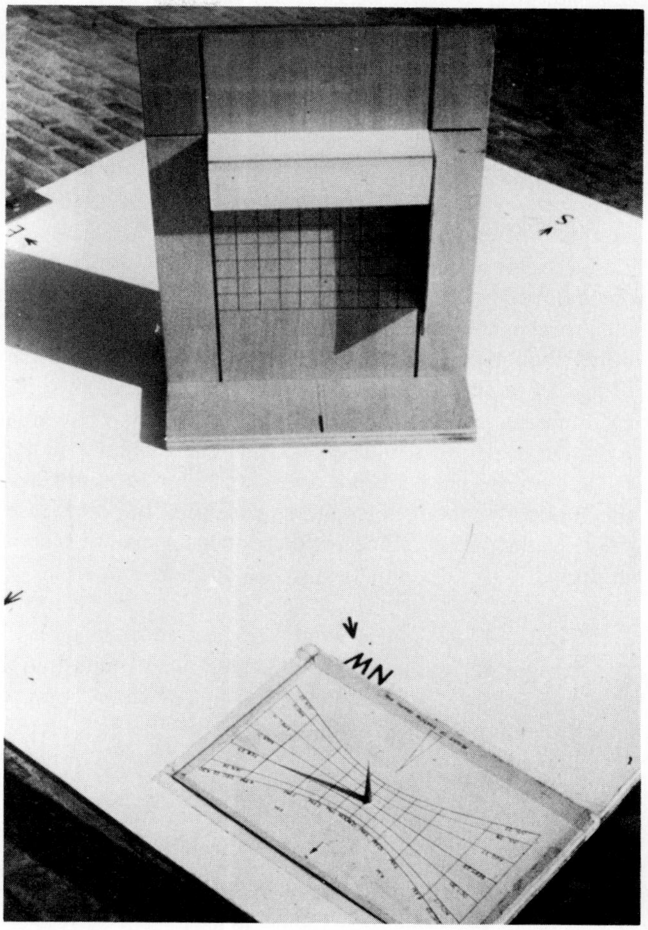

Figure 93 Solar table and sundial for the design of external shading devices to windows (Department of Building Science, National University of Singapore.)

Normal fenestration includes glazing bars, transoms, mullions, and window sashes, either in timber or metal. These may occupy a considerable percentage of the entire window area. Metal framing may conduct more heat than glazing, and the overall thermal performance of the fenestration may be hindered by a high percentage of framing area to glass area. Thermal isolation in large metal sections is useful to minimize excessive heat transfer.

Types. The type of fenestration to be used in buildings depends on the climate. In cold-winter zones, small windows with double or triple glazing are used, and infiltration is minimized to reduce heat loss. In temperate zones, both heating and cooling are required, and larger windows with casement or double-hung sashes are used. In warm humid zones, louvered windows are open below the level of the working plane to provide through airflow for thermal comfort. Hopper windows are used to provide ventilation in rainy weather. In air-conditioned buildings where the windows are seldom opened, the sashes are usually secured by locking devices so as to control operation.

Air locks should be provided for glass doors where isolation of the outdoor environment is needed. Revolving doors are used where pedestrian traffic requires distinct ingress and egress. The air curtain, a regular feature for openings without doors, is less popular due to its high consumption of energy.

Skylights are usually incorporated in the roofs of industrial buildings with large span to admit more daylight than reaches the interior through side windows. As the skylights should not admit direct sunlight, they face the part of the sky vault with the shortest sun path. Hence, skylights are more suitable in locations of high latitudes where solar radiation is also less intense and daylight hours are shorter in the winter. They are not appropriate for locations near the Equator without adequate sunshading. In some cases the skylights are combined with permanent openings so as to provide natural ventilation to the interior.

Wind. The effect of wind on fenestration depends on the strength of the glass, framing, and overall fenestration area exposed to the wind windward or leeward. The maximum bending stress f of the glass is given by Marcus's formula (Ref. 2):

$$f = (0.75\ W/h^2)\left[1 - \frac{5}{6}\frac{r^2}{1 + r^4}\right]\left[\frac{r(1 + \sigma r^2)}{1 + r^4}\right]$$

where W = total load on the glass pane
h = thickness of the glass pane
r = ratio of shorter side to longer side of glass pane
σ = Poisson's ratio for glass and equals approximately 0.21 for ordinary glass

The frames and assembly are usually designed to avoid excess deflection. The maximum deflection should be limited to 1/125 of the span for single glazing and 1/175 of the span for double glazing. To overcome excess deflection, buttresses or stiffeners are used to reinforce the framing. Glass stiffeners are sometimes attached to large spans of plate glass to give additional stiffening. A method of designing the fenestration assembly against wind load is presented in Ref. 3.

Infiltration. Sealing of window sashes and doors against air infiltration has been specified by various authorities. The acceptable airflow rate per unit length of opening joint depends on the pressure difference (Ref. 1). Testing of air infiltration is done in compression chambers where the difference between internal and external atmospheric pressures across a fenestration specimen is measured.

Acoustics. SOUND INSULATION of fenestration is chiefly governed by the mass law. Hence, single glazing has high sound transmission and low sound transmission class (STC) rating; this may be improved by double glazing. The introduction of an air cavity adds to the overall ability of the component to absorb sound energy before transmission. The wider the cavity, the more absorbent the component will be. Further improvement may be made by adding absorbent materials around the edges of the air cavity. The transmission losses of some doors and windows are given in Ref. 4.

Lighting. The methods of evaluation of DAYLIGHT in buildings have been developed since the 1930s. In the British system, daylighting is evaluated by the daylight factor method (Ref. 5). In the United States, the vertical windows are assumed to be luminous planes, giving light to the interior. Hence the U.S. method is analogous to electrical lighting (Ref. 6).

Deep rooms with windows on one side wall need permanent supplementary artificial lighting, and the apparent brightness of the room requires careful balance. GLARE from windows should be avoided by means of appropriate furniture layout. Reflectors outside windows are sometimes used to reflect sunlight through clerestories into the room.

References

1. *ASHRAE Handbook 1981, Fundamentals 1981* (Chapter 27, Fenestration). American Society of Heating, Refrigerating and Air-Conditioning Engineers, Atlanta, Ga., 1981.
2. H. MARCUS: *Die Theorie elastischer Gewebe* (The Theory of the Elastic Grid), Vol. 1, Springer, Berlin and New York, 1932.
3. B. P. LIM, K. R. RAO, K. THARMARATNAM, and A. M. MATTAR: *Environmental Factors in the Design of Building Fenestration*. Applied Science Publishers, London, 1979.
4. M. D. EGAN: *Concepts in Architectural Acoustics*. McGraw-Hill, New York, 1972.
5. R. G. HOPKINSON, P. PETHERBRIDGE, and J. LONGMORE: *Daylighting*. Heineman, London, 1966.
6. B. Y. KINZEY and H. M. SHARP: *Environmental Technologies in Architecture*. Prentice-Hall, Englewood Cliffs, N.J., 1963.
7. G. D. P. TURNERS: *Windows and Environment*. Pilkington Environmental Services, McCorquodale, Newton-le-Willows, 1969.

B.P.L

FIRE AND SMOKE ALARMS alert the occupants of a building to the existence of a fire. Severe losses occur mostly when a fire is detected too late.

Heat and Smoke Detectors are generally connected in electric circuits to a multizone fire alarm panel, which may sound an alarm, alert the fire department, and record the fire by latched-up indicator lights and/or hardcopy printout. The various circuits indicate the zone or floor of the building in which the detector is operating (Fig. 94). This minimizes the time required for locating the fire.

The monitoring of the circuits must be reliable; it should be continuously supervised electrically for faults by the alarm panel. A break in the wiring, which would interfere with the proper operation of the system, should actuate a trouble alarm.

Principles Used for Automatic Fire Detection. *Temperature detectors* sense either a fixed, critical temperature or a rapid rise of temperature. A *fixed temperature* is detected by bimetal thermostats, by elements melting at a low temperature, or by solid-state electronics. The *rate of temperature rise* is sensed by a detector operating when a specified rate is exceeded. This is achieved using special arrangements of bimetal elements, pneumatically operated switches, thermoelectric devices, resistance bridges, or solid-state electronics.

Smoke may be detected either by a photooptical device that ''sees'' the presence of smoke or by an ionization smoke detector. The *ionization chamber* monitors the electrical conductivity of ionized air in a sensing chamber and compares it with that of air in a reference chamber. The air space of both chambers is ionized, that is, made conductive, by the emission from a radioactive source, usually americium 243. When smoke enters the sensing chamber, it absorbs some of the radiation. The resulting electrical change is sensed by the electronic detector and translated into an alarm.

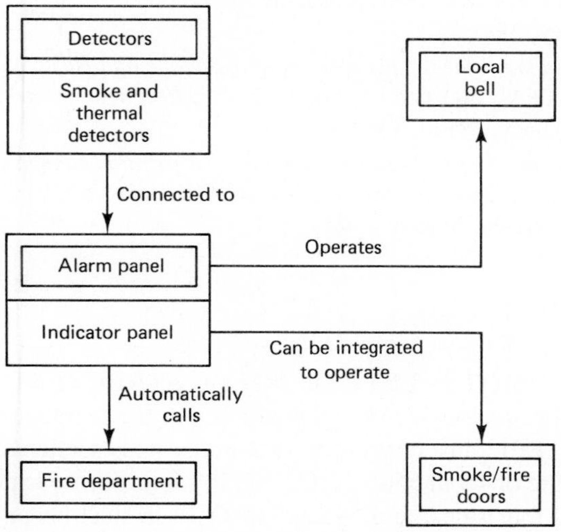

Figure 94 Schematic layout of a complete fire/smoke alarm system.

The *photooptical detector* has a light source and an electronic photoconductive cell in a chamber freely connected to the atmosphere, but constructed so that no outside light can enter. When a sufficient number of smoke particles enters the chamber, either the reduction in light or the scatter of light is sensed by the photocell and raises an alarm.

See also SPRINKLER SYSTEMS.

Reference

1. *Fire Protection Handbook*, 13th ed. National Fire Protection Association, Boston, 1969, *Section 14*, Detection, Alarm, and Fire Guard Services, pp. 1–44, 46, 47.

A.L.

FIREFIGHTING EQUIPMENT is provided in buildings for use by occupants and/or the fire department. ''Virtually all fires are small at origin and could easily be extinguished provided the proper type and appropriate amount of extinguishing agent were readily available and promptly applied. Portable fire extinguishers are designed to fulfill this need'' (Ref. 1).

Portable Fire Extinguishers. Portable fire extinguishers (successors to water and sand buckets) are designed to be carried and operated by hand. When fully charged, they should not be heavier than 23 kg (50 lb). They are commonly filled with water, foam-producing material, powder, carbon dioxide, or halon (halogenated hydrocarbon) and depend for their operation on stored pressure, chemical reaction, or a gas cartridge. Reversible or inverting types are obsolescent, and pumps are now used only for knapsack sprays. Fire extinguishers are marked alphanumerically; the letters designate the general class or classes of fire for which they have been found effective, and the numerals indicate the relative size of fire that an untrained operator can be expected to extinguish (Refs. 2, 3, and 4).

Fire Hose Racks and Reels. Small-bore fire hoses connected to a water supply are installed primarily for first-aid firefighting by building occupants. Two basic types are in use. The American system employs 15-m to 45-m (50-ft to 150-ft) linen or single-jacket rubber-lined hose, 40 mm ($1\frac{1}{2}$ in.) nominal diameter, on special reels (coiled) or racks (flaked on a quick-release mechanism). In the European system, hose reel assemblies consist of reel, inlet valve, hose, shut-off nozzle and, where required, a hose guide or fairlead (Fig. 95). European-type hose reels may be fixed (mounted on a wall with axis at right angles to the plane of the wall), swinging (mounted on a wall and capable of

Figure 95 Typical fixed-type fire hose reel with fairlead.

being swung through an arc of approximately 180° in a horizontal plane), or recessed (capable of swinging out of a recess through an arc of 180° in a horizontal plane). The hose of reinforced rubber or plastics is approximately 35 m (115 ft) long and has a 20 mm ($\frac{3}{4}$ in.) internal diameter. Nozzles may give a jet or spray, or a jet only.

Standpipe (or Hydrant) Systems are installed primarily for use by the fire department. They consist of piping, valves, and hose outlets connected to a water supply. They may incorporate an automatic or semiautomatic starting pump and a private water storage reservoir. Automatic starting pumps respond to a fall in water pressure and thus need pressure jacking with a package comprised of pump, electric motor, air chamber/pressure relief, pressure switch, pressure gauge, and nonreturn valves. Fire department booster (pumping-in) connections are fitted to the majority of hydrant systems near grade level in positions readily accessible to the fire department.

The number and arrangement of fire hydrants needed for the protection of a given building depends on the occupancy, character and construction of the building, the fire risk of any plant, exterior exposure, and accessibility. The water supply, system design, piping, valve, fitting, and testing are specified in national standards (e.g., Ref. 5). Standpipe risers may be dry, but the majority are permanently filled with water. Fire hoses, 63.5 mm ($2\frac{1}{2}$ in.) nominal diameter, may be preconnected to hydrant outlet valves, stored at such valves but not preconnected, or not provided at all, depending on local ordinances and fire department operational policy.

Special Hazard Systems. Many buildings are equipped with automatic and/or manually operated fixed installations to protect special hazard areas. All such systems incorporate manual release facilities for use by building occupants or plant operators. Water spray systems, frequently forming part of SPRINKLER SYSTEMS, may protect oil-filled transformers and flammable gas storage vessels. Gaseous systems, notably Halon 1301 (Ref. 6), may be installed for the protection of critically important areas such as those housing electronic data-processing facilities. Systems using liquid agents or dry chemicals are often fixed to protect kitchen ranges in restaurants, hotels, hospitals, and factories. Foam distributors may be installed in plant rooms, heat treatment departments, and other high-risk areas involving the storage or handling of flammable liquids.

References

1. *Fire Protection Handbook*. National Fire Protection Association, Quincy, Mass., 1981.
2. *Fire Extinguishers, Rating and Fire Testing of.* Standard for Safety UL 711, Underwriters' Laboratories, Inc., Northbrook, Ill., 1979.
3. *Specification for Portable Fire Extinguishers*. BS 5423:1980, British Standards Institution, London, 1980.
4. *Portable Fire Extinguishers; Classification, Rating and Testing*. AS 1850-1981, Standards Association of Australia, Sydney, 1981.
5. *Standard for the Installation of Standpipe and Hose Systems*. NFPA 14-1980, National Fire Protection Association, Quincy, Mass., 1980.
6. B. M. LEE: Fixed fire protection. *Australian Standards for Fire Protection*. Seminar Papers, Standards Association of Australia, Sydney, 1981.

B.M.L.

FIRE PROTECTION OF STEEL STRUCTURES has long been based on the proposition that the temperature in the STEEL must not exceed an average cross-sectional temperature of about 550°C (1022°F) in columns (fire exposure on all four sides) and 593°C (1100°F) in steel beams (fire exposure on three sides). As the temperature increases, the modulus of elasticity and yield strength decrease, and when this loss of strength is reduced to the design strength, collapses may occur depending on the loading of the column or beam. For this reason, many multistory steel structures require the steel to be protected by individual encasement using concrete, sprayed cementitious or mineral fiber, or precast mineral board to ensure that the critical temperature is not exceeded. Depending on the size and mass of the steel column, a thickness of 50 mm (2 in.) of concrete or sprayed cementitious or mineral fiber protection provides a fire-resistance rating of from 2 to 4 hours in severe fires.

Recent research has challenged the long-held assumption that bare or exposed structural steel cannot be used for all multistory buildings. The American Iron and Steel Institute (AISI) sponsored full-scale fire tests and statistical studies which show that fires in open automobile parking structures do not result in fire temperatures that do structural damage to bare steel (Ref. 1). AISI-sponsored research (Ref. 2) has shown that lightweight studs and joists of cold-formed steel can be protected by gypsum wallboard.

The fire-safe structural steel calculation method for fire protection of exposed exterior structural steel (Ref. 3) provides a step-by-step method for calculating the fire protection needed in a real fire. The fire exposure is conservatively simplified by describing it as a steady-state condition. The computer program FASBUS II (Fire Analysis of Steel Building Systems) developed by AISI (Ref. 4) provides a structural model of what happens to a steel-framed floor system in a fire. An evaluation of a 42-story building using FASBUS II produced fire protection material savings of approximately $250,000. All building codes in the United States now permit calculations for fire resistance of steel columns (Ref. 5), steel trusses (Ref. 6), load-bearing steel stud walls, and exposed exterior steel.

See also FIRE RESISTANCE OF BUILDING ELEMENTS; STEEL STRUCTURES; and VERMICULITE.

References

1. RICHARD. G. GEWAIN: Fire in an exposed steel parking structure. *Modern Steel Construction*, First Quarter (1973), pp. 12–14.
2. *Fire-Resistance Ratings of Load-Bearing Steel Stud Walls with Gypsum Wall-Board Protection with or without Cavity Insulation*. American Iron and Steel Institute, Washington, D.C., 1981.
3. *Fire-Safe Structural Steel, A Design Guide*, 1st ed. American Iron and Steel Institute, Washington, D.C., 1980.
4. D. C. JEANES: Predicting fire endurance of steel structures. *Reprint 82-033, presented at the ASCE Conference, Las Vegas, 26–30 April 1982*. American Society of Civil Engineers, New York, 1982.
5. *Designing Fire Protection for Steel Columns, 3rd ed.* American Iron and Steel Institute, Washington, D.C., 1975.
6. *Designing Fire Protection for Steel Trusses, 2nd ed.* American Iron and Steel Institute, Washington, D.C., 1976.

R.G.G.

FIRE RESISTANCE OF BUILDING ELEMENTS. The ability of building elements to function as barriers against the spread of fire is measured and interpreted according to the fire-resistance test standard (Ref. 1). In the standard test the conditions in a compartment fire are reproduced in an idealized way. A more or less full-sized specimen of the building element is exposed on one side (floors, ceilings, roofs, and beams on the underside, walls and partitions on one side or the other, depending on their use) or on all sides (columns) to a test furnace whose temperature is programmed to follow a temperature–time curve specified by the standard. Also specified are the design features and the permissible minimum size of the test specimen, the strength and moisture condition of the specimen before the test, the load or boundary restraint, or both, applied to the specimen during the test, the places where the temperature of the specimen is to be measured and the method of temperature measurement, the hose stream test to be conducted on certain types of specimens following furnace exposure, the criteria of failure of the specimen, and the procedure for reporting test results.

There are four kinds of failure criteria: (1) rise of temperature on the unexposed side (side opposite to the furnace) of the specimen by more than 139°C above that at the start of the test, (2) attainment of a critical temperature level (427°, 538°, or 593°C, depending on the nature of the building element) by a key load-bearing steel component of the specimen, (3) formation of through-holes large enough to allow the ignition of cotton waste held against the hole, and (4) partial or total collapse of the test specimen. The applicable failure criteria depend on the nature of the building element that the specimen represents.

The time taken to reach the first criterion applicable to the building element is the measure of fire resistance. The fire-resistance values found in listings published by testing laboratories are rounded down to match those specified in building codes.

The fire resistance of many kinds of building elements can be predicted by calculations. Some frequently used calculation methods and techniques for extending the use of test results by interpolation and extrapolation are reviewed in Ref. 2. Reference 3 is a guide applicable to concrete elements.

To ensure the satisfactory performance of a building during a fire, all its elements are required to possess certain minimum fire resistances. The current practice is to have these requirements assessed by the writers of building codes. A more accurate way of ensuring satisfactory performance is to match the fire resistance of the building elements with the severity of the fires that are expected to occur, using the concept of normalized heat load (see FIRE-RESISTANT COMPARTMENTATION).

References

1. Standard methods of fire tests of building construction and materials. ASTM Designation E119-81, in *Fire Standards*, 1st ed., American Society for Testing and Materials, Philadelphia, Pa., 1982.
2. T. Z. HARMATHY: Design to cope with fully developed fires, *Design of Buildings for Fire Safety*, ASTM STP 685, eds. E. E. SMITH and T. Z. HARMATHY, American Society for Testing and Materials, Philadelphia, Pa., 1979, p. 198.
3. AMERICAN CONCRETE INSTITUTE: *Guidance for Determining the Fire Endurance of Concrete Elements*, ACI 216R-81, Detroit, 1982.

T.Z.H.

FIRE-RESISTANT COMPARTMENTATION. In the concept of fire-resistant compartmentation, a building is pictured as consisting of compartments all perfectly insulated from one another and the spread of fire as taking place by the successive failure of the compartment boundaries. If the compartment boundaries are sufficiently fire resistant, it is argued, failure will not occur and fires will be confined to the compartments in which they start.

Fire resistance is a measure of the ability of building elements to withstand the spread of fires by destruction or (possibly) thermal conduction. It is determined by subjecting a specimen of the element to a standard fire test (Ref. 1). In that test the specimen is exposed on one side (except for building columns, which are exposed on all sides) in a furnace whose temperature is controlled to follow a prescribed temperature–time curve. The fire resistance is interpreted as the period of satisfactory performance. According to the test specification, collapse, development of through-holes, or rise of temperature at specified locations on or in the specimen mark the failure of the specimen.

Recent studies indicate that the potential of fires (real-world fires as well as test fires) for destructive spread can be quantified by the normalized heat load (Ref. 2). The normalized heat load is defined as the heat absorbed by a unit area of the compartment boundaries during the fire incident, divided by the thermal inertia of the boundaries. In real-world fires the normalized heat load depends primarily on three factors: (1) the amount of combustible material in the building compartment, (2) the ventilation level (determined mainly by the size and number of windows), and (3) the thermal inertia of the compartment boundaries (Ref. 3). In test fires, it is a function of the duration of the test.

A fire-resistant building consists of elements that, based on their fire resistances, are judged capable of enduring the normalized heat load that they are likely to be subjected to in real-world fires (calculated as described in Ref. 3).

There are indications that fires rarely spread by destruction or thermal conduction. The commonest mechanism of spread is convection, that is, the advance of flames and hot gases. Consequently, the design to counter the potential of fires for destructive spread is only part of the overall fire safety design.

References

1. Standard methods of fire tests of building construction and materials. ASTM Designation E119-81, in *Fire Standards*, 1st ed., American Society for Testing and Materials, Philadelphia, Pa., 1982.
2. T. Z. HARMATHY and J. R. MEHAFFEY: Normalized heat load: A key parameter in fire safety design. *Fire and Materials*, Vol. 6 (1982), p. 27.
3. J. R. MEHAFFEY and T. Z. HARMATHY: Assessment of fire resistance requirements. *Fire Technology*, Vol. 18 (1981), p. 221.

T.Z.H.

FIRE-RESISTING CONSTRUCTION. Fire resistance may be needed in elements of construction to confine a fire to a small compartment so that (1) occupants in adjacent risk areas can escape in good time, (2) the fire does not spread to cause large direct financial losses in terms of damage to the building and its contents and indirect losses such as lost production, and (3) the fire does not spread to adjacent buildings to cause a conflagration beyond the control of professional fire fighters. Compartments in multistory buildings are normally designed to resist a burnout of the contents in any one compartment so that there is no possibility of collapse should professional fire fighting be absent (Ref. 1).

Fire Resistance of Concrete. Reinforced and prestressed concrete elements have inherent fire resistance if the thickness of concrete cover to the steel needed to satisfy structural and corrosion-resistance conditions at room temperature is sufficient to keep the steel below its critical temperature for small periods of fire resistance. For separating elements, such as walls and floors, the designer needs to check, first, that the overall thickness of concrete is sufficient to ensure that the unexposed face temperature is within limits and, second, that the thickness of concrete cover shielding the steel bars or tendons from the fire is adequate to prevent the steel reaching its critical temperature. The critical temperature of reinforcing and prestressing steel is often taken as 550°C (1000°F) and 450°C (850°F), respectively, because at these temperatures the steel retains roughly 50% of its ambient strength so that failure is imminent assuming a safety factor of around 1.6 (Ref. 2).

The concrete cover must not spall away in fire or this will lead to rapid heating of the steel and premature failure. To prevent spalling, it is often advantageous to include a light steel reinforcing mesh within the concrete cover so that cracked concrete does not fall away. The type of aggregate has an important influence on the overall thickness and cover; lightweight aggregates (see LIGHTWEIGHT CONCRETE) have lower thermal conductivity than dense aggregates. For example, lightweight aggregate formed from, say, pulverized fuel ash has a low thermal conductivity, low thermal expansion coefficient, and good spalling resistance (Ref. 3).

Fire Resistance of Nonstructural Elements. Where walls and partitions are nonload bearing and tested as such in the standard fire-resistance test, the designer must ensure that in the actual building unacceptable loads and restraints arising from distortions of adjacent elements in a fire do not develop. This may mean, for instance, providing a sliding joint at the head of a metal framed partition to allow for upward expansion of the partition and downward deflection of the floor above (Ref. 1).

Lightweight suspended CEILINGS employing a metal grid supporting ceiling boards are often used to provide membrane fire protection to bare steel beams. It is fundamentally important that the grid should remain on station in a fire so that the ceiling boards are not dislodged, and this is achieved by restraining primary grid members at the boundaries and at changes of level of the ceiling and by providing expansion relief [20 mm ($\frac{3}{4}$ in.) for every 3 m (10 ft) length] at frequent intervals. The integrity of such ceilings in fire depends critically on good design, erection, and maintenance (Ref. 1).

Timber doors may have their fire resistance enhanced by means of a special strip that becomes intumescent at a temperature of approximately 150°C (300°F), thereby sealing any gap between the door and the frame. Steel rolling shutters are able to resist the passage of fire for long periods provided there is adequate expansion allowance for the various components, but single shutters transmit heat by reradiation from the unexposed face so that combustibles must be spaced away. This can be overcome with double shutters spaced apart within the same opening. Closed fire doors and shutters may provide an obstacle to users of the building, and there are magnetic and pneumatic hold-open devices that allow a door to close only when a fire occurs and fusible link devices to initiate closing of gravity-operated shutters. Fire doors used for means of escape purposes rarely require a fire resistance of more than $\frac{1}{2}$ hour, but it is advantageous to use smoke seals in door edges to limit the flow of cool smoke. (Intumescents will only prevent the flow of hot smoke.) There is as yet no internationally accepted method of test for assessing smoke flow past fire doors.

Silica-based sheet GLASS for use in buildings will crack within a few minutes when subjected to thermal shock because of stresses set up by temperature differences across the faces and between the edges and mid-zone. However, when reinforced with small-diameter wire mesh, glass of sufficient thickness (6 mm or $\frac{1}{4}$ in.) can be held in place for more than 1 hour in the standard fire-resistance test despite early cracking. Even better performance can be achieved if the edges of the wired glass are clamped so as to prevent sagging along the top and sides. Special grades of heat-treated glass have been developed that, in limited pane size of about 2 m (6 ft) high by 1 m (3 ft) wide are capable of providing 2-hour fire resistance when effectively clamped around the edges. Multilayer composites of unwired glass and transparent intumescent gel are also available that not only satisfy the stability and integrity criteria, as with the preceding glasses, but also satisfy the insulation criterion for periods up to $1\frac{1}{2}$ hours by virtue of the swelling of the insulating intumescent layers and associated cracking of the exposed face panes (Ref. 1).

See also FIRE PROTECTION OF STEEL STRUCTURES.

References

1. G. M. E. COOKE: Fire Protection, *Specification*, Vol. 1. Architectural Press, London, 1985.
2. *Fire Resistance of Concrete Structures*. Report of a Joint Committee of the Institution of Structural Engineering and the Concrete Society. Institution of Structural Engineers, London, 1975.
3. R. E. READ et al.: *Guidelines for the Construction of Fire Resisting Structural Elements*, Building Research Establishment Report. Her Majesty's Stationery Office, London, 1982.

G.M.E.C.

FLASHINGS are water-control devices that form a part of most wall and roof constructions. They are designed to prevent water from penetrating the exterior surfaces of the construction or, if water has penetrated, to intercept and lead it harmlessly back to the outside. Both functions are illustrated in Fig. 96. Base, counter, and cap flashings keep water out, and through-wall flashings control water that has penetrated the wall.

It is the designer's responsibility to carefully study all exterior surfaces and details of a building to determine where flashings are needed, how they must be detailed, and what materials should be used. They are generally needed where

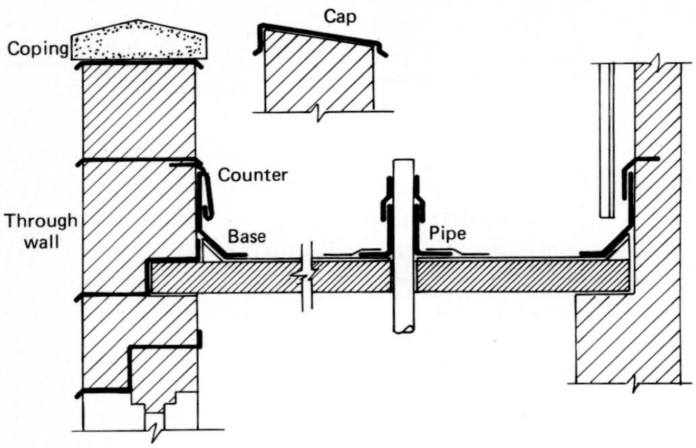

Figure 96 Typical flashings.

parapet and higher walls meet roofs and where pipes, chimneys, structural elements, and such items penetrate roof surfaces. They are also usually required at the top of walls as a capping or underneath masonry copings, at floor intersections with walls, and at the heads and sills of wall openings.

Flashings are made from a variety of materials, including bituminous saturated fabrics, rubber and plastic sheeting, and metals. Fabricating base flashings of the same material as the roofing or waterproofing element to which they are joined or of compatible rubber or plastic sheeting ensures the flexibility normally required. Cap, counter, and through-wall flashings, on the other hand, which usually must be stiff to hold their shape, are most often made of galvanized steel, copper, stainless steel, and aluminum sheet metal. Lead, zinc, terne, and Monel are used to a lesser extent. Through-wall flashings can also be made of copper foil bonded to asphalt-saturated fabric or other reinforcing. Many proprietary products made of preformed combinations of materials for special applications are also available. The choice of material is influenced by several factors, such as resistance to atmospheric pollution; staining of contiguous materials caused by products of pollution, corrosion, or color runoff of the material in question; and movement due to changes in temperature.

See also DAMP-PROOF COURSES and VAPOR BARRIERS.

References

1. M. C. BAKER: *Roofs*. Multiscience, Montreal, 1980.
2. C. W. GRIFFIN: *Manual of Built-up Roof Systems*. McGraw-Hill, New York, 1982.
3. H. PARKER, C. M. GAY, and J. W. MacGUIRE: *Materials and Methods of Architectural Construction*. 3rd ed. Wiley, New York, 1958.

M. C. B.

FLAT ROOFS are usually defined as those having a slope of less than 10° (1 in 5.7). Roofs with no intentional slope, sometimes referred to as dead level, comprise about half of all roofs in North America. Normal construction tolerances and structural deflections, however, prevent the achievement of a completely level roof. Flat roofs require a continuous watertight covering to keep precipitation out of a building and to hold water until it evaporates or drains slowly away. The minimum slope to assure reasonable drainage to outlets is about 1 in 50 (slightly more than 1°).

Flat roofs have come into use for a variety of reasons. They are sometimes considered to be more esthetically pleasing and can be easily designed to take vehicular traffic and landscaping. The high cost of land and the trend to greater population densities in urban areas has led to the increasing use of flat roofs as roof gardens, promenades, and parking areas. Buildings with complex floor plans, common in modern architecture, are more easily covered by such roofs, making them the obvious choice for most low-rise industrial and commercial buildings. Flat roofs also facilitate alterations and additions to factories to accommodate new equipment or production lines. In multistory buildings, the roof is structurally similar to the floors, except for the addition of thermal insulation and a watertight covering. Service conditions are usually more severe than for steeply sloping roofs, but maintenance is more easily done. Roofs with no slope seem to be a basic mistake in design, but low-slope roofs will continue to be used because of the advantages they provide.

Roof Systems. A flat roof system is a combination of components that has to perform a variety of functions in separating indoor and outdoor environments; it must control heat, air, and vapor flow and keep the building interior and its contents dry. Sound insulation, fire protection, and the ability to support traffic and landscaping may also be required.

The conventional combination of these components from inside to outside consists of a structural deck covered with a continuous air/vapor barrier, insulation, and a watertight roofing membrane. The membrane must be continuous without breaks or holes of any kind and must maintain its integrity throughout its service life, while subjected to heat and cold, temperature fluctuations, precipitation, ultraviolet radiation, and movements of the other components. A trap is formed by the combination of membranes and insulation that can hold construction moisture and water leaking or water vapor diffusing into the system, unless drainage at both membranes and venting between them is provided (Fig. 97).

A more recent but widely used system is the protected membrane. In this system a watertight membrane is laid directly on the structural deck, where it also acts as the air/vapor barrier. Insulation is placed on top of the membrane and is covered only by paint, gravel, paving, or landscap-

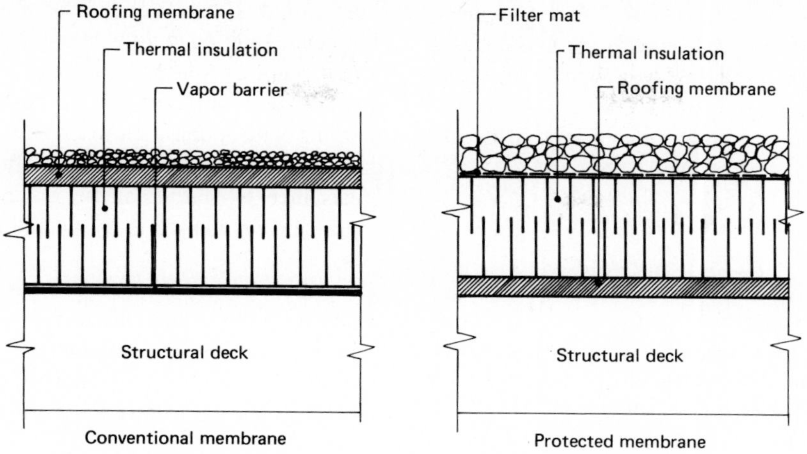

Figure 97 Two basic flat room systems.

ing. This provides protection to the membrane from traffic during construction and from direct weathering and traffic during service. Water from precipitation, however, usually drains through the top surfacing and insulation (Fig. 97).

It is usual, particularly in the conventional system, to attach components together and to the structural deck using combinations of adhesives and mechanical fasteners. For many modern roofs, especially those in which the protected membrane system is used, some or all of the components are loosely laid and ballasted against wind uplift and flotation.

Roofing Membranes. Until recently, watertight membranes were usually made of built-up bituminous roofing that consisted of three or four layers of bitumen-soaked felt, bonded together on site to form a continuous and somewhat flexible covering. The performance of such roofing, which still constitutes 80% to 85% of all flat roofing, has been generally good. On the other hand, for various reasons it has not always met the performance expectations of designers and owners and, in fact, sometimes gives problems from the outset, leading to replacement after only a few years.

As a result of such problems, new materials for membranes have been developed, such as modified bitumen combined with less moisture-sensitive felts for multiple-ply built-up roofing and plastic, rubber, or rubberlike sheet materials for single-ply roofing. They now constitute 15% to 20% of all flat roofing, but have created and will likely continue to create as many problems as conventional bituminous roofing. These problems can be minimized by careful selection of a roof system and its components, as well as attention to its construction.

References

1. M. C. BAKER: *Roofs*. Multiscience, Montreal, 1980.
2. C. W. GRIFFIN: *Manual of Built-up Roof Systems*. McGraw-Hill, New York, 1982.
3. *A. J. Handbook of Building Enclosure*. A. J. Elder, ed., Architectural Press, London, 1974.

M.C.B.

FLOOD-PROOFING OF BUILDINGS. Flood-proofing is a means of reducing potential damage to structures built in floodplain areas. It can protect new or existing buildings located in fringe flood zones or in areas where protection works, such as dikes, may not be adequate. It may be an acceptable alternative if flood-control works are not economically feasible or be part of a floodplain management program.

Owners, builders, architects, engineers, municipal officials, and planners must all appreciate the need for detailed study of many interrelated factors in assessing the feasibility of flood-proofing buildings. These factors include a study of flood characteristics, such as height of maximum flood level, duration and frequency of floods, and velocity of water during peak flows. Uplift and lateral loads on the structures during floods must also be determined so that the building can be designed to resist them. Economic factors are important to ensure that anticipated benefits exceed estimated costs.

There are several different methods of flood-proofing:

1. Constructing buildings on fill, piers, piles, columns, or bearing walls so that they are above design flood levels.
2. Surrounding buildings with flood-proof walls or beams.
3. Making lower levels of buildings watertight.

These methods are sometimes termed *dry* flood-proofing to distinguish them from *wet* methods in which it is accepted that the building interior will be flooded. (Damage is minimized by using special water-resistant construction materials in the lower levels of the building.)

The potential advantages of flood-proofing are greatest in metropolitan areas where low-hazard flood lands have already been developed or are being developed. Selective flood-proofing can be carried out in these areas with safety and often with greater economy than other forms of flood-plain management.

References

1. C. M. ANDERSON: *Manual for the Construction of Residential Basements in Non-Coastal Flood Environs.* Federal Insurance Administration, Department of Housing and Urban Development, NAHB Research Foundation, Inc., Rockville, Md., March 1977, 265 p.
2. DEPARTMENT OF FISHERIES and ENVIRONMENT CANADA: *Flood Proofing: A Component of Flood Damage Reduction.* Prepared by J. F. MacLaren Ltd., Vol. 1, Appendix A. *A Portfolio of Canadian Case Studies* is published separately as Vol. 2. Ottawa, 1978.
3. J. R. SHEAFFER AND ASSOCIATES: *Introduction to Flood Proofing: An Outline of Principles and Methods.* University of Chicago, Center for Urban Studies, Chicago, 1967, 61 p.

G.P.W.

FLOOR FINISHES are usually applied to structural floors to provide a level, smooth surface for foot and wheeled traffic and for furniture. The principal materials used are stone, wood, clay products, plastic, cork, and wool. The criteria for choice include wear, cleaning and surface protection, comfort, cost, and some that may only apply to particular situations, such as resistance to chemicals and sparking and behavior with underfloor heating (Refs. 1, 2, and 3).

Wear. Floor finishes receive more wear than any other internal building surface. Generally, the harder the surface, the greater its ABRASION RESISTANCE. Marble and slate show little wear after centuries of use, whereas cork may need resurfacing after a few years. Some materials, such as wood block, can be restored after wear by sanding and resurfacing.

Cleaning and Surface Protection. The durability of flooring is influenced by surface protection and cleaning methods. Most floor finishes in current use have background histories that show their likely performance over time.

Hard surfaces, such as marble, slate, clay tiles, and terrazo, can be maintained by sweeping and washing with warm water and detergent. Softer surfaces, such as linoleum, cork, vinyl, or rubber, may need an initial seal, renewed at intervals, with regular sweeping and/or mopping and/or polishing. Wooden surfaces usually require coarse and fine sanding before staining (if required) and sealing with a solvent-based seal (e.g., polyurethane) or a wax polish. Both surfaces can be maintained by sweeping and wiping with a damp cloth or wax polishing.

Carpets need no treatment after laying; however, their resistance to stains can be increased by a silicone-based spray. Maintenance is by vacuum cleaning and shampooing.

Floor finishes show more wear in heavily trafficked areas such as lobbies and the corners of corridors. Since some stones cease to be available and carpet dyes, plastic tile colors, and patterns may change, it is advisable to stockpile enough material to replace worn areas (Ref. 4).

The **Comfort** of a floor is partly objective and partly subjective. Hard floors and long-pile carpets can cause fatigue. The resilience of other carpets and the use of softer materials, such as cork or wood, lessens fatigue. Resilience is particularly important for gymnasiums and dance floors; it is usually achieved by supporting a wooden floor on springs, rubber pads, or air cushions.

Effective warmth of a floor will depend on its surface temperature, conductivity, thermal capacity, the temperature of the air near the surface, and, of course, the footwear used. However, the higher conductivity of stone, tile, or concrete floors will make them seem colder than wood or carpet.

Soft materials absorb airborne sound and reduce the transmission of impact noise. Harder finishes must be separated from structural floors by resilient underlays to reduce impact sound and, to a lesser extent, airborne sound. The floor finish must also be isolated from the surrounding walls.

Many hard surfaces are prone to slipperiness, particularly if polished. Wear can also produce slippery areas, but joints and patterns can reduce slipperiness. Stair treads of potentially slippery materials should have nonslip nosings or inserts.

Cost. Some finishes, such as marble or slate, can cost more than ten times as much as wood or vinyl. This is not only due to the higher cost of the materials, but also because more skilled labor is needed to place them. However, long-term cost may be more important than the initial capital cost; the LIFE-CYCLE COST of maintaining, and possibly replacing, a finish with a low initial cost may be greater than that of a finish with a higher initial cost.

References

1. H. KING: *Mitchell's Building Construction: Components and Finishes.* Batsford, London, 1971.
2. A. H. ELDER (ed.): *A.J. Handbook of Building Enclosure.* Architectural Press, London, 1974.

3. F. S. MERRITT (ed.): *Building Design and Construction Handbook*. McGraw-Hill, New York, 1982.

4. W. L. SALTER: *Floors and Floor Maintenance*. Applied Science Publishers, London, 1974.

G.E.H.

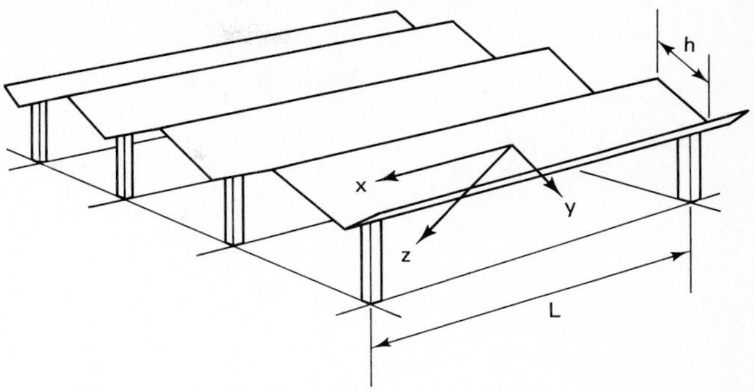

Figure 99 Common type of folded plate roof. L = span; x = horizontal axis parallel to span; y = axis in the plane of the folded plate; z = axis perpendicular to folded plate; h = slant height of folded plate, measured along y axis.

FOLDED PLATES, also known as hipped plates or prismatic structures, have many of the economic and esthetic advantages of curved SHELL ROOFS for covering large unobstructed floor areas. Formwork, steel fixing, setting out, concreting, the carrying of point loads and incorporation of large openings are all much easier than for shell roofs, but more concrete and reinforcement are required. The basic structural action is indicated in Fig. 98. A good practical system is shown in Fig. 99, and some other possibilities are shown in Fig. 100.

Design and Analysis of Folded Plates. The sizes of folded plates are assessed by simple approximate design methods and/or experience, followed by a structural analysis. The sizes are then altered if necessary and the process repeated.

Methods of structural analysis are reviewed in Ref. 1. Reference 2 compares several of these methods. The more accurate methods of analysis require computers for solution. A method of analysis not requiring a computer (but making more approximations) is given in Ref. 3.

The more approximate methods (such as Ref. 3) deal with fewer forces and moments. Resolve the loads in the y and z directions (see Fig. 99) for each plate; then each plate can be considered as a beam of span L and depth h. If everything were symmetrical, that is, if the folded plate system of Fig. 99 continued *ad infinitum* without terminat-

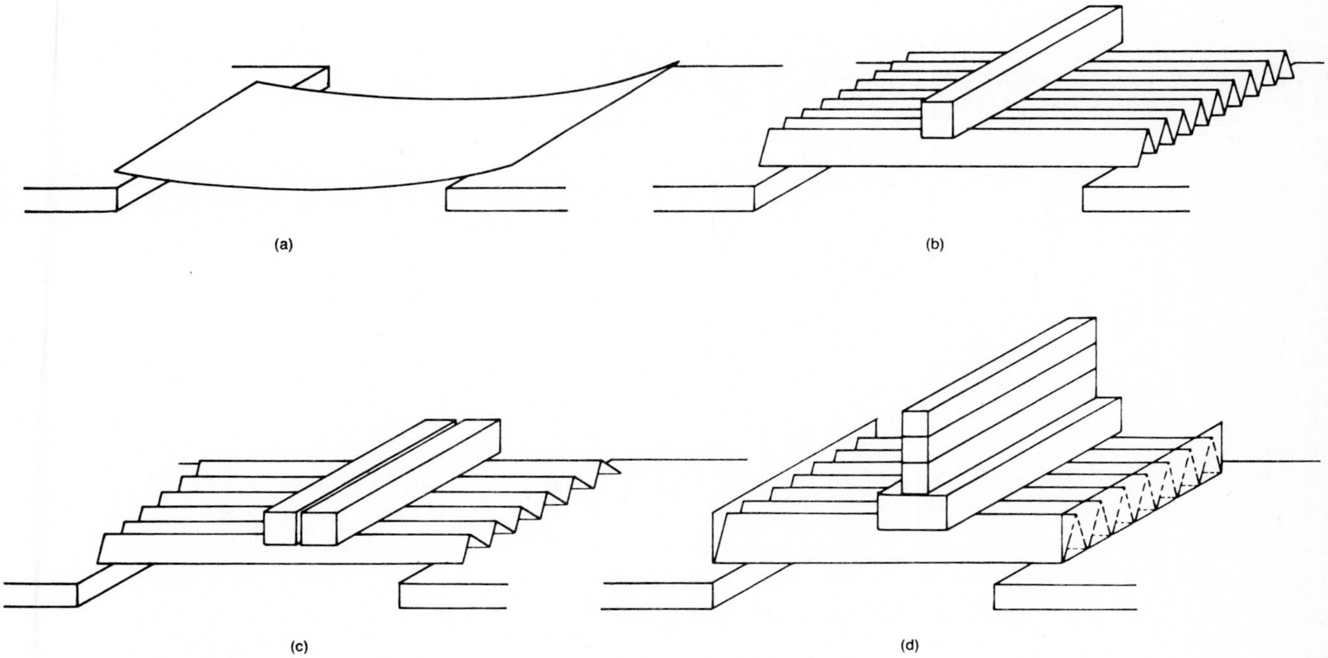

Figure 98 Folding a piece of paper greatly increases its strength; folding a reinforced concrete plate does the same. (a) A flat sheet of paper does not even carry its own weight. (b) A folded piece of paper can carry its own weight as well as a superimposed load. (c) As the load is increased, the folds straighten out, and the structure collapses. (d) Therefore, the structure can be greatly strengthened by gluing stifferers to its ends. (By courtesy of H. J. Cowan.)

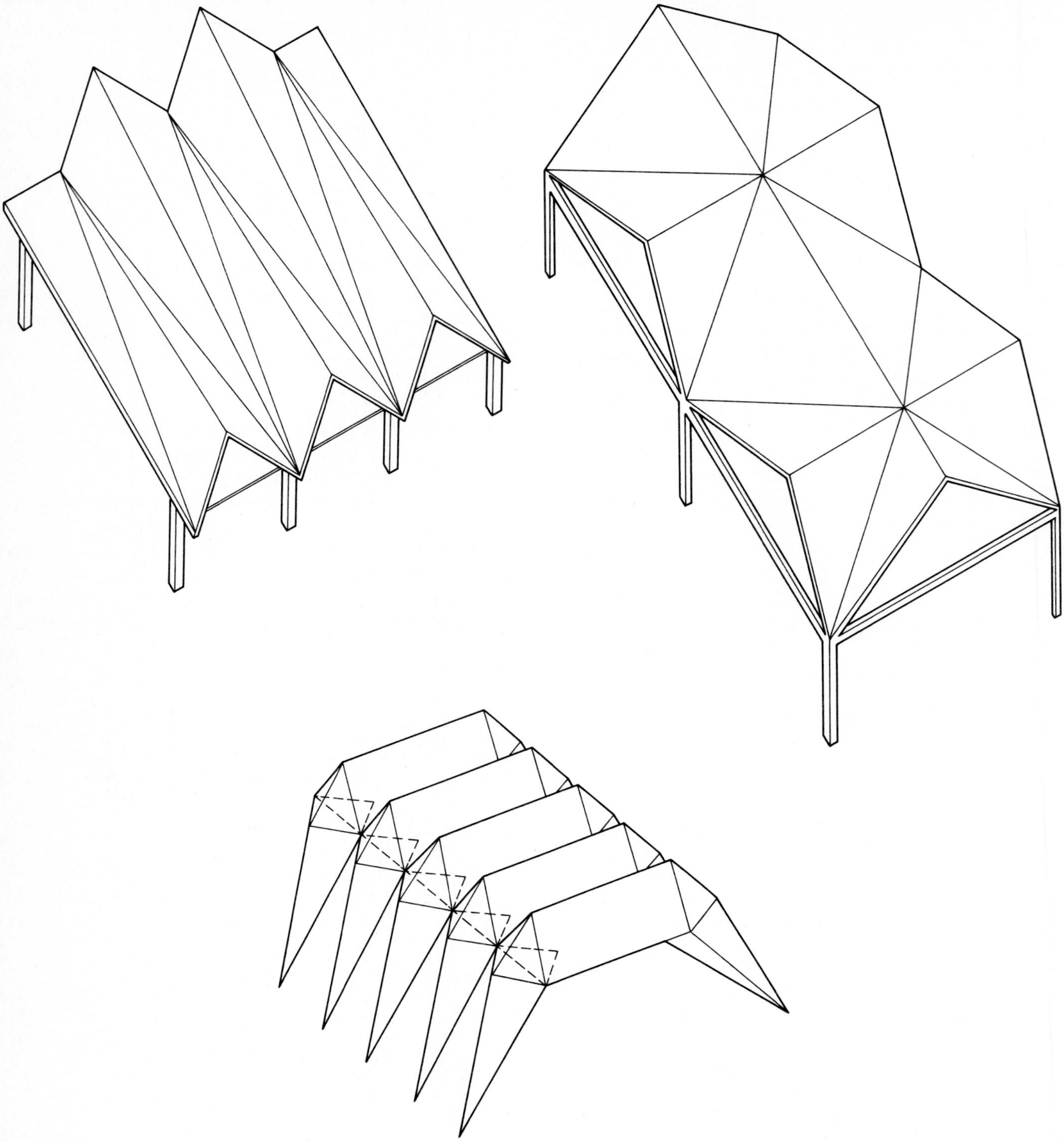

Figure 100 Some other types of folded plate roof. (By courtesy of H. J. Cowan.)

ing at either end, then the longitudinal flexural stresses in these plates would be the same where they touch one another, and the beam analysis for these stresses would be adequate and can be called a *beam analogy theory*. In addition, for this theory, each plate is very flexible in the z direction but relatively very stiff in the y direction. Hence each apex point and valley point is held in position, in that the two adjacent plates restrict its movement in their planes (i.e., the y directions). Hence, away from the end columns the slab can be regarded as spanning from apex to valley as a long continuous slab, assuming $L > 2h$, that is, the slab is not two-way spanning. If the roof consists of a large number of similar folded plates, this beam analogy theory is reasonable for the central bays. It should be added that, when each plate is analyzed for longitudinal bending and shear, its inclined reactions must be supported by end stiffeners; see Fig. 98d.

For a scheme like Fig. 99, where there are end plates, but not all the plates have the same dimensions h, the beam analogy would mean that the longitudinal stresses calculated at the junction of plates with different values of h would not be the same. Yet they must be, as the longitudinal strain at these junctions must be the same. Thus, more complex methods of analysis are required (Refs. 1 and 3).

Elastic versus Collapse Theory. The design methods described are based on the elastic theory. The folded plate should be designed for working loads and maximum allowable stresses, not for ultimate loads and failing stresses, because our main experience of shells and folded plates is based on the elastic theory. We know from experience that large cracks do not develop at working loads, so deflection calculations at working loads can be made; creep (see SHRINKAGE and CREEP) is allowed for by a suitable choice of the elastic modulus for concrete (Refs. 1 and 4). It is difficult in a laboratory to effect sufficient or even realistic research to justify ultimate collapse theories. A folded plate 18 m (60 ft) long and 100 mm (4 in.) thick, scaled down, say, 10:1 for a laboratory test would be 1.8 m (6 ft) long, but it would need to be 10 mm ($\frac{3}{8}$ in.) thick with several layers of reinforcement. Shrinkage could not be scaled down; in fact, it would be greater for the thinner section of the model.

Software. Internationally, many firms provide computer software services. If a designer does not use a computer package for folded plates, but has one for solving simultaneous equations, the author suggests he or she check the derivation of Parme's equations (Ref. 5) and calculate the coefficients of the simultaneous equations either with a computer or by hand. If computer facilities are not available, the designer could make an analysis using the relaxation methods given by Simpson (Ref. 3).

See also ELASTIC DESIGN OF STATICALLY INDETERMINATE STRUCTURES and COMPUTER METHODS IN STRUCTURAL ANALYSIS.

References

1. C. B. WILBY: *Concrete for Structural Engineers*. Newnes-Butterworth, London, 1977.
2. K. C. ROCKEY and H. R. EVANS: A critical review of the methods of analysis for folded plate structures. Proceedings of the Institution Civil Engineers, Vol. 49 (June 1971), pp. 171–192.
3. H. SIMPSON: Design of folded plate roofs. *Journal of the Structural Division of the American Society of Civil Engineers*, Vol. 84, No. ST1 (June 1958), pp. 1508, 1–21.
4. C. B. WILBY: *Structural Concrete*. Butterworth, London, 1983.
5. A. L. L. PARME: Computational arrangement for analysis of folded plate by direct solution. *Advanced Engineering Bulletin*, 3a, Portland Cement Association, Chicago, 1963.

C.B.W.

FORMWORK or falsework is the temporary supporting structure used to contain and provide shape to CONCRETE while in its plastic state. It is also used to a lesser extent to retain earth or clay mixtures for cottage wall construction and has its origins in this application.

The earliest known use of formwork with concrete is about the 2nd century B.C., when the Romans used it to contain *caementa* in wall and foundation construction (Ref. 1). Later it was also used for vault construction, but generally, because of the structural limitations of mass concrete, shapes were of necessity simple and progress was slow enough to make few demands of the formwork. Carpenter-designed and constructed timber forms were quite satisfactory and remained so until the invention of reinforced concrete early in the 19th century. The use of reinforced concrete permitted increased complexity of design shapes and faster rates of construction. Resulting reductions in the cost of concrete structures increased the relative importance of formwork until today, when it represents the major cost of most concrete construction.

Modern formwork construction has become highly specialized, and a complete industry has grown to service its needs. Specialist contractors, designers, suppliers, and a variety of proprietary formwork products are available to suit the majority of projects. Designs are engineered based on known performance of equipment and conditions (Ref. 2). The forming structures usually comprise either large fabricated mechanically handled sections or smaller lighter units, which can be manually handled. Larger forms in-

Figure 101 Circular wall formwork systems (crane handled).

clude crane-handled wall forms or floor forms (Figs. 101 and 102) of a type described as tableforms or flying forms. Smaller units are the proprietary systems comprising prefabricated panels, shoring frames, props, telescopic beams, and a whole range of small components known collectively as form hardware.

Of the materials used in formwork construction, timber is still the most popular in the form of PLYWOOD facing or sheathing mounted on sawn timber bearers (Ref. 3). Steel is now used generally for heavy structural support and almost exclusively as tubular vertical propping or shoring. There is limited use of other materials although structural aluminum is beginning to appear as an alternative to timber and steel. Plastics have their place, but so far with few exceptions have proved only suitable for nonstructural applications.

See also CONCRETE STRUCTURES and PLACING OF CONCRETE.

Figure 102 Table forms (crane handled).

References

1. NORMAN DAVEY: *A History of Building Materials*. Phoenix House, London, 1961.
2. *Code of Practice for Falsework BS 5975:1982*. British Standards Institution, London, 1982.
3. R. L. PEURIFOY: *Formwork for Concrete Structures*. McGraw-Hill, New York, 1964.

J.B.D.

FOUNDATIONS FOR BUILDINGS. The foundation of a building or structure is that part of the structure which is in direct contact with the ground and transmits the loads of the structure to the ground. In designing foundations, two criteria must be satisfied:

1. There must be an adequate factor of safety against bearing capacity failure of the supporting soil.
2. The settlements (vertical movements) and the differential settlements between adjacent parts of the foundation must be kept within acceptable limits for the structure.

For foundations on clay soils, either bearing capacity or settlement may govern the foundation design, but for foundations on sands or gravel, settlement is usually the controlling factor in design.

Foundations can be divided into two categories, shallow and deep. Shallow foundations (or footings) are most frequently used to support relatively light loads, whereas deep foundations can be used to support much heavier loads or to provide support in weak soils. The major types of foundations are described next.

Spread or Pad Footings. These are shallow foundations that support a single column or load (see Fig. 103). They are generally only used on relatively stiff soils where the loads are relatively light or where the structure is relatively rigid or is tolerant of differential settlement (vertical movements) between adjacent columns. They are usually located at the highest level at which the supporting soil is adequate. In cold climates, the founding depth should be greater than the depth of frost penetration (see also PERMAFROST). In arid climates where expansive clay soils may be present, pad footings should be founded below the zone of soil movement, typically 1 m to 3 m (3 ft to 10 ft) below the surface.

Continuous or Strip Footings. As illustrated in Fig. 104, strip footings are long shallow foundations that support a number of columns. They are frequently used in soils of medium to low compressibility and in circumstances re-

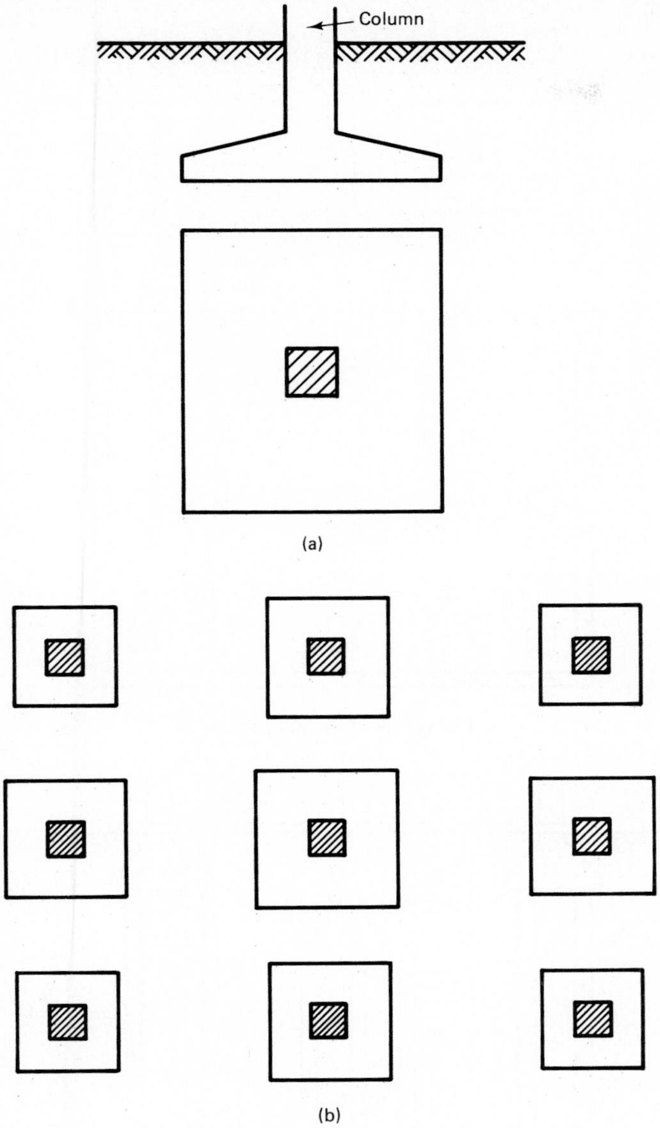

Figure 103 Spread or pad footings. (a) Single footing. (b) Group of footings (plan).

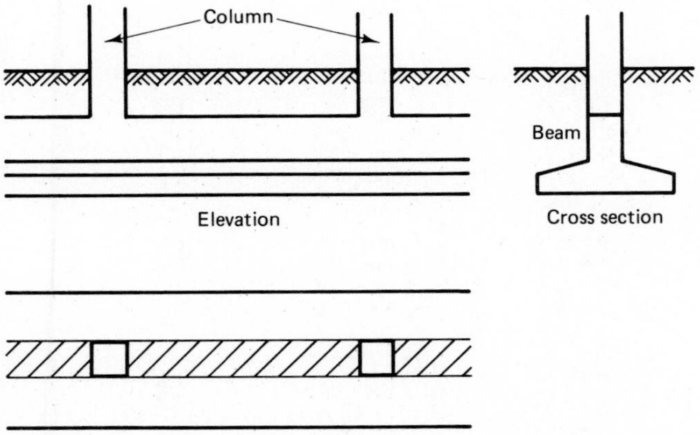

Figure 104 Continuous or strip footing.

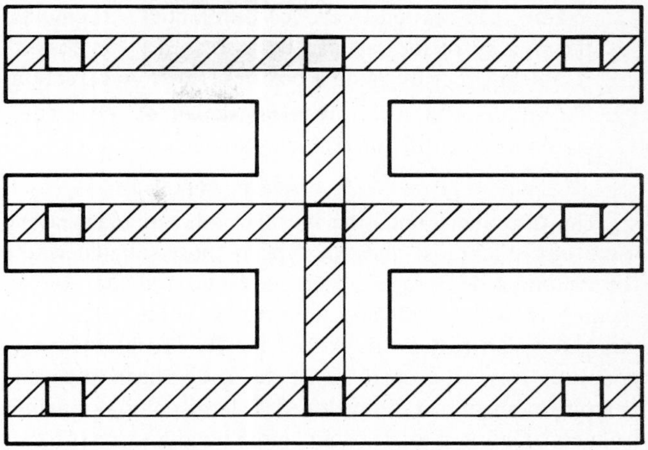

Figure 105 Example of two-way strip footing.

quiring some control of differential settlements; such settlements may be reduced by increasing the width and depth of the strip, thereby increasing the bending stiffness of the foundation. It is also possible to increase the stiffness by having strips running in both directions (Fig. 105).

Mat or Raft Foundations. These are continuous shallow foundations of relatively large extent that support the entire structure or large parts of the structure (Fig. 106). They are used in the following circumstances:

1. Where continuous or spread footings would cover more than about half the building area.

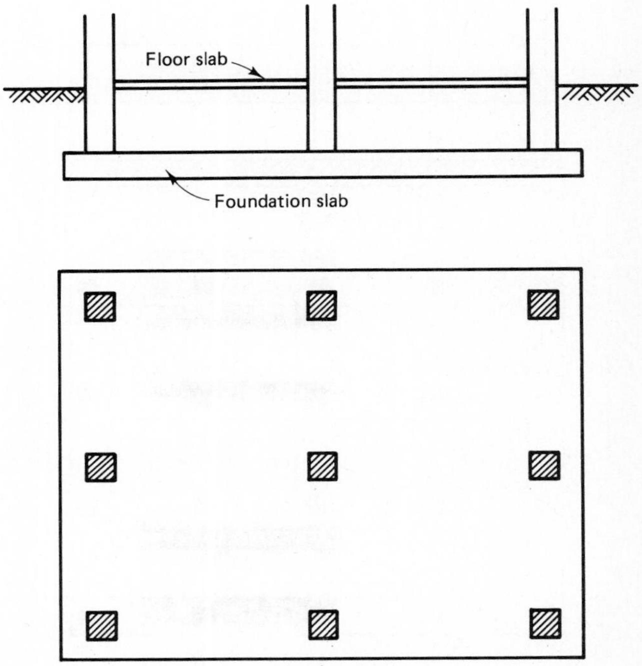

Figure 106 Mat or raft foundation.

2. Where it is desired to reduce differential settlements; the rigidity of the raft can be increased by increasing its thickness or by using a box or cellular construction.
3. To bridge over local areas of weaker or more compressible soil, fill, or rubbish dumps.

The true mat foundation is a flat concrete slab having a uniform thickness, and this type is most suitable where the column loads are small or moderate and the column spacing is fairly small and uniform. If column loads are large, the slab may be thickened locally or, alternatively, pedestals may be used to resist the large shear force and bending moment that is developed. If large column spacings or unequal column loads occur, large bending stresses may be developed, and it may be necessary to use thickened sections along the column lines (Fig. 107). In cases where very stiff rafts are required, a box structure of cellular construction or a rigid frame consisting of slabs and basement walls can be used (Fig. 108). PILES may be needed for soils of low beaming capacity (Figs. 109 and 110).

Factors Affecting Foundation Selection and Design. The selection and design of a foundation depends on many factors, and these can be grouped into four categories: structural requirements, engineering properties of the site, construction requirements, and economic and reliability considerations.

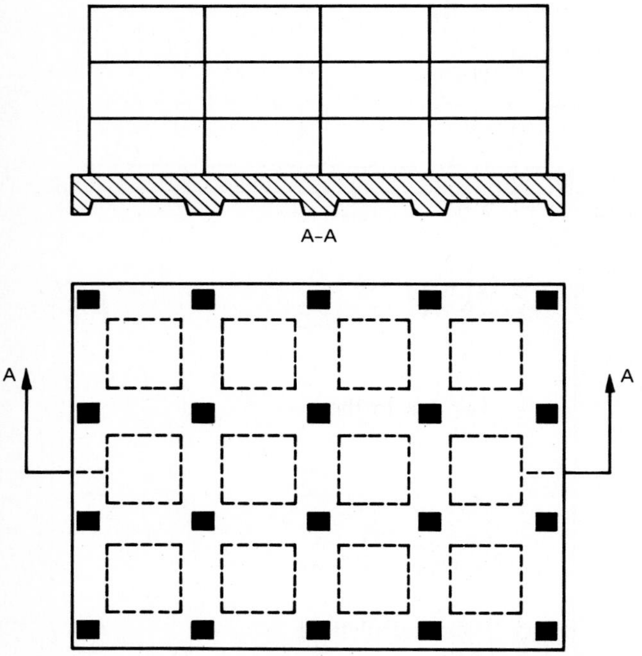

Figure 107 Mat foundation with thickened beams below columns.

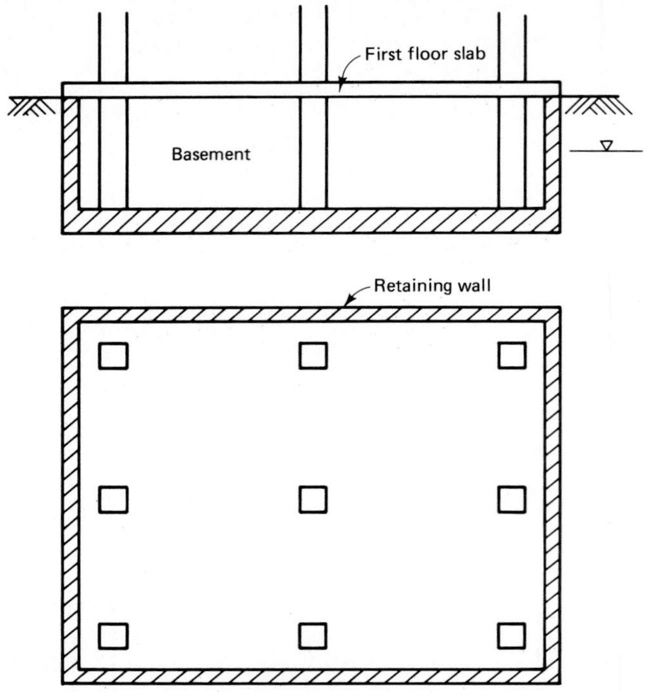

Figure 108 Floating foundation.

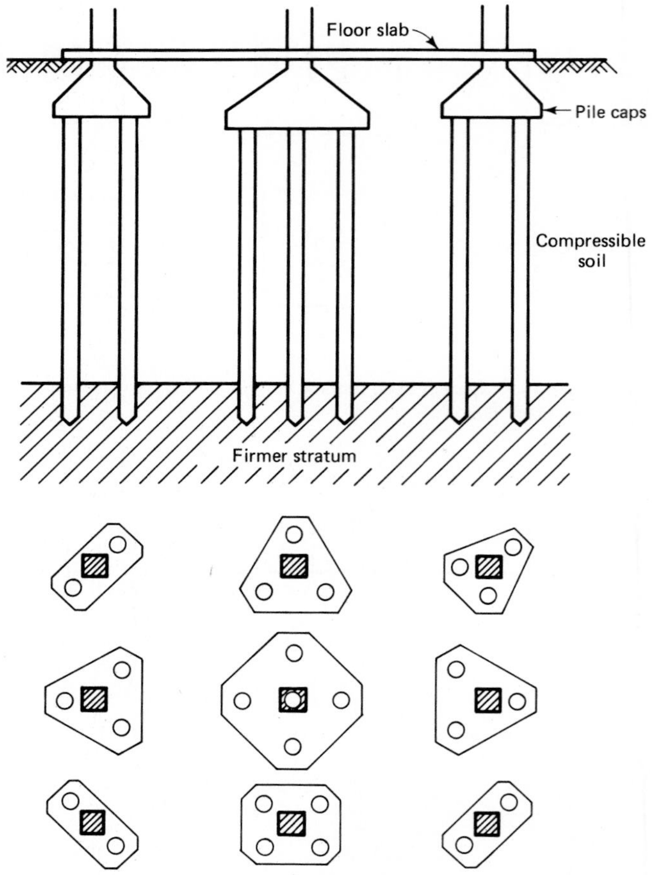

Figure 109 Pile groups.

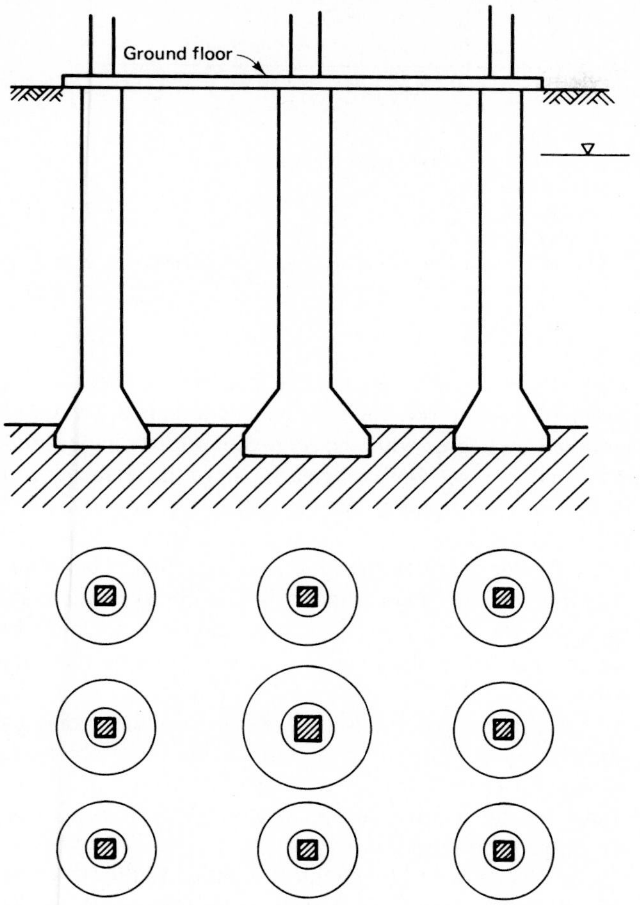

Figure 110 Bored piles with enlarged base.

Structural Requirements are generally determined by the architect and structural engineer and include the following:

1. The static and dynamic column loads of the superstructure.
2. The type of structure with respect to its flexibility, the amount of settlement that can be tolerated, the allowable differential settlements between columns or along walls, and the permissible tilt of the structure.
3. The allowable settlement in relation to adjacent construction and mechanical and hydraulic installations going into the building.
4. The effect of the construction of the building and foundation on existing adjacent buildings and structures.

Engineering Properties of the Site are generally of major concern to the foundation engineer, who requires the following information:

1. The stratigraphy of the site and the subsurface soil profile.
2. The mechanical properties of the subsoils, including the shear strength, compressibility, and permeability of the soils within the zone of influence of the structure (see also SOIL MECHANICS).
3. The location of the groundwater table and an appraisal of the possibility of future changes in groundwater conditions.
4. In seismic regions, the characteristics of the likely EARTHQUAKE (e.g., ground displacements and acceleration) and the dynamic properties of the subsoils. Particular problems with foundations in difficult ground conditions are discussed in Ref. 1.

Construction Requirements. These include the following:

1. The time available for construction.
2. The space available for construction, including access to the site, proximity to property lines, storage areas, and headroom for construction equipment.

Procedures for Design of Foundations. A prerequisite to any foundation design is adequate SUBSURFACE EXPLORATION. The extent of such investigation will depend on the cost of the project, but in all cases the information required includes the general geology of the site, any recent history of filling or construction, the nature of the subsurface deposits, the depth, thickness, and composition of each soil type, the location of the groundwater table, and the engineering properties of the soil and rock strata that may affect the performance of the building or structure.

On the basis of the soil data obtained from the subsurface exploration, the following design calculations are generally required:

1. A stability or bearing capacity analysis of the proposed foundation to determine the ultimate load capacity.
2. A stability analysis of any required excavation to determine the requirements for support and dewatering of the excavation.
3. Analysis of settlements (and horizontal movements if appropriate) due to the structural loads and the consequent differential settlements between various parts of the structure.
4. Structural design of the foundation system to withstand the stresses imposed by the loading and the interaction between the foundation and the soil.

Details of such calculations are given in Refs. 2, 3, and 4.

In assessing the suitability of a foundation, the customary procedure is to evaluate the overall factor of safety of the foundation as the ratio of the ultimate load capacity of the foundation divided by the design loading on the foundation. Typically, safety factors of 3 for shallow foundations and 2.5 for deep foundations are considered adequate. If the safety factor requirements are met, the computed settlement is then checked to determine whether it is excessive. Typically, acceptable settlements are of the order of 40 mm ($1\frac{1}{2}$ in.) for shallow foundations on sand and 60 mm (2 in.) for shallow foundations on clay. However, significantly larger settlements (e.g., 150 mm or 6 in.) may be tolerated if the structure or building is rigid. Allowable differential settlements vary with the type of structure and are generally specified in terms of the angular distortion (the ratio of the differential settlement to the distance over which it occurs). Detailed information is summarized in Ref. 1, but, typically, values of less than 1/500 are desirable to avoid significant cracking in steel and reinforced concrete frame structures. Where the building has brick walls, the angular distortion may need to be kept much lower (e.g., 1/2500).

In cases where the subsoil conditions are difficult or the structure is sensitive to foundation settlement, it is highly desirable to institute a program of observations of foundation movements and pore water pressures in the soil, during and after construction, in order to check whether the actual foundation behavior is similar to that predicted. If the observations are made during construction, it may be possible to modify the foundation design if the actual performance differs markedly from that predicted. This is the basis of the *observational method* (Ref. 5).

References

1. M. J. TOMLINSON: The design of foundation structures for difficult ground. In *Foundation Engineering in Difficult Ground*, ed. F. G. BELL, Butterworths, London, 1978.
2. M. J. TOMLINSON: *Foundation Design and Construction*, 4th ed. Pitman, London, 1980.
3. H. F. WINTERKORN and H. Y. FANG: *Foundation Engineering Handbook*. Van Nostrand Reinhold, New York, 1975.
4. J. E. BOWLES: *Foundation Analysis and Design*. 2nd ed., McGraw-Hill, New York, 1977.
5. R. B. PECK: Advantages and limitations of the observational method in applied soil mechanics. *Geotechnique*, Vol. 19, No. 2 (1969), pp. 171–187.

H.G.P.

FUZZY SETS, ANALYTICAL HIERARCHIES AND DECISIONS. Any group of elements must share one or more properties before we can think of it as a set. If we take one of the properties, such as the height of buildings, we find that all buildings have height, but it differs among them. In a fuzzy set, each element is assigned a value or membership degree with respect to the property in question. We can do this by comparing each pair of the elements with respect to that property. For example, we ask how much taller one building is than another. If we know the numbers we can enter them; furthermore, we note that if building i is five times taller than j, then j is $\frac{1}{5}$ as tall as i; this is the reciprocal property. If we have no measures for the property we can use the scale given in Table 1 to make the comparisons.

TABLE 1
Scale of Relative Importance

Judgment	Value
Equal importance	1
Moderate importance of one over another	3
Essential or strong	5
Demonstrated importance	7
Absolute importance	9
Intermediate values between two adjacent judgments	2, 4, 6, 8

The matrix of comparison A satisfies the reciprocal property $a_{ji} = 1/a_{ij}$, $a_{ii} = 1$. On solving the problem $Aw = \lambda_{\max} w$ (where $\lambda_{\max}$ is the largest eigenvalue of A), we obtain w, the relative degree to which the elements have the property in question.

We thus derive a hierarchic structure. Within it we derive an index of merit or degree of fuzziness with respect to all the properties taken together. For example, we can choose among several houses to buy by considering all relevant criteria.

The use of tangible and intangible criteria in decision problems removes most of the fuzziness in decision making and provides a way to strengthen rational decision making. There are many potential applications of fuzzy sets in different areas, but as yet there is no coherent theory with rules that one can follow in every situation.

References

1. T. L. SAATY: *The Analytical Hierarchy Process*. McGraw-Hill, New York, 1982.
2. T. L. SAATY: The analytic hierarchy process: a new approach to deal with fuzziness in architecture. *Architectural Science Review*, Vol. 25 (1982), pp. 64–69.
3. R. R. YAGER: An introduction to applications of possibility theory. *Human Systems Management*, Vol. 3 (1983), pp. 246–269.
4. L. A. ZADEN: Fuzzy sets. *Information and Control*, Vol. 8 (1965), pp. 338–353.

T.L.S.

G

GLARE is a form of light that causes visual discomfort and/or a reduction in the ability to see objects, due to an unsuitable distribution of luminance or to extreme luminance contrasts. Two separate forms of glare can be identified:

1. *Discomfort glare,* which causes discomfort without necessarily impairing the vision of objects.
2. *Disability glare,* which impairs the vision of objects without necessarily causing discomfort.

Typical examples of discomfort glare can be experienced in interiors in large rooms when rather bright LUMINAIRES are visible near to the normal line of sight or, under daylight conditions, when looking toward bright windows from the rear of a deep room.

A typical example of disability glare is when at night on an unlit road the headlights of an oncoming car impair a driver's ability to see details of the road. In interiors, disability glare is normally kept within bounds if discomfort glare is adequately controlled.

The sensation of **discomfort glare** is stronger the higher the luminance of the light sources (luminaires or windows), the larger the solid angle subtended by the light sources at the observer's eye, the smaller the angular displacement of the light sources from the observer's line of sight, and the lower the general field luminance controlling the adaptation level of the observer's eye. The interrelation between these parameters is fairly well known as a result of research carried out in several countries.

From the results of this research, the CIE (International Commission on Illumination) and several national illuminating engineering societies have worked out methods for evaluating discomfort glare in interiors. They make it possible to determine the degree of discomfort to be expected for existing situations and to choose luminance distribution, size, and arrangement of the luminaires for interiors to keep the degree of discomfort within a recommended value on a scale giving a precise description of several degrees of discomfort glare (Refs. 1 and 2, Chapter 3). Discomfort glare from windows in DAYLIGHT conditions can be avoided by providing an adequate level of the artificial lighting in the room (Ref. 3, and Ref. 2, Section 2.5).

In working areas, light from bright luminaires or windows is reflected into the eyes of the workers by glossy paper, thick writing in pencil, or the screens of VDUs (visual display units). This produces **reflected glare.** Apart from the discomfort caused by reflected glare, it reduces the contrasts of the task, and it is consequently a kind of **disability glare** (Ref. 4). The proper rendering of contrasts is therefore an important quality criterion of lighting for working areas. Good contrast rendering can be achieved by appropriate arrangement of windows and luminaires with respect to the work station and by adequate light distribution from the luminaires.

References

1. *Discomfort Glare in the Interior Working Environment.* Publications CIE No. 55 (TC-3.4.), Paris, 1983.
2. J. B. de BOER and D. FISCHER: *Interior Lighting.* Philips Technical Library, Eindhoven, Netherlands, 1978.
3. R. G. HOPKINSON and J. LONGMORE: The permanent supplementary lighting of interiors. *Transactions of the Illuminating Engineering Society (London)*, Vol. 24 (1959), p. 121.
4. D. FISCHER: Contrast rendering in office tasks. *International Lighting Review*, Vol. 34, No. 1 (Jan. 1983), p. 9.

J.B.d.B.

GLASS is a hard, transparent, brittle material that is insoluble in water and organic solvents and nonflammable. It is an inorganic fusion product that has been cooled to a rigid state without crystallization.

History. It is popularly supposed that about 3000 years ago Phoenician merchants discovered by accident how to make glass when they used blocks of soda from their ship's cargo on which to set their cooking pots over a fire on the beach.

The first method for making ordinary window glass was developed by the Venetians in the 13th century. It involved gathering molten glass on the end of a blow pipe, which was then blown and spun to make a disk or *crown.* After cooling and annealing, this was cut into small rectangular pieces for glazing windows.

The *blown cylinder process* (1841–1920s) was the next major development in glass manufacturing. Molten glass was blown into a cylindrical shape, which was then cut open and pressed flat. This process produced much larger sheets than could be made by the crown process, but the glass had a poor surface compared with the quality of crown glass.

Better-quality glass was produced by a process in which the molten glass was poured on to iron tables and then ground and polished. The finished product was known as *polished plate glass.*

The advent of the automobile created a need for greater quantities of polished plate glass at lower cost. This led to the development of the *Ford flow process*, in which the molten glass flowed between two rolls forming a ribbon of glass 2.4 m (8 ft) wide for later grinding or polishing. Continuous grinding and polishing was developed in 1930. A twin grinder, which ground both sides of the ribbon at the same time, was introduced by Pilkington in the United Kingdom.

The next major development was the *Pittsburgh flat drawn process*. Large sheets of flat glass were drawn vertically up an annealing tower, about 10 m (33 ft) in height; the sheets were then cut into convenient sizes. The drawn sheet process produced glass with a high-quality finish, but with considerable distortion.

Today, flat glass is produced by the *float process*, which combines the best features of both the polished plate and drawn sheet processes; it has the fire-polished finish and low cost of sheet glass and the distortion-free quality of polished plate glass. In the float process, invented by Pilkington in 1959, molten glass is poured continuously from a furnace onto a bed of molten tin. The molten glass floats on the tin, resulting in a smooth, flat, fire-polished surface.

Constitution and Properties of Glass. Glass for glazing buildings is generally of the soda-lime-silica type. Soda, Na_2O, and sometimes small quantities of potash, K_2O, are added as fluxing agents; they greatly reduce the viscosity of the silica, permitting the use of lower melting temperatures. The lime, CaO, and often small quantities of alumina, Al_2O_3, and magnesium, MgO, are added to improve the chemical durability of the glass. The proportions of the constituents are approximately as follows:

SiO_2	71%–73%
Na_2O	12%–14%
CaO	10%–12%
MgO	1%–4%
Al_2O_3	0.5%–1.5%

Provided the glass is cleaned at reasonable intervals, surface deterioration does not develop under normal conditions of use. The properties of soda-lime-silica glass are as follows:

Density:	Approximately 2560 kg/m^3 (160 lb/ft^3)
Softening point:	Approximately 730°C (1350°F)
Coefficient of linear expansion:	7.5–8.0 × 10^{-6}/K (4.2–4.5 × 10^{-6}/°F)
Specific heat:	830 J/kg · K (0.2 Btu/lb · °F)
Thermal conductivity:	Approximately 1.05 W/mK (0.60 Btu/ft · h · °F)
Refractive index:	Approximately 1.52
Young's modulus of elasticity:	69–74.5 × 10^9 Pa (10,000–10,800 ksi)
Poisson's ratio:	0.22–0.23
Strength (modulus of rupture):	19.3–28.4 MPa (2.8–4.1 ksi) for sustained loading
Coefficient of variation of strength:	0.20–0.25

Basic Glass Products. The bulk of the flat glass used in buildings is *clear glass* having a visible light transmittance ranging from 70% to 92%, depending on thickness. Where it is desired to have reduced solar transmittance, *tinted* or *heat-absorbing glass* can be used. This is made by adding various coloring agents to the molten glass batch. Four colors are made by the float process: blue, bronze, gray, and green. Their visible light transmittance varies from 14% to 83%, depending on the color and thickness.

Rolled glass is made by feeding molten glass from a furnace through rollers to produce a pattern on one or both surfaces. It is available for use in buildings as figured or patterned, wired, and art, opalescent, or cathedral glass.

Processed Glass Products. The following products are processed from the basic types:

1. *Reflective glass* is a clear or tinted glass coated with an extremely thin layer of metal or metallic oxide.
2. *Tempered (toughened) or heat-strengthened glass* is a clear or tinted glass subjected to heating and rapid cooling. This produces compression in the surface of the glass, resulting in greatly increased strength. Glass can also be tempered by a chemical process.
3. *Laminated glass* consists of two or more layers of glass bonded together by interlayers of the plastic polyvinylbutral (PVB).
4. *Insulating glass units* consist of two or more panes of glass that enclose hermetically sealed air spaces.

See also GLAZING.

References

1. R. MacGRATH, A. C. FROST, and H. E. BECKETT: *Glass in Architecture and Decoration.* Architectural Press, London, 1961.
2. E. B. SHAND: *Glass Engineering Handbook.* McGraw-Hill, New York, 1958.

3. D. P. Turner: *Windows and Environment.* McCorquodale, Newton-le-Willows, England, 1969.

I.J.C.

GLAZING is the installation of glass panes into suitable frames in the walls of a building in order to form a window.

History. From the Middle Ages up to the end of the 17th century, only very small panes of crown glass were available and this resulted in the use of leaded lights. In lead-light glazing, specially formed strips of lead, called *cames* (or *calms*) were used to fasten together the small panels of glass, and the assembled panels were then installed in a wooden or wrought-iron frame. As larger sheets of glass became available, the leaded light was replaced by sashes of wood in which the glass was held with wooden beads or putty.

Originally, the architectural conception of a window was a relatively small glazed opening in a thick load-bearing wall that admitted daylight into the building. This was altered, however, by the development of iron-framed buildings in the 19th century and of steel and reinforced concrete framing in the 20th century. In framed structures, the loads of the roof and floors are carried by the columns, and large areas of wall can be either glazed or filled in by other building materials. The availability of larger sizes of glass enabled these large areas to be completely filled with glass. In modern architecture, with the development of CURTAIN WALLS, the exterior cladding is hung from the structure, usually from floor to floor.

In more recent times, ''stopless'' or structural sealant glazing has become increasingly popular. The glass panels are glued to the framing members that make up the curtain wall or window wall, thus achieving a smooth exterior wall surface without protruding framing members.

General Guidelines. A great variety of glazing materials and glazing techniques is available, and only general guidelines for glazing can be given here. The purpose of the glazing system is to retain the glass adequately under the design load, provide effective weathering sealing, prevent loads or pressure points on the glass resulting from building movement, prevent glass-to-metal contact, and minimize glass breakage from mechanical or thermal stress.

Glazing should be carried out in accordance with the specific glazing recommendations of the glass manufacturer and the glazing material manufacturer(s). The chemical compatibility of all glazing materials and framing sealants with each other and with the glass must also be established before commencement of glazing.

In general, the glass should not directly touch the frame, and adequate edge clearance must be provided around all edges and also laterally. The edge clearance used should take into account the size tolerances of the glass and the opening to be glazed, as well as the deflection of the building structure under live and dead loads and thermal expansion and contraction of the glass and the frame.

The glass should always be set on two setting blocks of neoprene (having Shore A hardness of 80 to 90) or another suitable material, located preferably at the quarter-points of the sill. Edge blocking should also be used to limit lateral movement (side-walking) of the glass resulting from horizontal expansion or contraction, building vibration, and other causes. Where the nature of the glazing material is such that displacement may occur as a result of wind loading or other pressures, lateral shims should be used to center the glass within the glazing rabbett (rebate).

Glass glazed into frames is normally retained in the frame either by means of beads or by putty combined with clips or sprigs; sprigs are headless nails or triangular pieces of metal used to retain glass in timber frames. In most cases, the sprig is covered by face putty, and it is only required to retain the glass until the putty hardens. Clips are used to secure the glass in steel frames until the front putty hardens. Glazing beads are strips of metal, wood, or other suitable material used to retain the glass in the frame as an alternative to front putty or where the loading conditions are more severe than front putty glazing can accommodate.

See also FENESTRATION; GLASS; and SEALANTS AND JOINTING.

References

1. R. MacGRATH, A. C. FROST, and H. E. BECKETT: *Glass in Architecture and Decoration.* Architectural Press, London, 1961.
2. *Glazing Manual.* The Glass and Glazing Federation, London, 1978.
3. *Glazing Manual.* Flat Glass Marketing Association, Topeka, Kan., 1980.

I.J.C.

GYPSUM ($CaSO_4 \cdot 2H_2O$) is calcium sulfate dihydrate. This nonmetallic mineral was mostly formed 300 million years ago during the Paleozoic Era by precipitation from inland seas.

Properties. Gypsum is a monoclinic off-white to gray or pink crystalline solid, which can be easily scratched (Mohs hardness scale of 1.5–2.0). It is slightly soluble in water, has a specific gravity of 2.31 to 2.33, and loses water of crystallization at 128°C to form plaster of Paris (calcium sulfate hemihydrate).

Varieties. Alabaster is a compact, fine-grained rock gypsum prized by sculptors for its pleasing appearance and the ease with which it can be carved and polished. Satin spar is translucent fibrous gypsum, and selenite forms as clear crystals. Rock gypsums are dull colored and are often contaminated with clay, calcium carbonate, and silica.

Occurrence and Manufacture. Gypsum deposits are located in North America, South America, Europe, Asia, Africa, and Australia. During 1982, gypsum production in the United States and Canada was 20 million tons, and the world total was 82 million tons. It is mined by both open-pit and underground operations.

Uses. Most uses for gypsum are based on the unique ability of calcium sulfate to give up water or take in water of crystallization. When calcined (heated in the range 128° to 163°C), gypsum is converted to stucco (plaster of Paris), which, when mixed to a paste with water, hardens to reform gypsum, the least soluble hydrate.

Approximately 30% of all the gypsum produced in the United States is used either as a retarder for portland cement, a soil conditioner, or as mineral filler; the remaining 70% is calcined to stucco for the manufacture of gypsum wallboard and specialty plasters (see PLASTERING).

Gypsum Wallboard is the panel generally used in buildings as a lining material for walls, ceilings, and partitions to provide a surface suitable for decoration. It is a factory-made board consisting of a noncombustible gypsum core encased in a heavy natural finish paper on the face side and a strong liner paper on the back. The face paper is folded around the long edges to reinforce and protect the core.

References

1. S. J. LEFOND (ed.): *Industrial Minerals and Rocks.* Port City Press, Baltimore, Md., 1975.
2. W. C. RIDDELL: Physical properties of calcined gypsum. *Rock Products*, Vol. 53, No. 5 (May 1950), pp. 68–70.
3. K. SCHILLER: The course of hydration: its practical importance and theoretical interpretation. *Journal of Applied Chemistry and Biotechnology*, Vol. 24 (1975), pp. 379–385.

G.A.B.

H

HEAT ENGINES AND HEAT PUMPS. Heat engines extract mechanical work from a high-temperature energy source and discharge it to a lower-temperature energy sink. This is analogous to a turbine that produces mechanical work as water flows through it en route from a high-level reservoir to a lower-level reservoir.

A **heat pump** uses mechanical work to maintain a temperature difference between a low-temperature energy source and a higher-temperature energy sink during their interaction. A pump that transfers water back to the high-level reservoir from a lower level is analogous to a heat pump.

Common examples of heat engines are the gasoline or petrol engine (Otto cycle), the oil engine (Diesel cycle), the gas engine (dual or Otto cycle), and the steam boiler/turbine/condenser engine in a power station (Rankine cycle).

The domestic refrigerator and air conditioner (vapor compression cycle) are common examples of heat pumps (see REFRIGERATION CYCLES).

Background Thermodynamics. It is useful to retain the concept of the heat engine or heat pump operating between a "hot" and a "cold" reservoir (Fig. 111). The fact that two reservoirs of different temperatures must be involved is a consequence of the *second law of thermodynamics* (Ref. 1).

A heat pump operating between a cold reservoir and an ambient temperature hot reservoir is called a *refrigerator*. If it operates to maintain an above-ambient-temperature hot reservoir, it is called a heat pump. A *reverse-cycle heat pump* can be used in either of these modes by changing the reservoir role of the conditioned space.

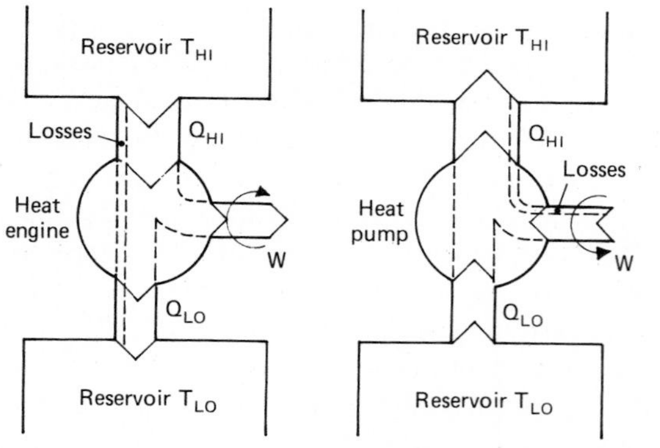

Figure 111 Energy (work and heat) flows for a heat engine and a heat pump.

The *first law of thermodynamics* tells us that energy is conserved, and we note that *work* and *heat* are both forms of energy. Thus, from Fig. 111, for both the heat engine and the heat pump,

$$Q_{HI} = Q_{LO} + W \tag{1}$$

where Q_{HI} and Q_{LO} are the quantities of heat transferred from the engine or pump to the higher and the lower reservoirs, respectively, and W is the quantity of mechanical work.

In any real system, friction and spurious internal and external heat transfers degrade the operation of the system from the ideal. For the heat engine,

$$Q_{HI} = Q_{HI,\,useful} + Q_{HI,\,losses}$$

and

$$Q_{LO} = Q_{LO,\,useful} + Q_{LO,\,losses}$$

Only the useful parts of Q_{HI} and Q_{LO} contribute to the generation of the work, W.

Similarly, for the heat pump, $W = W_{useful} + W_{losses}$. W_{losses} contributes to $Q_{HI,\,losses}$, and only W_{useful} and $Q_{HI,\,useful}$ contribute to the uptake of energy, Q_{LO}, from the low temperature reservoir, T_{LO}.

Carnot Cycle. When considering whether a heat engine or a heat pump system may be feasible in a given situation, it is useful to treat the devices initially as ideal and to assume that they operate on the most efficient possible *cycle* (repetitive sequence of expansions, heat exchanges, and compressions of a working fluid). If the system is not feasible or is only marginally feasible with such assumptions, it is certain that it will be infeasible in practice.

The most efficient *cycle* possible consists of an isothermal heat exchange (i.e., a heat exchange at a constant temperature) producing an isothermal expansion, an ideal adiabatic (no heat transfer) expansion, a second isothermal heat exchange causing an isothermal compression, and an ideal adiabatic compression back to the original state of the working fluid. The cycle is called the *Carnot cycle* in honor of Sadi Carnot (1796–1831), who showed in 1824 that the efficiency, η, with which heat can be converted to work in a heat engine is

$$\eta = \frac{W}{Q_{HI}} < \frac{T_{HI} - T_{LO}}{T_{HI}} = \eta_{Carnot} \tag{2}$$

where T_{HI} and T_{LO} are the temperatures of the hotter and the colder reservoirs, and η_{Carnot} is the efficiency of the ideal Carnot cycle, which is unattainable in practice.

The **coefficient of performance (COP)** is the inverse of the efficiency, and its use is convenient when considering a refrigerator. An energy transfer Q_{LO} is required during the cooling process, and a work input W is the price that must be paid. Thus

$$(\mathrm{COP})_{cool} = \frac{Q_{LO}}{W} < \frac{T_{LO}}{T_{HI} - T_{LO}} = (\mathrm{COP})_{Carnot} \tag{3}$$

For heat pumps working at above-ambient temperature it is more common for the COP to be referred to the hot reservoir, Q_{HI}. It is then called the *coefficient of performance for heating*.

$$(\mathrm{COP})_{heat} = \frac{Q_{HI}}{W} < \frac{T_{HI}}{T_{HI} - T_{LO}} = (\mathrm{COP})_{heat,\,Carnot} \tag{4}$$

As a consequence of the *first law of thermodynamics*,

$$(\mathrm{COP})_{heat} - (\mathrm{COP})_{cool} = 1 \tag{5}$$

The higher the temperature of the cold reservoir, the higher will be the temperature at which the hot reservoir can be maintained for a given work input. Thus, when using a heat pump for heating a space, it is possible to improve $(\mathrm{COP})_{heat}$ by augmenting the cold-reservoir temperature through recovery of waste heat or collection and storage of solar energy.

The **Chemical Heat Pump** is an alternative to the mechanical heat pump. The most common type is the *absorption cycle* heat pump, which may be used in an absorption refrigerator or an absorption heat pump. Both require two fluids. The primary fluid is evaporated by extracting heat from a cold reservoir. This heat is exothermally absorbed by the secondary fluid in the liquid phase. The pressure is increased through the solution pump before the mixture is boiled so that the primary fluid is vaporized before passing to a condenser in which heat is rejected to condense the primary vapor. The condensate is expanded through a throttle, cooled through partial evaporation, and returned to the evaporator for the next cycle. The system is shown schematically in Fig. 112, from which the internal heat recovery/exchange circuits, which serve to increase the COP, have been omitted for clarity.

The coefficient of performance (COP) is defined, for refrigeration of the cold reservoir, T_{LO}, as

$$(\mathrm{COP})_{cool} = \frac{Q_{LO}}{Q_{HI}} \tag{6}$$

Strictly, the work input from the pump to raise the pressure from the absorber to the generator should be added to Q_{HI}, but is usually omitted. For heating of a space through utilization of $Q_{condenser}$ (which includes $Q_{absorber}$), the COP for heating is used:

$$(\mathrm{COP})_{heat} = \frac{\text{heat released from condenser}}{\text{heat absorbed by generator}} \tag{7}$$

The maximum possible efficiency is still given by the

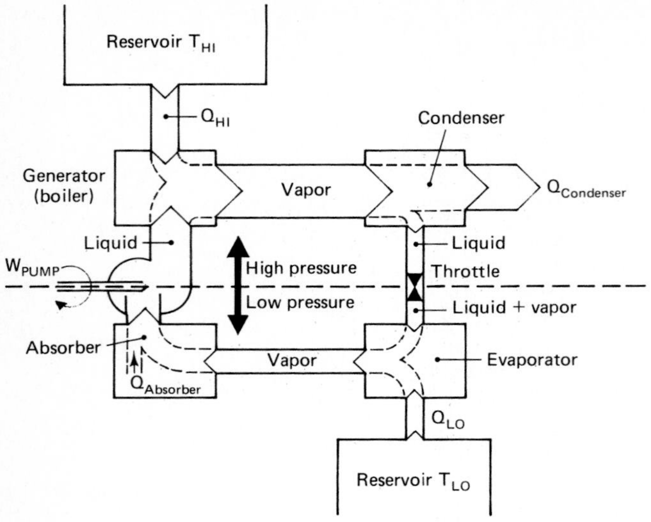

Figure 112 Chemical (absorption cycle) heat pump.

Carnot relationship, but the expression is more complex because two working fluids are involved. Thus

$$(\text{COP})_{\text{Carnot}} = \frac{T_{\text{evaporator}}}{T_{\text{condenser}} - T_{\text{evaporator}}} \times \frac{T_{\text{generator}} - T_{\text{absorber}}}{T_{\text{generator}}}$$

While many fluid pairs can be used in absorption systems, including hydrocarbons and fluorocarbons, the most common pairs are ammonia/water and lithium bromide/water. Several examples of ammonia/water systems are commercially available, and the Tokyo Sanyo Electric Company Ltd. manufactures a gas-fired industrial-scale LiBr/water heat-pump unit. More information on chemical heat pumps, including consideration of operating and maintenance problems, such as leakages, may be found in Ref. 2.

Location of Heat Engines and Heat Pumps in a network of heat exchange processes requires careful thought. Incorrect location can result in high-grade energy (electricity or fuel) simply being converted into waste heat discharged to the ambient reservoir. With proper design, however, the integration of heat engines or heat pumps can lead to very substantial increases in overall efficiency (Ref. 3, pp. 88–96).

Applications to Building. Heat engines are commonly used for regular or standby electricity generation or for driving refrigeration compressors. As discussed previously, all heat engines must reject heat to a reservoir. There is considerable potential for recovering this heat, say, from the cooling jacket water and the exhaust pipe of a diesel engine for use within a building or process, for example, for space or water heating, for drying (perhaps with the aid of a heat pump) in a hospital laundry, or for powering an absorption-cycle refrigerator or heat pump. Integrated systems of this type are known variously as *cogeneration, combined heat and power*, or *total energy* systems. The International Energy Agency has assigned a very high priority to such systems as a means of reducing the consumption of primary fuels.

Heat Pump Design Data. The primary design decision for the heat pump manufacturer is the choice of working fluid. Considerable attention has been paid recently to the establishment and tabulation of properties of candidate fluids (Ref. 4); supplementary data appear regularly in the *Journal of Heat Recovery Systems* (Pergamon Press). Design methods for heat pumps are clearly summarized in Refs. 5 and 6.

Equipment Suppliers. Many suppliers of heat engines now offer turnkey units that combine electricity generation in parallel with the utility grid with collection of waste heat for direct connection to a building or process heating system.

Suppliers of heat pump systems and components in North America are listed in *U.S. Heat Pump Research and Development*, U.S. Department of Energy Report DOE/CE-0035, 1983.

In the United Kingdom, the Marketing Department of the Electricity Council has published a comprehensive list of suppliers of packaged heat pump systems for use in buildings in *Heat Pumps and Air Conditioning—A Guide to Packaged Systems*, third edition, 1982.

A solar-boosted heat pump system is available in Australia from Siddons Industries Pty. Ltd.

See also AIR CONDITIONING AND HEATING; and REFRIGERATION CYCLES.

References

1. J. B. FENN: *Engines, Energy, and Entropy.* W. H. Freeman, San Francisco, 1982.
2. T. HEPPENSTALL: Absorption cycle heat pumps. *Journal of Heat Recovery Systems*, Vol. 3, No. 2 (1983), pp. 115–128.
3. *User Guide on Process Integration for the Efficient Use of Energy.* Institution of Chemical Engineers, Rugby, England, 1982.
4. F. A. HOLLAND, F. A. WATSON, and S. DEVOTTA: *Thermodynamic Design Data for Heat Pump Systems—A Comprehensive Data Base and Design Manual.* Pergamon Press, Oxford and Elmsford, N.Y., 1982.
5. D. A. REAY and D. B. A. MACMICHAEL: *Heat Pumps:*

Design and Application—A Practical Handbook for Plant Managers, Engineers, Architects and Designers. Pergamon Press, Oxford and Elmsford, N.Y., 1979.

6. R. D. HEAP: *Heat Pumps*, Spon, London, 1979, or Halstead Press, New York, 1979.

R.E.L.

A **HELIODON** or **SOLARSCOPE** is a machine for modeling the daily and annual motion of the sun at any latitude for the purpose of studying sunlight penetration into a building and for predicting the shadows it throws on its surroundings. There are two types. Figure 113 shows a heliodon with a fixed horizontal table on which the building model is placed, oriented correctly in relation to the north. The sun is modeled by a mirror reflecting light from the lamp behind the model. The position of the mirror and lamp are adjusted by three screws; one is set to the week of the year, the second to the time of day, and the third to the latitude of the place.

This type of heliodon is direct reading, but its accuracy is limited by the length of the arm on which the mirror is mounted because the rays from the lamp diverge slightly, whereas solar rays are parallel.

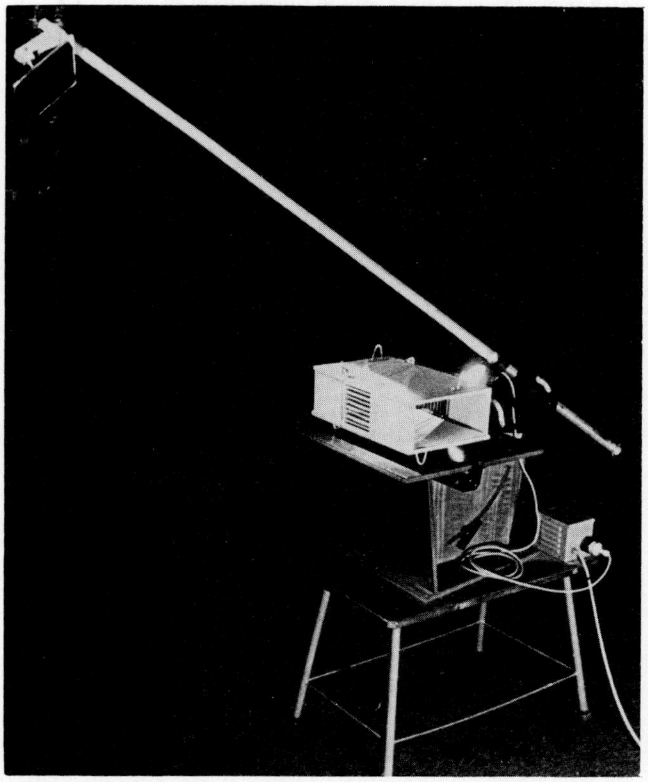

Figure 113 Heliodon with fixed model table and movable light source.

The second type of heliodon consists of a table that can be rotated about a horizontal axis to model the altitude of the sun. The azimuth (horizontal movement) of the sun is modeled by rotating the model on the table. The horizontally fixed lamp can be placed a long distance from the model, if space permits, so that the result is more accurate. However, the instrument is not direct reading, and the azimuth and altitude of the sun must be calculated for each time of day and year that is to be investigated.

If a table of a transparent material, such as Perspex, is used, it becomes possible to study sunlight penetration through the transparent base.

References

1. O. H. KOENIGSBERGER, T. G. INGERSOLL, A. MAYHEW, and S. V. SZOKOLAY: *Manual of Tropical Housing and Building. Part 1: Climatic Design.* Longman, London, 1974.
2. H. J. COWAN and P. R. SMITH: *Environmental Systems.* Van Nostrand Reinhold, New York, 1983.

H.J.C.

HELMHOLTZ RESONATORS are enclosed volumes of gas connected to external gas by small openings or passages. A bottle is the simplest form of Helmholtz resonator, with air in the neck acting like a piston on the enclosed air acting like a spring. In 1868, Helmholtz postulated that such volume/opening combinations would resonate at a specific frequency. The resulting pressure fluctuations near the opening reduce the level of external sound at this resonant frequency.

Simple formulas for calculating resonator/neck dimensions appear in many texts, but Alster (Ref. 1) shows that more complex formulas are needed to achieve greater accuracy. Van Itterbeek et al. (Ref. 2) show the effects of neck configurations; but, while complex calculations are needed in critical situations, many practical problems are better solved by simple calculations and in situ neck adjustments to obtain maximum effect.

Helmholtz resonators are often used to reduce low-frequency tonal noise where other treatments are less efficient. For very low frequencies, the resonators can become quite large, and sometimes spaces available incidentally can be converted into resonators at little cost. Similarly, tonal sources within enclosures can be enhanced if the volume/neck combination is resonant at a suitable frequency.

Sound-absorbing panels can be formed by perforated sheets of plywood or particle board and a backing of mineral wool. The space behind each perforation forms a Helmholtz resonator.

References

1. M. ALSTER: Improved calculation of resonant frequencies of Helmholtz resonators, *Journal of Sound and Vibration*, Vol. 24 (1972), pp. 63–85.
2. A. VAN ITTERBEEK, H. VAN ENGELEN, and H. MYNCKE: Influence of the shape of the neck of the Helmholtz resonator on its absorbing properties. *Acustica*, Vol. 14 (1964), pp. 213–215.

J.R.

HUMAN BEHAVIOR IN FIRES has been studied in order to provide guidelines for the EVACUATION OF BUILDINGS IN THE EVENT OF FIRE and the design of STAIRS and fire warning and sensing systems. The results of these studies also help to improve training for emergencies and to provide effective educational material on how to prevent fires and how to act if in a building on fire.

Panic a Misleading Concept. The need for detailed studies of what people do in buildings on fire is necessary because of the imprecise and sometimes misleading views that are presented in newspapers, derived from informal investigations looking at the *results* of fires. These views have, until recently, had a great impact on the formulation of building regulations that relate to fire prevention and evacuation in emergencies. This is unfortunate because the views are mistakenly based on the idea that all that people do when faced with a severe life-threatening situation is to panic.

Many studies have shown that not only is the concept of panic a confused and ambiguous one, but that most definitions of it do not lead to descriptions of human actions for which firm evidence can be found. Researchers have therefore attempted to develop clear descriptions of what people actually do in buildings on fire.

Empirical Research. Information on human actions in fires is usually collected by detailed interviews and questionnaire surveys with people who have experienced fires, although a few researchers have used laboratory and computer simulations. These studies have included detailed examinations of major conflagrations, such as the Beverly Hills Supper Club fire, which occurred in Kentucky in 1977, as well as much smaller scale domestic fires in which there were no injuries or fatalities.

A General Model. Out of all these studies a general model (Fig. 114) has emerged that many researchers now believe is a reasonable approximation of what people typically do when they experience a fire in a building. This model is based on the assumption that people usually attempt to cope as effectively as possible with what is a threat-

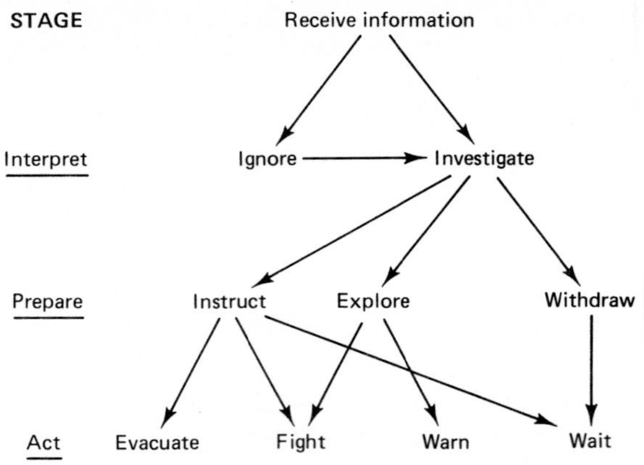

Figure 114 Model of human behavior in fire.

ening, ambiguous, rapidly changing situation. Their actions are thus often based on inadequate or confusing information. Thus fatalities are frequently the product of misinterpretation of what is happening and delay in responding to early indications of the presence of a fire.

The model represents people going through three broad stages in responding to the events associated with a fire. The first of these is an attempt to interpret what is happening. At this stage people frequently ignore early ambiguous cues, thus introducing potentially dangerous delay in their responses. Once they accept the likelihood of a fire existing, it is still common for people to investigate the source of the cues to satisfy themselves that it is not a false alarm. This can introduce further delays and directly expose people to danger. Because of the exponential growth of fires, these early delays can have effects that people caught in a fire do not appreciate. Thus preparations for dealing with the fire and actions relating to fire fighting or leaving are less effective than would otherwise have been the case.

Consequences for Fire Protection. Research has shown that the stages of action in a fire, if proceeded through quickly and effectively, can lead to safe evacuation of a building. However, if delays are introduced and people do not know what is occurring, conventional fire-protection hardware may not be sufficient. This case is emphasized by a tendency for people to act in ways that are normal for the building they are in, for example, leaving in a fire the way they would normally leave the building. Under-use of FIRE FIGHTING EQUIPMENT and special fire-escape routes is also a common finding.

Informative Fire-Warning Systems. Recognizing the importance of accurate information in the early stages of a fire and the ambiguity of many existing fire alarms, attempts are now under way to develop fire-warning systems that do give information about the fire, its location, and progress. Reconsideration of training procedures to take account of the natural stages in response to a fire is also being considered.

Relevance to Other Disasters. The findings from studies of human behavior in fires are also of relevance for other dangerous events such as bombing, flood, and EARTHQUAKES.

References

1. D. CANTER (ed.): *Fires and Human Behavior*, Wiley, Chichester, England, 1980.
2. D. CANTER: *Studies of Human Behavior in Fire*. Building Research Establishment, Watford, England, 1984.

D.C.

HURRICANE-RESISTANT DESIGN is employed in areas of the world that are subject to the effects of severe tropical cyclones (Fig. 115). Buildings require special attention if the effects of these storms are to be resisted.

Effects. Extreme winds, storm surges, and heavy rainfall may accompany tropical cyclones (Ref. 1). Winds that may exceed 60 m/s (135 mph) can persist for an hour or more as the storm passes a building site. These winds are often accompanied by windborne debris. Low-lying areas can be affected by storm surges, which can reach 8 m (26 ft) above sea level. Buildings at the coastline may also experience scour from wave action and battering from waves and wave-carried objects. Rain may cause flooding at the building site, and wind-driven rain can penetrate building envelopes, causing water damage to contents.

Designing for Wind. Buildings are usually designed for a hurricane-related wind speed that is expected to occur at the building site every 50 to 100 years (Ref. 2). Where windborne debris is a potential hazard, building designs must also consider the effects of missile impact, which can open the building to the effects of internal pressure (Ref. 3). Hurricane-resistant housing features special roof-to-wall and wall-to-foundation anchorages, shear-resistant walls, and missile impact resistant walls and windows (Ref. 4 and Fig. 116). Industrial and commercial buildings should contain wind-resistant doors and debris impact protection for windows. High-rise buildings should contain window glass that is resistant to high suction pressures and impact from small missiles such as roof gravel, where these hazards exist.

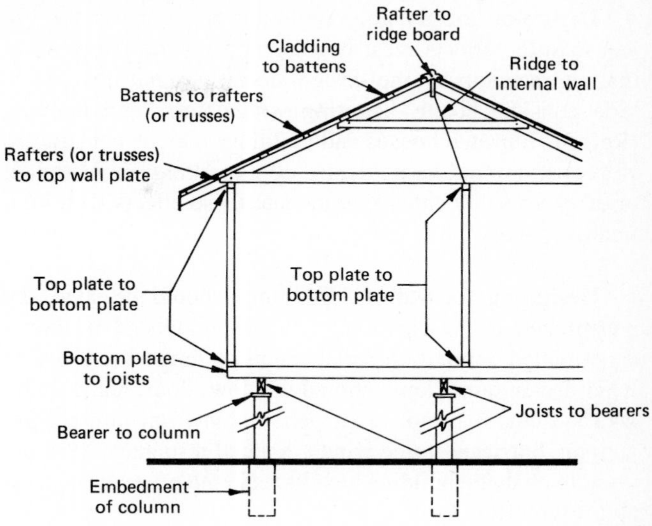

Figure 116 Hurricane-resistant houses contain connections between components. A chain of positive connections carries wind forces from roof, through walls, to floor and foundation.

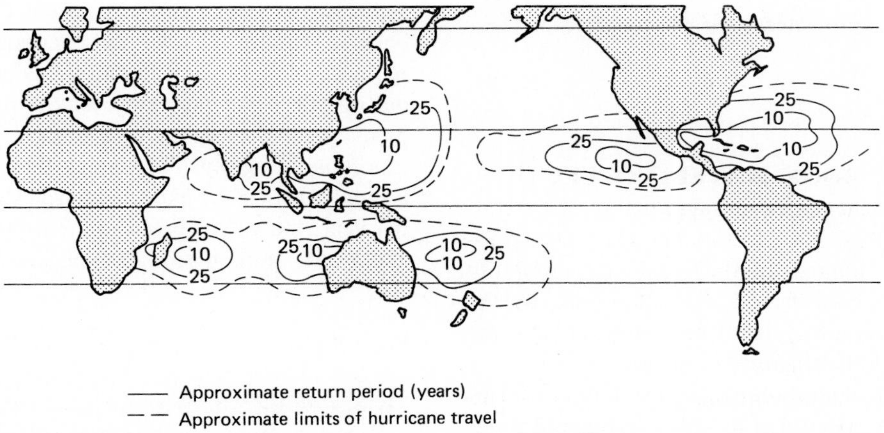

Figure 115 Areas of the world subject to hurricane-force winds. Contours indicate approximate return periods in years for winds exceeding 33 m/s (74 mph).

Designing for Surge. Buildings in coastal areas subject to surge effects must be designed for the rising water that can accompany the design storm. Special design criteria consider scour, battering with debris, and flooding (Ref. 5). Elevated houses and buildings may be constructed with open ground floors dedicated to parking, recreational, or other uses that can accommodate these effects with minimal damage.

Designing for Rain. Buildings should be sited and constructed in anticipation of flooding induced by hurricane-related rainfall. Special attention must be given to wind-driven rain. Conventional window, door, and curtain-wall systems may not resist penetration from rain accompanying hurricane-force winds. Special tests have been devised to evaluate building envelopes in the presence of wind and water (Ref. 6).

See also CLIMATE AND BUILDING and WIND LOADS.

References

1. R. H. SIMPSON and H. RIEHL: *The Hurricane and Its Impact.* Louisiana State University Press, Baton Rouge, 1981.
2. E. SIMIU and R. H. SCANLAN: *Wind Effects on Structures.* Wiley, New York, 1978.
3. J. E. MINOR: Preparing buildings for hurricane winds. *Postprints, Third Conference on Meteorology of the Coastal Zone*, American Meteorological Society, Boston, 1984.
4. *Home Building Code, Queensland*, Appendix 4 to the Standard Building By-Laws. State of Queensland, Brisbane, Australia, 1975.
5. *Model Minimum Hurricane-Resistant Building Standards for the Texas Gulf Coast.* Texas Coastal and Marine Council, Austin, 1976.
6. *Standard Test Method for Water Penetration of Exterior Windows, Curtain Walls, and Doors by Uniform Static Air Pressure Difference, ASTM E331-83.* American Society for Testing and Materials, Philadelphia, 1983.

J.E.M.

HYDROLOGY OF ROOF WATER COLLECTION.

The design of rainwater collection systems involves consideration of six factors. Two factors are fixed by the locality: the average annual rainfall and its variability throughout the year. Four factors are determined by the design: the demand on the system, the collection area, the storage volume, and the reliability of supply.

For domestic rainwater systems, it is common to start the design process by specifying the daily household water usage. Most households reliant on rainwater use about half as much water as the equivalent family on a reticulated supply. The minimum or basic catchment area may then be calculated from

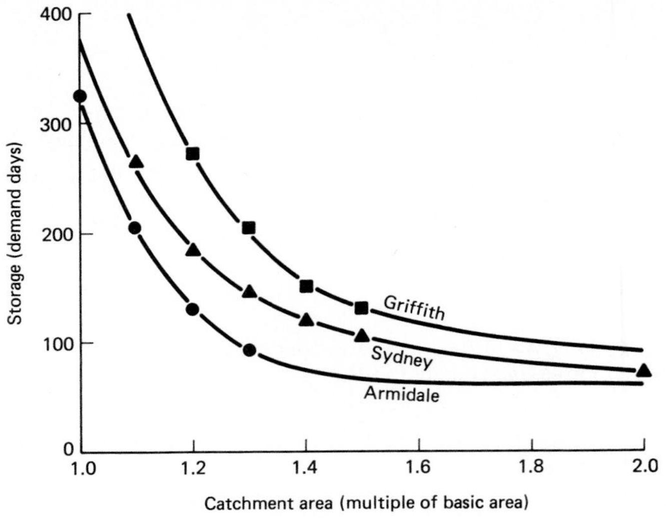

Figure 117 Relation between storage size and catchment area.

$$A = 365 \frac{D}{R}$$

where A = collection area (m^2)
D = daily demand (liters)
R = average annual rainfall (mm)

Note that this is a *basic minimum* area. The storage size required is a function of the variability of the rainfall throughout the year, the reliability of supply, and the actual catchment area available. Figure 117 shows combinations of catchment area (expressed as multiples of the basic area) and storage size (expressed as days of demand) for various locations in eastern Australia. There is a trade-off between catchment and storage. The final choice of the most suitable combination is dictated by economic factors.

Figure 117 relates specifically to conditions in which a shortage of water can be expected to occur, on average, once in 5 years. Under those conditions, most users of rainwater systems start to ration themselves before they run out or else make provision for the importation of supplementary supplies. If the household is prepared to tolerate a less reliable supply, the storage size may be reduced. For a supply that can be expected to run short, on average, once per year, the values given in the figure may be reduced by about 35% in high rainfall areas and 50% in arid areas.

References

1. S. J. PERRENS: Design strategy for domestic rainwater systems in Australia. *Proceedings of the International Conference on Rainwater Cistern Systems, University of Hawaii*, Honolulu, 1982, pp. 108–117.

2. P. J. HOEY and S. F. WEST: Recent initiatives in raintank supply systems for South Australia. *Proceedings of the International Conference on Rainwater Cistern Systems, University of Hawaii*, Honolulu, 1982, pp. 284–293.

S.J.P.

HYPOCAUST (Greek for "place heated from below"). In the 1st century B.C. the Romans invented an ingenious heating system for the hot rooms of their baths. It consisted of a tile floor raised on short brick pillars (Fig. 118) heated by smoke and hot air from a slow-burning furnace. The gases passed under the floor and were exhausted by a chimney at the far end. Hypocausts were later also used to heat houses in countries with colder climates. Remains of this heating system have been found in the ruins of many baths and some country houses.

After the fall of the West Roman Empire, hypocausts ceased to be used in Western Europe, but they continued to be employed in baths in Byzantine and Islamic buildings. They were independently invented by the Koreans and the Chinese, where they are still used.

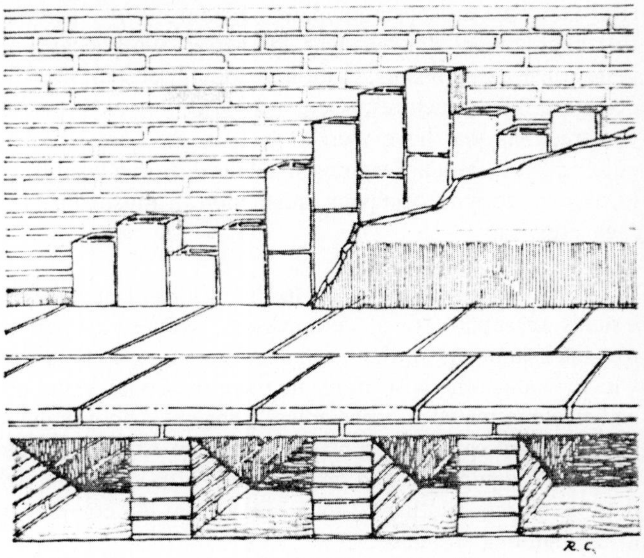

Figure 118 Hypocaust floor. The hot gases circulate between the brick pillars to heat the *bipedales* (two-foot tiles) that form the suspended floor.

References

1. MARCUS VITRUVIUS POLLIO (transl., M. MORGAN): *The Ten Books of Architecture.* Dover, New York, 1960.

2. H. PLOMMER: *Vitruvius and Later Roman Building Manuals.* Cambridge University Press, London, 1973.

H.J.C.

I

IMPACT AND VIBRATION ISOLATION have to do with attenuating the *structure-borne* sound in a building. The most common sources are footfalls from the room above or in abutting staircases, slamming of doors, and vibration of the heating and ventilation equipment for the building. This sound travels within the structure; thus the structure must be carefully detailed to provide effective breaks in the sound paths and to avoid bridging over any resilient vibration isolation elements.

Impact Noise refers to structure-borne noise radiated into a room when the overhead floor/ceiling structure is excited into vibration by impacts in the rooms above, either by footfalls, the scraping of furniture across the floor, or the dropping of solid objects. The problem is especially severe in kitchens and bathrooms, where there is often no floor covering to soften the impact. In high-rise apartment and office buildings with concrete floors, the provision of carpet and pad assures an effective solution for the impact noise (Ref. 1).

Wood-joist floors present a problem, because even when carpeted to attenuate heel impacts, there is a thump as the wooden structure is set into resonance by the body weight of the walker.

Concrete floors that *cannot* be carpeted must be provided with a floating slab in the upper room supported on resilient material such as a glass fiber blanket or neoprene pucks to decouple the impact vibrations in the upper slab from the lower slab, from which they would otherwise radiate as airborne noise into the room below.

Impact Test Method. There is not yet a fully acceptable test, although a standard method has been formally adopted, in which a standardized hammer machine administers a continuous series of standard impacts to the test floor while the sound level is measured in the room below

(Ref. 2). This has been a subject of controversy since the first international standard was adopted in 1960 because it yields misleading results. A more prudent procedure for the design of floor structures to attenuate impact noise is to use constructions that have worked well in the past, not according to the standard tapping-machine data, but according to experience with occupant satisfaction in finished buildings (Ref. 3).

Vibration Isolation refers to the provision of resilient mounts to support fans, compressors, pumps, and other rotating equipment in the mechanical rooms of a building, which would otherwise transmit their rotational vibration and create intrusive airborne noise. Hangers or floor mounts utilizing steel springs or rubber-in-shear are attached at one end to the equipment and at the other to the building, so there is no direct, ''hard'' sound path from the equipment to the building structure.

The amount of vibration isolation achieved depends on the relation between the excitation frequency of the machine and the resonance frequency of the system comprising the mass of the machine and the resilience of the mount. The basic rule is to select sufficiently resilient mounts so that the resonance frequency for the mass–spring system lies well below the rotational frequency of the machine. This rule often leads to springs with large deflections, and special means to maintain stability are needed during start-up and shutdown. If a number of machines must be vibration isolated in one room, it is often more economical to mount them on a single concrete block and isolate this block.

By comparison with the state of impact noise isolation, the field of mechanical vibrational isolation is well developed and a wide variety of suitable isolation devices is readily available.

See also SOUND INSULATION AND ISOLATION and VIBRATIONS IN TALL BUILDINGS.

References

1. JOHN H. CALLENDER (ed.): *Time-Saver Standards for Architectural Design Data*. McGraw-Hill, New York, 1982; chapter on Acoustics by R. B. NEWMAN.
2. *Impact Sound Transmission through Floor–Ceiling Assemblies Using the Tapping Machine, Laboratory Measurement of—ASTM E 492-77*. American Society for Testing and Materials, Philadelphia, 1977. The corresponding international standard is *Acoustics—Measurements of Sound Insulation in Buildings and of Building Elements—Laboratory Measurements of Impact Sound Insulation of Floors, Part VI—ISO R 140*. International Organization for Standardization, Geneva, 1978.
3. THEODORE J. SCHULTZ: Alternative test method for evaluating impact noise, *Journal of the Acoustical Society of America*, Vol. 60, No. 3 (1976), pp. 645–655.
4. *ASHRAE Handbook*, chapter on *Sound Control*. Published periodically by the American Society of Heating, Refrigerating, and Air Conditioning Engineers, Atlanta, Ga.

T.J.S.

INCANDESCENT LAMPS produce light by the electrical heating of a tungsten wire or filament to such a high temperature that radiation in the visible region of the spectrum is emitted. They form the oldest lamp family and remain the mainstay of home lighting, basically because they are easy to work with and cheap. Often an incandescent system is chosen because it is easy and reliable to dim.

The relatively short life and low efficacy (lumens emitted per watt of power consumed) of the incandescent lamp are the outcome of evaporation of the tungsten filament caused by its high operating temperature. Color rendering of incandescent lamps is excellent.

For general lighting purposes, screw and bayonet caps are available, which are identified by the letters E (Edison) and B (bayonet), followed by a figure indicating the diameter of the cap in millimeters.

General Lighting Service Lamps (GLS) are intended for general lighting use. They are made in sizes from 10 to 1500 W and are available in a range of bulb finishes (e.g., clear, inside frosted, silica coated) and bulb shapes (e.g., standard, candle, mushroom, tubular, globe). See Fig. 119.

Reflector Lamps have a reflecting metal coating directly applied to part of the inside bulb surface. This internal reflector is thus not subjected to damage, corrosion, or contamination. Consequently, cleaning costs are avoided

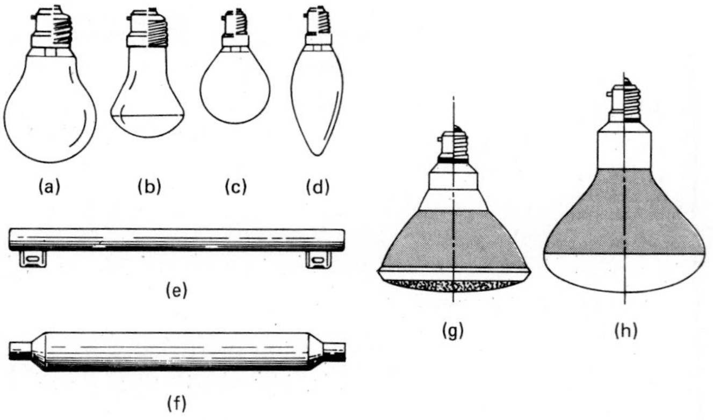

Figure 119 Incandescent lamps, typical shapes. (a) Standard. (b) Mushroom. (c) Luster. (d) Candle. (e) Linear decorative. (f) Show window. (g) Pressed-glass reflector lamp. (h) Blown-bulb reflector lamp.

and high output is maintained throughout the life of the lamp.

Pressed-glass reflector lamps are molded in tough, heat-resistant glass to achieve parabolic reflectors (PAR lamps). Their front element, or lens, is patterned to give a variety of beam width (e.g., spot, flood, wide flood).

Blown-bulb reflector lamps are available in flood, spot, and colored versions. The fact that small low-wattage versions are also available makes these lamps ideal for a wide range of indoor applications.

Tungsten–Halogen Lamps have a halogen (e.g., iodine, chlorine, bromine) added to the normal gas filling and work on the principle of a halogen regenerative cycle, which removes the evaporated tungsten from the bulb and redeposits it on the filament. The halogen cycle allows tungsten–halogen lamps to be designed for higher efficacy and/or longer life than normal incandescent lamps of the same wattage. Due to their compactness, halogen lamps approach the ideal point source needed for good optical control.

See also DISCHARGE LAMPS.

References

1. *IES Lighting Handbook*, Reference Volume. Illuminating Engineering Society of North America, New York, 1981.
2. J. B. de BOER and D. FISCHER: *Interior Lichting*. Kluwer Technische Boeken B.V., Deventer, The Netherlands, 1981.

D.F.

IRON ranks second in abundance among the metals on earth, but the extraction of the metal from its ores is harder than for copper, which consequently was used for tools at an earlier time in history. There are three main forms of iron.

Wrought Iron, which is almost pure iron, melts at a temperature unattainable before the 19th century. It is ductile at ordinary temperatures and easily forged when hot.

Steel, which contains 0.1% to 1.7% of carbon, has a lower melting point and is relatively soft when cooled slowly; but the higher-carbon steels become extremely hard when quenched, that is, cooled by plunging the metal into cold water (see METALLURGY). It can be forged at red heat.

Cast iron, which has a carbon content of 1.8% to 4.5%, is easily melted and cast into complicated shapes. It is a brittle material with little ductility (see TESTING OF MATERIALS).

Prior to the 19th century, steel was produced indirectly from wrought iron or cast iron, and it was consequently a costly material. However, wrought iron and cast iron are today little used in buildings, because STEEL is a better material for most of the purposes for which they were formerly employed.

Reference

A. K. OSBORNE (ed.): *An Encyclopaedia of the Iron and Steel Industry*. Technical Press, London, 1967.

H.J.C.

J

JOINTS FOR STEEL STRUCTURES. A steel structure is composed of an assemblage of structural members. Connections (or joints) provide structural continuity and load transfer to the foundations. Structural connections must be economical, constructable, and adequately proportioned to transfer forces.

Structural Joints. Since force transfer is critical to connection performance, identification of force types is pertinent. The forces associated with structural joints are shear, tension, and compression: these forces also produce moments. A combination of some or all of these may be occurring simultaneously on a joint.

Connections may be classified according to their rigidity and force transfer. A *rigid* connection develops the member

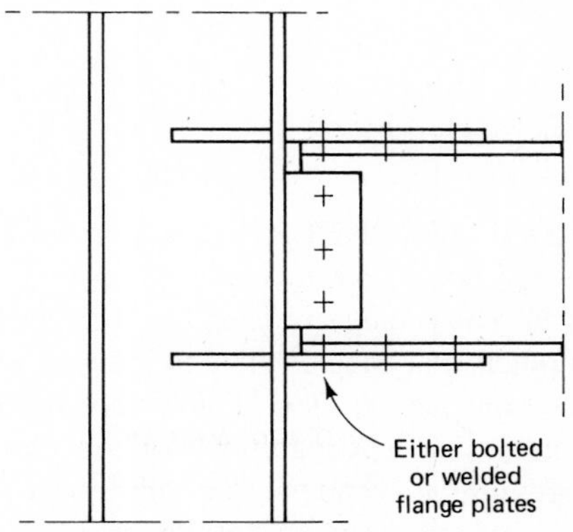

Figure 120 Flange plate connection. Either bolted or welded flange plates may be used.

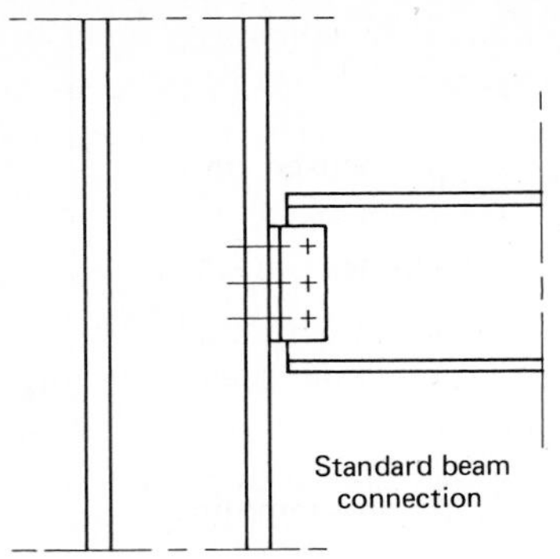

Figure 121 Standard beam connection.

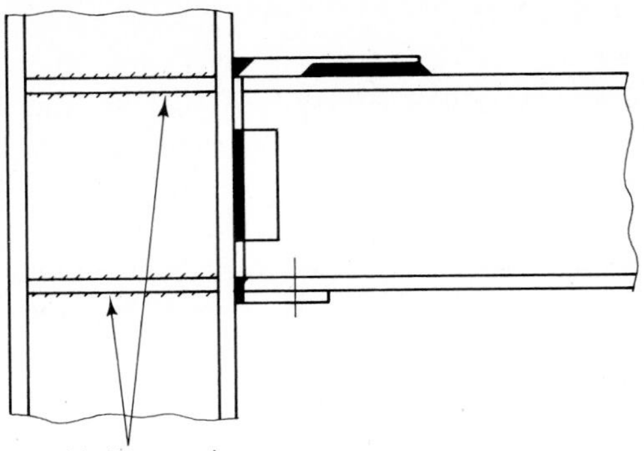

Figure 122 All-welded beam–column connection with column flange stiffeners.

moment capacity and can also transfer shear (Fig. 120). A *simple* connection assumes no moment transfer and transfers shear (Fig. 121). A *semirigid* connection can provide some percentage of the member moment capacity.

Methods of Fastening. The primary methods for fastening structural connections are by bolting or welding, or a combination of the two. Two general classes of bolts exist for structural applications in the United States (with similar counterparts in other parts of the world). The structural bolt is a general-purpose bolt whose use is limited to static conditions under relatively small service loads. The A-325 and A-490 high-strength bolts are used in the majority of structural connections under both static and dynamic conditions. High-strength bolts are generally preloaded to at least 70% of their tensile strength upon installation.

Welding consists basically of joining two pieces of metal by heating and melting them with connecting material. Almost all structural welding is electric, and the shielded metal arc process is the most common.

Structural joints may be made with groove, fillet, plug, seam, or slot welds; however, 80% to 90% of structural welds are fillet welds, and 15% are groove welds. There are equivalent bolted joints (see Fig. 122), and the most common practice is a combination of welding and bolting, such as illustrated in Fig. 120.

See also LIMIT DESIGN OF STEEL STRUCTURES; TIMBER JOINTS FOR STRUCTURES AND JOINERY; and WELDING, BRAZING, AND SOLDERING.

References

1. L. TALL et al.: *Structural Steel Design*. Wiley, New York, 1974.
2. J. W. FISHER and J. H. STRUIK: *Guide to Design Criteria for Bolted and Riveted Joints*. Wiley, New York, 1974.
3. *Manual of Steel Construction*, 8th ed. American Institute of Steel Construction, New York, 1980.

J.W.F.

K

KNOWLEDGE ENGINEERING is a branch of computing technology that seeks to make computers understand the worlds in which they are used, to recreate knowledge within computers that enables them to interpret and respond to demands from people. Two linked motivations help to define this field. One is recognition that conventional methods of programming computers are laborious, costly, and restrictive on what computers can do. The second is the recognition that people's ordinary activities and their consequent demands on computers are not amenable to orderly analysis leading to predefined sequences of operations; people do not normally provide the stable contexts necessary for conventional computer programs. This latter issue has relevance for architects and others engaged in the building industry.

Knowledge engineering is aimed at tasks that cannot be encompassed by algorithmic procedures that guarantee solutions but, essentially, at tasks whose goals are well defined. The general strategy is to develop techniques by which computers can extract meaning from expressions supplied by people. People should then be able to describe what they want a computer to do, relying on the computer to interpret what they are saying and expecting the computer to invoke appropriate machine responses. A long-term ambition is to lose the distinction between programming and using computers. Thus knowledge engineering refers to an attitude and an aspiration rather than a single definable activity.

History. Knowledge engineering spans the past decade, having emerged from the fields of artificial intelligence and computer science. Early work focused on relational database systems that provide access to data through descriptions of their content rather than knowledge of their physical locations in computer memory, thus supporting different views on stored information. Artificial intelligence's impatience with the difficulties of getting computers to do things prompted the development of new high-level programming languages that make it easier to build inference systems and knowledge bases, notably LISP in the United States. In Europe the same motivation led to development of Prolog, a software implementation of symbolic logic as a general-purpose programming language (Ref. 1). At the same time other efforts, often employing more conventional programming languages, have resulted in frame systems, production systems, and expert systems (Ref. 2). *Frame systems* make use of the stereotypical nature of people's perceptions of events to provide a methodical framework for a computer to receive information from a user. *Production systems* consist of production rules of the if–then variety, used to describe what inferences can be made from given data. *Expert systems* store in a computer experience from human experts that cannot be represented by algorithmic procedures, but operate on tasks that do have precisely defined goals. These techniques and their uses overlap.

Present and Future Trends. Japan's initiative on its fifth-generation computers project has helped to bring together previous efforts into a coherent field of study. This project identifies two main thrusts. One calls for radically new software advances needed to support a major expansion of the population of computer users to include ordinary people. The other is focused on new hardware, including knowledge base, inference and problem solving, and intelligent interface machines. Japan's own interest is in the opportunities for hardware manufacture that the project will create. This link between manufacturers' and users' interests is what makes this and similar projects in other countries so potent.

The Prolog logic programming language is widely recognized as a starting point for new efforts aimed at knowledge-based systems of the 1990s. These systems will build on new advances in formal approaches to natural language, leading to machine understanding of meaning intended by words and drawings. The systems then have to know how to assimilate received information and exploit it to respond to the demands of people. Work on knowledge bases that support graphical communication, which can extract meaning from drawings to arrive at object descriptions and can reflect changes to objects automatically in changed drawings, is still at an early stage of development (Ref. 4). Results of this work will have obvious significance for people associated with the design and production of buildings.

References

1. A. KOWALSKI: *Logic for Problem Solving*, Elsevier North-Holland, Amsterdam, 1979.
2. J. LANSDOWN: *Expert Systems: Their Impact on the Construction Industry*. RIBA Conference Fund, London, 1979.
3. *Preliminary Report on Study and Research on Fifth-Generation Computers 1979–1980*. Japan Information Processing Development Centre, Tokyo, 1981.
4. A. BIJL: Graphical input—Can computers understand people? To be published in a special issue of *Computer and Graphics*.

A.Bi.

L

LAMINATED TIMBER refers to timber elements, typically of rectangular cross section, fabricated by joining together several planks or strips of timber referred to as *laminae*. The joining process, referred to as *laminating*, may be achieved by the use of glue to produce a product termed *glulam* (Fig. 123) or by the use of metal connectors such as nails. The thickness of the laminae is typically in the range of 10 to 50 mm ($\frac{3}{8}$ to 2 in.). Glulam fabricated with wood veneer 2 to 5 mm ($\frac{1}{16}$ to $\frac{3}{16}$ in.) thick is sometimes termed *parallel laminated veneer*. It is a simple matter to fabricate laminated timber elements with curvature and also to sort and place the laminae so as to obtain an optimum lay up.

Glulam tends to have an enhanced strength and stiffness

Figure 123 Curved glulam portal arches of 41-m (135-ft) span, fabricated from *Eucalyptus marginata*, an Australian hardwood.

when compared with solid wood of the same cross section. Parallel laminated timber is particularly effective in this respect. By contrast, the properties of nail-laminated timber are generally poor compared with those of solid wood.

Laminated timber is fabricated from a great variety of timber species. The timbers of *gymnosperms*, often termed *conifers* or *softwoods*, tend to be easy to cut and glue, but require impregnation treatment with an oil-based preservative if they are to perform satisfactorily in outdoor locations. The timbers of *angiosperms*, often termed *hardwoods*, are difficult to cut and glue if their densities are too high: however, medium-density hardwoods are sometimes favored for glulam because of their high strength, natural durability, and esthetic qualities.

The primary benefit derived from the use of laminated timber is the potential of obtaining large structural elements fabricated from dry, high-quality timber. Thus, glulam is frequently used for medium- to large-sized structural members. Parallel laminated veneer is favored when particularly high strength is required. Nail-laminated timber is uncommon, but sometimes finds a use for medium-sized structural members when the available fabrication facility or technology is inadequate to achieve sound gluing of the laminae.

See also ADHESIVES and PLYWOOD.

References

1. U.S. FOREST PRODUCTS LABORATORY: *Wood Handbook: Wood as an Engineering Material*. USDA Agriculture Handbook No. 72. U.S. Government Printing Office, Washington, D.C., 1974.
2. AMERICAN INSTITUTE OF TIMBER CONSTRUCTION: *Timber Construction Manual*. Wiley, New York, 1974.

R.H.L.

LIFE-CYCLE COSTING (LCC) is an economic method used to evaluate various investment alternatives over a given period. It can be used, for example, to choose among different types of wall paints or to decide whether to renovate an existing building or build a new one. LCC takes into account not only initial capital costs (as has traditionally been the case), but also future costs. Initial costs are made up of all investment costs directly related to the project, including the costs of planning, design, engineering, construction, installation, and financing of future costs. Future costs mainly comprise operating and maintenance costs, repair and replacement costs, and property and capital gains taxes. The basic LCC equation can be expressed as follows: total life-cycle costs of a project = first or initial capital costs + present value of all future costs − present value of the resale price of the project at the end of the period covered by the analysis.

There are six requirements when applying LCC to any decision-making process:

1. Identify the objectives and constraints of the project in question.
2. Identify options to achieve the objectives.
3. Specify various assumptions regarding the discount rate, inflation rate, period covered by the analysis, types of costs to be considered, and tax treatment to be applied.

TABLE 1
Comparative Life-Cycle Costs[a] of Wall Painting, Constant (1978) Dollars

	Paint A	Paint B
Initial cost per m^2 (ft^2)	2.69 (0.25)	2.15 (0.20)
Present value of all repainting per m^2 (ft^2)	6.03 (0.56)	9.26 (0.86)
Total life-cycle cost per m^2 (ft^2)	8.72 (0.81)	11.41 (1.06)
Savings using paint A per m^2 (ft^2)	2.69 (0.25)	
Savings using paint A for a typical wall area of 350 m^2 (3767 ft^2)	941.50	

[a]Based on the following assumptions: Economic life of the building is 40 years. Discount rate is 11%. Frequency of repainting with paint A (latex) is every 5 years; with paint B (calcimine), every 3 years. Repainting cost with paint A after 5 years is \$2.37/m^2 (\$0.22/ft^2), thereafter increasing at an annual rate of 3% in real terms (exclusive of inflation rates) up to the 35th year, after which no repainting is done; preparation cost for repainting with paint A at the 15th year is \$2.15/m^2 (\$0.20/ft^2) and \$3.23/m^2 (\$0.30/ft^2) at the 30th year. Repainting cost with paint B is \$1.83/m^2 (\$0.17/ft^2), thereafter increasing at an annual rate of 3% up to the 36th year, after which no further repainting is done; the preparation cost is \$2.15/m^2 (\$0.20/ft^2) at the 9th year, \$2.69/m^2 (\$0.25/ft^2) at the 18th year, \$3.23/m^2 (\$0.35/ft^2) at the 27th year, and \$3.77/m^2 (\$0.35/ft^2) at the 36th year. All costs include labor and material.

4. Convert all future costs to their equivalent values in time, that is, to either present or annual values.
5. Compare the total life-cycle costs for each alternative and select the one with the lowest costs, other things being equal.
6. Test the reliability of the LCC analysis (sensitivity analysis) by changing the assumed values of various parameters.

Table 1, for example, indicates that paint A is preferred to paint B in spite of its greater initial costs on the basis of LCC.

See also VALUE ENGINEERING.

References

1. AMERICAN SOCIETY FOR TESTING AND MATERIALS: *Standard Practice for Measuring Life-Cycle Costs of Buildings and Building Systems*. Philadelphia, (in press).
2. AMERICAN INSTITUTE OF ARCHITECTS: *Life Cycle Cost Analysis: A Guide for Architects*. Washington, D.C., 1977.
3. H. E. MARSHALL, and R. T. RUEGG: *Energy Conservation in Buildings: An Economics Guidebook for Investment Decisions*. U.S. Department of Commerce, Washington, D.C., 1980.
4. A. S. RAKHRA: Economic aspects of the durability of building materials: An exploratory analysis. *Durability of Building Materials*, Vol. 1 (1982), pp. 15–22.

A.S.R.

LIFT SLAB CONSTRUCTION is a method whereby the floor slabs of a multistory building are all cast at ground level, one on top of another, around previously erected precast concrete or steel columns, with a separating medium between. The slabs are then raised to their final positions using hydraulic jacks, which are mounted on top of the columns and are connected to the slabs by two steel lifting rods (Fig. 124). The system was developed in America by Phillip Youtz and Thomas Slick in 1949, and the International Lift Slab Corporation was formed in 1950.

The floors are designed as ordinary reinforced or prestressed flat plates (see REINFORCED CONCRETE SLABS and LOAD BALANCING IN PRESTRESSED CONCRETE) using fabricated steel collars cast in each slab around each column for shear reinforcement; these are also used as anchorages for the lifting rods and in the final connection to the column. For structures up to five stories high, a single column length is used; thereafter, lifting is carried out in stages, with further lengths being added while the upper floor slabs are temporarily connected to the top of the existing columns.

Figure 124 Seventeen-story lift slab apartment block in Coventry, England.

Suitable applications include multistory residential and commercial blocks, parking garages, and mass low-cost housing in developing countries.

References

1. W. SEFTON: Multi-storey lift slab construction. *Journal of the American Concrete Institute*, Vol. 54 (Jan. 1958), pp. 579–589.
2. F. R. BENSON: Lift slab design and construction. *Reinforced Concrete Review*, Vol. 8 (Dec. 1960), pp. 495–527.
3. Prestressed Lift Slabs. *New Zealand Concrete Construction*, Vol. 5 (Apr. 1961), pp. 38–46.

S.B.

LIGHTING (ILLUMINATION), UNITS AND MEASUREMENT OF. Lighting is unique in both the SI and

Imperial systems of measurement because it uses a human response as the basis of its system of units. Lighting units are derived from radiant measurements modified by the response of the human eye at various wavelengths.

Luminous Flux, Φ, is the time rate of the flow of luminous energy. It is measured in *lumens* (lm) rather than watts, because photometric units were based originally on candlepower. Luminous flux is determined from radiant flux using a function standardized in 1924 by the Commission Internationale de L'Éclairage (CIE). There are two human visual systems: the *photopic*, which results in high-quantity color vision (as occurs by day and under good lighting by night), and the *scotopic*, which results in poor-quality achromatic vision (as at night). Similar definitions exist for the standard scotopic observer. Photometry cannot be performed in the region between the photopic and scotopic states.

Luminous Efficacy is the measure of the efficiency with which a light source converts electrical energy to light; it is the quotient of luminous flux and electric power and has the unit lumen per watt (lm/W). INCANDESCENT LAMPS have efficacies between 10 and 20 lm/W compared with low-pressure sodium vapor DISCHARGE LAMPS, which can have efficacies as high as 200 lm/W.

Luminous Intensity, I, is the measure of flux concentration in a particular direction; if the flux, Φ, emitted through a solid angle, ω, in a particular direction is uniform,

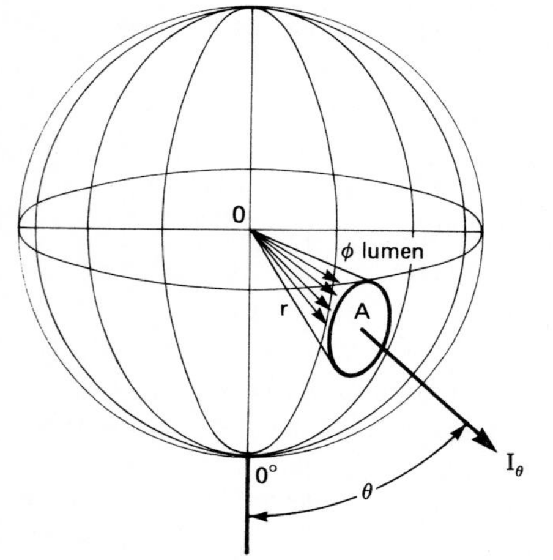

Figure 125 A sphere of radius r has a light source at its center. The source illuminates an area, A, on the sphere's surface. A is removed from the reference direction by an angle θ. The solid angle ω subtended by A is A/r^2. If Φ lumen are uniformly emitted through the cone formed by A, then the intensity, I_θ, is given by Φ/ω. The illuminance, E, on A is Φ/A lux, and it follows that $E = I/r^2$.

the intensity is Φ/ω lumens per steradian (the unit for measuring solid angles). It is called *candela* (cd). Figure 125 shows the definition of luminous intensity.

Luminance, L, is the measure of flux concentration leaving a source or surface in a particular direction. It is the ratio of the intensity in that direction and the projected area of the surface: $L = I/A$ in candelas per square meter (cd/m^2). Luminance is important for determining the flux transfer between surfaces and for predicting likely GLARE effects from lighting installations.

Illuminance, E, measures the flux density per unit planar area at a point on a given plane in *lux* (lx), which are lumens per square meter (lm/m^2). Illuminance is the quantity specified in lighting standards for the provision of adequate light for particular visual tasks.

Figure 125 shows that the illuminance on the surface, A, of a sphere is Φ/A lux, but since $\Phi = I\omega$ and $\omega = A/r^2$ (the definition of *solid angle*), it follows that $E = I/r^2$; this is the *inverse square law of illuminance*. The illuminance, on a plane whose normal is at an angle θ to the direction of the source, is given by $E = I_\theta \cos\theta/d^2$, where I_θ is the intensity in the direction θ (also called the *angle of incidence*) and d is the distance between the source and the point.

Lighting measurements are now based on the *photoelectric effect* (see PHOTOVOLTAIC CELLS), rather than using *visual photometry*. Certain semiconductors produce an electric current proportional to the flux density on their surfaces, which is utilized, with suitable filtering, to measure illuminance. It is necessary to ensure that the response is correct for other than normally incident light; that is, the photocell is *cosine corrected*.

The other lighting quantities are all derived from the ability to measure illuminance. Intensity is measured by multiplying illuminance by the distance squared. Luminance is measured with a constant aperture illuminance meter. Flux is measured indirectly, either by mathematical integration or by an integrating sphere whose nonselective white interior by interreflection produces a uniform illuminance on its surface. This is used to determine *lamp flux* and the *light output ratio* (LOR) of a luminaire.

The photometric and colorimetric qualities of materials are also determined by measuring flux as illuminance. The *reflectance*, ρ, of a surface is the ratio of direct to reflected flux. *Transmittance*, τ, is the ratio of transmitted to incident flux. Reflectance and transmittance are pure numbers less than or equal to unity. Colorimetric quantities are determined using spectrophotometric methods (see COLOR MEASUREMENT SYSTEMS).

See also LIGHTING LEVELS APPROPRIATE FOR VARIOUS VISUAL TASKS.

References

1. *Principles of Light Measurement*. CIE No. 18, Commission Internationale de l'Éclairage, Paris, 1970.
2. *Light as a True Visual Quantity: Principles of Measurement*. CIE No. 41, Commission Internationale de l'Éclairage, Paris, 1978.
3. J. E. KAUFMAN (ed.): *IES Lighting Handbook* (2 volumes). Illuminating Engineering Society of North America, New York, 1981.
4. P. R. BOYCE: *Human Factors in Lighting*. Applied Science Publishers, London, 1981.

W.G.J.

LIGHTING LEVELS APPROPRIATE FOR VARIOUS VISUAL TASKS should be related to the visual requirements of the task, the users' satisfaction, practical experience, and the need for cost-effective use of energy.

The Guide on Interior Lighting prepared by the International Commission on Illumination (CIE) and most national codes on interior lighting contain a schedule giving recommended illuminances for a large number of interiors and activities. The American Illuminating Engineering Society, in their *IES Lighting Handbook* (1981), has published illuminance ranges accompanied by a weighting-factor guidance system. Within Europe, lighting societies have published single-value illuminance recommendations allowing for modification factors, implying an increase or a decrease of recommended illuminances under certain circumstances.

Most of the recommended illuminances for offices, schools, and industrial activities lie in the range between 300 lux (30 footcandles) and 1000 lux (100 fc); 500 lux (50 fc) is the preferred value for normal office work and for medium machine and bench work in industry. Research based on the visual comfort criterion has shown that the minimum illuminance to be recommended in working interiors lies at about 200 lux (20 fc), irrespective of the visual task requirement. For circulation areas, a much lower minimum of around 20 lux (2 fc) has been found to be perfectly acceptable.

The recommended illuminances refer to the whole interior or specific part of the interior and generally to the horizontal working plane, which is assumed to be 0.85 m (34 in.) above the floor for standing workers and 0.75 m (30 in.) for those who are seated. Where the working plane is not horizontal, the recommended illuminance is referred to the appropriate inclination.

See also LIGHTING (ILLUMINATION), UNITS AND MEASUREMENT OF.

References

1. *Guide on Interior Lighting* (draft). Publication CIE No. 29/2. Bureau Central de la Commission Internationale de l'Éclairage, Paris, 1983.
2. *IES Lighting Handbook*. Illuminating Engineering Society of North America, New York, 1981.
3. *IES Code for Interior Lighting*. Illuminating Engineering Society, London, 1977.
4. D. FISCHER: Lighting levels for working interiors. *International Lighting Review*, Vol. 32 (1981), pp. 82–85.

D.F.

LIGHTWEIGHT CONCRETE is a cementitious material incorporating natural or manufactured lightweight aggregates. The concrete may be either structural grade or nonload-bearing and/or insulating grade depending on the particular aggregates used. The latter is made using perlite, VERMICULITE, or pumice, and its unit mass and properties are similar to those of CELLULAR CONCRETE.

For **Structural Grade Concrete,** the unit mass is between 1500 and 2000 kg/m^3 (100 and 125 lb/ft^3), and suitable aggregates include expanded shale and clay, foamed or granulated slag, sintered fly ash, and some scoria materials. It is usual for part or all of the fine aggregate to be a natural sand. Concrete strength will depend on mix proportions and on the strength of the aggregate. Mixes suitable for prestressed concrete can be designed. The indirect splitting-tensile strength is comparable with that of normal concrete if tested in a saturated condition; but when tested after air drying, it may be 25% lower. Reference 1 has detailed many parameters associated with lightweight concrete.

Shrinkage and Creep are generally higher than for traditional concrete, but it is noted in Ref. 2 that, where the lightweight aggregate had a continuous ceramic shell and provided natural fine sands were used, there was little difference in these properties.

Shideler (Ref. 3) evaluated the modulus of elasticity and found that it was from 44% to 82% of the value obtained with normal weight concrete.

Lightweight concrete may be pumped, but it is necessary to ensure that the aggregate has been presoaked. Its chief advantages lie in the reduced dead load of a structure, and the material has therefore been frequently used in multistory construction. Cost benefits can also be achieved in transportation of precast units.

References

1. H. S. WILSON: Lightweight aggregate: properties, applications. *Progress in Concrete Technology*. Canada Centre for Mineral and Energy Technology, Ottawa, 1980, pp. 141–187.
2. F. A. BLAKEY and R. K. LEWIS: A review of elastic deflection, creep and shrinkage of expanded shale concrete. *Constructional Review*, Vol. 37 (1964), pp. 3–6.
3. J. J. SHIDELER: Lightweight aggregate concrete for structural use. *Proceedings of the American Concrete Institute*, Vol. 54 (1957), pp. 299–328.

F.A.B.

LIMIT DESIGN OF STEEL STRUCTURES, also called plastic design, is a recent technique developed to represent the actual ductile behavior of indeterminate steel frames under increased loading and to take advantage of the fact that these frames have a greater load-carrying capacity than indicated by the allowable stress concept of ELASTIC DESIGN.

There are several causes of discrepancy between nominal and actual stresses, such as stress concentrations and residual stresses and (in statically indeterminate structures) secondary stresses due to FOUNDATION settlement, SHRINKAGE AND CREEP effects, and fabrication imperfections.

The practical behavior of a steel structure often departs strongly from the pure theory of elasticity, but remains safe only because of the material ductility. This ductility ought to be recognized in building codes. *In fact, there is no valid reason why the service stresses in a steel structure should under all conditions be kept below the yield point*, as long as there is no danger of fatigue or brittle failure. Fortunately, mild steel presents, in tension and compression, an extended yield plateau so that it can be admitted that a bent member cannot develop a bending strength larger than a definite value, M_p, called the plastic moment. In these conditions, the first-order theory (neglecting the effect of the deflections on the mode of action of external forces) shows that the frame will reach its maximum strength at a definite load called the *plastic limit load*.

In the plastic design methods, the frame is designed so as to give it a prescribed load factor against the attainment of the limit load. For a large majority of continuous beams and low building frames, the plastic limit load may be evaluated by the *simple plastic theory*. After unloading, the plastic deformations leave a set of residual stresses so that subsequent loadings produce a purely elastic response.

Knowledge of the ductile behavior of steel also permits the designer to simplify the design of connections by disregarding the local stress concentrations and to eliminate costly details intended to provide actual hinges that would be required if the construction material was brittle. The saving in fabrication costs that results from a design of details based on an understanding of ductile behavior is great.

In ELASTIC DESIGN any plastic deformations are avoided. Failure is therefore considered to occur as soon as, in the weakest place, the actual bending moment reaches its maximum elastic value (called *yield moment*), $M_y = \sigma_y S$ (where S is the elastic section modulus), and the dimensions of the members of the frame are so chosen that $M = M_y$ occurs at one place under the factored service loads. A design based on a reasonable load factor against this limit load provides a more appropriate structure than an elastic design based on allowable stresses.

Simple Plastic Theory: Behavior of Mild Steel in Bending. Most structural steels exhibit, in tension as well as in compression, an extended yield plateau followed by strain hardening (Fig. 126). The desirable postelastic properties of a steel for plastic purposes are a significant yield plateau and a good strain-hardening response. Neglecting strain hardening, the maximum bending moment that a mild steel beam can sustain is given by $M_p = \sigma_y Z$, where Z is the plastic modulus of the cross section. This moment produces a very large (theoretically infinite) curvature of the beam (Fig. 127).

Simple Plastic Theory: Concept of the Plastic Hinge. Consider a beam on two supports loaded by a concentrated force in the middle. The bending moment diagram is given

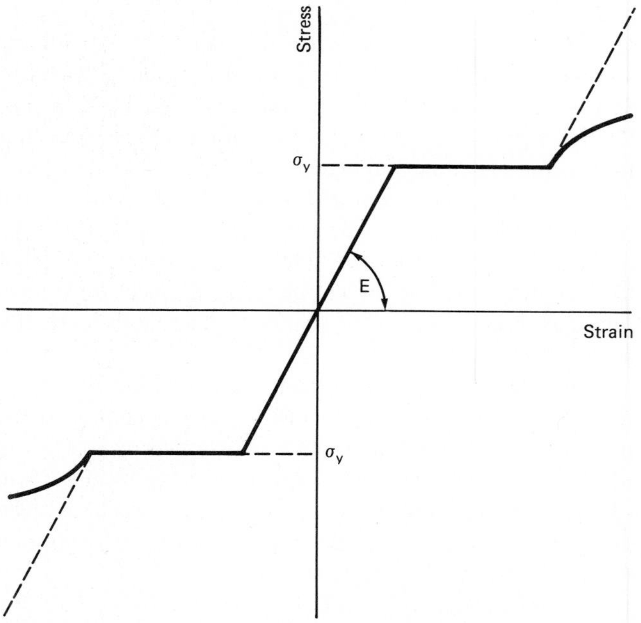

Figure 126 Structural steel has a yield plateau followed by strain hardening, both in tension and in compression.

 limit design of steel structures

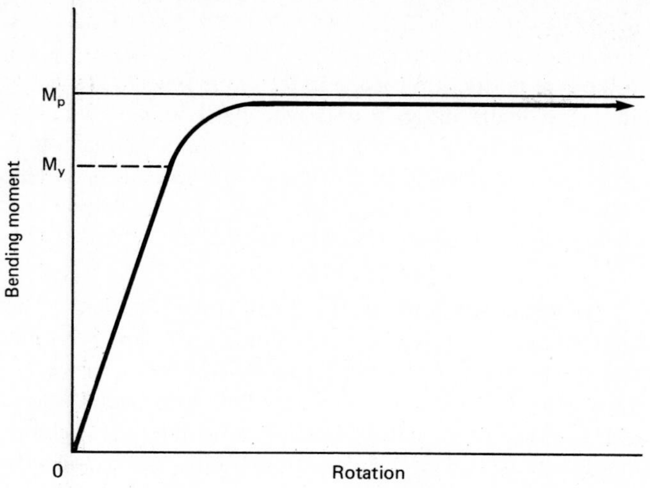

Figure 127 Maximum bending moment that a mild steel beam can sustain produces a very large curvature.

in Fig. 128. Increase the force P until the maximum bending moment $PL/4$ reaches the plastic moment M_p. Plastic deformations extend over the region where the bending moment exceeds the yield moment M_y (Fig. 128). Because of the shape of the moment–curvature diagram, the curvature remains very small near ends C and D of the plastic region, where at point E, where the force is applied, the curvature is theoretically infinite. Moreover, the elastic deflections of parts AC and DB are very small in comparison with these of part CD, and they can be neglected. *The beam therefore deforms as if it consisted of two rigid portions connected by a friction hinge in E* (Fig. 128). This approximation is valid for any bar of a statically indeterminate structure.

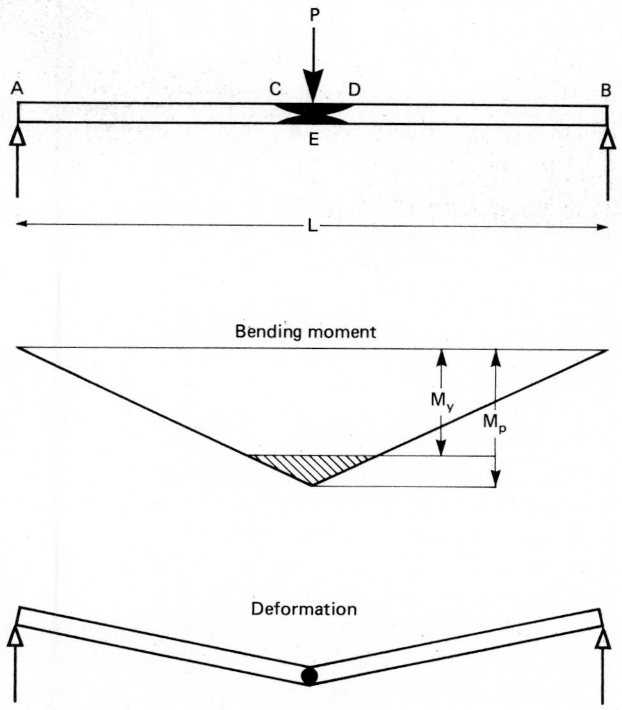

Figure 128 Plastic bending.

Behavior of a Statically Indeterminate Frame under Proportional Loading. A statically indeterminate structure, with degree of redundancy r, is subjected to set of loads that increase in the same proportion (Fig. 129). As the loads increase, plastic hinges appear in succession at sections at which the absolute value of the bending moment has a local maximum equal to the plastic moment. If the structure does not carry distributed loads, the only possible locations of plastic hinges are at the end sections of the members and at sections at which concentrated loads are applied.

As soon as a plastic hinge forms at a section, the magnitude of the bending moment at this section is stabilized at the known value $M_p = \sigma_y Z$, and the degree of redundancy is reduced by 1. The structure therefore becomes statically determinate when the rth plastic hinge appears. The next plastic hinge transforms this STATICALLY DETERMINATE system into a hinged mechanism with one degree of freedom, which can deform under virtually constant load. The appearance of the $(r+1)$th plastic hinge thus puts the structure out of service. The load at which the $(r+1)$th hinge appears is called the *limit* load.

The Theory is Valid provided:

1. The magnitude of the bending moment cannot exceed the plastic moment $M_p = \sigma_y Z$ in any section.
2. The plastic deformations in the vicinity of plastic hinges, because of the yield plateau of structural steel,

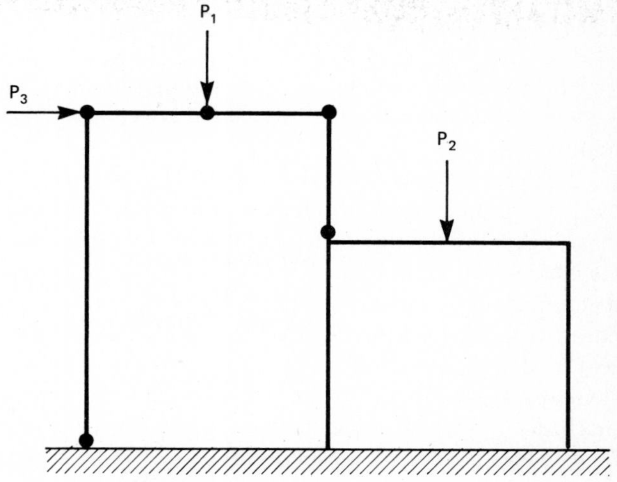

Figure 129 Plastic deformation of statically indeterminate frame.

are much higher than elastic deformations, which can be disregarded.

3. The ductility under the loading conditions allows large local strains in the narrow zones of high curvature without breaking.
4. The influence of normal and shearing forces on the plastic moment is not considered.
5. Despite large local plastic deformations, neither the structure nor a part of it becomes unstable before a mechanism is formed.
6. When the ultimate plastic hinge appears, the deformations are still sufficiently small so that the equilibrium and virtual work equations can be formulated in the undeformed structure (first-order theory).
7. The loads applied to the structure are assumed to increase in proportion.
8. The connections of the structure are able to transmit the maximum plastic moment of the connected members and exhibit enough plastic rotation capacity to allow for the formation of the collapse mechanism.

It is possible to fulfill these eight conditions by taking precautions that are discussed at length in Refs. 1 to 4.

Plastic Limit Theorems for Beams and Frames. The exact solution of a problem by the plastic theory must satisfy simultaneously three conditions:

1. *Mechanism condition:* A sufficient number of plastic hinges must form to allow the structure (or part of it) to deform as a mechanism that is compatible with the remaining constraints.
2. *Equilibrium condition:* The internal bending moment must be statically equivalent to the external forces.
3. *Plasticity condition:* The bending moment can nowhere exceed the plastic moment of the member being considered.

The two main methods for determining the limit load are based on the upper bound and on the lower bound theorem, respectively:

1. Upper Bound Theorem. A limit load computed on the basis of an assumed mechanism will always be greater than or equal to the true plastic limit load. In the *kinematic method*, which is associated with this method, a collapse mechanism is assumed; the corresponding virtual work equations are solved for the value of this load. As the theorem of virtual work is only valid for a structure in equilibrium and as equilibrium is not guaranteed for the assumed mechanism, the load value obtained in this way is an upper bound. It is the correct plastic limit load only if a bending moment diagram can be associated with it that complies with equilibrium and plasticity conditions.

2. Lower Bound Theorem. A load that satisfies the equilibrium and plasticity conditions is less than or equal to the true plastic limit load. In the *static method*, an equilibrium moment diagram is drawn that complies with the plasticity condition. The load equilibrating the moment diagram is a lower bound of the actual limit load. It is correct only if $M = M_p$ at enough places for the corresponding plastic hinges to create a valid mechanism.

General Problem of the Multistory Frame. The basic theorems of plastic analysis are sufficient for most single- and two-story frames. A multistory frame, however, must carry large gravity loads in addition to lateral loads. The secondary moments caused by the gravity loads acting through the lateral deflections of the frame can significantly alter the equilibrium relationships of the undeformed frame. The secondary overturning moments, however, reduce the strength and stiffness of the frame to resist gravity and lateral loads. A building may be called ''tall'' as soon as this effect cannot be neglected in the design.

The **Design of Connections** of plastically designed structures has a decisive influence on their ultimate strength, because plastic hinges most frequently occur where two or more members join. The connections must achieve the following:

1. Be able to support the limit load.
2. Be sufficiently stiff to maintain the relative positions of the connected members with the necessary degree of accuracy.
3. Have sufficient capacity for plastic rotation.
4. Be economical to fabricate.

See also JOINTS FOR STEEL STRUCTURES and LIMIT STATES DESIGN.

References

1. L. S. BEEDLE: *Plastic Design of Steel Frames*. Wiley, New York, 1958.
2. *Plastic Design in Steel*, 2nd ed. Manual and Reports on Engineering Practice of the ASCE, No. 41. American Society of Civil Engineers, New York, 1971.
3. B. G. NEAL: *The Plastic Methods of Structural Analysis*, 2nd ed. Wiley, New York, 1963.
4. C. MASSONNET and M. SAVE: *Plastic Analysis and Design, Vol. 1: Beams and Frames*. Blaisdell, New York, 1965.
5. C. MASSONNET and M. SAVE: *Plastic Analysis and Design, Vol. 2: Plates and Shells*. North-Holland, Amsterdam, 1972.

6. C. MASSONNET and M. SAVE: *Calcul plastique des constructions, Vol. 1*, 3rd ed. B. Nelissen, Liège, Belgium, 1978.

C.M.

LIMIT STATES DESIGN is a set of design procedures for engineering structures based on limit states design philosophy. This philosophy embraces the older methods of working-stress design and ultimate-strength or plastic design (see LIMIT DESIGN OF STEEL STRUCTURES). Its probabilistic basis enables account to be taken of the design life of the structure and of the inherent and inevitable variations in applied loads, properties of materials, workmanship, and dimensions of structures as built, in a way that leads to more uniform margins of safety and serviceability.

Limit states design was developed initially for concrete structures by the European Committee for Concrete, between 1953 and 1963 (Ref. 1). It is being extended to cover all structural materials. Its philosophy is the only one that has been accepted for use in international codes of practice, such as the seven Eurocodes. The first of these, EC1 (Ref. 2), is a comprehensive statement of the philosophy. Revisions to EC1 are to be expected as current difficulties in the application of limit states philosophy to the design of foundations are resolved and to take account of differences (mainly of terminology) that exist between North American and European work. The current draft EC1 follows closely the work of the Joint Committee on Structural Safety (Ref. 3).

In its simplest form, limit states design is based on loads that have a defined probability (typically 5%) of being exceeded (W_k, say) and strengths of materials that have a similar probability of not being reached (f_k, say). Most designs are governed by the *ultimate limit state*. The required low probability of failure is achieved by using design loads $\gamma_f W_k$ and design strengths f_k/γ_m, where γ_f and γ_m are *partial safety factors* exceeding 1.0, and by ensuring that load effects calculated from $\gamma_f W_k$ nowhere exceed the appropriate strength or resistance of the structure, calculated from f_k/γ_m.

See also SAFETY FACTORS IN STRUCTURAL DESIGN.

References

1. COMITÉ EUROPÉEN DU BETON: *Recommendations for an International Code of Practice for Reinforced Concrete*. American Concrete Institute, Detroit, and Cement and Concrete Association, London, 1964.
2. *Eurocode No. 1. Common Unified Rules for Different Types of Construction and Material*. Commission for the European Communities, Brussels. In preparation (3rd draft, August 1981).
3. JOINT COMMITTEE ON STRUCTURAL SAFETY: General principles on reliability for structural design. *Reports of the Working Commissions, International Association for Bridge and Structural Engineering*, Vol. 35 (1981), pp. 41–58.
4. P. THOFT-CHRISTENSEN and M. J. BAKER: *Structural Reliability Theory and Its Applications*. Springer-Verlag, Berlin and New York, 1982.

R.P.J.

LOAD BALANCING IN PRESTRESSED CONCRETE is a basic concept in designing tendons in such a way as to balance the dead load on a structure or its members. If the dead load on a slab, beam, or girder is balanced by the transverse upward component of the prestressing tendons, then that member will not be subjected to any moment or shear under this loading condition. Thus a structural member under flexure is transformed into a member under direct axial force due to prestress, and both the design and analysis of that member are greatly simplified.

Take, for example, a simple beam, prestressed with a parabolic tendon (Fig. 130). The upward uniform force exerted by the tendon on the concrete beam is given by

$$w_b = 8Fh/L^2$$

where F = prestressing force
L = length of span
h = sag of the parabola

Thus for a downward uniform load $w = w_b$ the net trans-

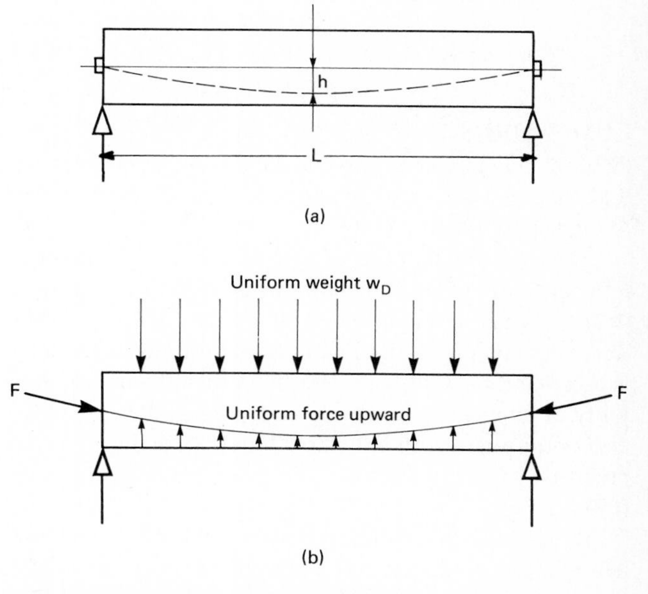

Figure 130 Load balancing of a prestressed concrete beam with a parabolic tendon. (a) Parabolic tendon. (b) Concrete as a free body.

verse load on the beam is zero. Hence, the beam is subjected only to the axial force F produced at the anchorages of the tendons. This F could result in a uniform compressive stress in the concrete of $f = F/A$. The change in stresses from this balanced condition due to live load can be computed by the ordinary formulas in mechanics, such as $f = Mc/I$, where the moment M is due to live load only or due to the unbalanced load.

For a flat slab or other two-dimensional structures, the load can be carried in either the x or the y direction and can be shared in any manner as may be desired. Again, for the unbalanced or additional load, calculations can be made by conventional mechanics; but since the unbalanced load is only one portion of the total loading, the accuracy in computation becomes less important.

The load-balancing concept for design and analysis becomes very useful when we deal with statically indeterminate structures. Their design as well as analysis will be much simplified. The amount of loading to be balanced can vary with each case. As a start, we could try to balance 100% of the dead load; but, if the live load is small, we can balance less than the full dead load; or if the live load is heavy, we may balance part of the live load in addition to the dead load.

See also PRESTRESSED CONCRETE.

References

1. T. Y. LIN: Load balancing method for design and analysis of prestressed concrete structures, *Journal of the American Concrete Institute*, Vol. 60 (June 1963), pp. 719–742.
2. T. Y. LIN and NED H. BURNS: *Design of Prestressed Concrete Structures*, 3rd ed. Wiley, New York, 1981.

T.Y.L.

LUMINAIRES are devices that control the distribution of the light given by lamps and include all the items necessary for fixing and protecting those lamps and connecting them to the supply circuit.

The primary purpose of light control is to direct the light in the required directions while reducing it in directions in which it might cause glare discomfort. The only luminaire commonly used without any form of light control is a batten luminaire with bare fluorescent lamps. The majority of luminaires, however, incorporate reflectors (enameled or mirrored, with or without louver shielding), louvers, prismatic panels (refractors), or opalescent diffusing panels (Fig. 131).

A reflector directs the light into the desired solid angle and so gives it a directional character. The directional effect is strongest where the reflector is of the mirrored-finish type. In some cases the reflector itself is designed to provide a certain degree of lamp shielding; but where more effective shielding is required, some form of louver is added. Louvers of the square-mesh and diamond-mesh types give both longitudinal and lateral screening. The lamellae louver, in which the screening strips run at right angles to the axes of the lamps, gives longitudinal screening only. A prismatic diffuser (or refractor) panel gives the light some slight directional character, while reducing the luminance of the luminaire in directions where glare could cause discomfort. The luminance of a luminaire fitted with an opalescent diffusing panel, shade, or globe is virtually uniform in all directions. Such a luminaire does not, therefore, afford the directional control of the light needed for efficient high-illuminance installations.

Where the space is air conditioned, lighting and air conditioning should, preferably, be combined in an integrated ceiling system (Fig. 132). Different types of air-handling luminaires have been developed for this purpose.

The vast majority of incandescent lamp luminaires used in interior lighting are designed for domestic use. Of these, spotlights and downlights are much used in interior lighting in general. Screening attachments can be used in combination with different lamp types to provide the desired degree of beam cutoff.

Luminaire Classification. Luminaires for general indoor lighting are classified by the Commission International de l'Éclairage (CIE) in accordance with the percentage of total luminous flux distributed above and below the horizontal. The six categories are direct, semidirect, direct–indirect, general diffuse, semiindirect, and indirect.

An important classification by the International Electrotechnical Commission (IEC) is made according to the degree of protection afforded by the luminaire against penetration of dust and moisture. The designation of the luminaire classes in this classification system is by means of two characteristic numerals preceded by the characters IP. For hazardous atmospheres, luminaires must additionally meet the requirements regarding protection against explosions, which are generally laid down in national regulations.

A classification made by the International Commission for Conformity Certification of Electrical Equipment (CEE) is according to the protection afforded against electric shock.

Luminaires may also be loosely classified according to the lamp they use, to the field of application, or to the method of mounting employed. Decorative luminaires are usually separated from those in which appearance is relatively unimportant.

Luminaire Photometrics. The optical performance of a luminaire is characterized by its luminaire efficiency (light output ratio), its luminous intensity distribution (candlepower distribution), and its luminances.

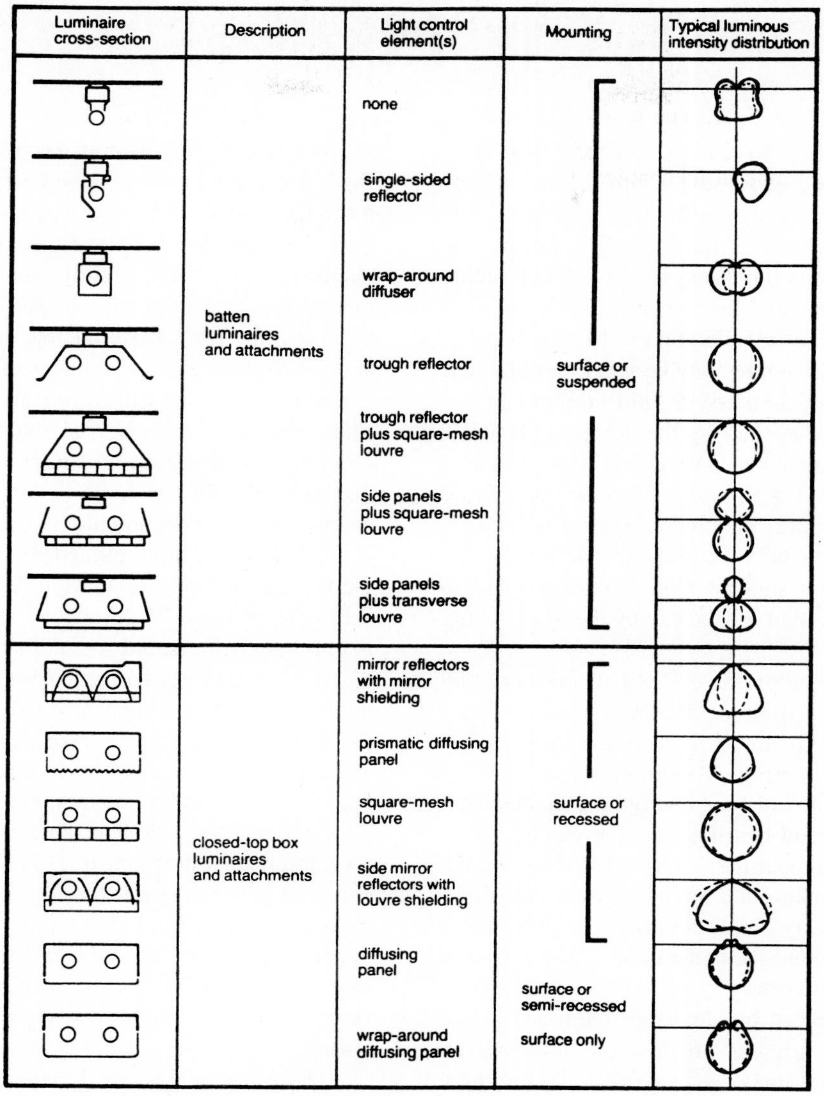
Luminaire cross-section
Description
Light control element(s)
Mounting
Typical luminous intensity distribution
none
single-sided reflector
wrap-around diffuser
batten luminaires and attachments
trough reflector
surface or suspended
trough reflector plus square-mesh louvre
side panels plus square-mesh louvre
side panels plus transverse louvre
mirror reflectors with mirror shielding
prismatic diffusing panel
square-mesh louvre
surface or recessed
closed-top box luminaires and attachments
side mirror reflectors with louvre shielding
diffusing panel
surface or semi-recessed
wrap-around diffusing panel
surface only

Figure 131 Some basic types of fluorescent lamp luminaires.

The luminaire efficiency is the ratio of the light output (lumens) of the luminaire to that of the bare lamps. An important factor when selecting a luminaire is the luminous intensity distribution, which represents the intensity of luminous energy from a luminaire in a particular direction. The luminaire luminances are important for evaluation of the glare to be expected in an interior. The average luminance of a luminaire in a given direction is calculated as the quotient of the luminous intensity in that direction and the projected area of the luminous surface area.

See also DISCHARGE LAMPS; GLARE; and INCANDESCENT LAMPS.

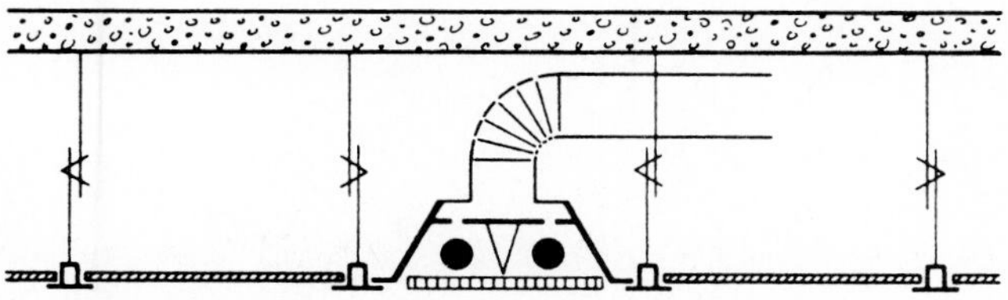

Figure 132 Cross-section of an integrated ceiling system.

References

1. *IES Lighting Handbook*, Reference Volume. Illuminating Engineering Society of North America, New York, 1981.
2. J. B. de BOER and D. FISCHER: *Interior Lighting*. Kluwer Technische Boeken B.V., Deventer, The Netherlands, 1981.

D.F.

M

MASONRY ARCHES, VAULTS, AND DOMES. Masonry is an assemblage of stones, or bricks, or, indeed, sun-dried mud (ADOBE) recognized by architects by various labels (Byzantine, Romanesque, Gothic), but by engineers by a common structural action. This action stems directly from the properties of the material.

Masonry cannot accept tensile forces. Mortar may be used in building, but this mortar may be weak initially and may decay; it cannot be assumed to add strength to the construction. Stability is assured by the compaction under gravity of the various elements. A general state of compressive stress can exist, but tension cannot be transmitted between two adjacent stones; they would simply pull apart. The compressive forces, however, allow friction to develop, locking the stones against slip. The shape of the whole structure is maintained in this way by the interlocking of the elements.

The mean level of compressive stress in masonry can be assessed by order-of-magnitude calculations or by direct study of old buildings. In round figures, the most highly stressed elements have average compressive stresses not more than one-tenth of the crushing strength of the material, the main portion of the load-bearing structure works at one-hundredth of the crushing strength, and infill panels and the like are subject to a background stress of, say, one-thousandth. Apparently, a very small compressive prestress is all that is necessary to avoid the dangers of slip and general loss of cohesion of the masonry.

A masonry structure can be idealized, then, as one for which (1) compressive stresses are low (so low that by comparison the material may be regarded as infinitely strong); (2) tensile stresses cannot exist; and (3) sliding failure does not occur.

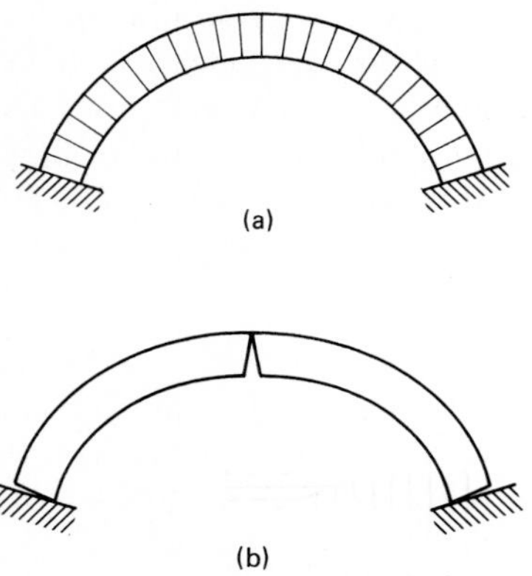

Figure 133 Voussoir arch. (a) The voussoirs, or perfectly fitted masonry blocks, of the arch. (b) Stable crack within the arch.

Structural Behavior of Arches. The voussoir arch of Fig. 133a is supposed to have been fitted perfectly between its abutments; the centering supporting the masonry is then removed, and the arch starts to thrust horizontally. The abutments give way slightly, and the arch must accommodate itself to an increased span. It does this, in general, by cracking; for a material that is strong in compression and weak in tension and that suffers no internal slip, the cracks of Fig. 133b will form (shown greatly exaggerated). These cracks, for an idealized voussoir arch, may be thought of as hinges; idealized or not, it is cracking that permits a masonry structure to alter its geometry in response to a change in the environment.

The cracked three-pinned arch of Fig. 133b is a perfectly satisfactory structural form. Such a stable arch is subjected to a single point load (Fig. 134), and the point load is slowly increased so that its magnitude becomes more and more important compared with the dead weight of the arch. How does the arch respond? It can be imagined that some preexisting cracks, such as those of Fig. 133b, may close and that new cracks will open, but none of this need lead to loss of structural integrity. Failure of the arch can only occur when sufficient hinges have formed, as in Fig. 134, to create a mechanism of collapse. The mechanism of Fig. 134 is indeed possible. There are some forms of arch, such as that of Fig. 135 (or Fig. 136), for which a mechanism of collapse cannot be constructed. The conclusion is that the arch of Fig. 135 is, within the present assumptions (low level of compressive stress, no sliding), infinitely strong.

These arguments are kinematic in nature. As always with a structural theory, there is an alternative approach through statics. For the arch, the statics may be investigated by the funicular polygon, leading to the construction of a line of thrust for given loading. (The shape of a flexible weightless chain subjected to the loads on the arch, but turned upside down, is the same as the shape of the line of thrust.) The theory of masonry arches lies within the plastic theory of structures (see LIMIT DESIGN OF STEEL

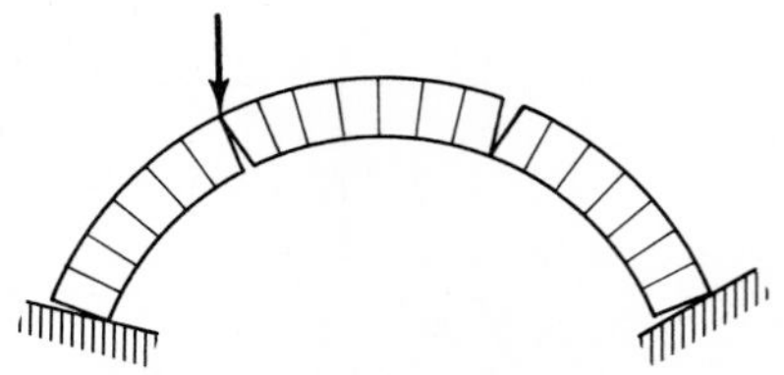

Figure 134 Failure of masonry arch, showing collapse mechanism.

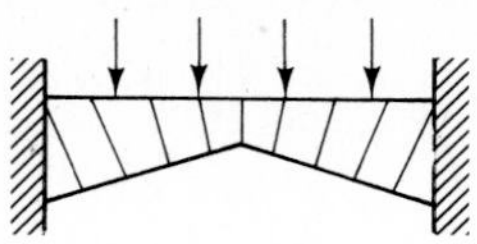

Figure 135 An arch that is infintely strong if there is no sliding action.

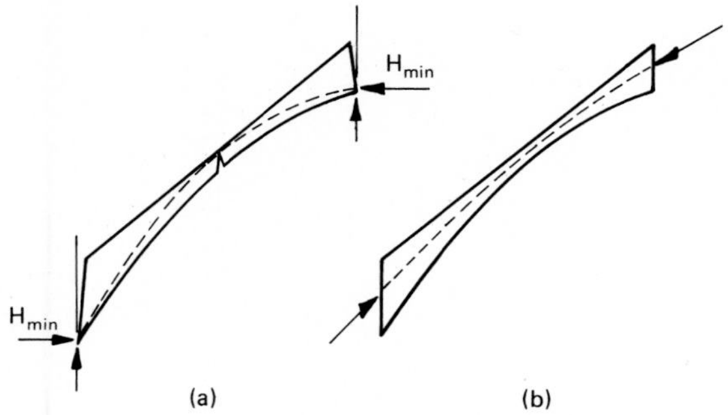

Figure 136 Flying buttress. (a) The minimum thrust, H_{min}, with which the buttress leans against the wall. (b) The buttress can absorb a large thrust from the high vault.

STRUCTURES), and the plastic theorems may be used. In particular, the powerful *safe theorem* applies: if a line of thrust can be found that lies within the masonry, the arch can never collapse under the given loads.

The safe theorem is important because the line of thrust found to satisfy the theorem need not be the actual line of thrust. Indeed, as the cracks shift within a masonry arch in response to changes in the environment, so the line of thrust must shift. It is of no use to ask for the actual state of the structure; that state is here today and gone tomorrow. Rather the analyst should find one possible state and be comforted by the knowledge that, if such a state can be found, then the structure itself certainly can. There are no conceivable settlements or other disturbances (always thought of as small compared with the dimensions of the structure) that can cause the line of thrust to shift outside the masonry.

Thus the flying buttress of Fig. 136a is in its *minimum* state; it has an easily calculable minimum thrust (say 10 tons) with which it will always lean against the nave wall. However, it can, if necessary, absorb very large thrusts from the high vault (Fig. 136b).

Cross Vaults. All these ideas can be extended easily to three-dimensional construction. First, any cracking that may be observed in an actual building can be ignored for the purpose of understanding structural action; the cracked state is the natural state of masonry. Similarly, jointing of the masonry is not important as long as compressive forces can be transmitted without slip. Cracks are papered over so they cannot be seen; the masonry is painted so that the joints cannot be seen. What is left is a structure having a certain geometry; what are sought are states of equilibrium conformable with that geometry.

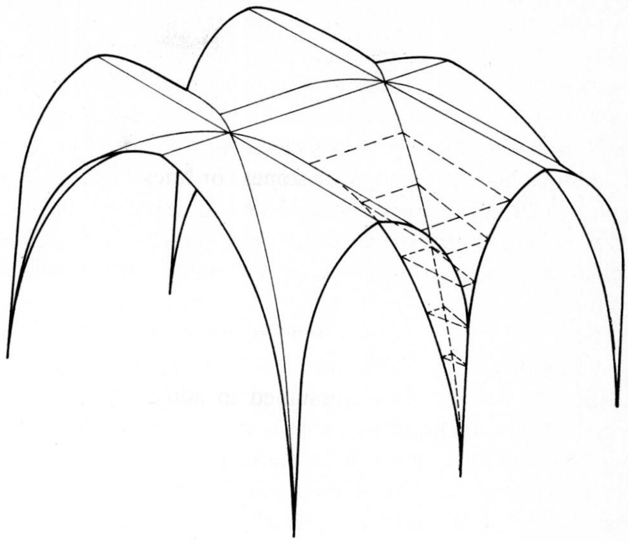

Figure 137 Quadripartite vault. (By courtesy of the University of Chicago Press. Copyright 1961, John Fitchen.)

Thus the basic quadripartite vault of Fig. 137 may be examined by the use of the membrane theory of SHELLS, a theory based solely on equations of equilibrium. The stresses in the smoothly turning web surfaces are well behaved and have magnitudes less than 1 MPa (1 N/mm^2 or 145 psi), or one-hundredth of the crushing strength of the material (hence the use of lightweight tufa whose strength is not of primary importance). However, a crease in the shell surface leads to a discontinuity, and the intersections at the groins define lines of stress concentration. If there are no reinforcing ribs (Romanesque), the stress in the groin will rise to, say, one-tenth of the crushing strength. The ribs are not strictly necessary structurally, but if they are used (Gothic), they will help to carry the vault; they serve also to cover awkward intersections between the webs, and they are of help as permanent centering during construction.

What is a structural necessity is the rubble infill to the vaulting conoid (Fig. 137). Equilibrium of the whole vault cannot be ensured by thrusts remaining within the diagonal ribs (or groins); the thrust escapes from the rib toward the springing and must be carried to the external buttress (by a flyer) through the infill. These same conclusions may be

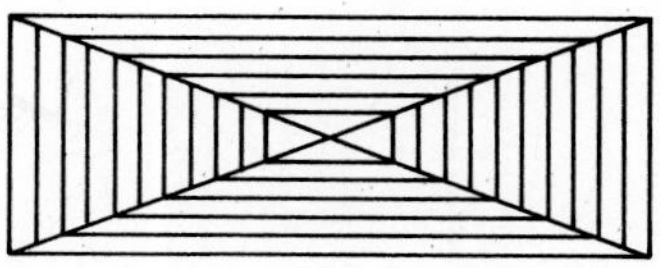

Figure 138 Plan view of quadripartite vault.

reached from a study of an even simpler model of the vault (Fig. 138). In plan, the vault webs have been sliced into parallel arches, springing from the diagonal ribs; each arch may be analyzed as a two-dimensional structure (Fig. 133) and the corresponding forces calculated.

More complex vaults, such as the later Gothic lierne vaults in England, may be analyzed in the same ways. A rib applied to a crease in a shell surface acts as reinforcement; a rib applied to a smoothly turning surface is decorative.

Fan Vaults. Thus the ribs on a fan vault serve to define the surface of the fan, but are not structural elements. Shell theory again gives a simple picture of the mechanics of the vault. It is possible to compute a profile (Fig. 139) close to those used in practice, for which only meridional stresses act and the hoop stresses are zero. No fan vault can stand without a compressive (meridional) load at its upper edge; this prestress is provided by the weight of spandrel material between adjacent fans. However, the fan vault of Fig. 139 can stand under the action of only the forces shown, and the vault thrust acts at a level lower than that of a normal cross vault; in practice, the conoid is filled with rubble masonry.

Domes. The membrane solution for a hemispherical DOME under its own weight is shown in Fig. 140. The circumferential stress is compressive from the crown round to 51.8°; large tensile stresses are developed toward the base. Tensile stresses are not admissible for masonry, and an alternative solution must be sought; but a general conclusion may be drawn from Fig. 140. An unreinforced dome requires buttressing at its base to resist the outward thrust, but encircling ties near the base help to stabilize the structure.

A slicing technique again gives a better picture of the forces in a masonry dome. The lune (orange slice) of Fig. 141 has been formed by meridional cuts, and an arch is formed by two such lunes leaning against each other. Figure 142 shows the line of thrust for a lune just thick enough (4% of the radius of the dome) for stability; from the weight W of the lune, the thrust H per unit of circumference can be calculated, and hence encircling ties and abutting masonry can be designed.

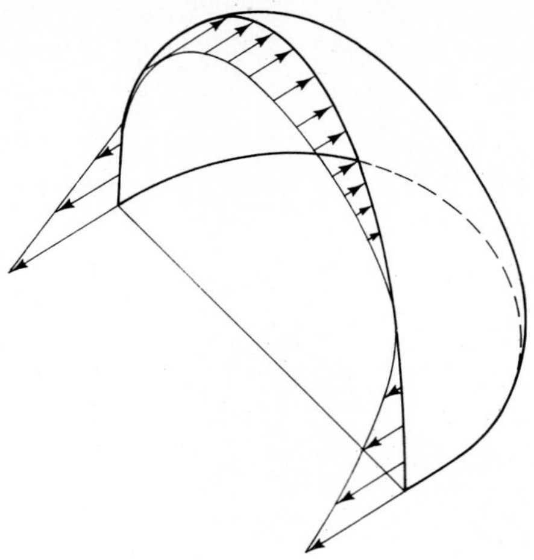

Figure 140 Dome.

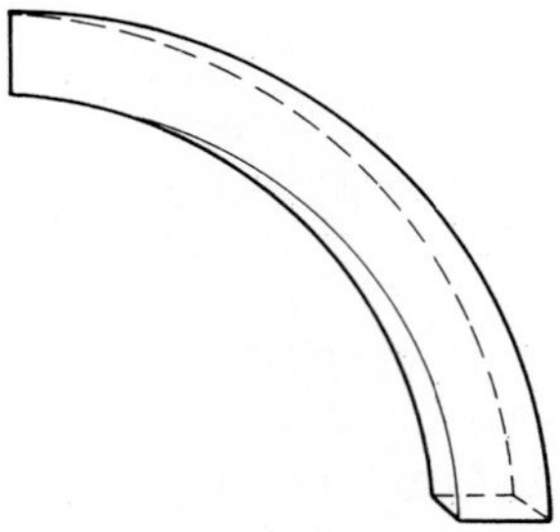

Figure 141 Orange slice of dome.

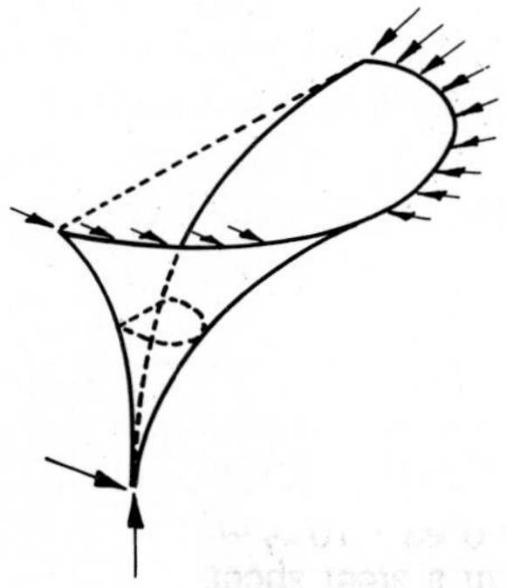

Figure 139 Fan vault.

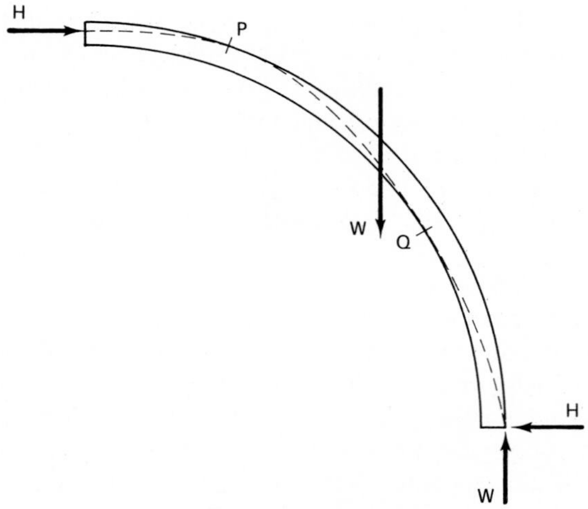

Figure 142 Line of thrust for dome.

A dome with an eye is perfectly satisfactory. A two-dimensional arch requires the placing of the keystone before decentering, but a dome can be open at the crown and can be built with little or no centering. Similarly, a lantern can be loaded onto the crown.

References

1. J. HEYMAN: *Plastic Design of Frames, Vol. 2, Applications*, Cambridge University Press, Cambridge, England, 1971.
2. J. HEYMAN: *The Masonry Arch*, Ellis Horwood, Chichester, England, 1982.
3. J. HEYMAN: *Equilibrium of Shell Structures*, Oxford University Press, Oxford, 1977.
4. J. HEYMAN: Chronic defects in masonry vaults: Sabouret's cracks, *Monumentum*, Vol. 26, (1983), p. 181.

J.H.

MEASUREMENT OF STRAIN is an important aid in experimental strength tests performed on structures or structural components. If a solid body is acted on by external forces, the material in it is subjected to a stress σ. This is the force, relative to the cross-sectional area, that must be withstood by the material at a certain point inside the body. The aim of the tests is to determine whether the material from which the body is made is capable of withstanding the stresses acting on it. Since it is not normally possible to measure the stresses directly, they are determined from the deformations that the body undergoes as a result of the force acting on it. Thus the stresses are obtained from strain measurements. There are a great number of materials in which the relationship between the stress and the deformations is linear in a certain range (Hooke's law).

Since the strain can only be calculated from the elongation Δl of a finite length l, the value $\epsilon = \Delta l / l$ is the mean value of the strain over the length l. Only when the strain is constant, or if the measuring point bisects the length l, is it also the same as the strain at the measuring point when the relationship is linear. In all other cases there is a greater or smaller deviation, especially when the strain gradient is very high. In such cases it is therefore preferable to work with the smallest possible gauge length. However, the smaller the gauge length, the shorter the elongation of the length Δl that is to be measured. This results in a considerable increase in the effort of obtaining a reliable measurement, because extensometers for shorter lengths require a higher magnification if they are to offer the same degree of sensitivity. The choice of this length largely depends on the task in hand, the degree of accuracy required, the stress gradient, and the external conditions (e.g., whether the measurement is in the laboratory or on a building site). At a stress of $\sigma = 140$ MPa (140 MN/m^2 or 20 ksi), the resulting strain in a steel specimen, according to Hooke's law, will be $\epsilon = 0.67 \cdot 10^{-3}$ m/m. To be sure of accurately recording a change in length of 10^{-4} mm, a gauge length of 20 mm ($\frac{3}{4}$-in.) is required.

Strain measurements on concrete provide a further example for choosing the right gauge length. This material consists of aggregates of various sizes, with an elastic modulus that is four to five times greater than that of the hardened cement paste. If the gauge lengths used are the same size as the diameter of the aggregates, stresses are measured that at certain points are unevenly distributed and that, because of the lack of homogeneity, may deviate considerably from the mean strain corresponding to the external deformation. With long gauge lengths, this is compensated for by averaging of the different strains occurring along the measured length. With a maximum particle size of 20 mm ($\frac{3}{4}$ in.) and a gauge length of 100 mm (4 in.), the deviations of the measured values from the mean strain may be approximately $\pm 8\%$. To prevent the deviation from exceeding $\pm 2\%$, a gauge length of 400 mm (16 in.) is necessary.

With mechanical strain gauges, the length l is marked between two knife-edges, one of which is fixed and the other movable. When an applied external load causes elongation of the specimen, the movable blade is moved around its fulcrum (Fig. 143). The change in length is shown by an indicator connected to the knife-edge. Various types of strain gauge differ only with regard to their indicators: either mechanical, mechanical-optical or electromechanical. The only types of strain gauge now in use are the following:

1. Detachable mechanical extensometers (used for long-term observations on structures and for shrinkage measurements).
2. Induction strain gauges (used for material testing);
3. Vibrating-wire strain gauges (used for long-term observations on structures).
4. Electric resistance strain gauges (the most commonly used recording devices at the present time).

Electric resistance gauges consist of a plastic foil on which a 3- to 5-μm-thick zigzag measuring grid is mounted (Fig. 144). They are fixed to the measuring point with adhesive, and the electric grid is connected to an amplifier. Strains affecting the test piece also strain the electric grid, whose electrical resistance changes accordingly. This is measured with a Wheatstone bridge and converted into a strain. The use of resistance strain gauges is a universally applicable method for experimental stress analysis because of the great advantages it offers. The gauges are simple to attach, are maintenance free, and can be operated by remote control. Analog or digital recording and digital processing of the measured values is possible. The digitally recorded measurements can be processed directly by computer.

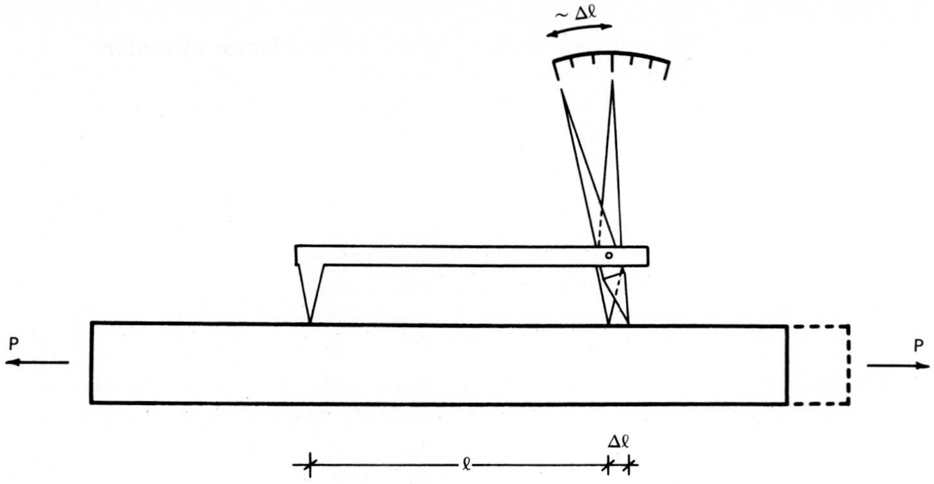

Figure 143 Basic principle of strain measurement with mechanical gauges.

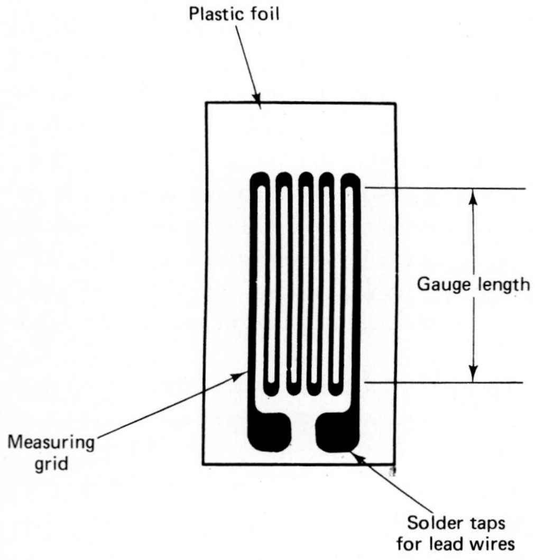

Figure 144 Electric resistance strain gauge.

See also TESTING OF MATERIALS FOR STRENGTH AND HARDNESS.

References

1. R. K. MÜLLER: *Handbuch der Modellstatik* (Handbook of Structural Model Analysis). Springer-Verlag, Berlin and New York, 1971.
2. R. K. MÜLLER: Einfluss der Messlänge auf Dehnungsmessungen an Beton (Influence of the measuring length of strain measurements on concrete). *Beton und Stahlbetonbau*, Vol. 14 (1964), pp. 205–208.
3. J. W. DALLY and W. F. RILEY: *Experimental Stress Analysis*. McGraw-Hill, New York, 1978.

R.K.Mü.

METALLURGY. There are now known to be more than 100 naturally occurring or artificially produced chemical elements of which 82 are generally classified as metals. This classification is based broadly on the physical and mechanical properties of the elements. A metal is an opaque, lustrous, chemically homogeneous element that also exhibits a high electrical and thermal conductivity and when highly polished is an excellent reflector of light. Most metals also deform elastically under load, and several show a marked tendency to deform plastically when stressed to higher levels. Although highly refined metals or those of commercial purity find ready use as metallic electric conductors, metal foil, surface coatings, and coinage, the majority of metals are utilized in the form of alloys where the restricted range of individual physical and mechanical properties of the component members gives way to a greatly enhanced utilization spectrum for the particular alloy.

An **Alloy** is a material with metallic properties that has been formed by the combination of at least two chemical elements, of which at least one is a metal. From this basic definition of an alloy there would appear to be the potential for creating a vast range of alloys since the additional elements added can be either metallic or nonmetallic. A listing of modern-day alloys would extend into some thousands, but many combinations are effectively merely slight variations on a common theme. Alloy compositions range from the simple two-component systems to extremely complex systems containing perhaps six or seven metals and two or more nonmetals. The very earliest alloys deliberately developed by people were almost certainly based on copper and tin, that is, the bronzes. The art of finding the relevant source materials (ores) and then blending them together in a liquid mass (smelting) prior to finally cooling to a solid (casting) and beating out to a useful shape (fabrication) developed slowly but steadily several thousand years ago. Since the days of the early Romans, iron and bronze have been used for weapons, tools, and machinery; copper, silver, gold, and tin for vessels and ornaments; lead for water reticulation; and brass and bronze alloys for coinage. The generic term metallurgy is now generally applied to a much expanded form of this activity.

If a pure metal A is melted along with a small percentage of an element B to form a homogeneous liquid and then cooled to ambient temperature, the physical characteristics of the resultant solid alloy depend on the relative sizes of the constituent atoms, their chemical properties, and the relative quantities involved. Similar elements may produce a *substitutional solid solution* in which the minority B atoms replace A atoms from their usual position on the atomic lattice (Fig. 145). Some degree of distortion of the parent lattice is always produced due to differences in atomic size or electrical charge. This results in a strengthening of the pure metal, often termed solid solution hardening, and the effect is well demonstrated by the hardening of gold by additions of other elements, such as copper and silver.

However, in many instances continued additions of the atoms of B cannot be accommodated within the distorted A parent lattice, and a new structural combination of A and B appears. The alloy then consists of two phases, the parent A phase fully saturated with B along with the new second phase consisting of a restructured arrangement of A and B. If the new phase can accept further additions of B and accommodate lattice distortion, the *intermediate solid solution* so formed may exist over a range of A and B relative composition ratios. If not, and A and B have such an affinity for each other that they tend to form an *intermetallic compound*, AxBy, then the new phase will have a very limited compositional range. Common brasses are alloys of copper and zinc, and two of the most important commercial alloys, 70% copper/30% zinc (cartridge brass) and 60/40 (Muntz metal) brasses, are appropriate examples of a single-phase solid solution alloy and a two-phase alloy.

A very common method of representing these phase relationships for two component alloy systems is shown in Fig. 146. This figure is an equilibrium phase diagram for the common brass alloys in which the various stable phase fields are mapped out as a function of the two important parameters, temperature and composition. Pure copper and pure zinc melting points of 1083°C and 419.5°C form the two extreme terminals on the composition axis, and between and below the liquid (L) phase regions, a succession of solid single- and two-phase regions are outlined. A common procedure is to label the succession of solid single phases from left to right as α, β, γ, and so on, noting that single-phase systems are always interspersed with two-phase fields.

A good example of the intermetallic compound AxBy alloy system would be the iron–carbon compound Fe_3C, better known as cementite, which is responsible for the high strength of many carbon STEELS. The distribution of this phase within primary iron–carbon solid solutions such as ferrite and austenite is of fundamental importance in establishing the characteristics of the many highly important commercial steel alloys.

If the atoms of element B are too large or too small to

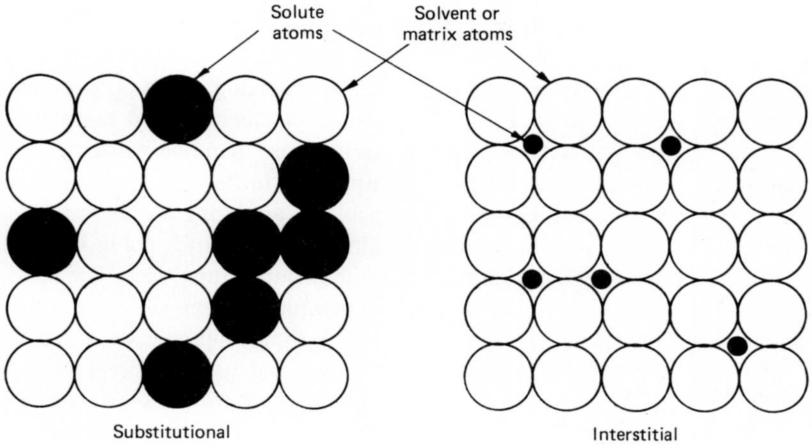

Figure 145 Schematic representation of solid solutions. (Redrawn from Charles O. Smith, *The Science of Engineering Materials*, 2nd ed., 1977, p. 204. Reprinted by permission of Prentice-Hall, Inc., Englewood Cliffs, N.J.)

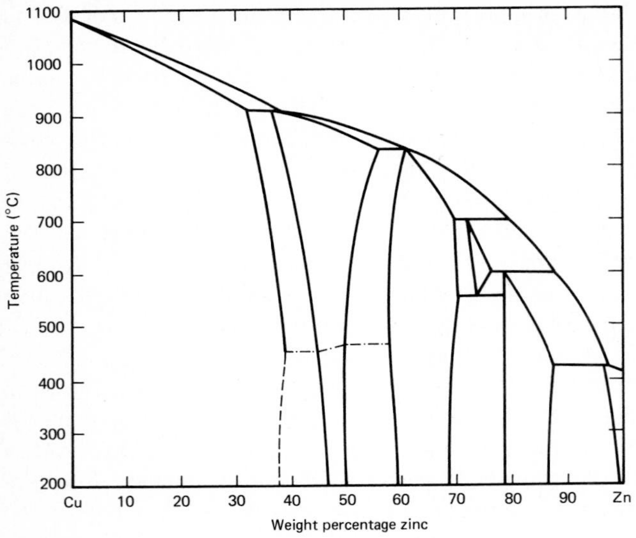

Figure 146 Phase diagram of copper–zinc system. (Redrawn from *Metals Handbook*, American Society of Metals, 1948.)

substitute for the parent A atoms on lattice sites, they may be able to find spaces between the A atoms, which they can occupy without causing too great an elastic distortion. This phenomenon gives rise to an important class of alloys known as *interstitial solid solutions* (Fig. 145b). The ferrite phase already mentioned is a typical example, the carbon atoms being sited between the iron atoms on a cubic lattice array.

Alloy systems need not be restricted merely to two-element components. In fact, an improvement in the qualities or properties of a binary alloy is frequently obtained by the addition of a third or even greater number of other elements. For example, the addition of nickel and chromium to steel gives rise to a class known as austenitic STAINLESS STEELS, and the presence of lead in copper–zinc brasses much improves the machinability of these alloys.

Alloy properties, however, react not only to the mere presence of the second (or third) phase. The physical distribution of the phases and their relative amounts are of prime importance to the final result. This physical distribution is generally determined by examining an appropriately prepared section of the alloy under a metallurgical reflected-light microscope or, in some instances, under an electron microscope. Once a suitable phase distribution has been developed for a particular alloy, this form of metallographic examination may be used to assess the likely performance of the alloy in service.

Heat Treatment. Heating and cooling a solid metal or alloy frequently alters the physical and mechanical properties of the system in a predictable manner. Commercially pure metals or simple solid solution alloys tend to become hard and less ductile when severely deformed at room temperature. Reheating and holding at a suitable temperature and then cooling at a suitable rate to ambient temperature can restore the original softness and ductility, a form of simple annealing. Such treatments may be varied in a complex manner, particularly with alloy systems, to provide a very wide range of end results.

For example, consider the aluminum-rich portion of the aluminum–copper phase diagram (Fig. 147). It is clearly possible to dissolve completely as an α phase 3% copper in aluminum at 500°C, the amount reducing significantly as the temperature drops toward ambient levels. The supersaturated solid solution produced on cooling can be manipulated to give varying results by applying an appropriate thermal cycle, giving rise to a precipitation or age-hardened alloy. The sequence could be as follows. The 3% copper in the ALUMINUM alloy is subjected to a solution heat treatment by reheating to a temperature within the α phase field indicated by point 1 on the diagram. If this alloy is then quenched to near ambient temperature (point 2), there would be insufficient time to reject copper from the aluminum-rich α phase, and a supersaturated situation is the result. In addition, a proportion of vacant lattice metal sites typical of the higher temperature formation would be retained in the quenched alloy. Even so, the quenched alloy is still reasonably soft and is capable of being mechanically worked without severe internal changes occurring. However, should the reworked alloy then be reheated to an elevated temperature (point 3) and held for a period of time, the excess copper would precipitate out of the α phase in any one of a series of very complex steps. The new precipitate phase, $CuAl_2$, is preceded by the formation of extremely small clusters of solute copper atoms. The result is an alloy of much greater strength and hardness than originally produced.

Heat treatment of an alloy may be particularly successful in changing the properties of the system in instances where the major metal component exhibits an ability to adopt more than one particular atom configuration (crystal structure) in the solid state. For example, pure iron is capable of arranging the dispersion of the iron atoms in two different forms of a cubic lattice configuration, according to the temperature level. The solid solution limit of carbon in one form is high and in the other, much restricted. Thermal cycling coupled with time variations of the cycle would be expected to give rise to great variations in the basic properties of iron. This is utilized widely in the form of steels and cast irons, where heat treatments such as annealing, normalizing, quenching, and tempering are all used to alter the properties of the basic iron–carbon alloy.

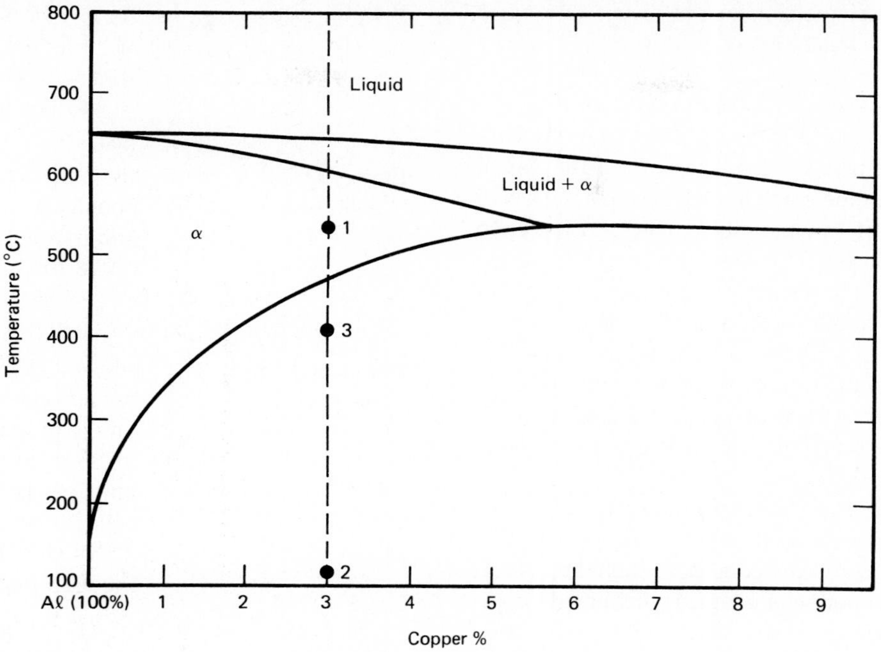

Figure 147 Partial aluminum–copper phase diagram. (Redrawn from *Metals Handbook*, American Society of Metals, 1948.)

References

1. CHARLES O. SMITH: *The Science of Engineering Materials*. Prentice-Hall, Englewood Cliffs, N.J., 1977.
2. Z. D. JASTRZEBSKI: *The Nature and Properties of Engineering Materials*. Wiley, New York, 1976.
3. *Metals Handbook*, American Society for Metals, Metals Park, Ohio, 1981.
4. R. W. K. HONEYCOMBE: *Steels*. Edward Arnold, London, 1981.

A.E.J.

METRIC AND CONVENTIONAL (BRITISH) AMERICAN SYSTEMS OF MEASUREMENT. The traditional systems of linear measurement are mostly based on the dimensions of the human body: the digit (width of a finger), the foot (length of a foot), the cubit (length of the forearm), and the yard (length of a step).

The meter was originally defined in 1793 by the new republican government of France as one ten-millionth of the distance from the north pole to the equator of the earth's meridian passing through Paris. It has since been redefined in terms of the wavelength of light at a particular frequency. The Napoleonic conquest spread the metric system to most of Europe.

Although the conventional systems of measurement related to the dimensions of the human body, they had many disadvantages. In the Middle Ages the length of the foot varied from city to city; it ranged from 295 to 350 mm, a variation of 18%. It was not uncommon for foot measures of different lengths to be used in the building of the same cathedral (Ref. 1). By the beginning of the 19th century the foot had become nationally standardized, but it differed from country to country. The metric system provided an international standard, although the same result could have been achieved if all countries had agreed to adopt, say, the British-American foot.

In addition, the metric system abolished the numerous traditional units that are well-known to the practitioners in a particular specialty, but are often a mystery to outsiders; for example, the standard and the cord of wood, the rod and the perch of masonry, the barrel of oil, and the ton of refrigeration.

The most important advantage of the metric system lies in its conversion factors, which are all multiples of 10. In the traditional systems of measurement, conversion factors such as 11, 12, 14, 16, and 20 were used. Some have no exact decimal equivalent, which creates difficulties when digital calculators or computers are used. This is the main reason for the recent conversion to the metric system in countries that had not already adopted it.

The SI System (*Système International d'Unités*), introduced in 1967 (Refs. 2, 3 and 4), differs from the conventional metric system in its definition of weight, force, pressure, and stress. In the conventional systems a mass of 1 pound has a weight of 1 pound, and a mass of 1 kilogram has a weight of 1 kilogram, because the acceleration due to gravity at the earth's surface is used in the conversion. In the SI systems, an acceleration of 1 m/s^2 is used.

The resulting unit of force and weight is called a *newton* (N). The acceleration due to gravity at the earth's surface is 9.807 m/s^2, and the weight of a mass of 1 kilogram at sea level is therefore 9.807 N; we may use 10 N as an approximation.

Stress and pressure are force per unit area. In most countries the *pascal* (Pa) is used, where 1 Pa = 1 N/m^2, but in several West European countries, including Great Britain, the pascal has not been adopted; this greatly complicates the international use of specifications and textbooks. The pascal is a very small unit, and most stresses used in building specifications are expressed in *megapascals* (MPa), where 1 MPa = 1,000,000 Pa. In Great Britain, N/mm^2 is favored:

$$1\ \text{MPa} = 1\ \text{MN/m}^2 = \text{N/mm}^2$$

The SI system replaces a number of established metric units by existing units of the same dimensions; notably the calorie (= 4.19 joules), and the metric horsepower (= 735 watts).

Multiples and Fractions are denoted by the prefixes kilo (one thousand, or 10^3), mega (one million, or 10^6), giga (10^9), tera (10^{12}), milli (one-thousandth, or 10^{-3}), micro (one-millionth, or 10^{-6}), and nano (10^{-9}). Thus the term "billion," which means a multiple of 10^9 in America and a multiple of 10^{12} in Europe, is avoided.

The **Principal Conversion Factors** between metric SI units and conventional American units, required in the design of buildings, are as follows:

Length	1 m = 3.281 ft = 39.37 in. 1 ft = 0.3048 m; 1 in. = 25.4 mm
Area	1 m^2 = 10.764 ft^2 1 ft^2 = 0.0929 m^2
Volume	1 m^3 = 35.315 ft^3 1 ft^3 = 0.02832 m^3
Capacity	1 liter (ℓ) = 0.001 m^3 = 0.220 Imperial gallon = 0.264 U.S. gallon = 1.057 U.S. quarts
Mass	1 kg = 2.205 lb 1 lb = 0.4356 kg
Weight, Force	1 kN = 224.8 lbf 1 lbf = 4.448 N
Stress, Pressure	1 MPa = 1 MN/m^2 = 1 N/mm^2 = 145 pounds per square inch (psi)
Energy (Work and Heat)	1 kilojoule (kJ) = 737.6 foot-pounds (ft · lbf) = 101.97 kilogram-meters (kgf · m) = 0.948 British thermal unit (Btu) = 238.85 calories (cal) = 0.0002778 kilowatt-hour (kWh)
Power and Heat Flow	1 kilowatt (kW) = 7376 foot-pound per second (ft · lbf/s) = 101.97 kilogram-meter per second (kgf · m/s) = 3412 British thermal units per hour (Btu/h) = 238.85 calories per second (cal/s) = 1.341 U.S. horsepower (hp) = 0.28435 ton of refrigeration

See also ACOUSTICAL TERMINOLOGY AND MEASUREMENTS; CONVECTION AND CONDUCTION; DIMENSIONAL COORDINATION; ELECTRICAL MEASUREMENTS; LIGHTING (ILLUMINATION) UNITS AND MEASUREMENTS; PSYCHROMETRICS; THERMAL RADIATION; and THERMAL TRANSMITTANCE.

References

1. J. JAMES: *The Contractors of Chartres*. Mandorla Publications, Dooralong, Australia, 1979.
2. *SI Units and Recommendations for the Use of Their Multiples and of Certain other Units*. International Standard ISO 1000. International Organization for Standardization, Geneva, 1981.
3. *The International System of Units (SI)*. NBS Special Publication 330. National Bureau of Standards, Washington, D.C., 1981.
4. *Metric Practice Guide*. E380-82. American Society for Testing and Materials, Philadelphia, 1982.

H.J.C.

MOBILE/MANUFACTURED HOMES. The U.S. housing industry has three major segments: traditional on-site construction, factory-built housing (packaged and modular), and mobile homes. As mandated by the U.S. Congress, mobile homes are now called *manufactured housing*.

Packaged homes (also known as prefabricated or panelized homes) are produced in factories and assembled on site by builder/dealers. *Mobile/manufactured homes* are transportable in one or more sections, built on a permanent chassis, and may be used with or without a permanent foundation when connected to utilities. *Modular homes* are similar to double-section mobile homes. Modulars, however, are designed for permanent foundations and conform to state factory-built housing codes, rather than to the federal performance code for mobile/manufactured homes. This article will deal only with mobile/manufactured homes.

The mobile home industry developed from the manufacture of recreational trailers in the 1930s. After World War II, mobile home producers separated from recreational vehicle makers, improved product quality, and developed uniform production standards. In the mid 1980s, mobile/manufactured homes represented about one-third of all new single-family houses sold in the United States. Their average price per square foot, excluding land, is roughly half that of site-built housing, and they dominate the new home market under $50,000.

The mobile/manufactured home is a highly standardized product manufactured in large volume from interchangeable components. It contains plumbing, central heating, air conditioning, and electrical systems and typically is equipped with major appliances, carpeting, and furniture (Fig. 148).

A single-section manufactured home usually is 14 ft (4.3 m) or even 16 ft (4.9 m) in width and up to 70 or 85 ft (21.3 or 25.9 m) in length. A multisection home consists of two or more sections combined at the site (Fig. 149).

Industry. Most firms in the manufactured home industry are specialists: manufacturers, distributors, or developers and/or operators of manufactured home subdivisions or parks. Manufacturers buy raw materials and components and mass-produce home units in assembly-line factories. Increasingly, these centralized producers are selling directly to subdivision developers, but most manufacturers will distribute their output to dealers who market the units directly to the more than 24,000 home park operators and to individual owners. Dealers also handle the complex tangle of local markets, land-use regulations, and politics.

Figure 148 Mobile/manufactured home assembly line: plumbing fixture installation.

Figure 149 Double-section mobile/manufactured homes.

Park operators provide sites, foundations, facilities, a community framework, and service for nearly half of all manufactured homes. Generally, the manufactured home owner rents the park site. Increasingly, however, the consumer buys the entire home package, including site, from a subdivision developer.

Design. Although conventional and mobile/manufactured homes serve many similar functions, they are physically very different. Manufactured homes must comply with highway regulations, require special strength during transit and while being put onto permanent foundations, and must be designed to resist overturning or sliding. Although fewer than 3% of all manufactured homes ever leave their original sites, all must be moved from factory to site. Most are hauled on their own running gear by truck.

Mobile/manufactured homes are designed to exploit the structural capabilities of construction materials and to meet the requirements of assembly-line production. Furthermore, the industry has developed an innovative systems analysis approach to product design, which enables producers to use materials, labor, and capital more efficiently than conventional home builders.

A mobile/manufactured home is engineered as a box beam supported by a semielastic steel chassis foundation, which transfers loads from walls and floor to the wheel assembly when the unit is in transit or to the foundation at the home site. This design principle permits thinner walls and smaller members. The box form also resists twisting and uplift.

The unibody design of the home firmly joins together four subassemblies, the chassis, floor systems, wall system, and roof system, into one structural unit. Mechanical service systems are housed in and integrated with these structural subassemblies.

Production. Mass production in off-site factories enables manufacturers to reduce costs through year-round production, mass purchasing and assembly, use of assembly-line machinery and techniques, quality control, efficient use of materials, and general economies of scale. Most manufactured home plants use a fully coupled main assembly line fed by subassembly stations at various points. Components are purchased or produced in the plant.

Assembly of each unit proceeds from the bottom up and from the inside out, starting with the chassis and ending with the roof. Approximately every 20 minutes, one completely finished and furnished home leaves the assembly line of the most efficient plants. Other factories produce as few as one unit every 8 hours.

See also PREFABRICATION: ORIGINS AND DEVELOPMENT and SYSTEM BUILDING.

Reference

1. A. D. BERNHARDT: *Building Tomorrow: The Mobile/Manufactured Housing Industry*. MIT Press, Cambridge, Mass., 1980.

Note: This article and Ref. 1 are based in part on research conducted by Program of Industrialization in the Housing Sector under a major contract with the U.S. Department of Housing and Urban Development.

A.D.B. and N.O.F.

MODELS are used for the three-dimensional representation of a proposed building to supplement the two-dimensional representation on drawings; in earlier times they have sometimes been used as a substitute for drawings. To gain a correct impression of the perception of the interior spaces, a small periscope, called a *modelscope*, can be inserted into a model at eye level from the top. An *urbanscope*, which contains a small television camera, conveys a perception of a person moving around a group of modeled buildings.

The only requirement for the correct scaling of an architectural model is that every dimension be reduced in the same ratio. This is also sufficient for models used on a HELIODON for the investigation of SUN SHADING AND SUN-CONTROL DEVICES or of sunlight penetration into buildings.

STRUCTURAL MODEL ANALYSIS requires additionally scaling of the modulus of elasticity of the material to be used in the structure and that used in the model and scaling of the loads. When loads are dynamic, as in EARTHQUAKES and WIND LOADS, several other variables are introduced that must be correctly scaled. The dimensional theory, used in all model tests, is briefly discussed in the article on STRUCTURAL MODEL ANALYSIS. Another type of structural model analysis is discussed in the article on PHOTOELASTICITY.

Another problem arises in ACOUSTIC MODELS. To match the time scale with the length scale, frequency of sound, which is inversely proportional to time, must also be changed.

Models for investigating illumination in an ARTIFICIAL SKY need a uniform linear scale reduction and correct reproduction of the reflectances.

Aerodynamic problems in and around buildings can be investigated in a wind tunnel, including the study of NATURAL VENTILATION. Scaling these models is complex, because it is generally not possible to satisfy all scale requirements simultaneously (see WIND LOADS AND WIND TUNNELS).

References

1. R. JANKE: *Architectural Models*. Thames and Hudson, London, 1962.
2. H. J. COWAN, J. S. GERO, G. D. DING, and R. W. MUNCEY: *Models in Architecture*. Elsevier, London, 1968.
3. D. J. SCHURING: *Scale Models in Engineering*. Pergamon Press, Oxford and Elmsford, N.Y., 1977.

H.J.C.

N

NATURAL BUILDING STONE. The earth's crust is composed of *rock*, which, when excavated for some useful purpose, is termed *stone*. This article deals with *building* or *dimensional* stone as distinct from crushed rock or *aggregate*, and thus with material to be used as masonry blocks, veneer, carved ornamentation, paving, structural units, and for monumental and statuary purposes.

Geological Classification of Rocks. Geologists recognize three classes of rocks according to their origins:

1. *Igneous*. Rock formed by the solidification of molten rock material (magma), which is familiar at the earth's surface as lava. Examples include granite, diorite, porphyry, and trachyte.

2. *Sedimentary.* Rock laid down in the sea or elsewhere and composed of the products (gravel, sand, mud) of the weathering of other rocks together with organic remains or chemical precipitates. Examples include sandstone and limestone.
3. *Metamorphic.* Products of the action of heat, pressure, or chemical changes on rocks usually deep below the earth's surface; they are recrystallized and/or deformed. Examples include marble, slate, gneiss, and quartzite.

Industry Classification of Building Stones. The classification within the stone industry concentrates on the appearance and properties of the final product, not on origin, and to only a minor extent on the composition. The five major types are as follows:

Granite. A hard, massive, coarsely crystalline rock capable of taking a polish. Composed of various proportions of silicate minerals (quartz, feldspars, and dark ferromagnesian minerals) producing varieties differing in color or texture. Ranges in color from pink or red (true *granite*), to gray (*granodiorite* or *diorite*), to dark gray or black (*dolerite*, *norite*, or *gabbro*). Includes textured variants such as *porphyry* (studded with large crystals), *pegmatite* (coarse and blotchy), *trachyte* (very fine and even grained to coarse and lustrous), and *gneiss* (banded).

Marble. A calcareous stone (i.e. composed of carbonate minerals, generally calcite) capable of taking a polish. Three main types are recognized:

1. *Crystalline marble:* true metamorphic rock formed by the recrystallization of limestone. This may be coarse grained (crystal) or fine grained (for statuary), massive, fragmental (breccia), or banded. Colors include white, pink, red, yellow, green, blue, black, and variegated.
2. *Sedimentary marble:* fine-grained, hard, dense limestone that has not been metamorphosed. This may be massive, fossiliferous, brecciated (consisting of angular or elongate fragments), stylolitic (showing jagged lines), or banded. Colors range from white to black through red, yellow, green, and variegated.
3. *Serpentine:* green stone consisting mainly of serpentine and carbonate (igneous ultrabasic rock or metamorphic ophicalcite). Structure is generally fragmental; some is schistone (foliated, cleaved).

Sandstone. A clastic sediment, that is, composed of small particles derived from the weathering of older rocks and deposited as sand grains. The particles are generally of quartz but may include other minerals such as feldspar, mica, rock fragments, clay aggregates, or carbonate. Sandstones are usually subdivided on grain size:

Fine grained	$\frac{1}{16}$ to $\frac{1}{2}$ mm; 0.0625 to 0.5 mm
Medium grained	$\frac{1}{2}$ to 1 mm; 0.5 to 1.0 mm
Coarse grained	1 to 2 mm

Most building sandstones have grain sizes of approximately 0.5 mm and are thus fine to medium grained.

An unconsolidated sand is converted to a sandstone by lithification (hardening) and cementation. The grains may be held together by a chemical cement or clastic matrix (commonly clay).

Limestone. Composed of the mineral calcite (calcium carbonate), some with other carbonates such as dolomite or argonite. It may be organic in origin (composed of the remains of shells or skeletons of organisms), clastic (composed of transported fragments of carbonate), or chemical (precipitates of carbonate from solution). They include hard dense fine-grained material (lithographic stone of Germany), oolitic (composed of small spherical particles, e.g., Bath limestone), coarse fossiliferous types, massive homogeneous (Indiana limestone), to soft, porous (travertine from Rome).

Slate. A fine-grained rock characterized by possessing a strong cleavage, that is, a structural direction along which the stone may be split easily. It is derived from mudstone by metamorphism and most are gray to black, green, or purplish. Thick slabs are used for veneer or paving but thin, platy material is utilized as roofing slates.

Properties, Use, and Durability. The correct use of stone (as massive masonry, veneer, carved ornamentation, paving, either internally or externally) depends mainly on its physical properties, which control performance in service and durability. Stone is a natural material and therefore variable in its properties; hence data on the more important physical properties, summarized in Table 1, are given as ranges.

References

1. *Code of Practice on Stone Masonry*. B.S. 5390:1976. British Standards Institution, London, 1976.
2. R. J. SCHAFFER: *The Weathering of Natural Building Stones*. D.S.I.R. Special Report 18. Her Majesty's Stationery Office, London, 1932 (reprinted 1972).
3. E. M. WINKLER: *Stone: Properties, Durability in Man's Environment*. Springer-Verlag, Berlin and New York, 1975.

A.H.S.

NATURAL VENTILATION is generally defined as the replacement of stale air by fresh air through the forces of

TABLE 1
Typical Physical Properties of the More Important Building Stones

Stone Type	Strength-Compressive Dry MPa	Strength-Compressive Dry psi	Strength-Compressive Wet MPa	Strength-Compressive Wet psi	Tensile Strength (Dry) MPa	Tensile Strength (Dry) psi	Modulus of Rupture Dry MPa	Modulus of Rupture Dry psi	Modulus of Rupture Wet MPa	Modulus of Rupture Wet psi	Apparent Porosity (% water absorp. by vol.)	Bulk Density t/m^3	Bulk Density lb/ft^3	Absolute Specific Gravity (grain density)	Expansion: Coeff. Thermal Expans. ($°C \times 10^{-6}$)	Expansion: Linear Expansion (in./ft · 160°F)
Granite																
Red,	60	9,000	40	6,000	7	1,000	8	1,200	7	1,000	0.02	2.56	160	2.60	7	0.13
gray	to	to	to	to	to	to	to	to	to	to	to	to	to	to	to	to
	150	22,000	120	17,000	25	35,000	120	3,000	15	2,000	0.05	2.9	180	2.80	8	0.15
Black	80	12,000	60	9,000	20	3,000	15	2,000	12	1,700	0.02	2.8	175	2.85	5	0.09
	to	to	to	to	to	to	to	to	to	to	to	to	to	to	to	to
	210	30,000	140	20,000	35	5,000	35	5,000	25	3,500	0.05	2.9	180	2.95	6	0.11
Marble	50	7,000	40	6,000	7	1,000	5	700	5	700	0.05	2.70	170	2.75	1	0.02
	to	to	to	to	to	to	to	to	to	to	to	to	to	to	to	to
	200	30,000	140	20,000	20	3,000	30	4,000	30	4,500	0.8	2.85	180	2.85	12	0.22
Sandstone																
Hard	40	6,000	20	3,000	8	1,100	10	1,500	2	300	6	2.25	140	2.60	10	0.19
	to	to	to	to	to	to	to	to	to	to	to	to	to	to	to	to
	80	12,000	60	9,000	25	3,500	20	4,000	8	1,200	10	2.50	160	2.75	11	0.22
Soft	5	700	1	150	4	600	3	400	1	150	10	2.0	125	2.60	9	0.17
	to	to	to	to	to	to	to	to	to	to	to	to	to	to	to	to
	40	6,000	20	3,000	8	1,100	10	1,500	2	300	30	2.25	140	2.75	10	0.19
Limestone																
Hard	30	4,000	20	4,000	15	2,200	10	1,500	6	900	1	2.56	160	2.65	8	0.15
(high	to	to	to	to	to	to	to	to	to	to	to	to	to	to	to	to
density)	80	12,000	60	9,000	25	3,500	30	4,000	20	3,000	7	2.65	165	2.75	9	0.17
(Mid-	12	1,700	5	700	5	700	5	700	3	400	7	2.16	135	2.65	7	0.13
density)	to	to	to	to	to	to	to	to	to	to	to	to	to	to	to	to
	30	4,000	12	1,700	15	2,000	10	1,500	6	900	12	2.56	160	2.75	8	0.15
Soft	5	700	2	300	1	150	1	150	0.1	15	12	1.76	110	2.65	4	0.07
(low	to	to	to	to	to	to	to	to	to	to	to	to	to	to	to	to
density)	12	1,700	5	700	5	700	5	700	3	400	50	2.16	135	2.75	7	0.13
Slate	70	10,000	50	7,000	7	1,000	40	6,000	40	6,000	0.01	2.6	165	2.75		
	to	to	to	to	to	to	to	to	to	to	to	to	to	to	9	0.17
	160	23,000	100	15,000	20	3,000	100	15,000	100	15,000	0.5	2.8	175	2.80		

nature. Satisfactory ventilation, however, means more than this because, besides providing a healthy and refreshing atmosphere, it must also keep it thermally comfortable, which does not necessarily require the replacement of stale air (Ref. 1). Fresh air in this context means air that is relatively free from harmful impurities and objectionable odors.

As far as ventilation requirements for health are concerned, the more important considerations are as follows:

1. To ensure an adequate supply of oxygen for respiratory purposes.
2. To prevent an undue rise in the carbon dioxide content of the air.
3. To minimize the risk of infection by airborne bacteria.
4. To avoid contamination of the air by hazardous gases (these include carbon monoxide, which may be given off by fuel-burning appliances or internal combustion engines, gases emanating from furnishings, and radioactive radon from concrete).
5. To reduce the concentration of unpleasant and other odors to acceptable levels.

Whereas some of these considerations may be of particular significance in industrial environments, the last mentioned is considered to be the most rational criterion on which to base minimum standards of ventilation for spaces devoid of industrial contaminants. Although a person's ventilation requirements vary with age and the air space allocated, a ventilation rate of 8 liters per second per person is normally accepted as a minimum standard for spaces in which moderate smoking occurs.

For industrial applications involving toxic substances, standards such as the threshold limit values (TLVs) of the American Conference of Governmental Industrial Hygienists (Ref. 2) are normally used for deciding whether a given airborne concentration of a toxic substance is acceptable. More often than not, positive ventilation using mechanical aids is required to control such substances.

For thermal comfort, ventilation is needed to remove excess heat. The amount of air required for this purpose can be many times that required for health, and it should be calculated separately for each circumstance (Ref. 1). Ventilation can also be used to control the temperature of a building. Thus, for example, in summer, ventilation by night can remove heat stored in heavy mass structural elements to give a cooler building during the day. Similarly, although less important, warm daytime air in winter can be used in the reverse process.

Air movement is an equally important factor in ventilation. High air speeds are very effective in cooling the body by the evaporation of perspiration, particularly under warm humid conditions. Lack of air movement on the other hand can result in a feeling of stuffiness.

Two forces are at work in the natural ventilation of buildings: temperature and wind (Ref. 1). Whenever a difference in temperature, and thus a difference in density, exists between the indoor and outdoor air, flow will take place if ventilation openings are provided at different levels. This is the well-known chimney or stack effect. When a wind blows across a building, pressure differences are set up, and if ventilation openings have been provided in the right places, air flow will result. Typical pressure distribution patterns over a single-roomed building are illustrated in Fig. 150 for two different roof pitches.

Since winds are not generally highly directional, ventilation openings, preferably of equal area, should be situated in as many external walls as possible to ensure effective cross-ventilation. In the case of industrial buildings, roof openings can be very effective if the roof has been correctly designed and the openings properly placed (Ref. 3).

The design and position of windows is also important for body cooling purposes. Openable sashes, for example, can be used to direct incoming air onto or away from the occupants, as demonstrated in Figs. 151a and b.

Natural ventilation has its limitations in practice because it is dependent not only on the vagaries of the weather but also on how well it is regulated by the tenants.

See also AIR CONDITIONING AND HEATING; ANEMOMETERS; PASSIVE SOLAR DESIGN; and WIND TUNNELS.

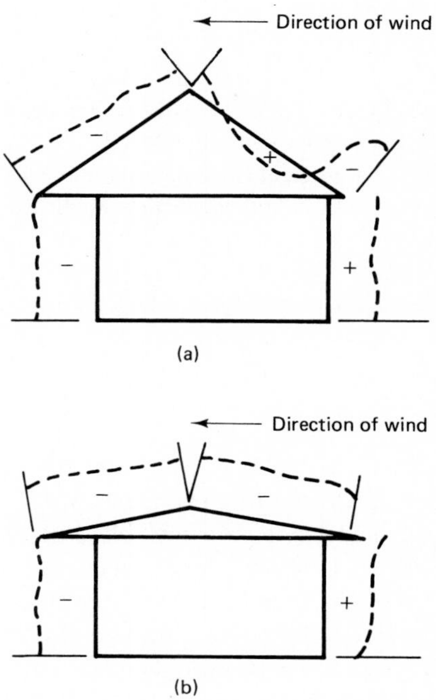

Figure 150 Typical static pressure distribution patterns over buildings with different roof pitches. (a) Roof pitch of 35°. (b) Roof pitch of 10°.

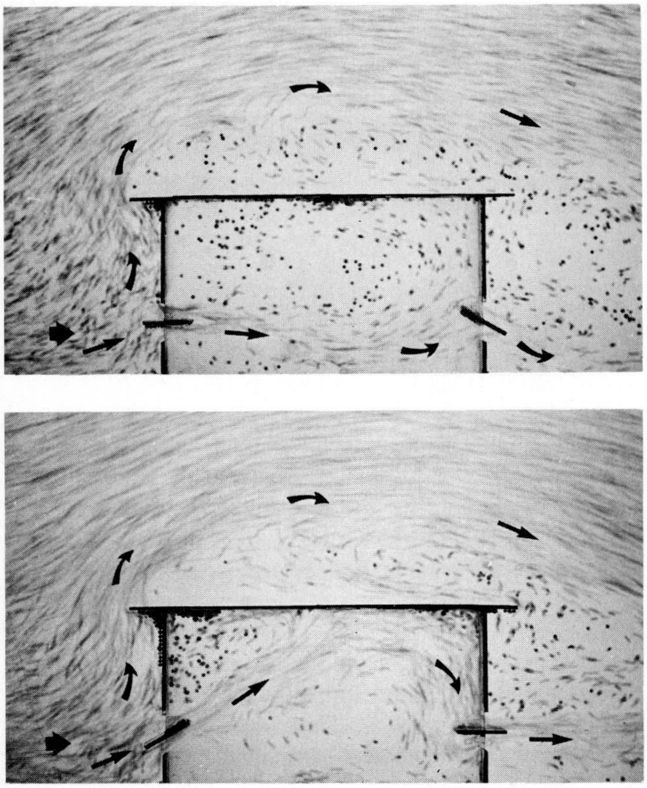

Figure 151 Influence of window design on the direction of air movement. (above) Sash set for summer ventilation. (below) Sash set for winter ventilation.

References

1. J. F. VAN STRAATEN. *Thermal Performance of Buildings*. Applied Science Publishers, London, 1967.
2. American Conference of Governmental Industrial Hygienists. *Handbook of Threshold Limit Values*. Published annually by the ACGIH.
3. *Guidelines for Factory Building Design*. K61, National Building Research Institute, Pretoria, South Africa, 1982.

J.F.v.S.

NETWORK ANALYSIS. The best technique for the planning and scheduling of construction work of all types is by network analysis, commonly called the *critical path method* (CPM); it provides more precise answers to all manner of problems than those available from the simple bar-chart form of construction program. It provides a diagram that is, in fact, a works program, but from which one can see at a glance which processes and operations are the vital ones controlling the smooth execution of the project.

The first step is to take each of the separate operations comprising the project and put them in sequence; for any specific operation, this depends on which operations must precede, which must follow, and which can be done at the same (or any other) time. The various processes, or parts of processes, are then arranged in loose sequence, and a network diagram or an arrow diagram is drawn up, as in Fig. 152. On this diagram each separate operation is represented by an arrow; related jobs are shown by arrows in consecutive series; jobs that can proceed concurrently are indicated by arrows in parallel. Arrows need not be drawn individually in proportion to time, but may be of any convenient length; they are merely drawn from left to right to indicate the flow of work in relation to each other. Hence the completed diagram at this stage shows all the operations required for the project in proper perspective and ensures that details are not ignored. The diagram is therefore as complicated (or as simple) as the project being planned. More important, it enables bottlenecks to be clearly seen.

Time Estimates. The next step is to add to the diagram the time estimates for each operation, so that on each arrow now appears the number of shifts required to carry out each process. The critical path is now revealed by simple addition; it is the chain of consecutive arrows anywhere on the diagram that requires the greatest time to complete. The total time along this path is the total time for the project, and any changes in the time required for any of the operations on this path must result in an equal change in the overall project time. These are therefore the operations requiring the most intensive planning so that they may be done within their estimated times, if the work is to be completed on schedule. Other operations in parallel with these are not as critical, as they have some leeway in starting and finishing times.

Critical Path. Figure 152 shows the critical path for the simple operation of installing a works water supply for any construction project. Examination of this diagram shows at once that the critical process is the erection of the tank stand and tanks and that the whole job will take 20 shifts to complete. It is also seen that the pipe trench setting out, excavation, and pipe laying have a leeway margin of 6 shifts, so a short delay is not vital here; the first task is not to set out the pipeline, but to start work on the tank stand. This example is a very elementary illustration intended merely to serve as an introduction to the subject; in practice, critical path diagrams of this type are prepared for every operation in the project, and then these can be correlated into one large comprehensive diagram for the whole works. At this point the starting and finishing dates for each item can be established, and alternative methods of executing parts of the work may be examined to see if the cost can be reduced without overrunning the contract time. This having been done, the truly critical operations are now obvious

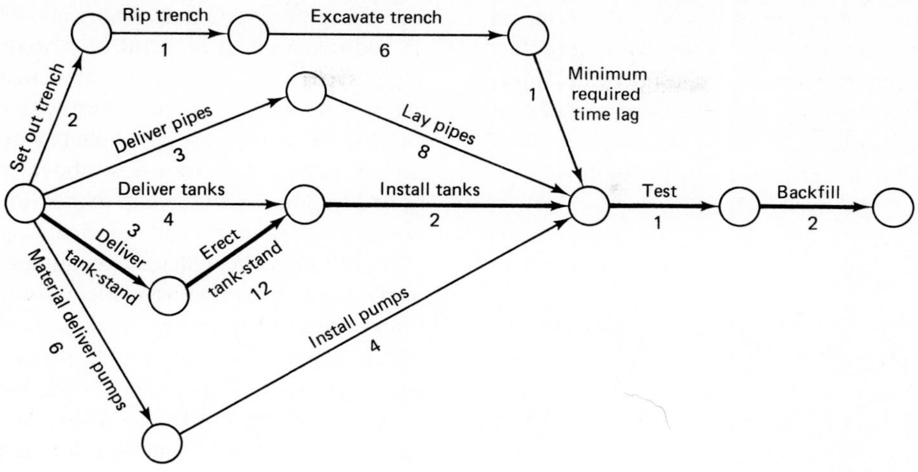

(a)

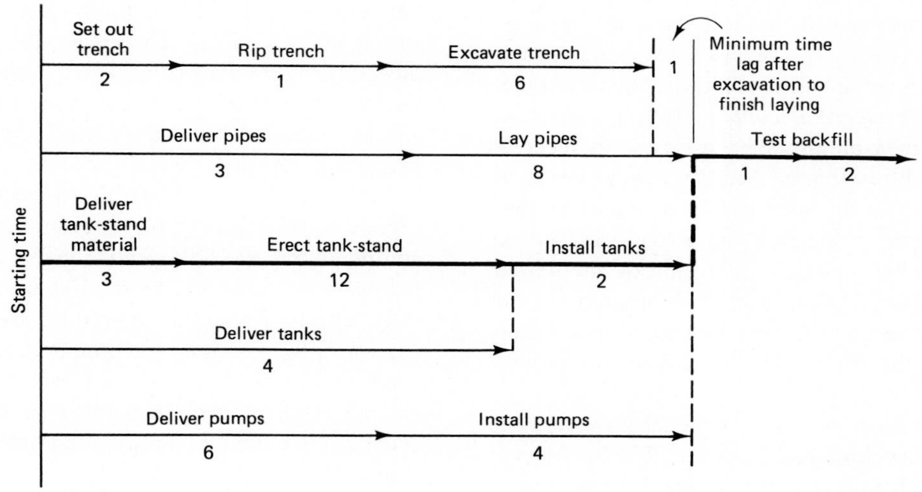

(b)

Figure 152 Critical path diagrams for the installation of a works water supply. (a) Network diagram. (b) Arrow diagram.

and may be thoroughly investigated when contingencies are under consideration. From this comprehensive diagram other planning schedules may also be drawn up with greater accuracy than was possible with the bar-type construction program. If desired, the critical path diagram of the arrow type may be drawn to a suitable time scale, so as to make it more understandable to people accustomed to bar charts.

Works Control. During construction, the critical path diagram for the project is also most important for works control; should any unforeseen delays occur, the construction manager may determine immediately whether or not they will affect the critical path for the project. If so, overtime may be warranted to retrieve the position; if not, advantage may be taken of the available leeway.

Thus the use of the critical path method of planning leads to a better understanding of any project, for it must not be forgotten that, as each operation on the critical path is shortened by the adoption of a more efficient approach, the path will possibly move to another part of the diagram, thus necessitating a reappraisal of other operations. The critical path is always the longest time path through the project; the works cannot be completed in less than that.

Cost Analysis. This technique can be extended to cover cost analysis also and to permit the computation of comparative costs for different time schedules or alternative construction methods. Manual diagrams and calculations are practicable for projects comprising up to about 400 operations; beyond this, recourse must be made to computers.

If the critical path method is applied to project costing in detail, the sheer number of cost combinations on a project necessitates that computers be used. This, however, can be a profitable application, for in this way the most economical duration of a project can be determined and a construction program built around this optimum construction time.

Reference

J. M. ANTILL and R. W. WOODHEAD: *Critical Path Methods in Construction Practice*. Wiley, New York, 1982.

J.M.A.

NOISE can affect the human in two ways, either through actions generated by mental stimulus or by a physiological reaction. Both are interrelated, but it is usual to separately attempt to quantify each. The acoustic characteristics of noise can be expressed in terms of three basic physical parameters: *Intensity*, *spectral content* by frequency, and *time*. Intensity is a measure of the acoustic energy received primarily by the auditory senses. Spectral content indicates the quality of the noise by identifying the high- or low-frequency components of the received sensation. Time allows discrimination between noise of a short or intermittent character and those noises that continually impinge on our senses. These parameters have been combined in a variety of units to measure the effects of noise on human response.

Loudness is perhaps the most elementary relationship between noise and psychological response. The loudness of a noise is defined in terms of the subjective response to each constituent frequency component judged against a standard sound level centered on 1000 Hz. The summation of these subjectively weighted components of the noise is defined as *loudness level*. The pioneering work of Stevens (Ref. 1) set the pattern for much of the subsequent psychoacoustic research. An attempt to broaden the scope of the relationship between noise and its effect was the introduction by Kryter in 1958 (Ref. 2) of the term *noisiness*. The experimental techniques adopted were similar to those of Stevens, but the results were presented in the form of noisiness units defined as the *perceived noise level*, which takes account of the tonal and temporal characteristics of the noise. This procedure has been adopted in the measurement of aircraft noise but has proved to be too complicated for universal application.

Response. Laboratory experiments have provided valuable data, particularly in relationship to the three basic physical parameters, but there is one important ingredient missing from this approach: the difficulty in simulating normal living or working environments. These factors provide an additional effect in terms of attitudes and social conditions that cannot be readily simulated in the laboratory; hence a parallel method of investigation has developed that attempts to relate noise and human response through social surveys. Questions are asked about environmental conditions and social attitudes as related to measured noise exposure. To provide valid information, great care must be exercised in the design and execution of these surveys. The results have provided valuable information, allowing environmental noise criteria and regulations to be derived. The data are usually presented in the form of a *dose–response function* (Fig. 153). The physical measure of noise exposure can be expressed mathematically in a variety of ways, but the most commonly adopted unit is the *equivalent continuous energy level* (L_{eq}) of a noise expressed in terms of a decibel unit (dBA) whose frequency response is called A-weighting. The A-weighting roughly corresponds to the loudness or noisiness of a noise dBA, and the noise exposure is defined as L_{eq}. The great value of the A-weighting procedure is its simplicity and the ability to make direct measurement of the value with relatively simple and inexpensive measurement instruments (see SOUND-LEVEL METER).

The data plotted against the subjective scales shown in Fig. 153 are derived from the results of the social surveys and relate the psychological response of a surveyed population to its noise exposure. Experience has shown that these effects are readily reproduceable for high levels of exposure, for example, in communities beside major highways or near busy airports. When levels of noise exposure are not dominated by a single type of noise, social and economic factors play an important role in determining response.

Audibility. It is difficult to discriminate between the psychological and physiological effects of noise, particu-

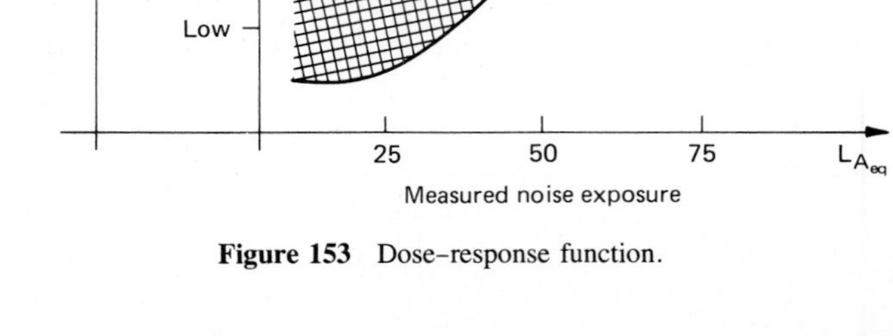

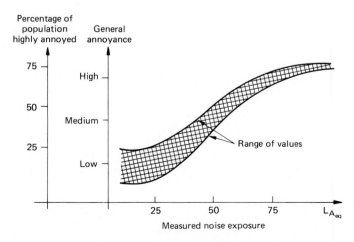

Figure 153 Dose–response function.

larly at low levels of intensity. Discrimination is also difficult between noise and vibration at frequencies below 10 Hz; such noise is primarily sensed through skeletal response rather than auditory perception. Noise to be heard must intrude above the *threshold of audibility*. Noises generated within the human body form the background noise that determines this threshold and is the zero base line for the decibel (dB). Audibility is not the only factor that can stimulate human response; for example, during periods of sleep the central nervous system can be aroused at around 20 dBA L_{eq} (Ref. 3), even though the sleeper will not hear the noise.

Damage to Hearing. Many laboratory studies have shown that exposure to noise can increase the heartbeat rate, change the surface temperature of the skin, interfere with sleep patterns, and produce many other physiological effects. But many other nonacoustic factors produce the same stressful effects; and in spite of much research, no conclusive evidence has been produced to indicate that the level of noise exposure experienced by most of us produces a sustained and clearly identifiable effect on our physiological well-being. There is one very important exception to this conclusion, the effect induced by prolonged exposure to high levels of noise. A significant proportion of industrial workers suffer the hazards of accelerated deafness due to excessive noise exposure. All of us lose some ability to hear high-frequency sounds as we become older, a natural effect called *presbycusis*, but exposure to exceptional levels of noise can increase this loss. Many investigations have been carried out (Ref. 4) to establish safe levels of noise exposure. These are called *damage risk criteria*. It is now generally accepted that workers over their normal working life should not be exposed to levels of noise during an 8-hour working day in excess of 85 L_{Aeq}. Higher levels of exposure require a shorter daily working period. Except for those sections of the population who are exposed to hazardous levels of noise, the majority of us are affected by noise primarily because it interferes with our normal daily activities. This disturbance is a social rather than a medical problem; but it is a social problem on a grand scale, and there is no excuse for ignoring or minimizing the effects.

See also ACOUSTICAL REQUIREMENTS FOR BUILDINGS; ACOUSTIC TERMINOLOGY AND MEASUREMENTS; SOUND ABSORPTION; and SOUND INSULATION AND ISOLATION.

References

1. S. S. STEVENS: Measurement of loudness. *Journal of the American Society for Acoustics*, Vol. 27 (1958), pp. 818–820.
2. K. KRYTER: *The Effects of Noise on Man*. Academic Press, New York, 1970.
3. H. BASTENIER, W. KLOSTERKOETTER, and J. B. LARGE: *Environment and Quality of Life*. Document EUR 5398e. European Economic Community, Luxembourg, 1975.
4. *Acoustics—Procedure for Describing Aircraft Noise Heard on the Ground. Addendum 1: Measurement of Noise from Helicopters for Certification Purposes. ISO 3891/DAD 1.* Draft. International Organization for Standardization (ISO), Geneva, 1981.

J.B.L.

NONFERROUS METALS. The great variety of metals and alloys commonly used in building and construction has traditionally been classified into the ferrous (iron and steel) and nonferrous divisions. Although making little sense from the purely metallurgical viewpoint, this division is based on the modern usage of metals and alloys, which is dominated in terms of sheer quantity by the ferrous group. The nonferrous group, about 10% of overall usage, is in turn dominated by aluminum, copper, lead, zinc, nickel, and magnesium alloys, in descending order. Since none of these alloys can compete with the iron-based alloys in cost and strength, they are used because of their special physical or chemical properties.

Copper Alloys have a unique combination of characteristics: high thermal and electrical conductivity, excellent corrosion resistance, good ductility and formability, and an attractive color for architectural purposes. It is possible to divide all the commercial copper alloys into single-phase and polyphase alloys. The former group is based on pure copper and consists essentially of strengthened versions of the solid solution of that attractive ductile metal, whereas the two-phase alloys rely more on the addition of other elements to obtain their particular useful properties.

The best-known copper-based alloys, the *brasses*, belong to the copper–zinc group. Alloys containing 30% and 35% zinc, for example, are often referred to as cartridge or yellow brass. However, the old terminology can often be confusing, because the term *bronze*, traditionally reserved for copper-based tin alloys, is also frequently used for certain copper–zinc alloys. This confusion has been overcome by a three-digit alloy classification series established by the Copper Development Association (Ref. 1). The digits identify cast and wrought alloys and the major alloying elements.

The simple alloys of copper and zinc are generally modified by adding manganese, tin, iron, and aluminum to increase strength, increasing in the order given. Lead is sometimes included to improve the machinability of brass, but at the expense of some strength and formability.

The term bronze has also been given to alloys with properties somewhat superior to the common brasses. For example, architectural bronze is a copper-based alloy containing 40% zinc and 3% lead. Two other commonly used materials, silicon bronze and aluminum bronze, are also tin free; they are alloys of copper with silicon and aluminum, respectively.

The Uniform Construction Index, adopted by the Construction Specifications Institute, the American Institute of Architects, the Consulting Engineers Association, and the Associated General Contractors of America, lists the products of copper and its alloys according to their major architectural end-use applications. This index is the basis of the CDA Building Products Source Book (Ref. 2), which provides a useful cross reference between the various alloys and their end use in an architectural environment.

Lead Alloys. The unusual physical and chemical properties of lead form the basis of its use in the building and construction industry. A combination of low melting point, high density, malleability, and an excellent resistance to corrosion provides an attractive basis for its use in sheet and foil, extrusions, castings, laminations, and other specialist forms. Coatings of lead over steel and other metals are now in widespread use. For example, steel sheet coated with an alloy of lead with 15% tin, commonly termed terneplate, finds widespread use as a roofing material. The Lead Development Association (Ref. 3) and the Lead Industries Association, Inc. (Ref. 4) both publish brochures on the use of lead and its alloys, which refer specially to their architectural uses. In addition to the more traditional uses in sheets, extrusions, and castings, special applications now include plenum barriers, antivibration mountings, radiation shielding, and the control of noise, which utilize the high internal damping characteristic of the metal.

Other Nonferrous Alloys. Although several wrought alloys suitable for use as rolled strip, sheet, or foil are available, *zinc* is not employed to any great extent as an alloy in the construction industry. Apart from brass, the major use of zinc as a structural material is in the form of alloys for pressure die casting. This development resulted from metallurgical advances and a clearer understanding of the effects of minor impurities on the mechanical and corrosion properties of zinc. The base alloy is now high-grade zinc of 99.99% + purity, with approximately 4% aluminum and 0.04% magnesium.

The other major use of zinc is in galvanized steel, which is steel with a thin coating of zinc. As the name implies, the zinc coating resists the galvanic or corrosive action of many otherwise aggressive environments on the underlying steel.

Magnesium and its alloys form a special class of materials. They exhibit extreme lightness coupled with a surprisingly high strength when appropriately alloyed. Technically, they are competitive with aluminum alloys, but their relatively high cost is a disadvantage. Their nomenclature is similar to the aluminum series.

The common wrought magnesium-based alloy utilized for sheet and plate contains 3% aluminum, 1% zinc, and 0.2% manganese and is designated AZ31B. The corresponding high-strength heat-treatable alloy, AZ80A, containing 8% aluminum, 0.2% zinc, and 0.2% manganese, is frequently used for forgings and extrusions.

See also ALUMINUM and METALLURGY.

References

1. Copper Development Association Inc., 405 Lexington Avenue, New York, NY, 10174.
2. *Copper Brass and Bronze Building Products Source Book No. 417/3*. Copper Development Association, Inc., New York, 1980.
3. Lead Development Association, 34 Berkeley Square, London W1X 6AJ.
4. Lead Industries Association, 292 Madison Avenue, New York, NY 10017.

A.E.J.

O

OPTIMIZATION. All of human endeavor is goal seeking. Although early Greek philosophers postulated certain perfect ideas, the concept of "doing the best with the available resources" has become deeply rooted. With the development of calculus in the 18th century, it became possible formally to prove the existence of an *optimum*, which matches the preceding concept. Thus, optimization is the process of locating the optimum. Examples abound in building design and performance: designing the minimum-cost building, or the maximum-area building, or the building that has the most daylight, or the minimum ingress of sound.

Modern optimization methods are almost invariably computer based. They can be categorized into two broad streams: those based on calculus and those based on numerical methods. The latter are called *mathematical pro-*

gramming. If both the goal to be optimized and the constraints on available resources can be described using linear algebraic relationships, the techniques associated with *linear programming* can be used.

When conflicting goals are to be optimized simultaneously, the notion of a single, unique optimum is inadequate; it must be replaced by a series of solutions each equally optimal in some way. The designer must trade off performance in one goal against performance in another in order to decide on a final solution. For example, a series of designs with differing capital costs and running costs may have been produced. There will be one design that represents the minimum capital cost and the maximum running cost and another the minimum running cost and the maximum capital cost. In between are designs that cost more than the minimum capital cost but less than the maximum running cost. The production of these intermediate solutions forms a subset of optimization, as does the trade-off process necessary to select the final solution. Optimization provides the potential to produce the best possible.

See also FUZZY SETS, ANALYTICAL HIERARCHIES AND DECISIONS; LIFE CYCLE COSTING; and VALUE ENGINEERING.

References

1. R. J. AGUILAR: *Systems Analysis and Design in Engineering and Architecture*. Prentice-Hall, Englewood Cliffs, N.J., 1973.
2. J. N. SIDALL: *Optimal Engineering Design*. Marcel Dekker, New York, 1982.
3. J. S. GERO (ed.): *Optimization in Computer-aided Design*. North-Holland, Amsterdam, 1985.

J.S.G.

P

PAINT is a formulated product applied in liquid form to a surface; when dry, it fulfills one or more of a variety of purposes. The principal roles of paint, however, are to decorate and protect.

Decorative paints can be produced in almost any color, gloss, and texture; some also resist wear and tear, staining, and dirt retention. As a protective barrier, paints may afford resistance to chemical or mechanical attack of the substrate, such as in the case of an anticorrosive coating on steel or as a sealing system to bind a friable surface, respectively.

Paints are also required for a variety of other purposes, including marking and identification, heat reflection or absorption, and protection against condensation, moisture penetration, or marine fouling.

History. Although paints were in wide use by the middle of the 18th century, until the 1880s they were invariably prepared by hand using binders from natural sources, including shellac, casein, vegetable oils, lime, wax, bitumen, and tars. Gloss paints were usually based on fish oil or vegetable oils such as castor and linseed oil; flat finishes were typically poorly bound products such as whitewashes and pigmented distempers. Pigments were mainly naturally occurring minerals, with a few selected manufactured compounds, notably white lead, lamp black, and Prussian blue.

House painting was popularized with the introduction of ready-mixed paints in the late 19th century; however synthetic binders did not appear until the early 1920s. The range of synthetic binders has expanded greatly in recent years; today binders are almost exclusively synthetic.

Constituents. Conventional liquid paints consist essentially of the vehicle, or liquid phase, comprising the binders dissolved or suspended in a volatile medium, and the solid pigments and extenders. Upon application of the liquid paint the volatiles evaporate from the film, and the binder, or film-forming constituent that remains, encapsulates the pigment and extender particles. The pigments and fillers are responsible for the opacity and largely for the decorative appearance of the paint. Products free from pigments are termed varnishes. Along with these key components of paint, there are a large range of additives, including driers, surfactants, antisettling agents, ultraviolet stabilizers, mold inhibitors, tinters, and colorants.

Binders. The binder largely dictates the physical and chemical properties of the paint and can be categorized in a number of ways; one categorization is into thermoplastic (nonconvertible) and thermosetting (convertible) types. Thermoplastic binders, by definition, may be softened by heat and cooled without undergoing chemical change. The dry film is formed solely by evaporation of the solvent or emulsifying liquid or by cooling the molten product. Common thermoplastic binders are acrylics, polyvinyl acetates, coal tars, and chlorinated rubbers. The thermosets solidify

TABLE 1
Thermosetting (Convertible) Coatings

Type	Principal Drying Mechanism(s)	Advantages	Limitations	Principal Uses and Dry Film Thickness
Drying oils	Solvent evaporation and reaction with atmospheric oxygen.	Very good wetability. Wide color/gloss range.	Slow drying, soft. Poor water immersion and chemical resistance, particularly alkalis and solvents.	Decorative painting, mainly of timber and ferrous metals. Largely superceded because of slow drying. DFT, 40 μm.
Alkyds (typical oleoresinous)	Solvent evaporation and reaction with atmospheric oxygen.	Good substrate wetability. Faster drying than oils. Wide color/gloss range.	Poor water immersion and chemical resistance, particularly alkalis and solvents.	Decorative painting of buildings and furniture. DFT, 40 μm.
Epoxies (2 part)	Chemical interaction of 2 parts.	Very hard, tough. Good chemical, heat, and abrasion resistance. Usually good water immersion. Wide color/gloss range.	Chalks on weathering. Slow cure below 10°C. Costly.	Chemically resistant floor screeds, paving paints, bench tops, and wear surfaces. DFT, 50 μm. High build, 150 μm.
Polyurethane (1 part)	Solvent evaporation and usually reaction with atmospheric moisture.	Harder and more resistant to chemicals and weathering than alkyds.	Oil-modified types are alkali susceptible.	Decorative paint for furniture and moderate wear surfaces. Clear interior finishes. DFT, 40 μm.
Polyurethane (2 part)	Chemical interaction of 2 parts.	Very hard. Chemical, abrasion, and weather resistant. Good gloss retention. Wide color/gloss range.	Epoxy primer required. Slow cure below 10°C. Costly.	Decorative weather and/or wear-resistant surfaces, e.g., bench tops, boats. DFT, 40 μm.
Zinc silicate	Solvent evaporation and reaction with atmospheric moisture.	Extremely hard. Heat and weather resistant.	Poor resistance to chemicals and solvents. Nondecorative. Flat gray spray paint only. Costly.	Primer system for steel. DFT, 80 μm.

Thermoplastic (Nonconvertible) Coatings

Type	Principal Drying Mechanism(s)	Advantages	Limitations	Principal Uses and Dry Film Thickness
PVA latex	Water evaporation.	Rapid drying. Low odor. Easy application. Wide color range.	Poor weather, wear and solvent resistance. Drying impaired below 8°C.	Interior decorative painting. DFT, 30 μm.
Acrylic latex	Water evaporation.	Rapid drying. Low odor. Good weather resistance. Wide color range.	Poor solvent and wear resistance. Drying impaired below 8°C.	Decorative painting. Particularly suited to exterior exposure. DFT, 30 μm.
Vinyl (solvent borne)	Organic solvent evaporation.	Good weather, water immersion, and chemical resistance. Wide color range.	Poor solvent resistance. Contains highly volatile solvents.	Protection of steel, especially in immersion service. DFT, 40 μm.
Bitumen or tar paints	Organic solvent evaporation.	Good substrate wetability. Moderate water and chemical resistance. Low cost.	Poor solvent resistance. Variable weather resistance. Black or silver. Soft.	Roof repairs, cast-iron pipes, underside of tank roofs, poorly prepared surfaces. DFT, 30 μm. High build, 200 μm.
Chlorinated rubber	Organic solvent evaporation.	Good chemical and weather resistance. Wide color range.	Poor solvent resistance. Contains toxic, volatile solvents.	Protection of steel and concrete in external, aggressive service, swimming pools, etc. DFT, 50 μm.

first by evaporation of any volatile solvents and then, more significantly, by the linking together or condensation of small molecules to form a much larger molecule (polymer). This reaction may variously be achieved by reaction with atmospheric moisture or oxygen or the mixing of two separate reacting components just prior to application. Thermosets include drying oils, alkyds, epoxies, and polyurethanes. Table 1 summarizes some of the properties and uses of the more common paint types.

Pigments are incorporated in paints for several reasons:

1. *Optical:* opacity, gloss, color.
2. *Protective:* durability and corrosion, fire, or mold resistance.
3. *Reinforcing:* thickness, durability, or hardness improvement.

The ability of a paint to hide the substrate, its opacity, is dependent not only on the amount of pigment and paint thickness, but also on the pigment color. For example, white and pure yellows and reds exhibit poor hiding power, while the addition of darker colors to these results in a marked improvement in their respective opacities. Some pigments have little inherent opacity, but are used in support of the more expensive hiding pigments. These, termed extenders or fillers, provide body to the paint, improve brushability, and increase the film build, cohesion, and, where applicable, the texture at a lower cost than would be achieved with more costly pigments. Typical extenders are talc, clays, silica, gypsum, whiting, and barites.

Pigment type and proportion also control gloss. When pigment content is low, all the pigment is below the surface of the film with a resulting planar, glossy surface. With increasing pigment content, the pigment particles increasingly disrupt the surface and reduce the gloss. At the critical pigment volume concentration (CPVC), the point at which there is just sufficient binder to wet the pigment particles, and coincidentally the maximum PVC that permits a continuous relatively impermeable coating, the finish is regarded as flat. Textures are similarly produced by varying the coarseness and content of pigment.

Classification of pigments is usually according to color, and a very wide range of inorganic, organic, and organometallic types is available. However, pigment selection is not just a matter of esthetics, and the resistance of a given pigment to light, weathering, selected chemicals, and bleeding, its compatibility with the binder, and particle size can also be relevant. The characteristics of the more popular pigments are well established and readily available through pigment manufacturers and texts.

Typical common pigments include titanium dioxide and zinc oxide whites, carbon and lamp blacks, toluidine, para and iron oxide reds, hansa, chrome and iron oxide yellows, phthalocyanine and Prussian blues, phthalocyanine green, molybdate and chrome oranges, and metallic pigments, principally aluminum and zinc. Specialized anticorrosive pigments such as red lead and zinc chromate, fire-retardant antimony oxide, and marine antifouling pigments such as cuprous oxide are also important.

Pigments protect both the binder and substrate. Organic coatings are invariably degraded by moisture, light, and heat, and pigments retard this influence. The effect of pigments on the weathering performance of paints can be seen in the poor durability of unpigmented paints of the same generic type. Similarly, light-colored products tend to be superior to their darker counterparts outdoors because of the greater tendency of light colors to reflect the damaging incident radiation. Laminar pigments, notably micaceous iron oxide and flake aluminum, are particularly successful in improving resistance to weathering because of the multiple layering of these pigments, which acts more effectively than rounded particles to shield the binder and substrate. Some pigments inhibit deterioration of the substrate. For example, zinc metal pigment reacts sacrificially to protect steel; other pigments produce alkaline or oxidizing environments that retard the corrosion that would otherwise occur when moisture diffuses through the coating to the substrate.

Pigments are also important in modifying the physical properties of the binder. They provide body, enabling greater film thickness and increased hardness and abrasion resistance, and can improve film cohesion.

Solvents are necessary to control the consistency of the paint and allow it to be readily applied as a homogeneous mixture. The solvent must also evaporate at an acceptable rate and in a manner that retains the nonvolatile phase as an even film. The principal solvents are aliphatic and aromatic hydrocarbons, alcohols, esters, ketones, glycolethers, and water, the latter being used either as a solvent or as an emulsifier.

In recent years there has been a marked increase in the use of water-based formulations, due both to the increasing cost of organic solvents and because of statutory restrictions, which limit the use of organic solvents that generate photochemical smog. The toxicity, dermatitic properties, and flammability of many organic solvents are additional hazards.

Additives. Conventional paint essentially comprises pigment, binder, and volatile solvent, but commercial products are usually quite complex and typically contain perhaps 15 or more different constituents, although in quantity the minor additives would represent only a few percent of the formulation. Typical formulations include a balance of pigments and extenders, a blend of solvents, the binder and perhaps a modifying plasticizer, tinters, driers, preservative, stabilizers, and thickening agent.

For conventional solvent-borne alkyds, driers are perhaps the most important additive. These are usually soaps of metals such as lead, calcium, cobalt, or manganese; they accelerate drying, which occurs by reaction with atmospheric oxygen. Without them drying would be unacceptably slow.

Despite all these complexities, a wide range of established formulations is readily available in appropriate texts and through raw material suppliers.

Film Formation. Paints cure by a number of different mechanisms, and these are often used as a way of categorizing products. The principal modes are as follows.

Evaporation of the Continuous Phase of an Emulsion. This is the mechanism by which latex paints cure; as the water evaporates, the small resin droplets coalesce to form a continuous film.

Solvent Evaporation and Reaction with Atmospheric Oxygen. Binders that form film by this mechanism include the natural drying oils (linseed, tung, and fish oils) and the synthetic oil-modified resins, such as the alkyds and oil varnishes. The mechanism of film formation involves the cross-linking of largely linear molecules with oxygen to form a continuous, rigid, three-dimensional structure.

Solely by Solvent Evaporation. Nitrocellulose lacquer is the traditional example of this type of film formation; however, chlorinated rubbers, vinyls, and indeed any thermoplastic resin dissolved in a solvent system cures by this physical mechanism. With all solvent-borne paints, solvent evaporation is involved in the drying process, albeit often complementary to the other mechanisms, such as oxidation and chemical combination.

Chemical Combination. This involves chemical interaction of different compounds that are mixed together immediately prior to application to form a solid polymer, usually at ambient temperature. Two-pack epoxies and polyurethanes are examples.

Although these modes of film formation account for the majority of traditional finishes, curing can also be achieved by other means; for example, cement-based coatings and some zinc silicates and polyurethanes cure by reacting with moisture, whereas fusion-bonded polyolefins and stoved epoxies are cured by the use of heat.

Function. Painted finishes generally comprise a number of separate coats. The prime or seal coat is designed to key to the substrate, sealing porosity, encapsulating friable material, and/or providing a barrier to prevent reaction between the finishing coats and a chemically reactive substrate. Primers for metal surfaces are usually formulated to include corrosion inhibitors; those for nonferrous substrates often include metal etchants such as phosphoric acid.

Undercoats provide additional thickness, which, for decorative finishes, facilitates uniform opacity, fills small discontinuities, and provides a surface suitable for sanding to a smooth even finish. Undercoats or intermediate coats for anticorrosive paint systems or high-build architectural finishes provide the required film build, largely to enhance the protective qualities. The finishing coat or coats provide the requisite color, gloss, and chemical, wear, and stain resistance.

There are, however, many other paint functions. For example, water-repellent paints, typically silicone or acrylic based, are applied as thin films to inhibit surface wetting and capillary absorption of water, while road-marking paints must combine the features of rapid drying and wear resistance.

Latex and Solvent-Based Paints. In terms of conventional, decorative painting, the fundamental choice is usually between the solvent-based and latex paint systems. Each has its particular advantages and limitations. For solvent-borne types, the tendency to a higher achievable gloss, superior wetting and brushing properties, enhanced wear and stain resistance, and lower moisture permeability favors their use where these parameters are important. They therefore tend to be preferred for metal substrates, timber trim, and surfaces subject to moderate wear and tear and dampness, such as in kitchens and bathrooms. In contrast, latex types are preferred where rapid drying, low and short-lived paint odor, damp or alkaline substrates, and mold growth are factors. Exterior latex paints, notably the acrylics, exhibit superior durability, particularly on dynamic substrates, such as timber, that respond to changes in humidity.

Testing. Quality control and evaluation of liquid paints involve examining properties such as application and film-forming characteristics, viscosity, drying time, fineness of grind of the pigments, and density and proportion of nonvolatile material. Dry films are assessed for color, gloss, thickness, opacity, resistance to mechanical damage and chemical agents, and weathering.

Defects. Paints are essentially thin, degradable, semipermeable plastic membranes that rely on intimate contact with the substrate for adhesion and a minimum thickness to provide the requisite decorative and protective value. Although there can be deficiencies in formulations and quality control of manufacture, premature paint failures are usually attributable to poor surface preparation, adverse climatic conditions during application or drying, or incorrect film thickness. Indeed, the increasing sophistication and specialization of contemporary coatings is, to a degree, paralleled by a reduction in tolerance to deficient application procedures, and there is a consequent need for greater emphasis on correct selection and use of paint products.

References

1. G. E. WIESMANTEL: *Paint Handbook*. McGraw-Hill, New York, 1981.

2. OIL AND COLOUR CHEMISTS ASSOCIATION: *Surface Coatings*, Vol. 1. New South Wales University Press, Sydney, 1983.
3. H. F. PAYNE: *Organic Coating Technology*. Wiley, New York, 1961.
4. P. NYLEN and E. SUNDERLAND: *Modern Surface Coatings*. Wiley, Chichester, England, 1965.
5. M. HESS: *Paint Film Defects*, 2nd ed. Chapman and Hall, London, 1965.
6. *Painting Testing Manual, STP500*. American Society for Testing and Materials, Philadelphia, 1972.

D.B.

PARTITIONS are the nonstructural divisions of internal spaces. They are not assumed to support the building structure, but impose a load on it. The main types are masonry partitions, demountable partitions, and movable partitions.

Masonry Partitions can be divided into the following categories.

Clay Brick Partitions have excellent SOUND and THERMAL INSULATION. They may expand, but the movement is less than the shrinkage of concrete blocks or bricks. It is difficult to get two good faces on a fired brick; if an exposed brick face is required on both sides of a single skin wall, it is better to use an unfired brick, such as a sand-lime type.

Concrete Blocks or Bricks also have good thermal and sound insulation qualities. The selection of the correct unit to provide the relevant properties is important. Some concrete units have a high shrinkage.

Lightweight Concrete Panels are light and easy to handle, even in floor to ceiling modules. However, sound insulation properties are not as good as for an equivalent thickness of normal brickwork. See also BRICKS AND BLOCKS and LIGHTWEIGHT CONCRETE.

Demountable Partitions vary widely in their degree of demountability. Some partitions, notably those faced with flush-jointed plasterboard or ceramic tiling, while certainly demountable, are unlikely to be reusable. Door frames and doors are generally integrated into the design of these partitions; this may influence the selection of the type to be used. The main classes of demountable partitions are as follows.

Full Panel Type Partitions, using GYPSUM plaster as the core and facing to other core materials, may be reused if visible jointing with a cover batten or tongue is adopted. With flush set joints between panels, the units are not usually re-erectable.

Stud and Lining Type Partitions (Fig. 154) may be constructed with a pressed steel stud and rail system, usually sheeted with plasterboard, using two sheets to each face when fire ratings are required. Similarly sheeted, timber-framed stud partitions are also widely used where fire-resistance ratings are not required. Sound insulation ratings may be improved by placing in the voids in the framing a blanket of sound-absorbing material, such as rock wool. The use of heavy fixings to a plasterboard-sheeted wall is possible only if provision has been made in the framing to accommodate the loads. The cavity in hollow partitions may be usefully employed for running service pipes and conduits.

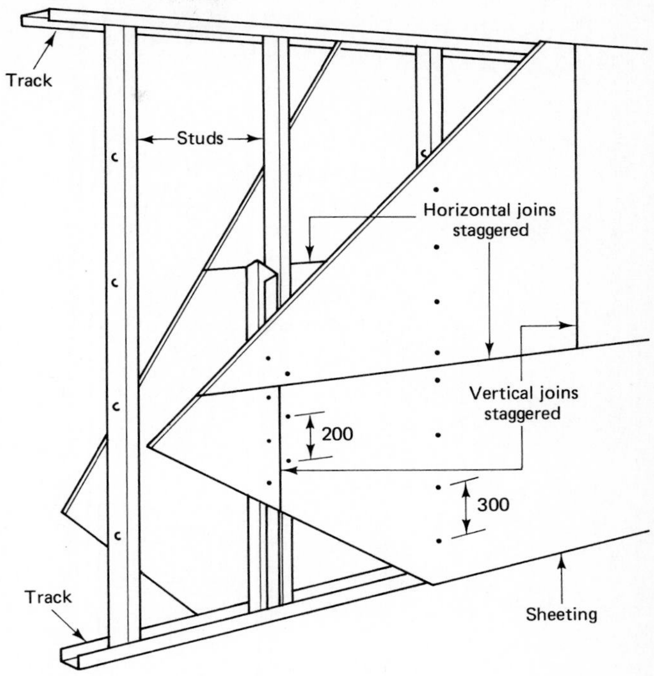

Figure 154 Stud-framed sheeted partitions from technical literature of A.C.I. Fibreglass, Melbourne, Australia.

Post and Panel Partitions (Fig. 155) are available in a wide range of proprietary systems, utilizing posts of hardwood, pressed steel, or extruded aluminum. Hollow posts permit the insertion of wiring runs and switches. Panel thicknesses vary considerably; 50 mm (2 in.) is an average thickness. Post and panel partitions usually incorporate a skirting duct at the floor; this sometimes includes jack-bolts, which wedge the partition between floor and ceiling.

Glazed Screens are usually employed in foyers and shops, and they incorporate narrow-stile or frameless doors. Frames are made from box sections, using stainless steel or extrusions of aluminum, finished by anodic methods, colored or clear. Glass may be clear or tinted, annealed, tempered, or laminated.

Toilet Cubicle Partitions can be supported from the floor, suspended from the ceiling, or cantilevered from the wall. In each case they can be constructed with plastic

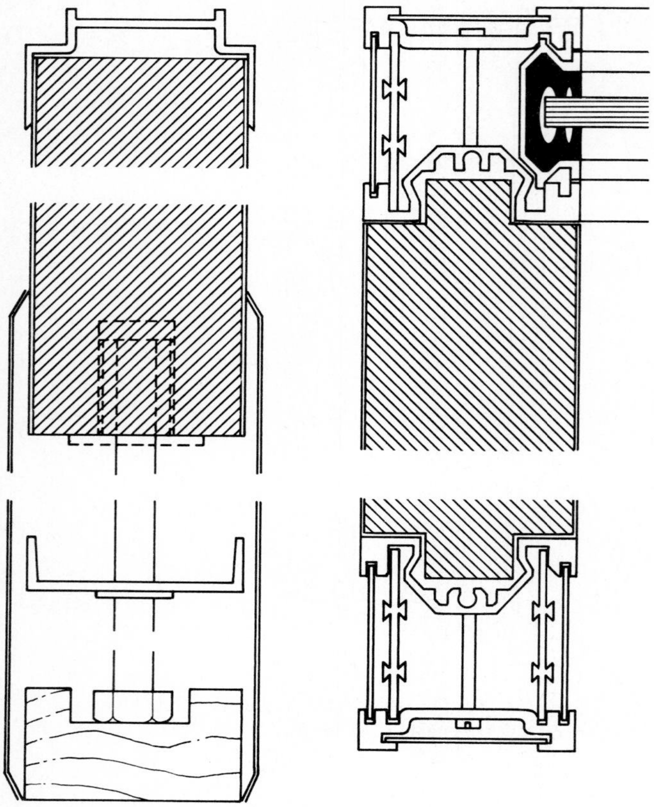

Figure 155 Post and panel partitions from technical literature of Moyle, Sydney, Australia.

laminate facing, on a core of particle board or timber, with matching doors. Two other materials sometimes used for panels are terrazzo and asbestos cement; both should be in floor-supported systems.

Low-Level Screens are available in most of the materials already discussed. They are used mainly to subdivide space in offices. Since these partitions are not secured at the top, it is necessary to provide cross walls or right-angle junctions to achieve stability.

Movable Partitions are divided into the following categories.

Sliding Doors on tracks that may be linked or unlinked are available. A linked door can either slide as a unit behind a covering wall or fold concertina style at the end of its travel. An unlinked door travels in divided tracks at the end of the run, and the leaves are stored side by side in a cupboard or recess.

Portable Panels (Fig. 156) are often used in convention centers, hotels, and restaurants where rapid subdivision of areas is required. The panels have rebated or grooved edges and are designed to be lifted individually and carried away. Retractable edge fixings can be fitted to the panels to provide an acceptable sound barrier.

Reference

B. LAUBCHBERRY, Partitions, Section 8 in *AJ Handbook of Building Enclosure* (eds., A. J. ELDER and M. VANDENBERG). Architectural Press, London, 1975, pp. 314–336.

R.G.Co.

PASSIVE SOLAR DESIGN attempts to reduce the energy requirements of buildings by the utilization of solar radiation for winter heating. It is distinguished from active heating with SOLAR COLLECTORS by the fact that passive solar buildings do not depend on mechanical equipment, and therefore "parasitic" energy, to function. Energy collection is usually by normal elements of the building envelope (such as windows, skylights, glazed porches); heat storage is usually within the structural mass of the building, and energy flow from collection to storage and utilization zones is by natural processes of THERMAL RADIATION, CONVECTION AND CONDUCTION. The performance of some passive systems may be enhanced by an auxiliary mechanical device: for instance, a small fan may be used to augment airflow that is primarily by natural convection and to increase the degree of control over the operation of the system. But such a device would always be smaller than those required by an active system.

The building itself and most of its design details, from site planning through location of the various functions within the building volume, to choice and detailing of the materials of its construction, constitute the "heating plant." As a result, the design of solar heating by passive systems is integral to the architectural design process and should be taken into account in the earliest design decisions.

Passive Solar Heating and Energy Conservation. Reductions in the energy consumed in heating may be achieved by both passive solar heating features and improvements in the general thermal quality of the building envelope (see ENERGY CONSERVATION). In practice, the appropriate combination is determined by CLIMATE. In cold, cloudy climates, energy conservation measures are usually more effective, until the total heating load is substantially reduced. In milder, sunny regions, solar heating may be more cost effective in reducing fuel bills.

However, when incorporating solar heating features in a building, it is important to consider not only the efficiency in energy terms, in other words the financial saving, but also to ensure general improvements in THERMAL COMFORT. Not by chance, several features of successful passive solar designs, notably the raised mean radiant temperatures, the dispersion of heat sources, the lower working temperature of circulated heating air, and the general emphasis on better control of heat losses, all tend in that direction.

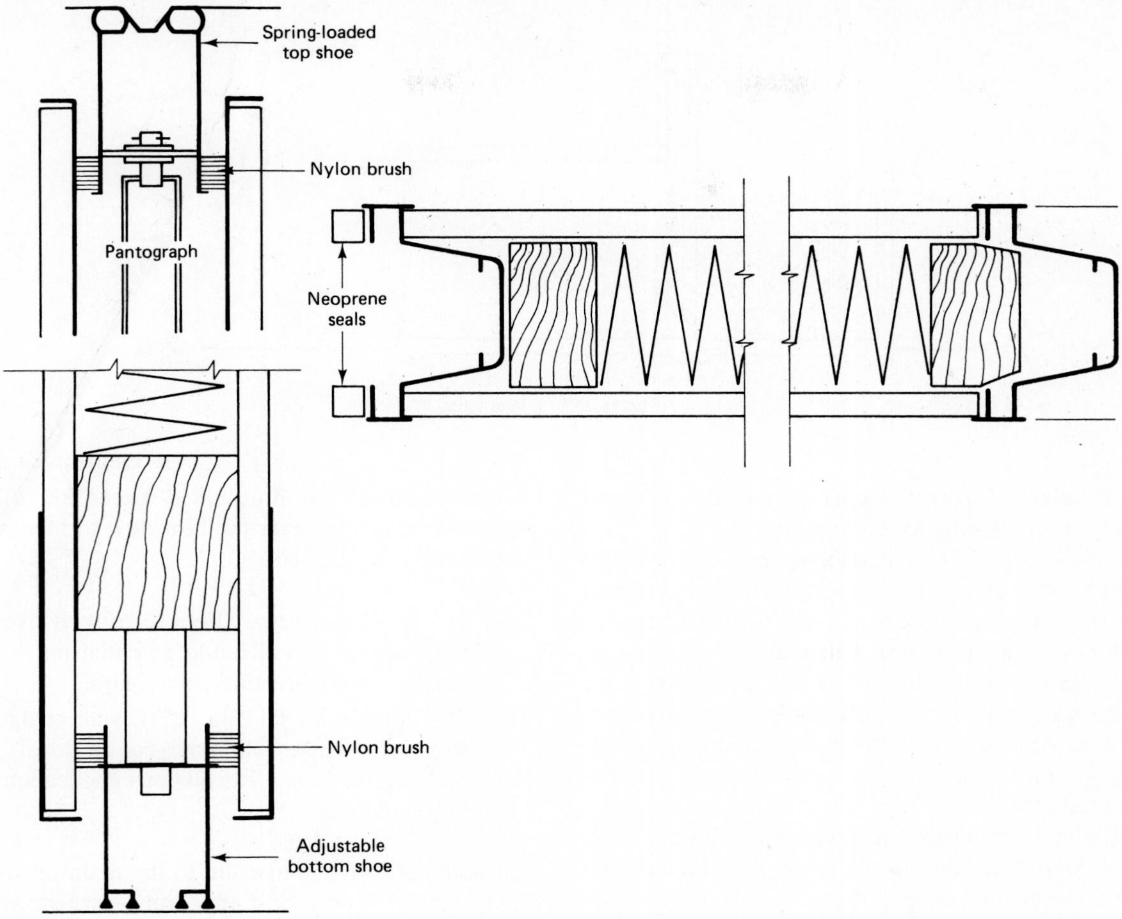

Figure 156 Portable partitions from technical literature of Indeng Architectural Products, Melbourne, Australia.

The successful use of solar energy for winter heating depends on the following factors:

1. Sufficient means of admitting solar energy to the building in the heating season.
2. The prevention of excessive heat loss from the inhabited spaces during times when such energy is unavailable.
3. The provision of means to store the excess energy from sunlit periods to be utilized for heating at other times, most importantly at night.

Early attempts at building solar houses, especially in the United States in the 1930s, failed to provide for any but the first factor. Insufficient controllable insulation over large areas of glazing and the absence of significant thermal storage in the traditional lightweight timber-frame construction combined to allow these houses to cool down rapidly in the absence of sunshine. As a consequence, no significant solar savings were achieved, and undesirable overheating often occurred during sunlit periods.

Solar Collection through Glazing. The basic mechanism of solar collection is the admission of direct radiant energy through GLAZING, making use of the fact that glass, while transparent to solar radiation, is virtually opaque to infrared of longer wavelengths, such as the radiation from warm surfaces inside a building.

This one-way action, known as the *greenhouse effect*, may be incorporated in buildings in a number of ways. Each system has particular implications for the habitable rooms of which it is part. In some cases these involve functional problems, such as potential overheating or glare, and in others the requirement for regular actions by the occupants, such as the operation of shutters or curtains. In practice, a building rarely employs only one basic system, but in fact is a composite of a number of them applied to spaces with differing functional requirements or to achieve complementary effects for the building as a whole.

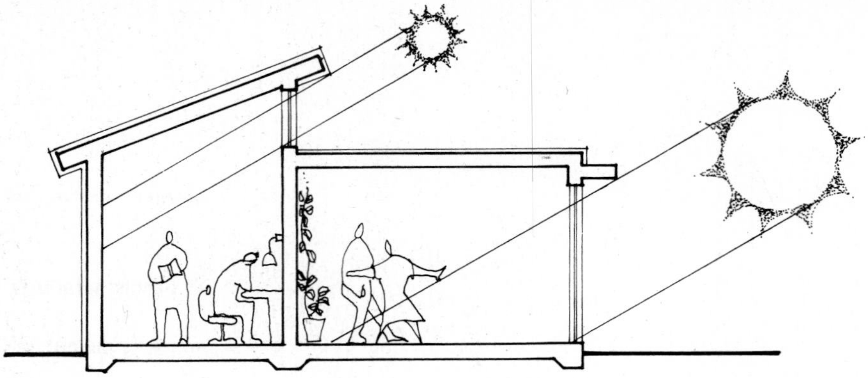

Figure 157 Direct gain.

Passive Solar Heating Systems are usually classified according to the following loose definitions.

Direct Gain (Fig. 157). Inhabited spaces are heated by sun admitted through conventional windows, skylights, and the like. The building fabric itself acts as the necessary thermal storage. Nighttime heat loss may have to be regulated by operable insulation at the windows. There are several potential problems with "living in the solar collector," such as obstruction of the sun-patch by furniture, fading of furnishings and damage to artworks, and discomfort from excess heat and glare.

Direct gain is applicable in almost all climates with appropriate control of night insulation in cold regions and of summer overheating in hot regions. It is most easily incorporated in massive construction; in lightweight buildings, specialized storage (such as water walls) would need to be introduced.

In single-story construction, direct sunshine may be introduced to the majority of the dwelling with relative ease, but in high-rise construction, the depth of the building away from the south facade may have to be limited.

Attached Sun-spaces. Solar radiation heats some auxiliary space such as a greenhouse, enclosed balcony, or corridor that is physically divided from the rest of the dwelling. The intention is to remove the major disadvantages of excessive direct sun in occupied spaces and to introduce a measure of control over the heat flow to them. Two types are usually distinguished:

1. *The greenhouse type* (Fig. 158), with overhead glazing, maximizes collection potential but is difficult to integrate with structural sun control.
2. *The sun-porch type* (Fig. 159), with opaque roof construction, is a much more conventional part of the building fabric and has obvious applications in multistory buildings.

In each case, the connection to the main inhabited spaces may take a variety of forms, such as a massive storage wall, an insulated wall with vents (which may be conventional door openings), a glazed opening, another collector wall with glazing, or any combination of these.

Collecting Storage Wall, often referred to as a *Trombe wall* (Fig. 160), in its simplest form is glazing placed in front of a dark-colored, thermally massive wall. Thus the functions of collector and thermal storage are combined in one external wall element. While a small proportion of the collected energy may be transmitted to the inhabited spaces by circulating air through vents, the effectiveness of this system is dependent on an accurate selection of material and thickness for the thermal mass to ensure the correct time lag between peak collection at the outer face and heat given off at the inner.

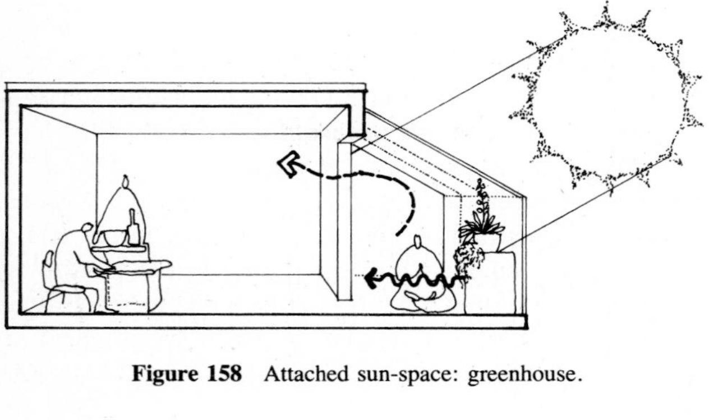

Figure 158 Attached sun-space: greenhouse.

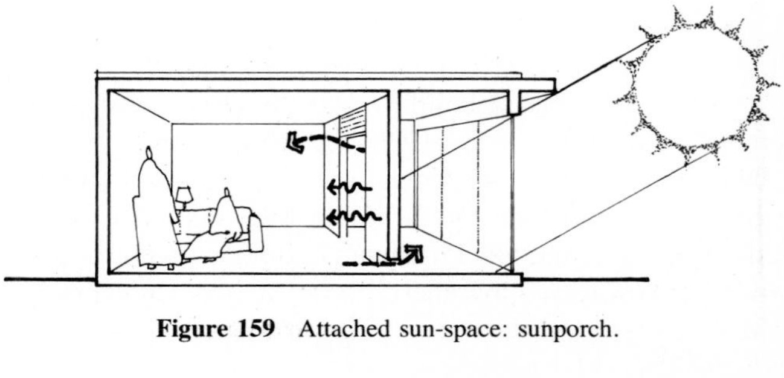

Figure 159 Attached sun-space: sunporch.

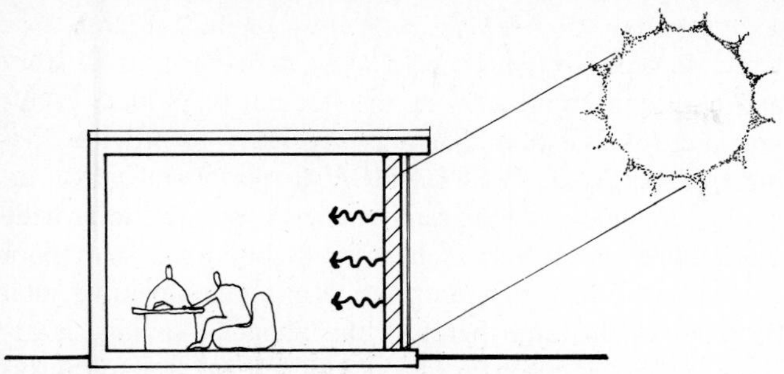

Figure 160 Collecting storage wall.

Convective Loops. Using air as the circulating medium, heat is collected at an insulated solar collector and transferred to either the inhabited space or specialized heat storage by convection. Convective loop air-heating collectors are best suited to structures where the heating load is large compared to the collector area or where the building use is intermittent and predominantly in the daytime. Schools and office buildings are suitable applications. In residential construction, convective loops may be particularly useful in supplying heating air to spaces remote from south-facing walls.

Generally, two types are distinguished:

1. *Insulated collecting walls*, such as the Barra–Constantini system (Fig. 161). Here the glazing is placed in front of an insulated wall, effectively forming a solar collector integrated into the normal building fabric. Heat transfer may be enhanced by the details of the collecting element. The circulating air may be passed through thermal storage or vented directly to adjacent inhabited spaces.
2. *Remote collectors* (Fig. 162). In some locations, the collector may be removed from the conventional building fabric. This can have advantages, such as the ability to tilt the glazing to optimize collector efficiency or to locate the collector below the level of the inhabited space and convect heated air through subfloor storage.

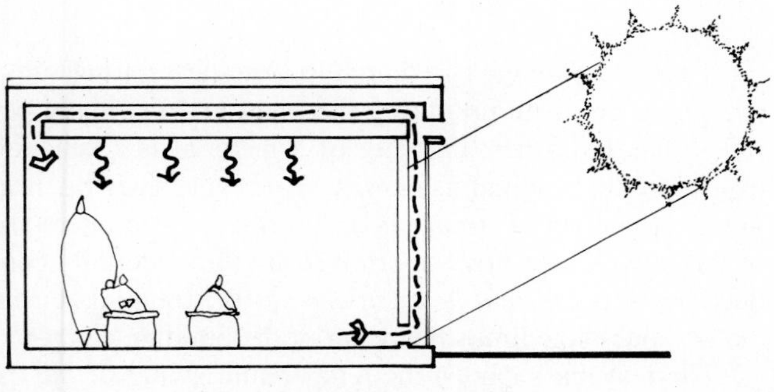

Figure 161 Insulated collecting wall.

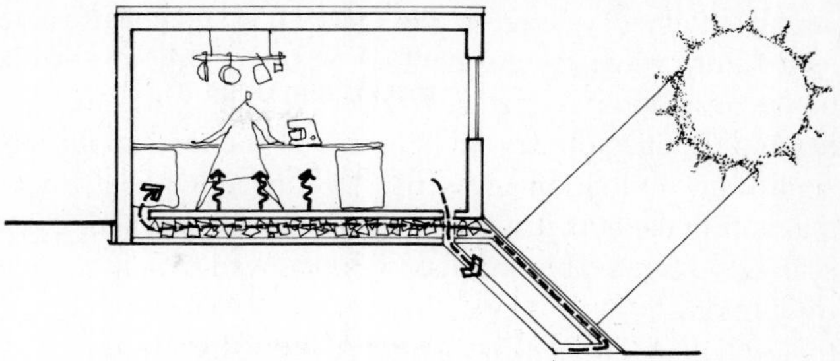

Figure 162 Remote collectors.

Collecting Storage Roof (Fig. 163). The energy balance of a massive horizontal roof freely exposed is normally negative during the winter. However, if heat losses are greatly reduced by insulation placed over the roof at night, useful energy may be transmitted to the interior by conduction. The best known example of this system is Harold Hay's Skytherm installation. The roof consists of water in plastic bags supported on a metal structure whose soffit forms the radiant ceiling in the interior. The operable insulation consists of horizontal panels positioned by a motor drive. A simple radiation sensor opens the panels during sunlit periods and closes them at night or when cloudy.

The system is completely free of the orientation con-

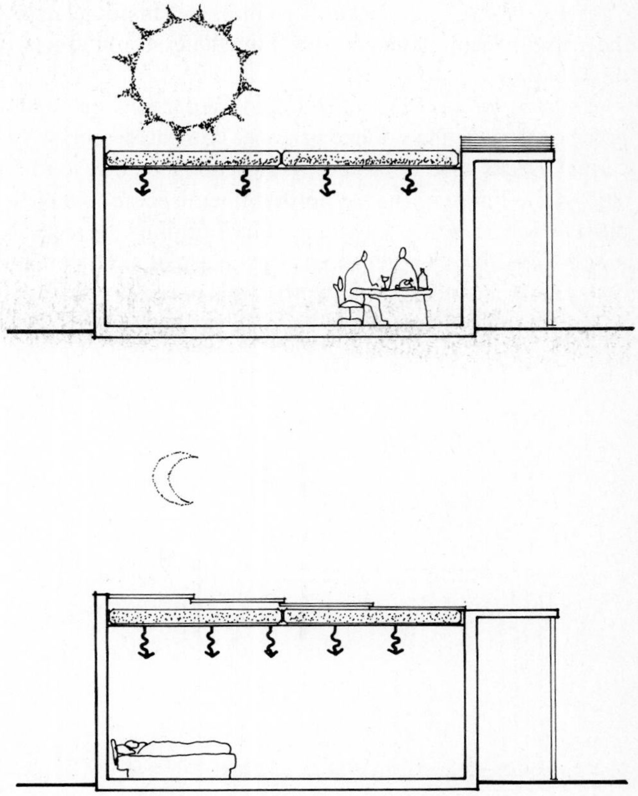

Figure 163 Collecting storage roof.

straints of other systems, and in theory it is equally effective as a summertime natural cooling system, if the operation of the insulation is reversed. However, its efficiency for heating depends on the air-tightness of the insulation; any significant infiltration gives rise to convective heat losses that nullify the bulk insulation. The technology for adequate sealing with movable insulation has proved expensive and troublesome.

Radiation Traps (Fig. 164). A special collector, with south-facing, slightly tilted glazing, insulated roof and sides, and operable internal insulation to the solar glazing, is positioned above a thermally conductive, massive ceiling. Solar radiation penetrating the glazing irradiates the ceiling element, which transmits the heat to the interior. Heat transfer may be further enhanced if a small fan is used to draw down the warmed air from the collector as well. At night, heat losses are minimized by insulating the glazing.

The system can be positioned at a different orientation to the spaces that it serves to optimize solar collection on otherwise unsuitable sites. With little modification it can also be utilized for summer cooling. However, this system is an extra to the conventional built volume, and it has to be justified by its cost/benefit in energy terms.

Design Factors. In passive solar design, the building itself is an integral part of the energy system; the following factors should be taken into account.

Ensuring Sun Access to the building. This is related to the natural features and to the immediate surroundings of the building.

Incorporating Possibilities for Heat Flow by natural processes, from the collecting zones to heated rooms. Locating spaces that need the most heating to the southern side of the building (in the northern hemisphere) simplifies passive solar heating. In single-story buildings, direct gain may be possible to other rooms by the use of skylights, and under certain conditions (Fig. 161) it is possible to circulate warm air to northern rooms via ceiling ducts.

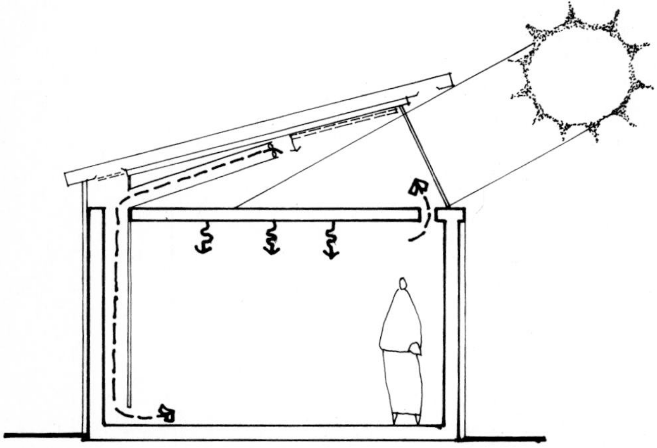

Figure 164 Radiation trap.

Providing High Thermal Quality to the building as a whole in order to reduce the heating demand. In multistory and multiunit configurations the internal units have excellent thermal isolation, but it is necessary to provide adequate THERMAL INSULATION to peripheral units, including top floors. At the same time, it is possible to provide worthwhile passive solar heating to buildings with poor insulation; in fact, the contribution of a given passive solar heating system, and therefore the financial saving, is actually higher as the heat loss coefficient of the building is greater, because more heating is needed in the first place.

Defining the Functional and Thermal Comfort Requirements in the particular building. This covers times of use, lighting, visual contact with the outside, and ventilation.

Choice of Suitable Passive Systems according to building type, local climatic conditions, and functional and comfort requirements.

Design of the Passive Solar Heating Systems. The most important factors are the collecting elements, the thermal storage, the heat transport mechanisms, and, of course, the ability to evaluate the expected performance of the proposed design.

Solar Glazing. The design of the collecting elements is largely a matter of determining the location, orientation, geometry, type, and size of the solar glazing. For solar collection in the northern hemisphere, orientations within 20° of due south provide maximum available solar radiation, the most favorable energy balance over summer and winter, and ease of automatic, fixed sun control (see SUN SHADING). Tilted glazing can maximize penetrating radiation by being nearer normal to winter sunshine, but may incur penalties in heat losses and complicated shading, insulation, and condensation control. The type of glazing is determined by the energy balance in a particular location and the transmission and thermal conductivity of the assembly (see GLASS and FENESTRATION). From the pure energy viewpoint, quite large glazing areas would often be justified, but the recommended maximum size in any system should be determined by comfort considerations, specifically the need to limit the maximum temperature variations within the occupied space and to avoid summer overheating.

Thermal Storage is required to store heat during sunlit periods to be given up during the night and in cloudy periods. The gross heat capacity of a material is a function of its specific heat and its density. However, since the heat flow into and out of storage is cyclic, we are also interested in the rate of heat flow determined by the material's conductivity. The *diurnal heat capacity* of the building during the 24-hour cycle limits the depth of the building elements.

Thermal mass damps the temperature swing of the interior; but as this depends on the surface area available for rapid absorption of excess heat, it fixes a minimum ratio between the mass surface area and that of the solar glazing.

Some systems (Fig. 160) have integral storage capacity. In heavy-weight construction, the structure and partitions can provide the required heat storage without additional mass beyond structural needs. In lightweight construction, however, nonstructural masonry, rock beds, or water containers may be needed.

Energy Transfer between the collecting elements and storage or utilization zones is by natural conduction, long-wave radiation, and convection. Its efficiency depends on the spatial relationships and the design of the connections.

In direct-gain systems, heat distribution should be mainly by radiation, as heat transfer to storage elements by convection is only about 40% as efficient as by radiation (Ref. 2, Vol. 1). In sun-spaces, convective heat transfer by the thermosyphonic loop between the sun-space and the cooler northern spaces (Ref. 3) is most efficient.

Evaluation of Expected Performance depends on the development of suitable computation systems to make possible this evaluation under different simulated environmental conditions based on readily available meteorological data (see CLIMATE AND BUILDING).

Simulation Models can give accurate prediction of the behavior of both complex building configurations and variable environments. The PASOLE Code 3, developed by the Passive Solar Group at Los Alamos National Laboratory, is an example. The simulation approach requires hourly data and large amounts of iterative computation. As this needs much computer power and time, simulation is used principally for scientific research and especially for the validation of simpler *calculation models*, for example, the LASL model (Ref. 2, vols. 2 and 3), and the *predictive model for direct gain* (Ref. 4).

See also EARTH-COVERED ARCHITECTURE.

References

1. E. MAZRIA: *The Passive Solar Energy Book*. Rodale Press, Emmaus, Pa., 1979.
2. R. W. JONES (ed.): *Passive Solar Design Handbook*, Vols. 1, 2, and 3. U.S. Department of Energy, Washington, D.C., 1980, 1981, and 1982.
3. F. D. BALCOLM: *Heat Storage and Distribution inside Passive Solar Buildings*, in Passive and Low Energy Architecture, Second International PLEA Conference, Pergamon, Oxford and Elmsford, New York, 1983.
4. B. GIVONI: Performance of direct gain solar buildings in relation to their heat capacity. *Passive Solar Journal* (in press).

B.G. and S.K.

PERMAFROST, or perennially frozen ground, refers to "the thermal condition in soil or rock of having temperatures below 0°C (32°F) that persist over at least two consecutive winters and the intervening summer" (Ref. 1). Moisture in the form of water or ground ice may or may not be present. Permafrost is found under about 25% of the land area of the world and is particularly widespread in the northern hemisphere. It underlies about 50% of Canada and the Soviet Union, more than 75% of Alaska, about 22% of China, extensive areas in northern Europe, and most of Greenland and Antarctica. It may also occur under rivers, lakes, and oceans and at high elevations in mountainous areas. It may be discontinuous, that is, existing as islands surrounded by unfrozen ground, or continuous. The thickness of permafrost may vary from about 1 m (3 ft) to more than 1000 m (3300 ft) (Ref. 2).

Permafrost gives rise to many difficult engineering problems because of its negative ground temperature regime, its rather unpredictable nature and distribution, and, in particular, the large quantities of ice often found in it. Careful site and route investigations are essential, and special design and construction procedures must be used to ensure satisfactory performance of engineering works. The basic approaches in most cases are to design and construct to (1) preserve the frozen condition of the ground or (2) control or reduce the rate and depth of thaw of ice-rich ground to minimize settlement and ground instability problems (Refs. 3 to 5).

References

1. R. J. E. BROWN and W. O. KUPSCH: *Permafrost Terminology*. Technical Memorandum No. 111, Associate Committee on Geotechnical Research, National Research Council Canada, Ottawa, 1974.
2. A. L. WASHBURN: *Geocryology: A Survey of Periglacial Processes and Environments*. Edward Arnold, London, 1979.
3. G. H. Johnston (ed.): *Permafrost: Engineering Design and Construction*. Wiley (Canada), Toronto, 1981.
4. O. B. ANDERSLAND and D. M. ANDERSON (eds.): *Geotechnical Engineering for Cold Regions*. McGraw-Hill, New York, 1978.
5. N. A. TSYTOVICH: *The Mechanics of Frozen Ground*. McGraw-Hilll, New York, 1975.

G.H.J.

PHOTOELASTICITY is an optical method of stress analysis based on the phenomenon of *birefringence* or *double refraction* of polarized light in certain stressed transparent materials (usually an epoxy resin). In two-dimensional analysis, a geometrically similar model of the component

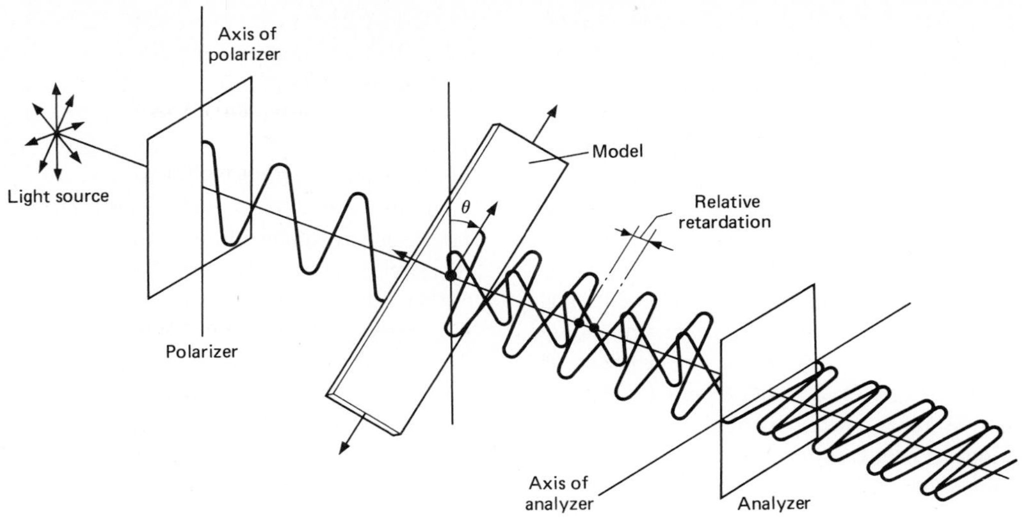

Figure 165 Optical principle of photoelasticity.

under test is set up between two polarizing filters, as indicated in Fig. 165. The first filter (the *polarizer*) passes a beam of light polarized, say, in the vertical plane, while the second (the *analyzer*) is set with its axis at right angles to this and thus does not transmit light from the polarizer unless there is a stressed model between the two.

Principles. Polarized light entering the model at a particular point emerges polarized in directions parallel to the principal stresses, that is, the greatest and the smallest stresses at that point in the material. These two components have different transmission velocities and therefore emerge with a *relative retardation* between them. The analyzer passes components of both, which combine to form a single wave on the observer's side of the analyzer. The relative retardation is found to be directly proportional to the difference of the principal stresses for a given specimen thickness, and if it is equal to one wavelength of the light used, the transmitted components cancel each other and no light is observed through the analyzer.

Assuming, therefore, that monochromatic (single wavelength) light is used, a uniformly stressed specimen appears alternately dark and light in the field of view as the stress is increased, and the relative retardation becomes equal to successive multiples of a wavelength. If the stress field is nonuniform, points at which the difference between the principal stresses, and thus the relative retardation, is the same are joined by dark lines, referred to as *fringes*. The specimen thus displays a *fringe pattern* from which the stress distribution can be inferred. Figure 166 shows a fringe pattern for a symmetrically loaded beam. The use of white light in place of monochromatic light results in a corresponding pattern of colored bands (*isochromatics*); but as these become paler and less distinct as the relative retar-

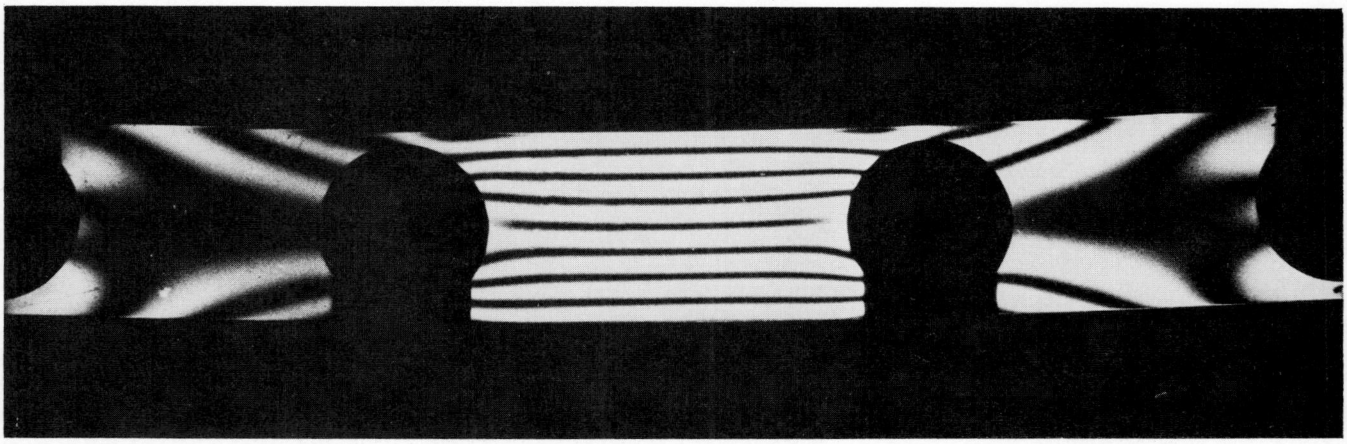

Figure 166 Fringe pattern in a beam with four-point loading.

dation increases, the use of monochromatic light is preferred for quantitative work.

If at any point the plane of polarization of the incident light coincides with the direction of one of the principal stresses, the wave is not resolved, and the model remains dark at that point. The loci of such points are termed *isoclinic lines* and may be used to determine the directions of the principal stresses at all points in the model by rotating the polarizer and analyzer together and tracing the position of the *isoclinic lines*; they may also be used to determine the directions of the principal stresses at all points in the model by rotating the polarizer and analyzer together and tracing the position of the isoclinic lines at a series of settings. To examine the fringe pattern only, the isoclinics can be eliminated optically from the field of view by using circular polarized rather than plane polarized light.

Interpretation of Results. At an unloaded boundary, the principal stresses normal to the boundary are zero, and therefore the edge stress distribution can be determined directly from the fringe pattern. At internal points, the fringe pattern gives the difference between the principal stresses, which equals twice the maximum shear stress. A variety of experimental and numerical methods is available for finding the separate principal stresses.

Plane photoelasticity has been applied to the investigation of a variety of structures and components, including portal frame connections, deep beams, dams, perforation patterns in bricks, and many situations involving stress concentrations and complicated geometry difficult to analyze by mathematical or numerical methods.

Three-Dimensional Analysis by photoelasticity is possible by making use of the *frozen stress* method. A loaded model is heated in an oven and gradually cooled with the leads still applied. When cold, the model is carefully cut into slices that, when examined in polarized light, show fringe patterns similar to those observed in two-dimensional models. In general, separation of the principal stresses is more difficult in this case except on planes of symmetry.

The Determination of Stresses in Actual Structures or Components has been attempted by applying a coating or by cementing pieces of photoelastic material to the part concerned and observing the optical effects as it is stressed.

See also STRUCTURAL MODEL ANALYSIS.

Reference

1. A. W. HENDRY: *Elements of Experimental Stress Analysis*. Pergamon Press, Oxford and Elmsford, N.Y., 1977.

A.W.H.

PHOTOVOLTAIC CELLS are electronic devices that convert light falling on them directly into electricity. They are used for detecting and measuring light and as a source of electricity, particularly when exposed to sunlight.

Operation. Photovoltaic cells are made from thin layers of *semiconductors* whose electronic properties are between those of metals (conductors) and insulators (nonconductors). Silicon is commercially the most important semiconductor, and it forms the basis of the present microelectronics industry. Other semiconductors used in photovoltaic cells are selenium, cadmium sulfide, and gallium arsenide. When suitable light enters a semiconductor, its energy is absorbed by creating electrical current carriers. The photovoltaic cell is designed to move these carriers in a preferred direction and produce an electrical current (Ref. 1).

Structure. Figure 167 shows a cross section of a silicon photovoltaic cell (Ref. 1). Normally, cells are only a fraction of a millimeter thick. Their lateral dimensions depend on the particular application; they range from 1 mm up to 10 cm for a solar photovoltaic cell. In the regions labeled *n*-type and *p*-type in Fig. 167, different impurities have been introduced into the silicon to control its electronic properties. The junction between these regions forces the carriers created by illumination to move off in a preferred direction. The back contact covers the entire rear surface, while the top contact has an open grid pattern to let light through. When illuminated, electrical current flows in any electrical load connected between these contracts. A thin antireflection coating is frequently applied to the top surface of the cell to increase its sensitivity.

Light Measurement. For light measurement, photovoltaic cells are operated at low output voltages to maximize the linearity of their response. However, their color sensitivity differs from that of the human eye. For example, the peak sensitivity of some silicon cells is to infrared radiation, which the eye cannot even detect. At the same time, these cells can also respond to ultraviolet radiation, which lies beyond the opposite extremity of the eye's response. If cells are used to measure light properties as seen by the human eye, filters have to be placed in front of the cell to correct for this difference. Photovoltaic cells made from the semiconductor selenium have the closest response to that of the human eye and are sometimes used in light measurements on this account (Ref. 2), despite lower sensitivity and a poorer stability than silicon cells.

Since photovoltaic cells generate their own electrical power during light measurement, they can be used as the basis of very simple and rugged instruments. All that is required is the cell, a meter that measures electrical current, and a range switch to change the meter sensitivity.

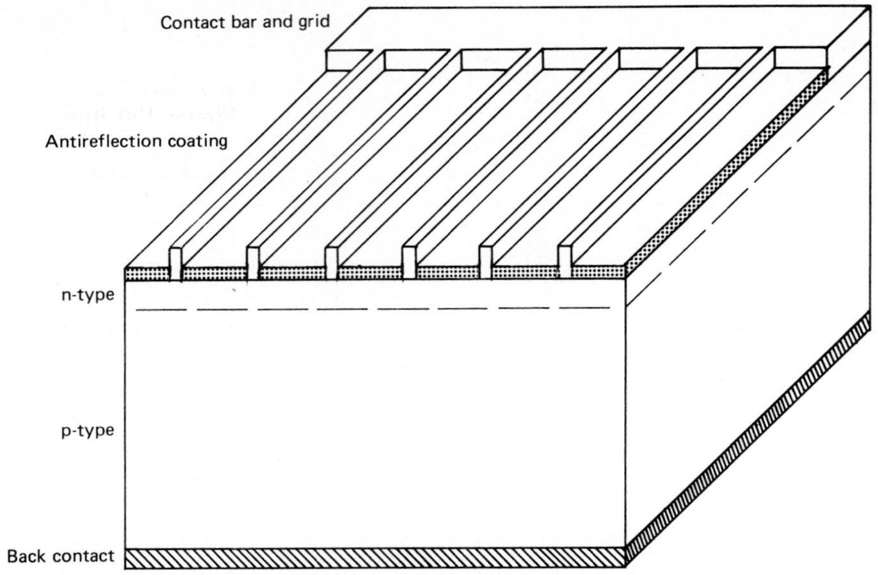

Figure 167 Cross section of a silicon photovoltaic cell.

Electricity Generation. Photovoltaic cells find their most widespread use in converting sunlight to electricity. A number of solar cells are electrically connected together and encapsulated into weatherproof packages known as solar cell modules. Under bright sunlight about 100 W of electrical power is generated for every square meter of module area (about 10 W/ft^2).

The first major application of solar photovoltaic cells was on spacecraft. In the early 1970s, the terrestrial use of cells began to increase, particularly for providing small amounts of electricity in regions remote from conventional sources. Cell costs have since decreased, and many experts believe that electricity generated by solar cells may become as cheap as that from any other source by the turn of the century. Large power stations such as the 100 peak megawatt station being constructed by the Sacramento Municipal Utility District may then be more common (Ref. 3).

Figure 168 The Carlisle House, completed in 1981, a residence near Boston with 7 kW of silicon photovoltaic cells mounted on its roof. These supply a large portion of its electricity requirement.

One use of the cells that will increase in the future is the generation of electricity on site to supplement the normal grid electricity supply for buildings and private residences (Ref. 3), as shown in Fig. 168. An inverter converts the direct-current output of the solar cells to the alternating current normally used within homes; it also automatically sends excess electricity back down to the electricity supply network for use by others during periods of bright sunshine. At night, electricity flows in the opposite direction. Even though this use of photovoltaic cells cannot be justified on purely economic grounds at present, associated attractions have since encouraged the building of several privately financed homes of this type (Ref. 3).

See also ARTIFICIAL SKY; FIRE AND SMOKE DETECTORS; and SOLAR ENERGY.

References

1. M. A. GREEN: *Solar Cells: Operating Principles, Technology and System Applications.* Prentice-Hall, Englewood Cliffs, N.J., 1982.
2. A. STIMSON: *Photometry and Radiometry for Engineers.* Wiley, New York, 1974.
3. *Photovoltaics in the United States*. Special Issue of *Sunworld*, Vol. 6, No. 3 (June 1982), pp. 78–86.

M.A.G.

PILES AND PILING. Piles are relatively long and slender members whose major function is to transmit foundation loads through relatively weak or loose soil strata to stiffer underlying soil or rock strata. Figure 169 illustrates some of the situations in which piles may be used (Ref. 1). Their main use is to provide FOUNDATIONS FOR BUILDINGS. As well as carrying compressive loads, piles may be used to carry uplift loads, resist lateral loads, carry loads below scour level in marine situations, support structures adjacent to future excavations, and provide a stable foundation in expansive or swelling soils that undergo significant movement when their moisture content changes. As indicated in Fig. 169, piles are frequently used in groups as well as singly.

When carrying vertical loads, the resistance of a pile is derived partly from end-bearing resistance at the pile tip and partly from skin friction developed between the pile shaft and the surrounding soil. The relative proportions of end-bearing and shaft load will vary, depending on the nature of the soil surrounding the shaft, the soil or rock near the tip, the length and stiffness of the pile, and the applied load level. At normal design load levels, a significant resistance is derived from skin friction, even if the pile tip rests on rock. Figure 170 illustrates a typical load–settlement relationship for a pile and distributions of load along the pile shaft at various applied load levels.

In contrast to vertically loaded piles, piles subjected to lateral loading obtain the majority of their resistance from the soil near the ground surface. This soil is frequently disturbed or may be subjected to seasonal moisture changes; consequently, the prediction of pile behavior under lateral loading is often less accurate than that under vertical loading.

Design Considerations. The design of pile foundations includes the following stages:

1. Selection of the type of pile and the method of installation.
2. Determination of the size and number of piles to ensure

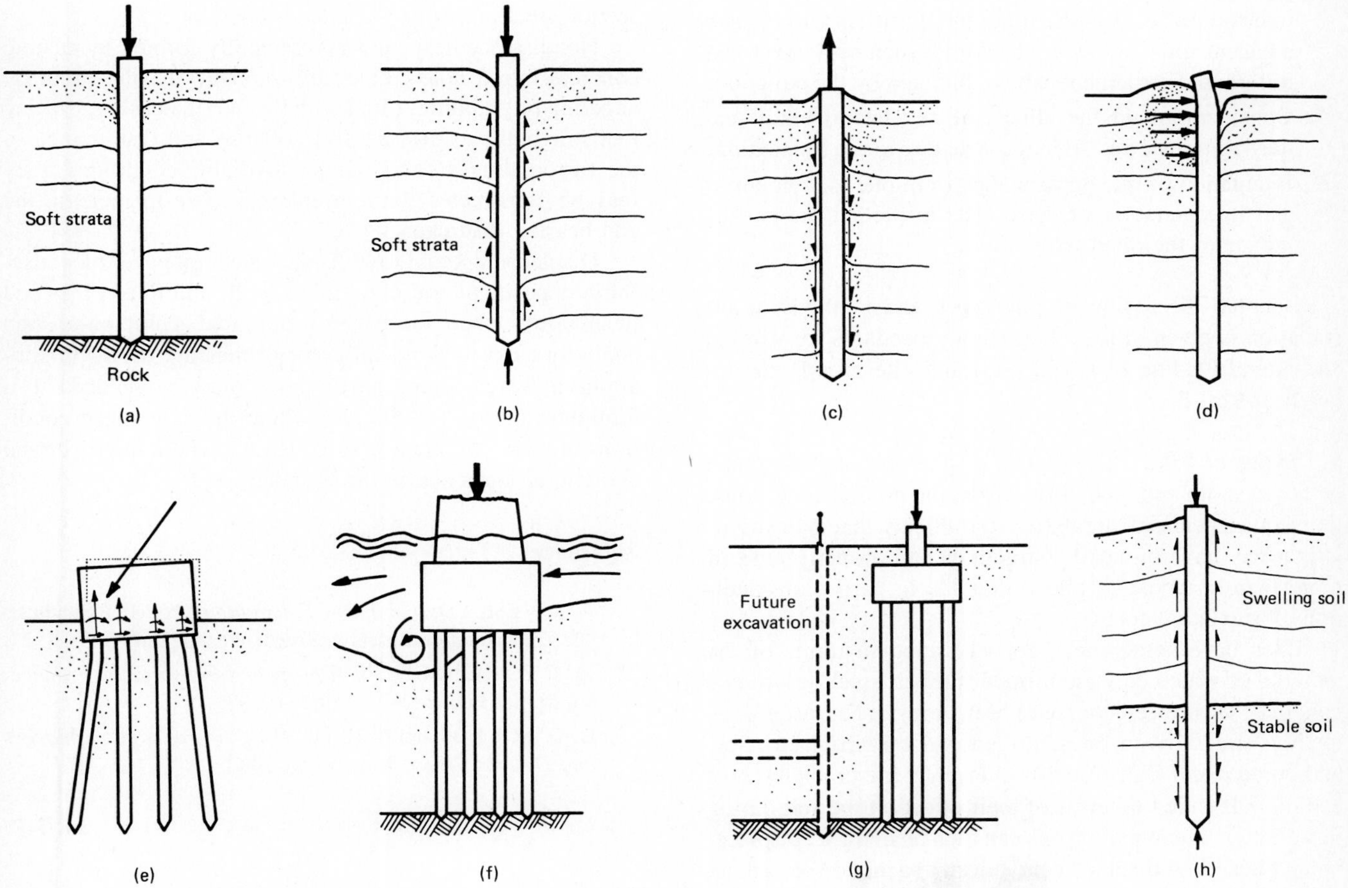

Figure 169 Applications of pile foundations (Ref. 1). (a) End-bearing pile. (b) Friction or floating pile. (c) Pile resisting uplift force. (d) Laterally loaded pile. (e) Pile group subjected to combined and lateral vertical loading. (f) Pile group in scour area. (g) Piles supporting structure adjacent to future excavation. (h) Pile in expansive soil.

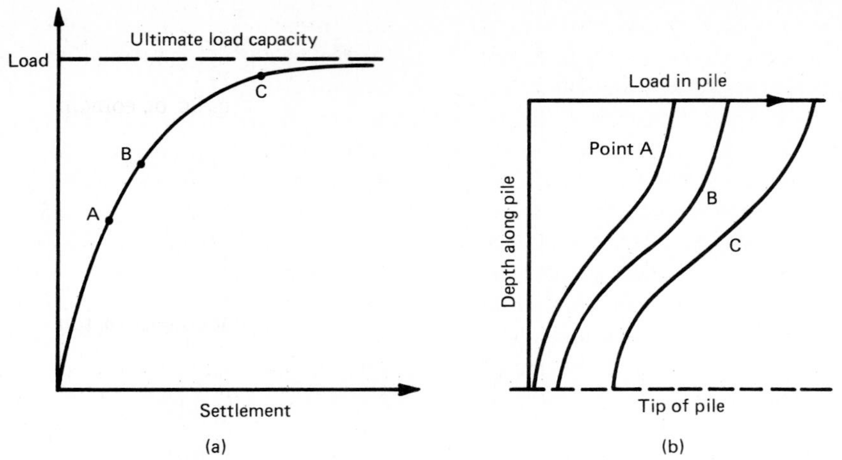

Figure 170 Behavior of vertically loaded pile. (a) Load–settlement curve. (b) Load distribution along pile shaft.

an adequate factor of safety against failure to the supporting soil and the pile material.

3. Estimation of the settlement (vertical movement) of the pile foundation and the differential settlement between adjacent foundations to check that such settlement and differential settlement can be tolerated by the structure.
4. Consideration of the effects of any lateral loads that may be transferred from the structure to the foundation.
5. Evaluation of pile performance from pile loading tests, and interpretation of these tests to predict the performance of the piled foundation.

Detailed discussion of pile types and methods of installation is given in Ref. 2. Design procedures for vertical and lateral loading of single piles and pile groups are detailed in Ref. 3.

Types of Pile. The choice of pile type is determined by the location and type of structure, the ground conditions, the requirements of durability, the effects that pile installation may have on nearby structures and property, and, if more than one type of pile is suitable in particular conditions, by relative costs.

Piles have sometimes been classified in terms of the material of which they are formed (timber, steel, reinforced concrete, prestressed concrete) or in terms of the installation method (i.e., driven piles, driven and cast-in-place piles, and bored piles) (Ref. 2). However, they are probably most usefully classified in terms of their effect on the soil during installation. Three main types can then be defined: displacement piles, low-displacement piles, and nondisplacement piles.

Displacement and low-displacement piles are installed in the ground by driving, jacking, or vibration. In each case, the pile displaces the surrounding soil; the extent of this displacement depends on the installation procedure and the pile cross section. Solid sections or closed-end steel pipe piles cause larger displacements than low-displacement H-section steel piles or open-ended tube piles.

Nondisplacement piles are generally formed by casting concrete into a hole created by removal of soil, either by augering or grabbing, often with the aid of a casing, drilling mud, or both, to prevent soil collapse and disturbance of the surrounding ground. Under favorable conditions, they may be constructed with an enlarged base to increase the end-bearing load capacity.

Displacement and low-displacement piles are most useful in deposits of soft clay, loose sand, and filled areas and in situations where the groundwater table is high or artesian conditions exist. Nondisplacement piles are useful in situations in which vibration or noise must be reduced to a minimum during installation, where the subsurface conditions involve stiff clay or soft rock, or where large vertical and lateral loads act on the foundation.

References

1. A. S. VESIC: *Design of Pile Foundations*. NCHRP Synthesis 42, Transportation Research Board, Washington, D.C., 1977.
2. M. J. TOMLINSON: *Pile Design and Construction Practice*. Viewpoint Publications, London, 1977.
3. H. G. POULOS and E. H. DAVIS: *Pile Foundation Analysis and Design*. Wiley, New York, 1980.

H.G.P.

PLACING OF CONCRETE techniques are key factors in its subsequent performance. To provide sound durable concrete, it is essential that it be free of segregation and that the mortar within it be intimately in contact with coarse aggregate, reinforcement, and any adjacent faces to which it is to be bonded. Because of the nonhomogeneous character of concrete, segregation of the product is a continual hazard. It is essential to avoid dumping at an angle or depositing continuously at one point and allowing the concrete to flow. Similarly, if the concrete is confined laterally, as in forms, there is a tendency for the coarser and heavier particles to settle and for the fine and lighter particles to rise. The concrete should be deposited as nearly as practicable in its final position to preclude the use of placing methods that might cause the concrete to flow in the forms. To ensure that the structural integrity is maintained in large concrete placements, concrete should not be placed in layers greater than 500 mm (20 in.) thick. Each layer should be shallow enough to ensure that the layer below is still soft, and the two layers can be integrated during compaction. Each layer should be thoroughly spaded close to the formwork so as to close voids in the surface skin of the concrete.

There are many mechanical aids for use in the placing of concrete. Barrows, manually or motor powered, are economical on small work; most recently pumping (Fig. 171) has been used extensively on large and small work. With the use of these portable concrete pumps, many of them fitted with booms, it is possible to place concrete directly in position. Ready-mixed concrete can be discharged over an extended range using the folding chutes on most truck mixers. Hoppers are commonly used at the top of narrow columns or walls. Immersion vibrators are important for compaction of structural concrete.

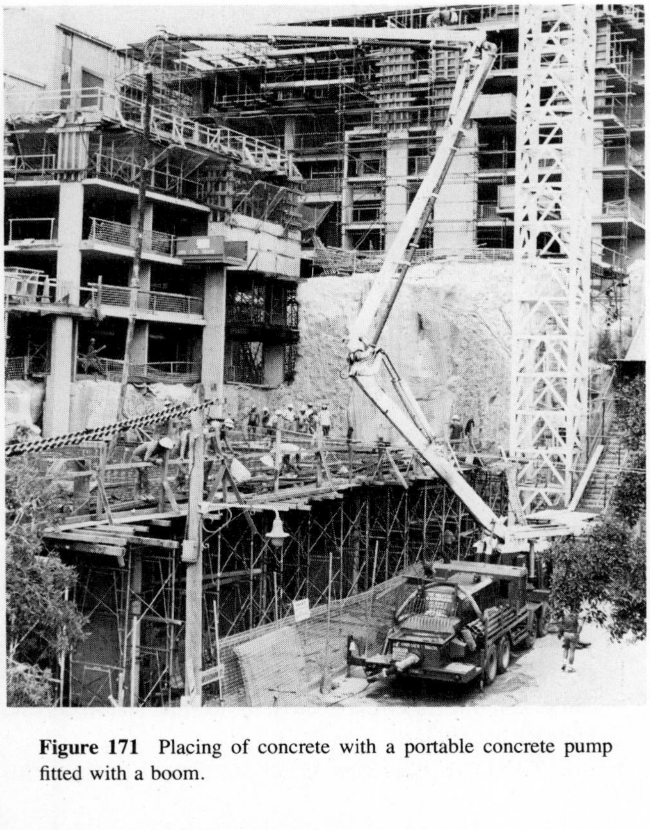

Figure 171 Placing of concrete with a portable concrete pump fitted with a boom.

See also CONCRETE MIX PROPORTIONING and FORMWORK.

References

1. AMERICAN CONCRETE INSTITUTE: *ACI Manual of Concrete Practice MCP Part 2,* Construction Practices and Inspection. American Concrete Institute, Detroit, 1983.
2. W. H. TAYLOR. *Concrete Technology and Practice.* McGraw-Hill, New York, 1977.
3. Bureau of Reclamation, U.S. Department of the Interior: *Concrete Manual,* U.S. Government Printing Office, Washington, D.C., 1975.
4. W. G. J. RYAN. *Pumped Concrete—Mix Technology.* Proceedings of a Seminar on Pumped and Pneumatically Placed Concrete. Association of Consulting Structural Engineers and National Ready-Mix Concrete Association, Sydney, 1971, pp. 47–59.

W.G.J.R.

PLASTERING is the application to a surface of a plastic mass or paste consisting of a cementitious material or a combination of cementitious material, aggregate, and water, which adheres to the surface and subsequently hardens, preserving in a rigid state the form or texture imposed during the period of plasticity (Ref. 1).

Plaster, also known as *stucco* or *rendering*, may be applied either to improve the appearance of the surface, to prevent rain penetration, or to obtain a surface that can be cleaned easily. The two *types* of plaster mainly used for plastering in building are portland-cement-based plasters and gypsum-based plasters.

Portland Cement Plaster is a mixture of portland cement, natural or manufactured sand, and water and can contain either lime or an admixture to improve workability. Portland cement plasters can be used internally as well as externally as undercoats or finishing coats. The most commonly used mix proportions are 1 part of portland cement and 5 to 6 parts of sand. When lime is added, mixes of 1 : 1 : 6, 1 : 1 : 8, and 1 : 2 : 9, portland cement : lime : sand are common.

Gypsum Plaster is made from plaster of Paris (calcium sulfate hemihydrate or sometimes anhydrite) and includes additives to control the setting time and in some cases other aggregates, which may be natural or manufactured sand or lightweight aggregates such as VERMICULITE or perlite.

When no aggregate has been added, gypsum plaster produces a very smooth finish, often applied over an undercoat of Portland cement plaster. With aggregates, it is used instead of Portland cement plaster for undercoats or finishing coats. Gypsum plaster should only be used internally since it can be adversely affected by moisture. For the same reason and because of the possibility of poor bonding, it should not be applied over green portland cement plaster.

The **surface** to be plastered should be clean and somewhat rough and should have a degree of suction to obtain a proper bond. Smooth surfaces should first be roughened and/or treated with a bonding agent. Undercoats in multilayer plastering should be scratched before they harden to ensure a good bond.

Method of Application. Both types of plaster can be applied by hand with a trowel (Fig. 172) or by using mechanical aids (Refs. 2, 3, and 4). It is common where the plaster is to be less than 15 mm thick for a single coat to be applied. (For thicker applications, plaster is best applied in one or more consecutive undercoats or scratch coats, followed by a finishing coat.) Leveling is carried out with a rod or straightedge and finishing with a wooden float or steel trowel, depending on whether a rough or smooth surface is desired. Various decorative finishes can be achieved (Ref. 5), such as a *rough texture* (Fig. 173) obtained by the spraying of a wet coarse sand and cement mix on a fresh finish, *trowel texture*, obtained by strokes of the trowel held at different angles, *brush finish*, obtained by raking the fresh finish with a broom or brush, or *exposed aggregate* finish made by partly embedding small stones in the surface. *Colored finishes* can be obtained by the intermixing of pigments with the materials; however, painting over plaster is more common because, besides giving the desired color, it improves the resistance to rain penetration. Both types of plaster, if of good quality, provide durable surface finishes, which are easily cleaned and can be renovated by repainting.

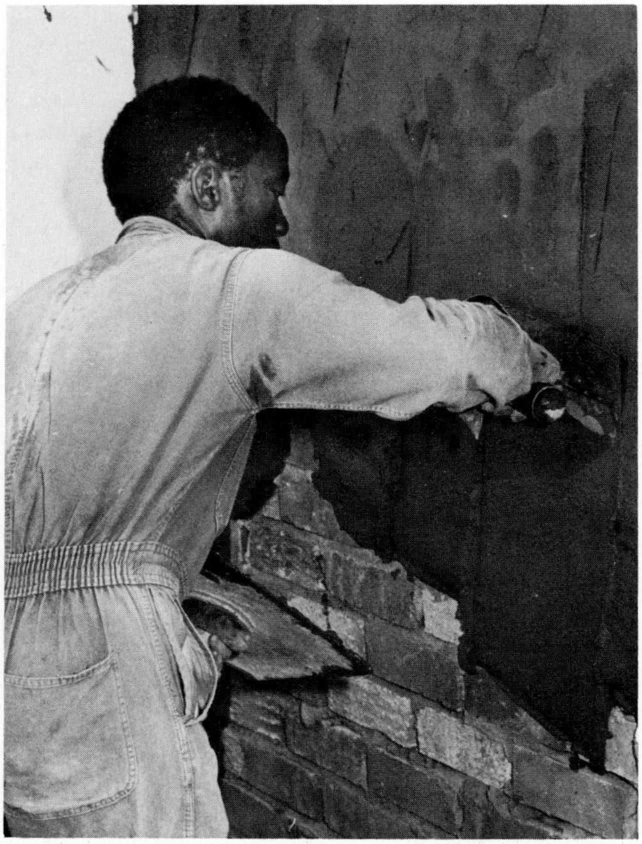

Figure 172 Plaster being applied manually.

Figure 173 Rough-textured finish.

Cracks caused by excessive shrinkage may affect the durability of the plaster. These can be caused by the use of plaster sands that are too fine or too clayey, mixes that are too rich in portland cement, the addition of too much mix water, or too rapid drying of the plaster.

See also COLORED CONCRETE; GYPSUM; and PORTLAND CEMENT.

References

1. *Cement and Concrete Terminology*. Report by Committee 116. Publication SP-19 (1978). American Concrete Institute, Detroit, 1978.
2. *Standard Specification for Application of Portland Cement-based Plasters*. Designation C926-81. Annual Book of ASTM Standards. Part 13. American Society for Testing and Materials, Philadelphia, 1982.
3. *Standard Specification for Application of Interior Gypsum Plaster*. Designation C842-79. Annual Book of ASTM Standards. Part 13. American Society for Testing and Materials, Philadelphia, 1982.
4. J. B. TAYLOR: *Plastering*. George Godwin, London, 1970.

5. JOHN J. BUCHOLTZ: *Stucco Textures and Finishes*. California Lathing and Plastering Contractors Association, Van Nuys, Calif., 1973.

J.J.J.v.R.

PLASTICS are relatively new materials with properties that are very different from those of traditional materials such as wood, metal, concrete, and glass. The great diversity of plastics has resulted in their use by many industries, such as construction, transportation, shipbuilding, aviation, and aerospace. Each plastic has a particular combination of properties, fabrication procedures, and economics that makes it ideal for certain applications, yet unsuitable for others.

In a technological sense, a plastic is a material containing polymer (a high-molecular-weight substance) and various ingredients (stabilizers, plasticizers, fillers, and reinforcements) that are fabricated into a finished product. The current definition of plastics is further restricted to those materials based on organic polymers that are not considered to be elastomers or fibers.

Polymers. Plastics are based on organic polymers. Although some polymers exist in natural substances such as cellulose and natural rubbers, almost all of those used in plastics are synthetic. Organic polymers are made from substances consisting of small molecules that combine under certain conditions to form a pattern of repeated groups of atoms (repeat units) in the structure of a large molecule in much the same way as links make up a chain. The substances from which polymers are made are known as *monomers;* the process by which polymers are produced is called *polymerization*. If only one kind of monomer is used, the resulting polymer is called *homopolymer;* if more than

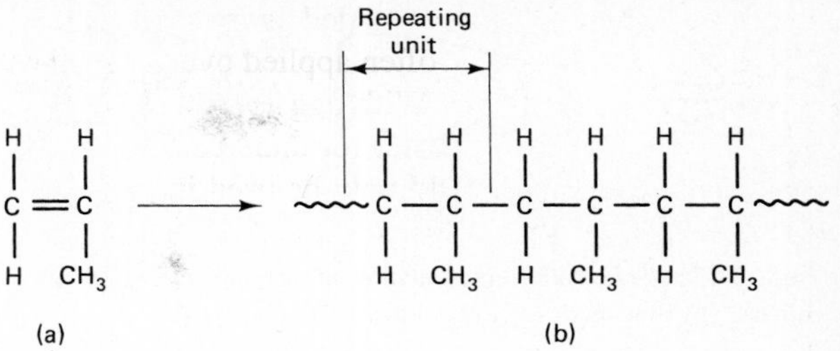

Figure 174 Formation of polypropylene. (a) Propylene. (b) Polypropylene molecule.

one kind of monomer is used, the product is a *copolymer*. A good illustration of a homopolymer is polypropylene, which is made by polymerizing propylene, a gaseous substance (Fig. 174). The molecules of polypropylene consist of long chains of carbon atoms linked to each other; each carbon atom, in turn, is linked to two hydrogen atoms or one hydrogen and one carbon atom (Fig. 174b). The repeating group of atoms,

$$\begin{array}{ccc} & \text{H} & \text{H} \\ & | & | \\ -\text{C}- & \text{C} \\ & | & | \\ & \text{CH}_3 & \text{H} \end{array}$$

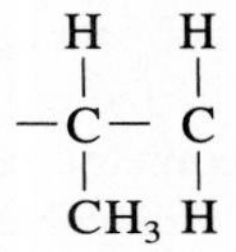

has the same kind and number of atoms as the starting material, but this is not the general rule. For example, in a synthetic polyamide, such as a nylon 66, the repeating unit in Fig. 175b has fewer atoms than the sum of the atoms in the two monomers (Fig. 175a). Polymerization, in this instance, involves elimination of a substance of low molecular weight, water (Fig. 175d). Depending on their chemical nature and the method used to produce them, molecules may be linear and chainlike, with or without appendages called branches, or they may be tridimensional net-

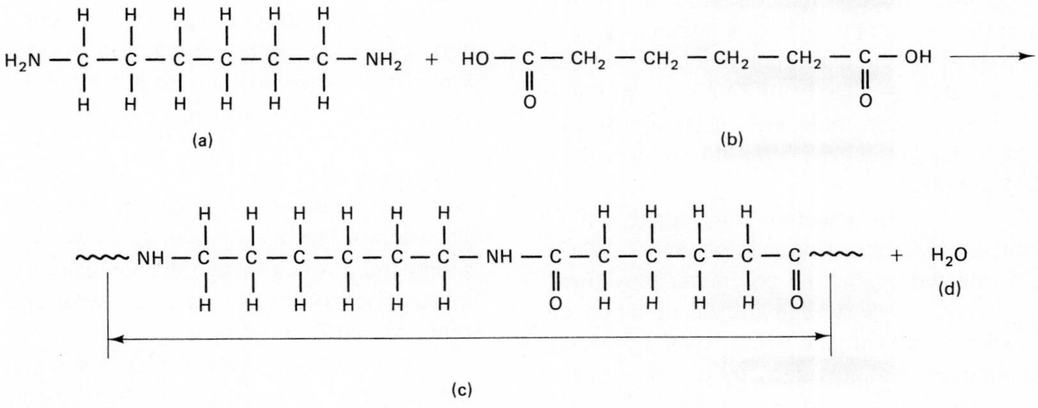

Figure 175 Formation of nylon 66 (a synthetic polyamide). (a) Hexamethylenediamine. (b) Adipic acid. (c) Repeating unit of nylon 66. (d) Water.

Figure 176 Schematic representation of polymer molecules: (a) Linear. (b) Branched. (c) Cross-linked.

work structures resulting from permanent chemical cross-links between the linear chains (Fig. 176).

Physical Structure of Solid Polymers. The chainlike molecules of a solidified polymer mass may form an amorphous, crystalline, or oriented state, according to the arrangement they assume. In the amorphous state the molecules are arranged in a completely random manner, somewhat like a mass of cooked spaghetti. Polymers in an amorphous state usually have irregularly shaped molecules. They are transparent and glasslike, and hence this solid state is sometimes referred to as the glassy state. Two commonly used amorphous polymers are poly(methyl methacrylate) and polystyrene. If the polymer molecules possess sufficient chemical or geometric regularity, the solid mass may assume an ordered arrangement and is said to be in a crystalline state. In this state, however, the amorphous regions usually range from 5% to 95% of the total volume of the polymer mass. There are hundreds of crystallizable polymers, of which the best known are polyethylene, polypropylene, nylon, polytetrafluoroethylene, and polycarbonate.

Finally, a polymeric mass (amorphous or crystalline), when subjected to uniaxial or biaxial tensile stress, may assume an oriented state in which the molecules are aligned in the direction of the applied load. In an oriented material, the properties in the direction of molecular alignment (or orientation) differ from those at right angles to the alignment; that is, the material in the oriented state exhibits anisotropy. Thus, an oriented polymer sample is considerably stronger in the direction of orientation than a sample of unoriented material of the same polymer. The tensile strength can be increased by a factor of 2 or 3 by unidirectional orientation of the chain molecules of an amorphous polymer; at right angles to the orientation it is weaker than in the unoriented sample.

Both crystallization and orientation can strongly influence the mechanical properties of organic polymers. When combined, as in oriented crystalline polymers, they form the strongest materials. High strength in the direction of orientation is imparted to synthetic fibers, most of which are highly crystalline oriented polymers, by stretching them after spinning.

Plastics. In a pure state, the polymers available do not meet the requirements of industry, despite the wide spectrum of properties they provide. Therefore, to extend the range of properties, to reduce deterioration during fabrication and use, and to facilitate processing, polymers are generally mixed with various materials (additives). The main classes of additives used in plastics are lubricants, stabilizers, plasticizers, fillers, reinforcement, fire retardants, and colorants (dyes and pigments). Lubricants reduce friction and improve flow characteristics during processing; stabilizers provide protection against deterioration by heat, oxidation, and solar radiation; plasticizers reduce brittleness of the end product; and fillers are inert materials added (in amounts ranging from 5% to 60% or more) to improve hardness and abrasion resistance, modify electrical characteristics, and reduce cost.

Although the terms polymer* and plastic are often used synonymously, they refer to different materials. Plastic is usually the physical mixture of polymers and additives. The precise nature and amount of each additive used in this physical mixture, or plastic compound as it is also called, depend on the polymer, the processing method used to convert the plastic compound into a finished article, and the properties required in the final product.

Heat Behavior and Transition Temperatures. When heat is applied to a mass of polymer or plastic having linear or branched molecules, the material softens at a certain temperature and then flows because these molecules are free to slide over one another; they harden again when cooled, reassuming their solid or rubbery state. The process of softening with heating and hardening with cooling can be repeated as often as required. Thus, plastics based on such polymers can be made to take on new shapes repeatedly by application of heat and pressure, followed by cooling. They are referred to as *thermoplastics*, the most common examples being polyethylene, polystyrene, and nylon. Polymers with highly cross-linked molecules do not soften with heat or show any significant flow because the individual chain segments are chemically attached to each other, forming rigid, network-type structures. Products made with such polymers, once formed, cannot be reshaped. These polymers are called *thermosetting* (thermoset) polymers. The principal commercial thermosetting polymers are the unsaturated polyesters, epoxies, alkyds, phenolics, and polyurethanes.

Because they are organic materials, both linear and cross-linked polymers decompose eventually upon excessive heating. The linear polymers, however, generally soften at temperatures below their decomposition point; the cross-linked polymers produce charred residue and smoke without softening and flow.

Two important characteristic temperatures associated with polymers and plastics are the glass-transition temper-

*The term *resin* is used by industry to designate a commercial polymer.

ature (T_g) and the melt temperature (T_m). The glass-transition temperature is the temperature below which the polymer mass has many of the properties of an inorganic glass, including hardness, stiffness, and transparency. Above this temperature, the material has plastic or elastic properties and is said to be in a rubbery state; hence, the glass transition is also called glass–rubber transition (temperature). The position of T_g with respect to room temperature determines the type of application or the usefulness of the polymeric material. For example, polymers such as polystyrene and poly(methyl methacrylate), with a T_g of 100° and 110°C (212° and 230°F), respectively, which is well above room temperature, are used in the glassy state. Such polymers and derived materials are brittle. Polymers with T_g below room temperature, such as polyisoprene (−83°C; −181°F) and polyisobutylene (−65°C; −85°F), are used in rubbers. If the T_g lies near room temperature, the polymers are generally not very useful because they are rubbery in the summer but glassy in the winter. Poly(vinyl acetate) (T_g = +28°C; 82°F) is a major exception to this rule.

The T_m, or first-order transition, of amorphous polymers is generally very indefinite and determines the rubber–liquid transition; it is sometimes referred to as flow temperature. In crystalline polymers, T_m is the temperature range at which crystalline volume elements in a polymer system are in equilibrium with the molten state.

Appearance. Many objects made from polymeric materials are inherently transparent, some are translucent, and a few are opaque. Rigid amorphous polymers, free of fillers and other impurities, are almost as clear as glass, particularly when they are fabricated so as to avoid orientation of molecules and flow. Poly(methyl methacrylate) is a good example of this type of polymer.

Similarly, derived plastics show a range of optical properties, depending on the compounding ingredients. Blends of impact-grade plastics and those based on crystalline polymers are usually composed of intermixed solid phases of different densities that can make them translucent or opaque. Polymers and plastics can also be made translucent or opaque by deliberately incorporating appropriate amounts of certain pigments or fillers.

The **density** of most plastics is considerably lower than that of metals, a useful feature where reduction in weight is required; consequently, a number of plastics are stronger than metals on a weight basis. On a volume basis, however, the opposite is normally true. Polymers and plastics generally have densities in the range 0.83 g/cm^3 (52 lb/ft^3) to 2.5 g/cm^3 (160 lb/ft^3), although some, such as foamed plastics, have densities as low as 0.01 g/cm^3 (0.62 lb/ft^3); filled plastics have densities as high as 3.5 g/cm^3 (220 lb/ft^3).

Thermal and Electrical Properties. Because they are organic materials, polymers and plastics have, with few exceptions, much lower heat resistance than do metals, particularly in the presence of oxygen. Among the common polymers, one exception is tetrafluoroethylene, which has very high heat stability because it has only C—C and C—F bonds, both very stable. When heated through a range of temperatures, thermoplastic materials slowly change from rigid solids to high viscous liquids. Although thermosetting materials do not appreciably soften with heat, excessive or prolonged heating results in overhardening, contraction, charring, or disintegration. Plastics have coefficients of thermal expansion (4 to 20 × 10^{-5}/°C; 2 to 11 × 10^{-5}/°F), considerably higher than those of common metals (1.0 to 2.5 × 10^{-5}/°C; 0.6 to 1.4 × 10^{-5}/°F). Polymers and derived plastics are generally good electrical insulators; some, such as polytetrafluoroethylene, are excellent.

Mechanical Behavior. Polymeric materials are chiefly valuable for their mechanical properties. In comparison with metals, polymers and plastics have low moduli and high strength-to-weight ratios. Products made from plastics show a wide variation of impact strength (resistance to brittle failure), from very tough to very brittle.

Polymeric materials do not obey Hooke's law (law of elasticity), but exhibit viscoelastic behavior. The prefix "visco" means that the material has some of the features of a viscous liquid, implying that its reaction to an applied force depends on the duration of the application of the force. According to this behavior, the deformational response of amorphous polymeric materials to nondestructive (relatively small) stresses may be of three kinds, differing in the way they depend on time under load: (1) instantaneous (or Hookean) elasticity, (2) delayed or retarded elasticity, and (3) if not cross-linked, viscous flow (irreversible flow in which entire polymer molecules are displaced relative to neighboring molecules).

Ultimate Mechanical Properties. Engineering applications of plastics are governed by strain considerations to a greater extent than are applications of other materials. As for most materials, a simple tensile stress–strain curve (Fig. 177) illustrates the basic mechanical behavior of a particular plastic or polymer. The initial slope provides a value for the tensile modulus of elasticity, which is a measure of stiffness. The curve also shows yield stress and strength and elongation at break. The area under the curve is a rough indication of the toughness of the polymeric material. The stress at the knee in the curve (known as *yield point*) is a measure of the strength of the material and of its resistance to permanent deformation. The stress at the breaking point is a measure of the force required to fracture the material completely. Stress–strain behavior for four typical classes of polymeric material is shown in Fig. 178.

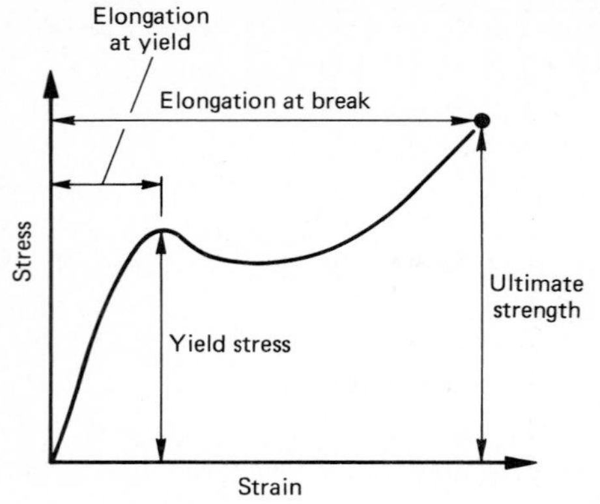

Figure 177 Generalized tensile stress–strain curve for polymeric materials [based on Winding (Ref. 7)].

A hard, brittle material such as an amorphous polymer at temperatures far below its T_g usually has an initial slope indicative of very high modulus of elasticity, moderate strength, a low elongation at break, and a low area under the stress–strain curve (Fig. 178). Typical values of modulus of elasticity and tensile strength are 3.4 GPa (500,000 psi) and 69 MPa (10,000 psi), respectively; typical elongation is about 2%. Generally, such materials exhibit elastic deformation up to the point of fracture, which is a brittle fracture. Polymeric materials showing hard, brittle behavior at room temperature or below are polystyrene, poly(methyl methacrylate), many phenolformaldehyde resins, and most other thermosets.

Hard and strong polymers have a high modulus of elasticity, high strength, and elongation at break of approximately 5%. The shape of the curve often suggests that the material has broken where a yield point might be expected. This type of curve is characteristic of some rigid polyvinyl chloride formulations and polystyrene polyblends.

Hard, tough behavior is shown by polymers, such as cellulose acetate, cellulose nitrate, and nylons; they have high yield points, a high modulus of elasticity, high strengths, and large elongations. Their stress–strain behavior is very closely depicted by a curve similar to that of the generalized curve (Fig. 177).

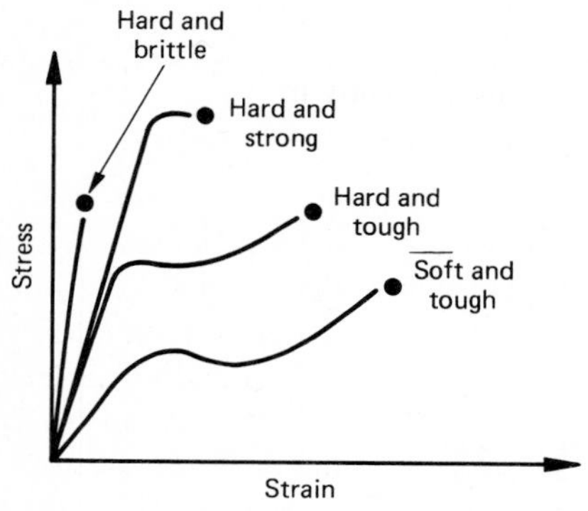

Figure 178 Tensile stress–strain curves for four types of polymeric materials.

Polymeric materials that are soft and tough show low modulus of elasticity and yield values, moderate strength at break, and very high elongation ranging from 20% to 1000%. This type of stress–strain curve is characteristic of plasticized polyvinyl chloride (PVC) and rubbers (elastomers).

Chemical and Weathering Resistance. The chemical resistance of polymeric materials depends on the nature of the polymer constituent and formulation (Ref. 1); hence, specific information on their resistance in a given environment is best obtained from the manufacturer. In general, polymers and plastics are resistant to weak acids, weak alkalis, salt solutions, and water, although some polyesters and polyamides may hydrolyze in acids and alkalis. Strong oxidizing acids may attack plastics, resulting in discoloration or embrittlement. Most polymers and plastics are affected by organic liquids; for example, fuels, oils, and various organic solvents may attack plastics, causing swelling, softening, and dissolution. Resistance to these agents depends on the prevailing temperature of the agents and the plastic when they come in contact and on the composition of the plastic. Most plastics may be used as corrosion-resistant materials. In view of the considerable variation in their resistance to various environments, great care should be taken to select the best possible plastic material for specific anticorrosive applications.

Polymeric materials vary widely in their resistance to outdoor weathering, in particular to solar radiation (Refs. 1 and 2). Some polymers and the plastics based on them have displayed outstanding weather resistance, whereas others have performed very poorly. Most commercial plastic compounds, however, can now be formulated to have fairly good weathering resistance.

Fire Resistance. Plastics show a wide range of behavior in fire; some ignite, some are self-extinguishing, others are slow to fast burning (Ref. 3). Flammability depends on the polymer and other constituents, such as fillers, reinforcing materials, plasticizers, or fire-retardant additives.

Halogen-containing polymers, like PVC or chlorinated PVC, are inherently flame retardant; with heating, they liberate chlorine gas that interrupts the free radical oxidation chain reaction. However, adding plasticizers to PVC may make it flammable. The fire-resistant properties of plastics can be improved by incorporating appropriate additives (Ref. 4) or using polymers with built-in fire resistance.

The products of combustion of most plastics are similar to those of wood, paper, and fabrics because their chemical constituents are essentially similar. Products of combustion, however, depend not only on the chemical nature of the material but also on the conditions under which the material burns. For example, with sufficient air the main combustion products of most plastics, woods, and fabrics are harmless carbon dioxide and water; but if there is an oxygen deficiency, large volumes of toxic carbon monoxide and smoke are formed. Fire-retardant plastics rapidly produce dense smoke that is not easily cleared by ventilation; and if plastics contain combined chlorine, fluorine, nitrogen, and sulfur (or their derivatives), these elements or their derivatives will also be present in the smoke.

Some Typical Applications in Building and Construction. Because plastics are available in many types and shapes, they are increasingly used in a variety of building applications (Refs. 1, 4, and 5), for example, inner partitions, luminous ceilings, roofing, vapor barriers, glazing, window and door frames, insulation, cladding and sidings for exterior walls, and piping materials (Ref. 6).

See also ADHESIVES; PAINTS; and SEALANTS AND JOINTING.

References

1. N. M. BIKALES (ed.): *Encyclopedia of Polymer Science and Technology*, Vols. 1–16. Wiley, New York, 1964–1972.
2. A. BLAGA: *Deterioration Mechanism in Weathering of Plastic Materials.* STP 691, American Society for Testing and Materials, Philadelphia, 1980.
3. NATIONAL MATERIALS ADVISORY BOARD: *Fire Safety Aspects of Polymeric Materials*, Vols. 1–10. Academy of Science, Washington, D.C., 1977–1980.
4. ENGINEERING EQUIPMENT USERS ASSOCIATION: *The Use of Plastics Materials in Buildings.* EEUA Handbook No. 31:1973, Constable, London, 1973.
5. M. VANCE: *Plastics in Building: A Bibliography.* Vance Bibliographies, Monticello, Ill., 1981.
6. D. A. CHASIS: *Plastic Piping Systems.* Industrial Press, New York, 1976.
7. G. C. WINDING and G. D. HIATT: *Polymeric Materials.* McGraw-Hill, New York, 1961.

A.Bl.

PLENUM SYSTEMS AND DUCTS are used to force fresh air into a building by a fan. The vitiated air is thus expelled through windows, doors, or any cracks in the construction so that drafts and the ingress of dust are prevented. In a warm-air plenum system, the fresh air is passed across a furnace before it enters the rooms of the building, and therefore it provides heating as well as ventilation (Fig. 179). In an AIR CONDITIONING system, the air is filtered, either heated or cooled as the climate requires, and its moisture content is controlled before it is delivered to the rooms of the building.

Since it is wasteful of energy to allow heated or cooled air to escape after it has passed through the rooms, some of it is led back to the fan through return air ducts. It is then mixed with a proportion of fresh air and goes through the filtering and heating or cooling process again before being sent back to the rooms. The amount of fresh air added must be sufficient to satisfy the requirements of health and freshness laid down in building codes. The proportions are commonly about 80% recirculated and 20% fresh air for rooms that do not produce objectionable odors. In the case of kitchens, toilet rooms, and rooms where heavy smoking is expected, recirculation should be avoided.

In the spring and fall, when the temperature of the outside air is more suitable than that of the return air, the system should be capable of operating on 100% fresh air. This operation mode is called an *economy cycle* or *free cooling.*

The **Velocity of Air** in the ducts usually ranges from 5 to 10 m/s (1000 to 2000 ft/min) in a low-velocity system, up to 20 to 25 m/s (4000 to 5000 ft/min) for a high-velocity system. The higher velocities permit smaller duct sizes, but they require more fan energy to circulate the air, require stronger ducts, and cause more noise than low-velocity systems.

The ducts for the supply or return of air are commonly made of sheet metal, but masonry is used for large vertical ducts and flexible or glass-reinforced plastics for small ducts are used to connect with the supply or return grilles. Ducts for low-velocity air are usually rectangular, as this shape fits easily into confined spaces. For good airflow in the duct, the ratio of the longer to the shorter sides of the rectangle should not exceed 3 : 1, although ratios up to 5 : 1 are used when there is a space problem. For higher air velocities, circular cross sections are more usual, as these resist more easily the higher pressures needed to maintain the higher velocities, and airflow is more efficient in a circular duct.

The supply air is usually introduced into rooms through grilles or registers that control the speed and direction of the air as it enters the space to ensure that it mixes thoroughly into the space. The return air can be collected by another set of grilles. The position of the return air grilles influences the pattern of airflow within the space, but does not control its speed or direction. The grilles may be integrated with the CEILING and especially with the LUMINAIRES.

Plenum Ceiling. Sometimes the space above a false ceiling is used as a plenum for supply *or* return air. The

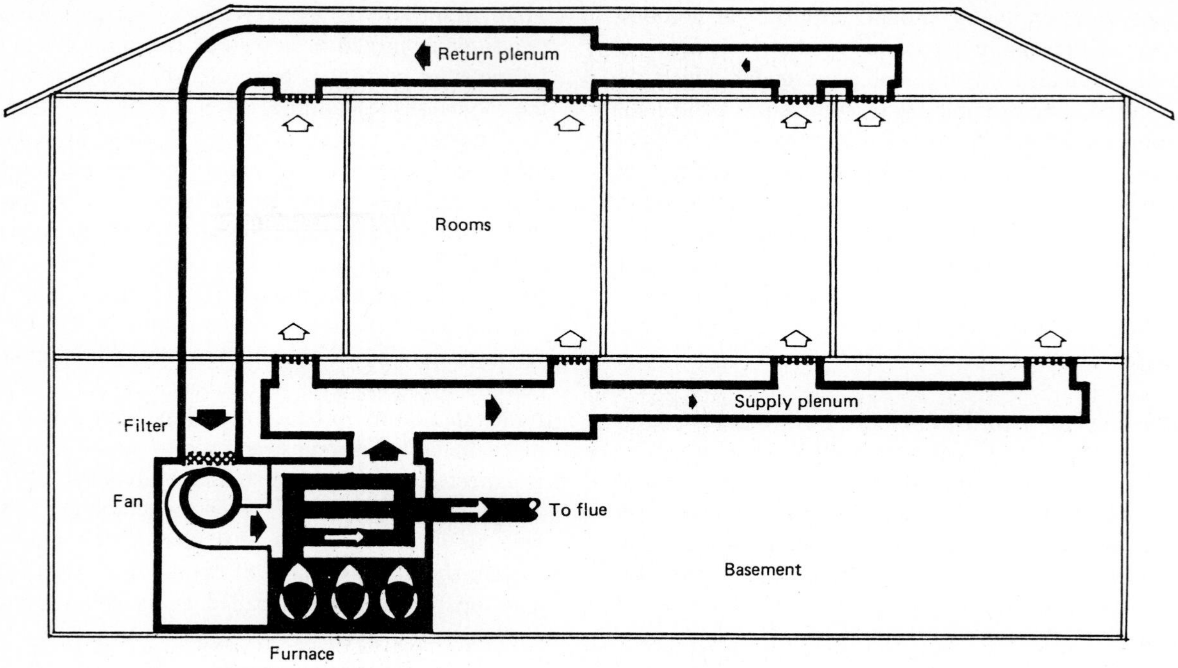

Figure 179 In a warm-air plenum system, a fan circulates air around the furnace and distributes it to the rooms through ductwork. It is recirculated through a return grill back to the furnace.

air travels between the ceiling plenum and the occupied space by small perforations or by grilles in the ceiling. Better control of the air distribution in the room is obtained by the use of supply registers and a ceiling plenum return, rather than the reverse.

Supply ducts that contain air significantly hotter or colder than the surroundings should have THERMAL INSULATION, usually fibrous glass wrapped around the outside of the duct, covered with reflective foil. Insulation exposed to the weather or to abrasion must be protected, for example by galvanized steel.

Air ducts transmit sound from the fans to the occupied rooms or from one room to another (see SOUND INSULATION). This can be reduced by lining the portion of the duct near the outlets with an acoustical absorber, such as a sheet of perforated metal backed by fibrous packing.

Fire Isolation. Air ducts passing from one room to another or from one floor to another provide a path that transmits fire between otherwise fire-isolated compartments. Therefore, in a building designed to isolate a fire within compartments, air ducts must be fitted with fire dampers at the position of the fire partition. The dampers are normally held open, but close by gravity or by a spring when hot gases cause a fusible link to melt.

Ducts for plumbing, electrical, and other servies should have the space around the pipes or cables sealed with a fire-resistant material at each floor.

References

1. *ASHRAE Handbook.* American Society of Heating, Refrigeration and Air Conditioning Engineers, Atlanta, Ga.
   (a) *Systems Volume and Product Directory*, 1980.
   (b) *Equipment Volume*, 1983.
2. W. J. McGUINESS, B. STEIN, and J. S. REYNOLDS: *Mechanical and Electrical Equipment for Buildings*, 6th ed. Wiley, New York, 1980.
3. H. J. COWAN and P. R. SMITH: *Environmental Systems.* Van Nostrand Reinhold, New York, 1983.
4. N. S. BILLINGTON: *Building Services Engineering; a Review of its Development.* Pergamon, Oxford and Elmsford, N.Y., 1982.

P.R.S.

PLYWOOD is defined as a panel consisting of a number of thin sheets of wood of uniform thickness, normally called *veneers*, bonded together with a rigid adhesive. The direction of the wood grain in alternative layers or plies is usually at right angles. Thus plywood consists of a series of cross laminations of veneer.

Plywood can be as thin as 1 mm ($\frac{1}{20}$ in.) and usually has a maximum standard thickness of 25 mm (1 in.), although thicker plywoods can be manufactured where required. It may consist simply of three veneers in thin assemblies or of many layers in thicker plywood. Usually, plywood is manufactured with an odd number of plies balanced in terms of thickness, species, and grain direction each side of a central or core veneer. When even numbers of plies are used, the two innermost are glued parallel.

The veneers, can be produced by rotating the log in a lathe against a fixed knife, causing a continuous ribbon of rotary veneer, or sliced off the log in pieces by moving the log backward and forward over a knife. This sliced veneer is usually decorative, as the log is cut to enhance the natural figure and hue of the wood.

Types of Plywood fall into two categories, constructional and decorative. The line of differentiation lies in the type of adhesive used to bond the veneers together. This makes it possible to apply basic working stresses to the plywood. All constructional plywoods are bonded with permanent, durable rigid adhesives of the synthetic phenol formaldehyde or natural phenolic type, and they have specified structural properties. The species of timber and the quality of the veneer are also defined. This type of plywood is an engineering material; it is used for domestic and industrial flooring, bracing of house frames against wind or earthquake forces, marine applications such as boat hulls and marina decks, plywood webbed beams, pallets, containers, agricultural uses, and other engineered applications involving exposure or the application of long-term stress.

The decorative plywoods are bonded with less durable urea formaldehyde adhesive. Veneers are either rotary cut or sliced decorative. This plywood is used in interior nonstructural applications, such as decorative wall paneling, furniture, and internal fitments in buildings.

See also ADHESIVES and LAMINATED TIMBER.

References

1. *Glossary of Terms Used in the Plywood Industry—AS2289*. Standards Association of Australia, Sydney, 1979.
2. *Technical Information Manual*, Plywood Association of Australia, Brisbane, 1977.
3. R. F. BALDWIN: *Plywood Manufacturing Practices*. Forest Industries Books. Miller Freeman, San Francisco, 1975.

K.J.L.

PNEUMATIC STRUCTURES (air houses) carry load by compression of an enclosed gas or fluid, such as air or water, the pressure difference between the inside and outside being carried in the skin. They are the most basic form of structure in nature. The first amoeba was a pneumatic structure; the human not only grows as one, but still largely distributes load on himself in that way. It is an extremely efficient structure in its use of materials.

History. The idea of utilizing pneumatic structures is not new. Three thousand years ago rafts were made of goat skins; the sail of a ship is probably the best known example of their use in the past, the car tire probably the best known example in the present.

F. W. Lanchester, the British car pioneer, patented in 1918 the design of a tent supported by air (Ref. 1). It comprised a "skin" reinforced with cables and anchored round its edges to ensure small air leakage; the increased internal air pressure required for support was provided by a centrifugal fan, and entrance to the tent was achieved through an air lock with double doors; balancing the air within the lock so that the doors could open was achieved by small shutters in the door, like paddles in a canal lock. Movement of the flexible air house skin relative to the stiff air lock structure would be allowed for by bellows of fabric around the lock, and additional cable reinforcement would be provided around the opening. This patent describes all that is included today in a single skin air house, yet Lanchester never made one; a suitable fabric was not available in his day.

It was not until the late 1950s that pressure-supported domes to cover equipment to detect intruder aircraft and missiles in the Arctic were developed by Walter Bird in the United States. Walter Bird's firm, Birdair, started using isotropic sheet materials such as neoprene and hypalon, but subsequently changed to using coated fabrics. The use of woven fabric is important because, although when woven it has the threads at right angles with certain extension qualities, when pulled at 45° to this warp and weft it is much more extensible. In other words, the Poisson's ratio of the material at 45° axes, arising from the lack of in plane shear stiffness in the material, used wisely allows for a change of shape that enables loads to be distributed in a manner suitable for the material. Coating of the fibers then provides both airtightness and protection against deterioration for the base fabric.

Design. Like most developments in the building industry, this type of building largely evolved in a craft sense, made by tent or cover manufacturers looking for new materials, so design has been largely empirical and theory has been relatively underdeveloped. The first fabrics were woven nylons and terylenes coated with PVC (polyvinyl chloride) usually with stitched seams. The life of the cloths was relatively short, although today advances in the chemical formulation of the coatings, together with the welding of seams, give lives of 10 years or so.

Since the only normal loads are the weight of the cloth together with that required to combat wind pressure, the air pressure required to maintain uplift is relatively low (typically on the order of 30 mm water pressure); it has long been the habit to run airhouses with electrically driven fans coupled to a pressure gauge and, in the event of a drop in pressure due either to primary fan failure or to increased loads, provide a gasoline-driven emergency fan motor.

Lack of research has inhibited development. Codes produced in the various countries contain considerable oversimplifications. Airhouses were viewed as static structures, and the wind pressure assumed to exert fixed pressures. So analysis was assumed as static and fixed factors against failure were applied. Stress with fabrics was regarded as being the determining criterion. Deformations were considered to be of secondary importance. The result was that stability was assumed to be assured by a given fixed level of pressure, rather than by considering the fans together with the enclosed volume of air as the structural support. There was a desire to provide a rigid support system in the event of fan failure, which ignored the problems of compatible deformations. Far too little attention was given to the nature of the materials apart from a fear of their deterioration.

Failures. A survey of known failures (Ref. 2) showed that almost the only cause was puncturing, leading to tearing, from an external or internal pointed object (either roof tiles, fork lift trucks, etc.) or internal supports (sometimes the emergency support system) when deformations became too great. Put another way, most of these collapses came from insufficient air pressure when external loads were large, but the actual failure was tearing of the fabric following puncturing.

For airhouses that were just dying of old age, the points of weakness were in areas where the stresses were low in one direction and, due to incorrect fabric cutting patterns, compression wrinkles were caused; thus cracks occurred in the coating, dirt collected in the wrinkles, and bacteria attacked the skin.

These lessons are slowly leading to better performance. Toughness and resistance to tear propagation are now recognized as the most important material characteristics and a good shape to the skin is the prime quality in ensuring a long life and freedom from problems. Since airhouses perform in a nonlinear manner, concepts of designing for excess fan capacity require development. A strategy for regular maintenance is essential.

Current Practice. By 1980 it was estimated that more than 60,000 airhouses had been erected throughout the world for exhibition, office, sports, storage, shopping, or other purposes (Fig. 180). The International Expo at Osaka in 1970 marked a watershed. Not only were many of the national or commercial pavilions air supported, but the U.S. pavilion was a low profile 142 m × 83 m (465 ft × 273 ft) oval-shaped, cable-reinforced roof supported on cantilevers so that the wind pressure on the roof was largely uplift and, in the event of a collapse, the roof was still supported over the occupants (Ref. 3). This was the first practical example of what designers had been arguing for some time—that the future of the airhouse lay in its ability to cover large spaces cheaply since the reservoir capacity of air within made it such an economical space (Ref. 4).

Figure 180 Common type of air house, designed by Buro Happold.

The engineering practice of Geiger, Berger, and others has gone on to design a number of stadium roofs on this same principle (Fig. 181), largely using new fabrics such as teflon-coated fiber glass and, more recently, silicon-coated glass. Even metal foil has been used. These materials have the problem of being extremely stiff and so susceptible to fatigue failure if subjected to much flexure. Yet they have an extremely long theoretical life, desirable reflective and translucent properties, and, since the coating is dielectric, do not collect the dirt. Again in these developments Walter Bird's firm, Birdair, now called ChemFab Structures, has been the leader, although there are other designers and manufacturers elsewhere in the world.

Figure 181 Pontiac Stadium in Michigan, U.S.A. This is one of the new low-profile, polytetrafluorethylene-coated, glass-fiber-covered roofs (built in 1975, 65,000 seats).

Of course, the main quality of an air-supported roof is to keep the weather out, and insulation, both thermal and acoustic, is limited. Recently, thermal performance has begun to be studied, and double skins and more sophisticated air movement systems and shading have begun to be used. Major studies have already been carried out for their use in covering living places in areas of extreme climatic conditions such as the Arctic (Ref. 5). It is only a matter of time before such roofs are attempted on a very large scale.

See also TENSILE ROOFS.

References

1. F. W. LANCHESTER: *An Unproven Construction of Tent for Field Hospitals, Depots and the Like Purposes*. British Patent No. 119339 (accepted 1918).
2. E. HAPPOLD and M. J. COOK: A review of airhouses in use. *IABSE Colloquium on Inspection and Maintenance*. Cambridge, England, 1978.
3. D. H. GEIGER: Developments in incombustible fabrics and low profile air structures including those with thermally active roofs. *IASS World Congress on Space Enclosures*, Concorde University, Montreal, July 1976.
4. *City in the Arctic, IL2*. Institute for Lightweight Structures, University of Stuttgart, Stuttgart, Federal Republic of Germany, 1968.
5. FULLERTON, WILKINSON, BURO HAPPOLD, ATELIER FREI OTTO: *Climate Tempering Structures for Central Areas*. A report for the Government of Alberta, Canada 1982.

E.H.

POLYHEDRA are spatial configurations with polygonal faces, edges, and vertices, which combine to form a continuous envelope with exactly two faces meeting along every edge. *Polyhedral envelopes* are divisions of the three-dimensional Euclidean space into two subspaces.

Polyhedra are *finite* and topologically related to the sphere. They may be divided into convex plane-faceted polyhedra (Fig. 182a, b, and c), star polyhedra (Fig. 182d), and saddle polyhedra (Fig. 182e). Polyhedra envelopes are *infinite*, and they may be divided into plane (Fig. 182f) and cylindrical tesselations (Fig. 182g), and infinite skew polyhedra, which are locally saddle-shaped envelopes, subdividing space between two interwoven space networks (Fig. 182h).

History. The *regular polyhedra* have identical vertex configurations, regular face polygons, and equal edge lengths. Their symmetry has given them an esthetic and intellectual attraction and a hold on human imagination since

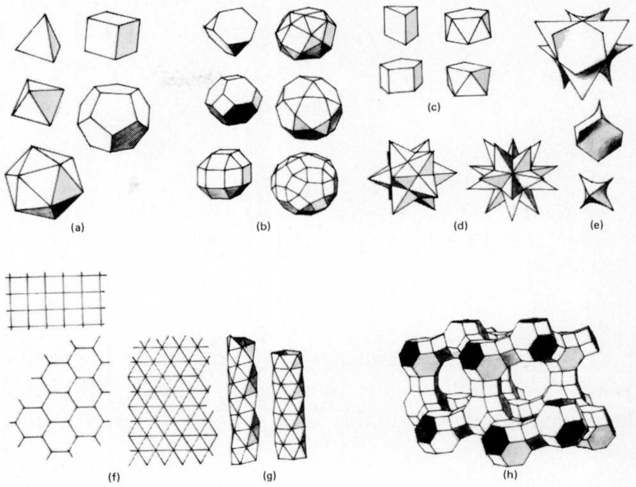

Figure 182 Classification of polyhedra. *Finite-convex polyhedra*: (a) The five platonic polyhedra. (b) Archimedean polyhedra. (c) Prisms and antiprisms. (d) Star polyhedra. (e) Saddle polyhedra. *Infinite polyhedra*: (f) Plane tessellations. (g) Cylindrical and helicoidal tessellations. (h) An infinite skew polyhedron. (b), (c), (d), (e), (f), (g), and (h) are just representative polyhedra of much larger groups.

the dawn of civilization. Some were studied by the ancient Egyptians and the Pythagoreans. Plato referred to the five regular polyhedra, known today as the *Platonic solids*. Euclid devoted his final book to their discussion. The 13 semiregular convex polyhedra are attributed to Archimedes, and Kepler associated them with the cosmic order. Many other prominent artists and mathematicians, such as Leonardo da Vinci, Dürer, Descartes, and Euler, contributed to our knowledge of the subject.

Application. In modern times, polyhedra play an important role in mathematics, molecular chemistry, biochemistry, and crystallography, as well as in engineering and architecture. Polyhedral envelopes and networks, especially

Figure 183 Geodesic domes can be based on several types of polyhedral geometries. This one is derived from the icosahedron, shown top left, which is a platonic polyhedron formed by 20 equilateral triangles.

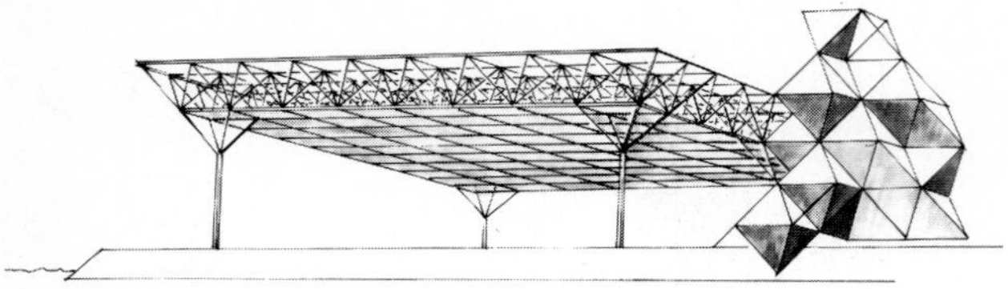

Figure 184 Bar-and-joint space structure with flat top and ceiling, whose geometry is derived from a close packing of tetrahedra and octahedra.

those with *triangular geometry*, inspired many highly industrialized solutions for lightweight SPACE FRAMES displaying structural rigidity and stiffness (Figs. 183 and 184).

References

1. ROBERT WILLIAMS: *Natural Structure—Toward a Form Language*. Eudaemon Press, Moor Park, Cal. 1972.
2. ARTHUR LOEB: *Space Structures—Their Harmony and Counterpoint*. Addison-Wesley, Reading, Mass., 1976.

M.B.

PORTLAND CEMENT is typically manufactured by proportioning limestone, a source of calcium oxide, with clay, a source of silica, alumina, and ferric oxide. This mixture is ground and then heated in a rotary kiln to a temperature of approximately 1450°C to form the compounds that comprise portland cement clinker. The clinker emerges from the kiln as particles having an average size of about 3 cm. The cement clinker is then mixed with gypsum and ground until the average particle size is reduced to approximately 5 to 10 μm with an average surface area of 350 m^2/kg.

Portland cement is composed of five major compounds as shown in Table 1. These compounds are not chemically pure. Impure forms of tricalcium silicate and dicalcium silicate are often referred to as alite and belite, respectively. The relative amounts of these compounds are determined by controlling the composition of the raw materials. This depends on the application for which the cement is to be used. These applications include general construction and specific applications where high early strength, low heat evolution, or sulfate resistance may be required. Typical weight percentages of these components in cement used in ordinary construction are listed in Table 1.

Each cement compound reacts exothermically with water to yield hydrated solids. However, it is the hydration of the silicate bearing phases that imparts a major contribution to the mechanical properties required in concrete. Tricalcium silicate reacts far more rapidly than does dicalcium silicate, and it is the hydration of the former that results in early strength gain in concrete. Accordingly, cements used in applications where early strength is important have high tricalcium silicate contents and are finely ground. In massive concrete, such as large dams, the rate of strength development is less important than is minimizing the rate of heat evolution. Decreasing the rate of heat evolution can be accomplished by using cements having high dicalcium silicate and low tricalcium silicate contents.

Both tricalcium silicate and dicalcium silicate react to form calcium hydroxide and a calcium silicate hydrate of variable composition, but are often represented by the formula $3CaO \cdot 2SiO_2 \cdot 3H_2O$. The formation of calcium silicate hydrate is principally responsible for strength development. The formation of calcium hydroxide is not believed to contribute significantly to the strength of concrete. However, the formation of calcium hydroxide allows the addition to cement of various pozzolanic materials, such as fly ash, or latent hydraulic material, such as blast-furnace slag. Pozzolanic materials are those that contain a source of reactive silica that is available to combine with calcium hydroxide to form calcium silicate hydrate. The hydration reaction of latent hydraulic materials is promoted by the presence of calcium hydroxide. Cements containing these materials are referred to as portland–pozzolan or portland–blast furnace slag cements.

Both calcium hydroxide and calcium silicate hydrate are basic, and the pH of the pore solution in concrete is approximately 12.6. Steel develops a passive iron oxide layer on its surface at this pH. It is for this reason that, in the absence of corrosive anions such as chloride, concrete can effectively be reinforced with steel for extended design lives.

TABLE 1
Composition of Portland Cement

Compound	Common Chemical Name	Composition: Range (%)	Composition: Typical (%)
$3CaO \cdot SiO_2$	Tricalcium silicate (alite)	25–75	55
β-$2CaO \cdot SiO_2$	Dicalcium silicate (belite)	10–50	25
$3CaO \cdot Al_2O_3$	Tricalcium aluminate	0–18	10
$4CaO \cdot Al_2O_3 \cdot Fe_2O_3$	Tetracalcium aluminoferrite	0–18	6
$CaSO_4 \cdot 2H_2O$	Calcium sulfate dihydrate (gypsum)	0–7	4

The reaction of tricalcium aluminate with water is extremely rapid. It is undesirable that this reaction occur prior to concrete placement, as it results in premature stiffening, called *flash set*. Flash set is eliminated by adding gypsum to cement clinker prior to grinding. The hydration of tricalcium aluminate in the presence of gypsum results in the formation of ettringite. Ettringite coats the surfaces of the unhydrated tricalcium aluminate particles, thereby slowing the rate of tricalcium aluminate hydration. The hydration of tetracalcium aluminoferrite is generally similar to that of tricalcium aluminate, and this reaction is also retarded by the addition of gypsum.

The amount of gypsum added in the manufacture of cement is controlled such that only the amount of ettringite required to control flash set is allowed to form. The formation of ettringite is an expansive reaction. If this expansive reaction occurs in hardened concrete, cracking tends to occur. However, if the concrete is in a plastic state, this reaction is not disruptive. Advantage of this expansive tendency is taken in applications where the normal shrinkage that occurs in concrete is undesirable. In this instance, extra gypsum or other compounds, such as $3CaO \cdot 3Al_2O_3 \cdot CaSO_4$, are added to cement to promote controlled expansion. These cements are called expansive or shrinkage-compensating cements.

In numerous instances, concrete is used in applications where it is exposed to environmental sulfate, such as in sewage systems. Under these conditions, sulfate ion slowly intrudes the cement paste matrix and reacts to form ettringite, resulting in disruption. The tricalcium aluminate content of cement manufactured for use where environmental sulfate is present is limited. Depending on the amount of compound present, these cements are referred to as sulfate resisting or moderately sulfate resisting cements.

References

1. F. M. LEA: *The Chemistry of Cement and Concrete*. Chemical Publishing Co., New York, 1971.
2. H. F. W. TAYLOR (ed.): *The Chemistry of Cements*. Academic Press, London and New York, 1964.
3. S. MINDNESS and J. F. YOUNG: *Concrete*. Prentice-Hall, Englewood Cliffs, N.J., 1981.

P.B.

PRECAST CONCRETE is reinforced concrete constructed by industrialized methods. It is characterized by dividing the building into components, which are manufactured in a precast concrete plant, transported by special transportation equipment to the site, and erected by cranes to form a building.

Precasting reduces construction time because the manufacturing of the elements is carried out beforehand. The precast elements have a better finish than site-cast concrete work. Consequently, doors and electrical and hydraulic conduits are often embedded in the elements, and the surface of the precast walls is ready to receive wall paper. The manufacturing plant allows the use of highly industrialized techniques and makes the production independent from the weather.

Precast concrete is used in the construction of the following:

1. Single-story buildings: industrial halls with various forms and spans.
2. Multistory buildings, which can be partially or totally precast by using a structural frame.
3. Large-panel buildings whose walls are load bearing without a structural frame.

The **Basic Elements** of precast concrete are double tees, single tees, channel sections, and hollow core slabs. These elements can be used for various dimensions and spans and also external walls with or without thermal insulation (Fig. 185).

Architects appreciate that concrete can be used for decorative purposes. Many contemporary buildings have precast concrete external walls that produce a pleasant architectural design (Fig. 186).

Types of Building.

Single-Story Buildings. The roof slabs and external walls of single-story buildings are often constructed with single

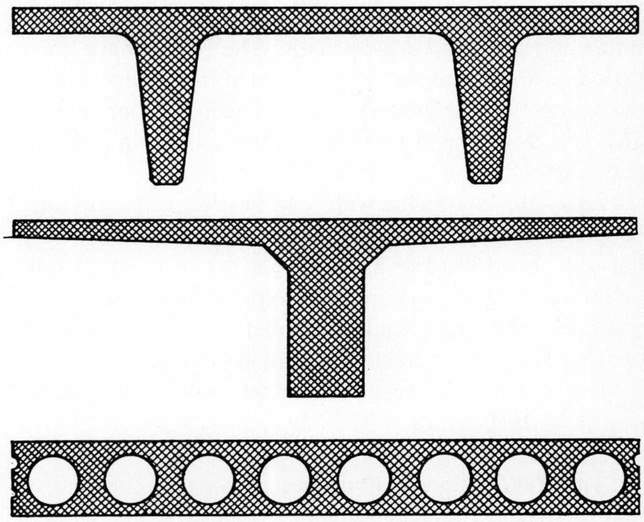

Figure 185 Shape of basic precast concrete elements for floor slabs and walls.

Figure 186 Architectural concrete in Toronto, Canada.

Figure 187 Erection of a single-story building with folded-plate north-light roof.

or double tees, or the roof slabs are carried by prestressed precast concrete girders of T shape.

North-light roofs are becoming popular again because of the energy saving. A roof with FOLDED PLATES produces a pleasant architectural solution (Figs. 187 and 188).

Multistory Buildings are constructed with a structural frame and prestressed floor slabs. The beams have an inverted T form, and the external walls often have a decorative function. Multistory columns reduce labor, but can give rise to technical problems in transportation and erection (Fig. 189).

Large-panel systems with load-bearing walls provide a structural solution for housing. Floors are hollow-core or flat-slab. Load-bearing walls have a thickness of 140 mm to 200 mm ($5\frac{1}{2}$ in. to 8 in.). In cold countries the external walls are multilayer sandwich panels. If the load-bearing panels are properly connected, the system has high earthquake resistance, because the structure is stiff in any direction (Fig. 190).

The **Production Techniques** of precast concrete are industrialized. The T and double T slabs, beams, and columns are manufactured in steel forms and compacted by high-frequency vibrators. Slabs and beams are generally prestressed. The prestressing strands are deflected at one or two points to save steel and to give a better shear stress distribution at the supports.

The **Prestressing** of precast concrete is important especially for big spans or large live loads. For small spans,

Figure 188 Bus garage with folded plate roof in Switzerland.

Figure 189 Erection of laboratory building with multistory columns for the University of Riyadh, Saudi Arabia.

Figure 190 Apartment buildings with large panels in Ratzeburg, Germany.

and especially in housing construction, nonprestressed floor slabs are better. Prestressed hollow core slabs are produced by extruders, with machines that receive the concrete directly from a batching plant and form the elements on a long prestressing bed.

Nonprestressed hollow core slabs, up to six elements per hour, are manufactured in a continuous production process. The steel tubes that form the cores are pulled out immediately after vibration is finished. The elements are then cured on a pallet by steam. The entire process is highly industrialized.

Large Panels can be manufactured either vertically or horizontally. For flat-floor slabs and internal walls, precasting is more economical vertically, and for external walls, horizontally. Vertical pouring produces smooth surfaces, which can be covered with wallpaper without further treatment.

Machines for the vertical production of elements employ a conveyor belt system with circulating molds. Manufacture is divided into preparation, pouring, steam curing, and demolding. The concrete can be poured on the same station three to four times a day.

Changes of the element's dimensions in the circulating pallet system do not influence the production, as some of the pallets can be taken out of the production circuit to change the dimensions while the other pallets continue to produce elements.

Transportation limits the maximum possible dimensions of the elements. Bridge clearance is generally not higher than 4.50 m (15 ft), so an element's dimension cannot exceed 4.20 m (14 ft) in one direction. The length of transported elements may be 36.0 m (120 ft) and more in the United States, but in Europe it is generally no more than 30.0 m (100 ft).

Erection. Mobile cranes are generally used for erection. Multistory structures, especially large-panel buildings, can be erected more economically by a tower crane. Tower cranes can lift elements up to 20 tons in weight with an arm of 20 to 30 m (65 to 100 ft). A combination of mobile and tower cranes is also possible.

Precast concrete construction has great flexibility, and the same elements can be used for single- and multistory buildings for different purposes like floor slabs, roofs, or wall elements, and also for different spans and building heights. There is also flexibility in the production, as the same machine can produce elements of various dimensions.

The development of precast concrete depended on the production of equipment for transportation and erection. By using the technology of prestressing, large spans can be reached.

See also CONCRETE STRUCTURES; PREFABRICATION: ORIGINS AND DEVELOPMENT; PRESTRESSED CONCRETE; REINFORCED CONCRETE SLABS; and SYSTEM BUILDING.

References

1. T. KONCZ: *Manual of Precast Concrete Construction*. 3 volumes. Bauverlag, Wiesbaden and Berlin, 1978.
2. T. KONCZ: Assembly line speeds panel construction for low cost housing. *Journal of the Prestressed Concrete Institute*, Vol. 24 (Mar./Apr. 1979), pp. 38–57.
3. T. KONCZ: Precast concrete construction correctly suited to design and manufacture. *Betonwerk + Fertigteiltechnik*, Vol. 12 (Dec. 1978), pp. 697–704, and Vol. 1 (Jan. 1979), pp. 46–53.
4. T. KONCZ: Various types of prefabrication plants for large panel components. *Betonwerk + Fertigteiltechnik*, Vol. 4 (Apr. 1983), pp. 242–249.

T.K.

PREFABRICATION: ORIGINS AND DEVELOPMENT. In 1787, Samuel Wyatt manufactured 12 "moveable hospitals . . . for his Majesty's distant possessions." From this modest beginning, the fabrication of buildings in component form for later assembly on adjacent or remote sites developed into a 19th century industry of some substance. It produced buildings described as "temporary" or "portable"; the term prefabrication was probably not in general use before the 1930s. From the workshops of Europe and the United States, using the new tools and techniques of the Industrial Revolution, came the components of wooden and iron buildings of every type, from utilitarian structures to churches and theaters. Most found their way to the expanding frontiers of Asia, Africa, the Americas, and Australasia (Fig. 191).

Timber Prefabs. Houses were perhaps the most important product, from the pioneer "Portable Colonial Cottage" produced by Manning of London in the 1830s, to the precut timber and panel houses to be found in the California goldfields and in the settlement of the American Midwest. Wooden barracks, premade in England, housed the British and French armies in the Crimean War, and Brunel's 2200-bed hospital at Renkioi must rank (with Paxton's Crystal Palace) as one of the outstanding achievements of 19th century prefabrication. The 1850s saw the climax of timber prefabs in Britain; in the United States, Scandinavia, and Germany, with such firms as Christoph and Unmack, they continued well into the 20th century.

Iron replaced timber as the more advanced industrial nations such as Britain or France turned to more sophisticated techniques and materials. *Corrugated iron* buildings were to be found not only in California, but in the mining towns of Australia and South Africa; iron prefabs provided a palace for an African king, a ballroom for Victoria's Prince Consort, and an officers' club later to be converted into General Smuts's residence. The extensive range of iron buildings could be examined in the catalogs of John Porter, Morewood and Rogers, and Samuel Hemming. There were also *cast-iron* structures, solid and architecturally more pretentious, made by such Scottish ironworks as Macfarlanes or C. D. Young, whose century-old products still stand today in daily use, in Geelong, Kimberley, or Manaus.

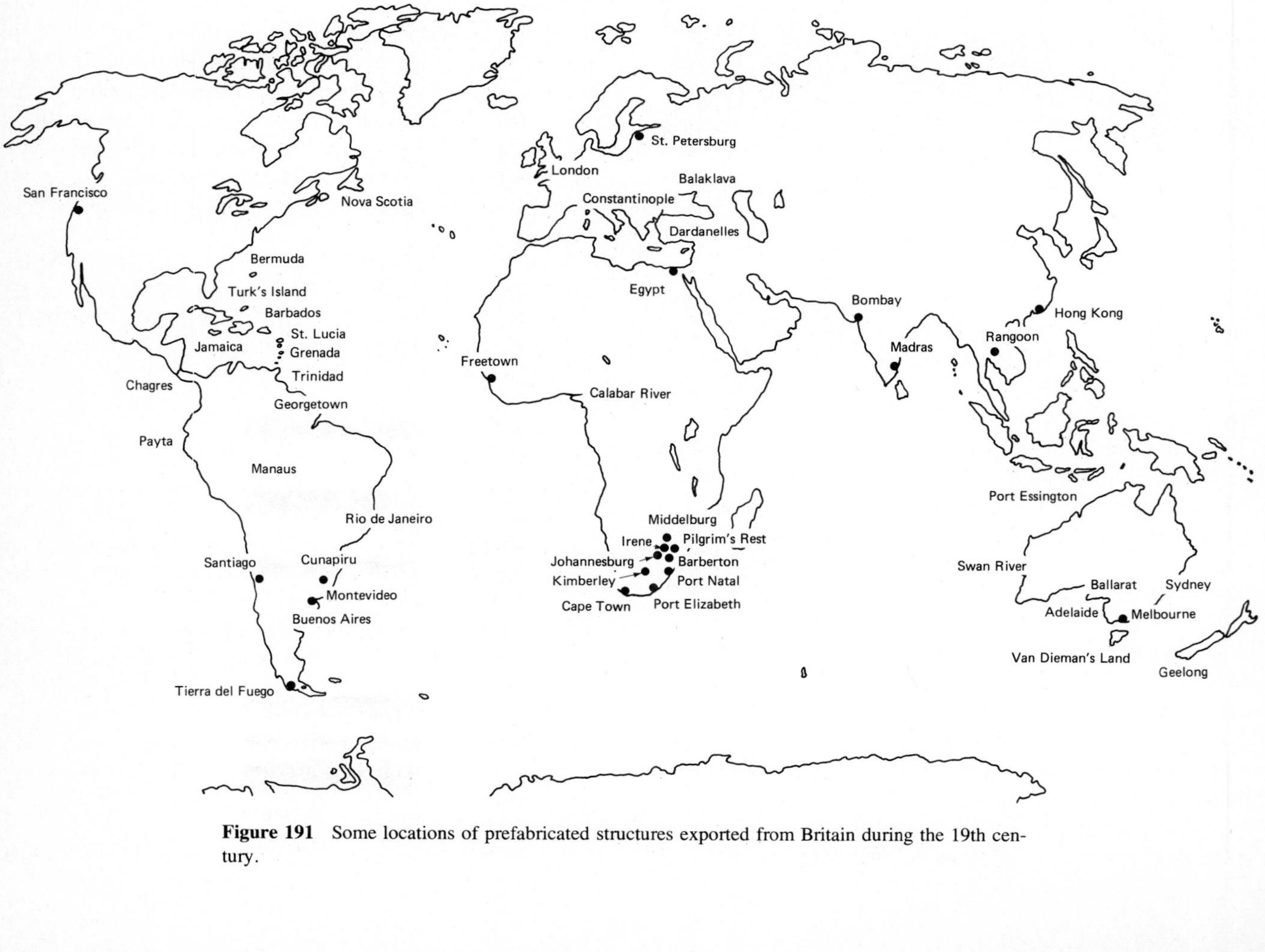

Figure 191 Some locations of prefabricated structures exported from Britain during the 19th century.

Glasgow ironfounders produced the Jamaica Street warehouses, and in New York Bogardus and Badger competed in the erection of multistory buildings with handsome cast-iron fronts.

Reinforced Concrete, as used by the end of the 19th century by Hennebique in France or Ransome in America, had obvious potentialities for prefabrication. In 1905, J. A. Brodie of Liverpool constructed a 12-apartment block of flats in a panelized system of his own devising; inspired by this and supported by the Russell Sage Foundation, the New York architect Grosvenor Atterbury built several hundred dwellings of hollow-cored precast units in 1910–1920. Edison experimented with on-site mechanized procedures for casting concrete houses, to which Walter Gropius referred when he presented his seminal memorandum on the industrial production of dwellings, the first explicit statement of a theory of prefabrication, to Emil Rathenau in 1910.

Between the Wars, in the 1920s and 1930s, the housing crisis was endemic; this spurred perhaps the greatest strides in the development of concepts and techniques of prefabrication. In Britain, after the war, prefabricated dwellings were constructed in the thousands, not only in reinforced concrete, but also in *steel*. Inspired by Britain's Weir house, experiments in steel and other metals were carried out in Germany. Gropius acted as consultant in the development of the Hirsch copper house and exhibited two variants at the Berlin "Growing House" exhibition of 1932, the apogee of the factory-built house in Europe. All these were technically advanced systems, but not a radical rethinking of the problem, as in Fuller's Dymaxion House. In any event, the emphasis in Europe was shifting from the house to mass housing, to the *siedlungen* whose repetitive blocks of flats lent themselves ideally to industrialized systems of construction, whether through rationalized on-site mechanization, as developed by Gropius at Toerten-Dessau or Taut at Berlin's Hufeisensiedlung, or the fully prefabricated systems of May at Frankfurt or, later, Beaudoin and Lods at Nancy.

Postwar Reconstruction, after World War II, depended on techniques and concepts established by the 1930s. Extensive state-sponsored housing schemes were undertaken both in the socialist East and the welfare states of the West, generally in precast concrete panel, frame, or box systems. In addition, lightweight, highly industrialized systems were explored, as in Britain's Aluminium House at the war's end or the highly successful Hertfordshire schools program. In the United States, prefabrication aimed at the private house market failed, in the main, with the notable exception of the MOBILE HOMES industry, to compete with the increasingly efficient on-site construction methods of the merchant builders. Nor did technologically sophisticated prefabrication systems such as the Lustron house or Gropius' and Wachsmann's "Packaged House" (Fig. 192) prove economically viable. A decade later, Operation Breakthrough, sponsored by the U.S. Department of Housing and Urban Development (HUD), failed to bring forth the expected fruit: the realization of the dream of the fully industrialized, mass-produced closed building system. However, more adaptable open systems using compatible premade elements continue to be used extensively and with success.

Figure 192 General Panel Corporation, California. "Packaged House" wall panels, about 1947. (Photograph by courtesy of Mrs. Judith Wachsmann.)

See also MOBILE/MANUFACTURED HOMES; PRECAST CONCRETE; and SYSTEM BUILDING.

References

1. B. KELLY: *The Prefabrication of Houses*. Technology Press, Cambridge, Mass., 1951.
2. R. B. WHITE: *Prefabrication: A History of its Development in Great Britain*. Her Majesty's Stationery Office, London, 1965.
3. G. HERBERT: *Pioneers of Prefabrication: The British Contribution in the Nineteenth Century*. Johns Hopkins University, Baltimore, Md., 1978.
4. G. HERBERT: *The Dream of the Factory-Made House: Walter Gropius and Konrad Wachsmann*. M.I.T. Press, Cambridge, Mass., 1984.

G.H.

PRESERVATION OF TIMBER. The type and species likely to be used, the timbers available and otherwise suitable for the application, the suitability of the proposed preservative treatment for a particular timber, and the conditions of exposure likely to be encountered should be known in order to design timber in the most satisfactory and economical manner.

Timber Structure. We distinguish between pored and nonpored timbers. The first group embraces the wide range of broad-leafed trees that produce timber containing vessels and fibers. It includes, however, some species with needle-shaped leaves, resembling those of the other group.

The second group consists of the conifers, whose timber is composed of tracheids. In some species of this group, resin canals can produce openings which could be mistaken as pores. This may cause a nonpored timber to behave like a pored timber in the presence of some destructive agencies, but this is unusual.

In every tree the stem is laid down in concentric cones, starting from the pith at its center and terminating at the junction between the wood and the bark, which is the position of the *cambium*, the layer in which the wood is formed. Each concentric ring is termed a growth ring and consists of sections formed in the fast-growing period, known as *early wood*, and that formed later in the growing season, known as *late wood*.

The growing tree may contain a greater weight of water in its stem than the dry weight of its wood substance. The water is present in two forms: (1) free water in the cell cavities and (2) combined water held by the material of the cell walls. The water combined with the wood substance is usually on the order of 25% to 30% of the moisture-free weight of the wood in the stem. The *fiber saturation point* is the condition at which all the free water has been removed, and the evaporation of combined water is about to commence.

Destructive Agencies. Depending on its species, moisture condition, and the conditions under which it is to be used, timber in service may be attacked by fire, weathering, abrasion, borers, termites, marine organisms, or decay, or by several of these destructive agencies. The conditions necessary to cause destruction of the timber and the rate at which it progresses vary considerably, and a large number of treatments have been developed to extend the life of timber. However, many problems can be solved by a careful analysis of the factors that tend to reduce it. As an example, timber maintained at a condition of saturation does not decay. Thus the piles for the construction of a house should be driven to below the permanent water level and then cut off at a point below that level. They should be capped with a concrete headstock to raise the base above the high-water level.

Other piles can be based on the headstock, with a suitable damp-proof course under them, to carry a timber structure, since timber maintained at a moisture content below 18% does not decay. In buildings, the main causes of timber destruction are attacks by borers or termites, fungal decay, and fire.

Borers. Timber in its unseasoned condition, and also in the living tree, may be attacked by a range of borers of various sizes. These cause damage such as shot hole or pupal cavities; but it is unusual for the attack to continue after the moisture content has been reduced to below the fiber saturation point. The effect of their activity can be seen in the sawn timber, and it can be accepted or rejected according to the assessment of the extent and the effect of the damage present.

Borers that damage timber at moisture contents below the fiber saturation point can be considered in two main classes, typified by the *Lyctus* species and the *Anobium* species. Each of these groups develops in a life cycle commencing with the deposition of an egg in the timber, hatching of a larva, feeding of the larva on the wood and substances in the wood, pupation of the larva, and eventually emergence from the wood of a beetle. The beetle bores its way out of the wood, forcing out ahead of it a finer powder. After mating of the adult beetles, more eggs are deposited in the wood and the life cycles continue with the next generation.

Unless an effective treatment is applied to the timber, borer attack will progress from one season to the next until the timber susceptible to attack is completely destroyed. Borer attack may be prevented by a suface coating that completely seals the timber after having effectively sterilized the timber by heat treatment or by the application of suitable insecticides. Such a method relies on the effective sterilization of the timber and on the protective barrier, paint, or similar treatment remaining effective. This is difficult to achieve, and the method can only be recommended for those applications where the use of other more permanent treatment is not considered practicable.

The prevention of borer attack can be effected by the use of nonsusceptable timbers or by the application of suitable preservative treatments. The *Anobium* borer is more commonly associated with nonpored woods, but occasionally a pored timber in which the internal cavities and fiber structure can support anobium attack is experienced. *Lyctus* attack usually occurs in pored timbers containing starch, but attack has been found in a few species of nonpored timber.

Timber Treatments consist of the application of specially developed preservatives, using diffusion or pressure impregnation. These are usually compounds of salts or oil-based preservatives applied in specially designed timber-treatment plants. Each preservative is formulated to permit its placement in the timber, to become fixed in the wood so that it will not leach out, and to provide protection to the timber against the organisms expected to attack it. Suitable treatments have been developed for the purpose of increasing the ability of timber to resist fire.

Protective Coatings. Protection against weathering is normally achieved by the application of protective coatings, sometimes in conjunction with a preservative treatment to prevent the breakdown of the timber immediately under the coating. A similar approach is adopted for protection against abrasion. Each treatment has a finite life, which in most cases can be estimated. If performance beyond that life is expected, provision must be made for appropriately scheduled maintenance treatments.

References

1. *Manual of Recommended Practice*. American Wood Preservers Association, (a loose-leaf handbook revised annually), New York.
2. FREDERICK S. MERRITT: *Standard Handbook for Civil Engineers*, 3rd ed. McGraw-Hill, New York, 1983.
3. P. B. CORNWELL: *Pest Control in Buildings*. Hutchinson, London 1973.
4. AMERICAN INSTITUTE OF TIMBER CONSTRUCTION: *Timber Construction Manual*. Wiley, New York, 1966.

E.B.H.

PRESTRESSED CONCRETE is concrete in which internal stresses of such magnitude and distribution have been introduced that the stresses resulting from the given applied loading are counteracted to a desired degree. Concrete is strong in compression but weak in tension; prestressing is an effective way to reduce or eliminate undesirable tensile stresses. Two methods of prestressing are in general use: pretensioning and posttensioning. In *pretensioning*, the high-strength steel tendons pass through the mold or molds for a number of similar members arranged end to end and are tensioned between external fixed abutments, by which the tension is maintained when the concrete is placed. When the concrete has hardened sufficiently, the tendons are slowly released from the abutments; the tensile force in the tendons is thereby transferred to the concrete by *bond* stresses. In *posttensioning*, the concrete member is cast incorporating ducts for the tendons. When the concrete has hardened sufficiently, the tendons are tensioned by jacking against one or both ends of the member, and are then anchored by means of *anchorages* that bear against the member or are embedded in it. Pretensioning is more suitable for the mass production of standard members in a factory. Posttensioning is generally used for large special-purpose units, which may be cast on site or off site; within limits, tendon profiles of any shape can be used.

Prestressed concrete has several advantages over reinforced concrete: reduced deflections and the absence or near absence of cracks under service load; increased shear strength; increased fatigue resistance; and efficient use of materials. In general, the same applied load can be carried by a lighter section in prestressed concrete; this yields more clearance where it is required and enables longer spans to be used. On the other hand, prestressed concrete construction requires special equipment, high-quality materials, and a skilled labor force.

History. The concept of prestressing structures dates back at least to 1886 when P. H. Jackson of San Francisco applied for U.S. Patent No. 375999. In 1919, Karl Wettstein of Austria conceived the idea of prestressing concrete planks with highly tensioned piano wires and received Patent No. 384009 in Bruex, Bohemia, in 1921. Wettstein's ideas and tests subsequently enabled E. Hoyer to manufacture prestressed concrete units on a large scale in 1938. A breakthrough, which was to make practical prestressing a reality, occurred in 1926–1927 when Eugene Freyssinet of France became the first person to grasp the significance of the *creep* of concrete (see SHRINKAGE AND CREEP) and to appreciate the futility of using mild steel tendons. His French Patent No. 686547, applied for in 1928, referred specifically to the use of high-strength steel. Today Freyssinet is accepted as the father of prestressed concrete; other pioneers who made important contributions include G. Magnel of Belgium, P. W. Abeles of the United Kingdom, and T. Y. Lin of the United States. The rapid development of prestressed concrete after World War II owes much to their ideas and efforts.

Applications. Prestressed concrete finds wide applications in bridges, nuclear vessels, offshore structures, low-temperature storage tanks, towers and masts, railway sleepers, piles, floor slabs for buildings and pavement slabs for roads and airports, and, especially, precast concrete construction. It also has useful applications in rock and soil anchors.

Sufficient is now known about prestressed concrete that serious problems in its use are unlikely to arise, provided what is currently accepted as good practice is applied in design and construction. Efforts should now be directed not so much at discovering new concepts as at further advancing recent progress in techniques such as pressure grouting, crack sealing, and, in bridge construction, the use of match-cast glued joints and balanced cantilevers. Encouraging support is developing for *partial prestressing*, by which is meant the application of prestress of a limited magnitude so that only a part of the service load is counteracted. Partial prestressing offers decided economic advantages over full prestressing and may be expected to achieve growing importance in the near future.

Looking into the future, it is likely that prestressed concrete will find applications in ocean energy conversion plants such as OTEC and DAM-ATOLL, large reactor pressure vessels for coal conversion, large desalination plants, and in structures in the Arctic regions. Another growth area is in low-temperature storage structures. Today most concrete bridges, long or short, are prestressed; this trend will continue. Developments in cable-stayed bridges look promising.

Further Information. For history see Ref. 1; for analysis and design, see Refs. 2 and 3; for recent developments see Ref. 4. Reference 5 covers many topics associated with prestressed concrete, including chapters on full and partial prestressing, construction, maintenance, and demolition, as well as lessons from structural failures.

See also CONCRETE STRUCTURES; LOAD BALANCING; PRECAST CONCRETE; and REINFORCED CONCRETE DESIGN.

References

1. C. DOBELL: Patents and Codes Relating to Prestressed Concrete. *Proceedings of the American Concrete Institute*, Vol. 46 (1950), pp. 713–724.
2. T. Y. LIN and N. H. BURNS: *Design of Prestressed Concrete Structures*. Wiley, New York, 1981.
3. F. K. KONG and R. H. EVANS: *Reinforced and Prestressed Concrete*. Van Nostrand Reinhold, Wokingham, England, 1983.
4. F. SAWKO: *Developments in Prestressed Concrete*. Applied Science Publishers, London, 1978.
5. F. K. KONG, R. H. EVANS, E. COHEN, and F. ROLL: *Handbook of Structural Concrete*. Pitman, London, 1983.

F.K.K.

PROJECTIONS. The most common method of representing a three-dimensional object on a sheet of drawing paper is the *orthographic* projection, which shows the object by means of three separate drawings: the *plan*, the *elevation*, and the *side elevation*. In the *first-angle* projection (commonly used in Great Britain), each view is placed so that it represents the side of the object remote from it in the adjacent view. In the *third-angle* projection (commonly used in America and Australia), each view is placed so that it represents the side of the object near to it in the adjacent view.

Pictorial Projections. While the orthographic projection is true to scale in every respect, it fails to give a *pictorial* representation of the object. In the *oblique parallel* projection, the elevation is drawn as for the orthographic projection, and the plan and side elevation are then attached to the same picture at an angle of 45°. The object invariably looks too deep if drawn this way, and consequently an artificially foreshortened scale is sometimes used along the 45° lines.

In the *isometric* projection, all three orthographic projections are drawn at an angle. Vertical lines remain vertical, but all lines that are horizontal in the orthographic projection are drawn at 30° to the horizontal in the isometric projection. While the plan, elevation, and side elevation are all given equal prominence, the object frequently appears distorted. However, all dimensions on the vertical and on the 30° axes are accurately to scale.

The *axonometric* projection is used to overcome the pictorial limitation of the isometric. Vertical lines remain vertical; the other axes are drawn at angles other than 30°, and they consequently have different scales. If the two horizontal axes are drawn at the same angle, the projection is *dimetric*; if they are drawn at different angles, it is *trimetric*. The latter gives a better pictorial representation, but all three axes have different scales.

Perspective Drawings. An even better picture is obtained by the use of a *perspective* projection, which is the way the eye sees an object. Lines that are parallel in plan are drawn to converge on a vanishing point, and lines that are parallel in the elevation and side elevation are also drawn to converge on their vanishing points. The complete or *three-point* perspective is difficult to draw precisely, and it is sufficient to use a *two-point* perspective (in which the parallel lines of the elevation and the side elevation only converge on two vanishing points). Even a *one-point* perspective is often sufficient; in this only the parallel lines of the side elevation converge on a vanishing point.

Although perspectives could be used to scale dimensions with specially drawn scales, this is rarely required. Because of the complexity of the projection, perspectives are therefore frequently sketched without undue concern for dimensional accuracy.

Stereographic Projection. The two-dimensional representation of the surface of a sphere is of particular importance for the drawing of maps and of charts showing the movement of the sun and the stars. In building technology it is employed for the design of SUN SHADING AND SUN-CONTROL DEVICES. The most commonly employed projection is the *stereographic*. This is a perspective projection, whose perspective center is on the point of the sphere diametrically opposite to the point where the pictorial plane touches the sphere. It has the remarkable property that all arcs of *great circles* and *small circles* are shown either as arcs of circles or as straight lines.

References

1. J. D. BETHUNE: *Essentials of Drafting*, Prentice-Hall, Englewood Cliffs, N.J., 1977.
2. F. E. GIESECKE, A. MITCHELL, H. C. SPENCER and I. L. HILL: *Technical Drawing*, 6th ed., Macmillan, New York, 1974.
3. R. W. GILL: *Creative Perspective*, Thomas and Hudson, London, 1975.
4. F. W. SOHON: *The Stereographic Projection*. Chemical Publishing Company, New York, 1941.

H.J.C.

The **PSYCHROMETRIC CHART** (from the Greek *psychros*, "cold") shows the variation of the humidity and heat content of air with temperature (see TEMPERATURE AND HUMIDITY MEASUREMENT). It is used for the evaluation of THERMAL COMFORT CRITERIA and for the design of AIR-CONDITIONING systems.

The horizontal coordinate is the *dry-bulb temperature*

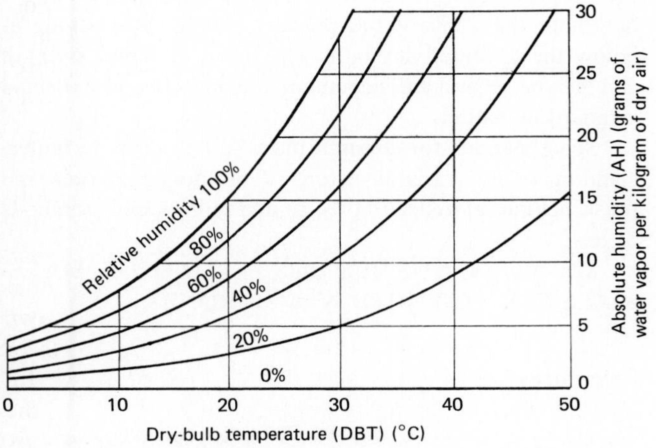

Figure 193 Relation between dry-bulb temperature, absolute humidity, and relative humidity, as plotted on a psychrometric chart in metric units.

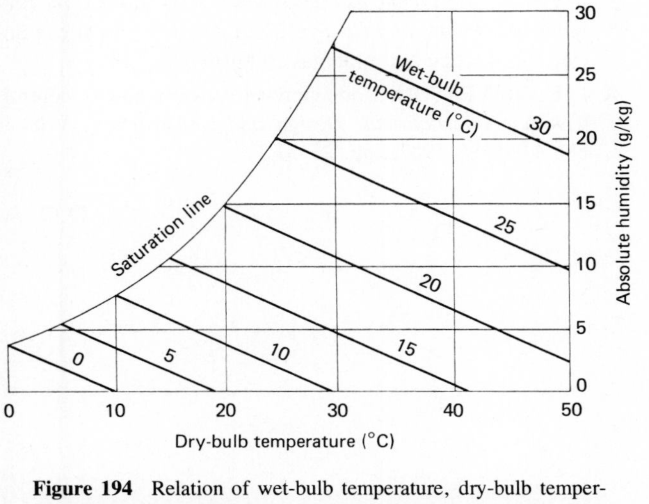

Figure 194 Relation of wet-bulb temperature, dry-bulb temperature, and absolute humidity, as plotted on a psychrometric chart.

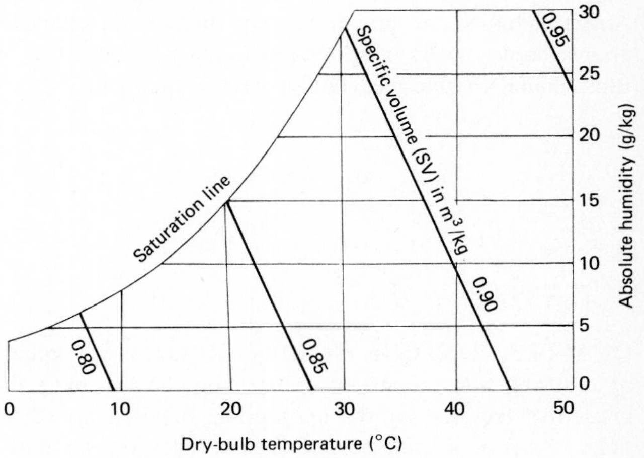

Figure 195 Relation of dry-bulb temperature, absolute humidity, and specific volume, as plotted on a psychrometric chart. It is necessary to know the density of air, which changes with temperature and humidity, to determine the volume of air to be handled by the air-conditioning plant. The specific volume, which is the reciprocal of the density, is plotted in place of the density because it yields straight lines.

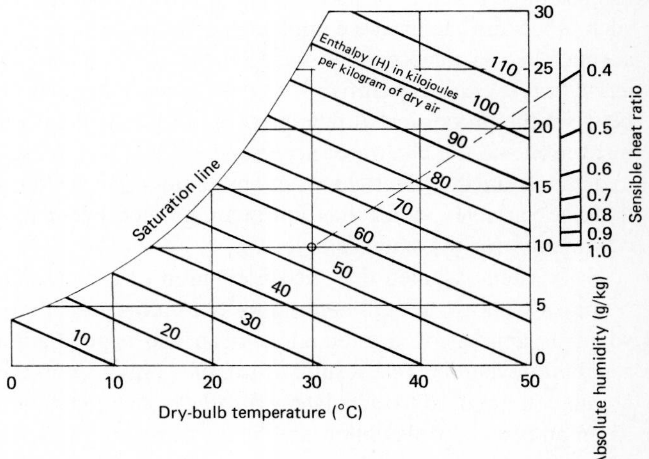

Figure 196 Relation of dry-bulb temperature, absolute humidity, and enthalpy (heat content of the moist air), as plotted on a psychrometric chart. The enthalpy is the sum of the *latent heat*, the heat required or released without a change of temperature when a fluid changes its state (evaporates or condenses) and the *sensible heat* due to a change of temperature without a change of state. The ratio of the sensible heat to the total (sensible plus latent) heat is shown by the additional scale on the right side. Thus at a dry-bulb temperature of 30°C and an absolute humidity of 10 g/kg the sensible heat ratio is 0.4. (Figures 193 to 196 by courtesy of Van Nostrand Reinhold.)

in degrees Celsius (or Fahrenheit). The vertical coordinate is the *absolute humidity* in grams of water vapor per kilogram of dry air (or pounds of water vapor per pound of dry air). Plotted on these coordinates are four sets of curves or straight lines, all on the one psychrometric chart. The metric versions only are shown separately in Figs. 193–196. The components of the psychrometric chart in conventional American units are similar, except that kilojoules, kilograms, cubic meters, and degrees Celsius are replaced by Btus, pounds, cubic feet, and degrees Fahrenheit.

References

1. H. J. COWAN and P. R. SMITH: *Environmental Systems*. Van Nostrand Reinhold, New York, 1983.
2. W. J. McGUINESS, B. STEIN, and J. S. REYNOLDS: *Mechanical and Electrical Equipment of Buildings*, 6th ed. Wiley, New York, 1980.

H.J.C.

Q

QUALITY CONTROL OF MATERIALS is the process of ensuring that the materials used on the job have the properties required by the appropriate material specifications. Properties may be specified by description (attributes), such as "Steel shall be free from pipe, lamination, harmful segregation and other defects," or by measurement, such as determination of compressive strength (variables).

As all materials have variations of properties from piece to piece and as strength tests involve destruction of test pieces, the determination of the properties of a material in use has to be based on a statistical sampling procedure. No such procedure can ensure complete discrimination between acceptable and unacceptable material. A curve known as an *operating characteristic curve*, which shows the relation between the proportion of defectives in a lot submitted for inspection and the chance of acceptance of that lot, is used to demonstrate the effectiveness of any sampling plan. Plans that use variables rather than attributes provide better discrimination for the same sample size.

It is often assumed that the distribution of a measured property such as strength approximates to a normal curve, which is defined by a population mean and a population standard deviation. Test results of samples drawn from the population are used to calculate estimates of the population mean and standard deviation and hence to determine what proportion of the material population lies below some specified or characteristic strength. It is usually accepted that not more than 5% of the product should have strengths below the specified strength. The mean or target strength can then be set and will depend on the measured or expected standard deviation.

Specifications for strength must be based on an understanding of the variable nature of building materials and must include reference to precise and reproducible methods of test.

See also STATISTICS and TESTING OF MATERIALS FOR STRENGTH AND HARDNESS.

References

1. *Australian Standard AS 1199—Sampling Procedures and Tables for Inspection by Attributes.* Standards Association of Australia, Sydney, 1972.
2. *Australian Standard AS 1399—Guide to AS 1399.* Standards Association of Australia, Sydney, 1973.
3. *ASTM Special Technical Publication 15-D—Manual on Presentation of Data and Control Chart Analysis.* American Society for Testing and Materials, Philadelphia, 1976.
4. J. E. OLIVERSON: Computerized concrete quality control. *Concrete International: Design and Construction*, Vol. 4, No. 11 (Nov. 1982), pp. 23–31.

D.C.-A.

R

REFRIGERATION CYCLES. Of the numerous systems of refrigeration and air cooling, vapor compression, absorption, evaporative cooling, air cycle, thermoelectric, steam jet, and ultralow-temperature systems, only the vapor compression, absorption refrigeration, and evaporative cooling systems are discussed since they are predominantly used in buildings.

The cooling load of modern buildings has greatly increased, because of lighting requirements, electric office equipment, computer installations, larger glass exposure, lightweight building materials, and higher standards for environment control.

The cost of energy and the value of space in the modern building have both risen sharply. Service space for refrigeration plant and piping and spaces in below-ground-level subbasements and high on rooftops have become valuable for shopping arcades, car parking, or restaurants. Space can be saved if the refrigeration plant is located close to the air-conditioning systems it serves and the cooling tower located near the condenser it serves, because shafts connecting basement to roof level and intermediate-level equipment (e.g., chilled water and condenser water supply and return piping) then become unnecessary.

The Centrifugal Compressor and the Absorption Refrigeration System. With the advent of the centrifugal compressor and the absorption refrigeration system, it became possible to locate the refrigeration plant close to the air-conditioning system and the cooling tower. Previously, large reciprocating compressors necessitated the location of the refrigeration plant in the basement.

The absorption refrigeration system had previously been used mainly for domestic gas refrigerators. When the lithium bromide–water absorption system was developed, it became popular in sizes up to 5.6 MW (1600 tons of refrigeration). Being vibration free, it comfortably fits on top of large modern buildings. Unlike the aqua–ammonia absorption refrigeration plant, it requires no rectifying columns. These plants operate well under a very low part load. They can utilize steam where it is available either through boilers required for purposes of winter heating or, as is the case in downtown New York City, distributed from a public utility plant that supplies it at low summer rates. It is frequently used in hospitals where both medium- and high-pressure steam is normally available. In time, absorption refrigeration systems using solar energy as a heat source may become a viable proposition.

The **Evaporator,** common to both the vapor compression and the absorption refrigeration cycle, is the most important component of the refrigeration system. Through this heat exchanger, the air is simultaneously cooled and dehumidified to offset the heat and moisture loads. In small installations this is done directly by passing the air to be treated over the evaporator or direct expansion coil, where it is both cooled and dehumidified. In medium and large buildings it is done indirectly due to the presence of long runs of coolant serving a number of air-conditioning systems; chilled water is used as an intermediary fluid and cooled in an evaporator. The total refrigerating system also includes a compressor, condenser, expansion valve, controls, and auxiliaries, which together make up a packaged chiller. From the evaporator, the chilled water is pumped to the various air-conditioning systems where the air is cooled and dehumidified. There is considerable scope for further optimization (Ref. 1).

Evaporative Cooling entails minimum running costs. The major expenditure of energy in the vapor compression system is the work of compression; in the absorption refrigeration system it is the heat input to the generator. These components are absent from the evaporative cooling system. In the ideal situation, there is a minimal energy requirement since the process of evaporative cooling entails an intrasystem exchange where the energy content of the air remains substantially unchanged. The air is cooled due to the evaporation of water vapor into the treated air. The reduction of the energy of the moist air due to cooling is approximately equivalent to the increase of energy due to the addition of water vapor. This system is limited in application to arid environments. It is misapplied if there is no proper disposal of the treated air, as it becomes laden with water vapor under climatic conditions that are too humid.

See also AIR CONDITIONING AND HEATING and HEAT ENGINES AND HEAT PUMPS.

References

1. A. SHAW: Exploration of air velocity across air-conditioning system dehumidifiers—an energy conservation project, *ASHRAE Transactions*, Vol. 88, part 2 (1982), pp. 142–156.
2. *ASHRAE Handbook, Equipment Volume*. American Society of Heating, Refrigerating and Air-Conditioning Engineers, Atlanta, Ga., 1983.
3. *ASHRAE Handbook, Applications Volume*. American Society of Heating, Refrigerating and Air-Conditioning Engineers, Atlanta, Ga., 1982.
4. *ASHRAE Handbook, Fundamentals Volume*. American So-

ciety of Heating, Refrigerating and Air-Conditioning Engineers, Atlanta, Ga., 1981.

5. *ASHRAE Handbook, Systems Volume and Product Directory*. American Society of Heating, Refrigerating and Air-Conditioning Engineers, Atlanta, Ga., 1980.

A.S.

REFUSE DISPOSAL is corequisite with SEWAGE DISPOSAL in the preservation of community health and amenity. *Refuse* comprises the solid wastes of households, commercial buildings, and industries. Animal wastes and liquid wastes, which are too strong, corrosive, toxic, or hazardous for sewer discharge, are usually excluded from refuse disposal systems and require special treatment and disposal. The residues and sludges from municipal and some industrial wastewater treatment plants may, however, be acceptable for disposal with refuse.

Refuse disposal systems may involve on-site storage, collection, processing, transport, and ultimate disposal. Major factors affecting disposal are refuse production rate (both weight and volume), accessibility to collection vehicles and traffic conditions, remoteness and capacity of ultimate disposal sites, and markets for reclaimed materials or products of refuse processing.

Refuse Production per capita in industrialized countries varies from 0.5 to 2 kg (1 to 5 lb) per day, made up of mixed paper (25% to 40%), foodstuffs (25% to 40%), glass (5% to 15%), metals (5% to 8%), plastics, rags, broken toys, and furnishings (5% to 10%), and inert materials such as ashes (3% to 10%). On-site storage may be in 40- to 50-liter (10- to 13-US gallon) plastic or metal bins, 200-liter (50-US gallon) wheeled containers, or various sizes of individual and community containers, depending on the method and frequency of collection. Receptacles must have tight-fitting lids to keep out flies, insects, and rodents. Storage areas in multistory buildings, commercial centers, and industrial premises must be convenient for users and for access by collection vehicles.

Collection accounts for over 80% of the cost of most refuse disposal systems. Purpose-built compactor vehicles are preferred to open-bodied trucks because of their greater load capacity and better control of nuisance from odors, flies, and windblown paper. Storage receptacles are emptied manually or with the assistance of a machine lift. When larger receptacles are employed, an empty may be left in place of the full container, which is then taken to the disposal site for emptying.

Transfer Stations located centrally to the collection area are economical where travel time or distance from collection to disposal areas is excessive. The special-purpose collection vehicles deliver to the transfer station, where the refuse is transferred to large road-haulage *container vehicles*, often fitted with compactors, for transport to the disposal site. Materials may be *classified and recovered* at transfer stations if markets exist for recovered materials, such as paper, glass, or metals; but separation of materials into separate containers at the source can be more effective.

Composting of refuse, especially with digested sewage sludge to provide nitrogen and moisture to balance the high carbon content of refuse, produces a useful soil conditioner; but this is only worthwhile if there is a market for the compost. Furthermore, metals, glass, and other inert materials separated in composting plants still require disposal.

Refuse Processing reduces the volume and weight of materials for disposal. Baling under pressure and *pulverization* followed by compaction increase the bulk density of material to be used as landfill in a disposal area. Pulverization also improves separation efficiency in materials recovery; it also reduces the volume and mass of the material for disposal.

Incineration reduces refuse volume to less than one-tenth and weight to less than a quarter. Combustion is self-sustaining because of the high energy value of municipal refuse. The heat can be recovered and used for heating buildings or in electricity generation. Incinerators require (1) careful location with respect to potential users of the energy produced, (2) remoteness from residential areas and yet proximity to the collection areas to reduce collection cost, and (3) control of air pollution.

Sanitary Landfill. Ultimate disposal of refuse is usually to sanitary landfill. Landfill requires sound management to ensure effective compaction in layers up to 2.5 m (8 ft) thick, daily covering and effective sealing with earth, protection from windblow papers and dust, diversion of stormwater away from the landfill, and collection, treatment, and disposal of the leachate escaping from the landfill (Fig. 197).

References

1. J. L. PAVONI, J. E. HEER, Jr., and D. J. HAGERTY: *Handbook of Solid Waste Disposal*. Van Nostrand Reinhold, New York, 1975.
2. G. TCHOBANOGLOUS, H. THIESEN, and R. ELIASSEN: *Solid Wastes, Engineering Principles and Management Issues*. McGraw-Hill, New York, 1977.
3. D. G. WILSON (ed.): *Handbook of Solid Waste Management*. Van Nostrand Reinhold, New York, 1977.

P.J.B.

REINFORCED CONCRETE DESIGN is an iterative process that usually involves a preliminary (or conceptual) design stage, in which the general layout of the structure is

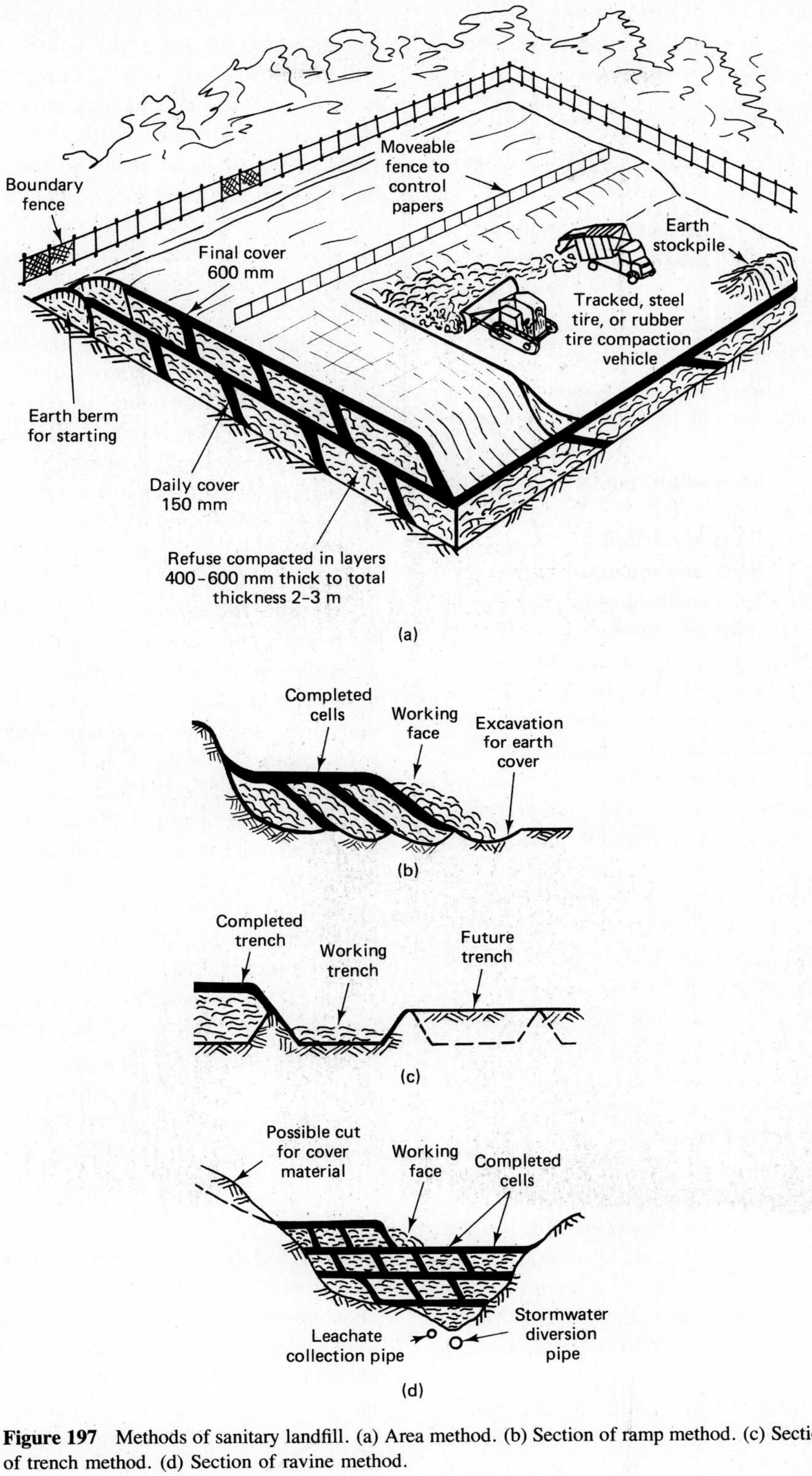

Figure 197 Methods of sanitary landfill. (a) Area method. (b) Section of ramp method. (c) Section of trench method. (d) Section of ravine method.

established, and a detailed design stage, in which the proportions and reinforcing details of the component members are determined. The overall design objective is to ensure that the structure satisfies the following requirements:

Functionality: It should serve its intended (usually nonstructural) purpose.

Serviceability: under normal in-service conditions it should perform acceptably, without excessive deflection, excessive cracking, or unacceptable vibration.

Safety: It should have adequate reserves of strength to resist unexpected overloads.

Durability: It should remain safe and serviceable over its design life without the need for excessive expenditure on repair or maintenance.

Economy: All the preceding requirements should be achieved at a competitive total cost.

Preliminary Design. In the preliminary design stage, the structural concept is developed and the manner in which the applied loads are to be channeled through the structure into the supports is established. The general layout is thus decided on, at least provisionally, so that the location and structural function of the component members such as columns, walls, beams, and floor slabs are determined. The members are also sized very approximately by means of simplified design calculations (Ref. 3). The layout chosen for the structure must be compatible with the requirement of functionality.

Detailed Design. In the detailed stage of design, the preliminary sizing of the individual members is checked and adjusted as necessary so that the requirements of serviceability, safety, and economy are satisfied. Full details are determined for each member, including size and shape, quantity, and distribution of the reinforcement and properties of the component steel and concrete (see DETAILING OF REINFORCED CONCRETE).

Iterative Nature of Design. The initial decisions concerning the structural layout have an important effect on the proportions of the individual components. For example, a decision to use widely spaced columns in a building will usually result in deep floor beams and heavy slabs. Modifications to the layout and indeed to the structural concept may become necessary during the detailed design stage in order to achieve an economical and structurally acceptable overall design. Also, in the detailed design the procedure is essentially iterative, with the member proportions chosen on a trial basis and checked for adequacy in strength (safety) and serviceability and adjusted and rechecked as necessary.

Design Methods and Their Codification. Minimum requirements for safety and serviceability of concrete structures are usually stipulated in national codes of practice. Codes in current use in the United Kingdom (Ref. 1), the USSR, Canada, and various European countries are written in *limit states* format. In other countries, including Australia (Ref. 2), the changeover to limit states format has now (1984) reached the draft code stage. In the United States, the *ACI Building Code Requirements* are formulated in terms of *design for strength and serviceability*. Differences between the limit states approach and the strength and serviceability method tend to be linguistic rather than fundamental.

In earlier concrete codes, the structural design requirements were expressed in terms of the *working stress method*. The stresses in the steel and concrete were calculated on the assumption of elastic behavior under the working design loads and were limited to safe, permissible values. In the 1960s and 1970s this method was replaced by the *ultimate strength method*, which in turn has evolved into the strength and serviceability method and the limit states method.

Design for Strength. In the ACI procedure for strength design, and also in the design procedures for the strength limit states, a check is made to ensure that the strength of each critical cross section is sufficient to resist, with an appropriate margin of safety, the stress resultants (i.e., moments, shears, and compression forces) that are produced by the design loads (dead, live, WIND, EARTHQUAKE). The stress resultants are calculated for each design load on the assumption of overall elastic structural action, multiplied by partial load factors and then combined together to give the design stress resultant that has to be resisted by the section. For example, according to the ACI code the moment to be carried by a critical cross section in a flexural member when subjected to combined dead and live load is $1.4M_d + 1.7M_l$, where M_d and M_l are the moments produced by the dead and live loads, respectively, and 1.4 and 1.7 are the partial load factors. The section must be so proportioned that its design bending strength is adequate to carry this combined moment:

$$\phi M_u > 1.4M_d + 1.7M_l$$

The term ϕ is called a strength reduction factor. It is a partial safety coefficient, introduced to allow for unavoidable variations in the strength of the member, just as the load factors allow for variations in the magnitude and distribution of the loads. For bending strength calculations, the ACI value for ϕ is 0.9. The nominal moment capacity of the section, M_u, is determined from *ultimate strength theory* (Ref. 4) for reinforced concrete members. Other load combinations, including wind and earthquake, also have to be considered with appropriate load factors. Strength checks are made in a similar manner for sections subjected to shear, axial force, and torsion. Where the trial design properties of the member are inadequate to satisfy the design requirement as expressed in the preceding equation, or its equivalent, either additional reinforcement must be used or the dimensions must be increased.

Design for Serviceability. In design for serviceability, attention is concentrated on the control of deflections and cracking in flexural members. Deflection control can be undertaken directly by calculating instantaneous and long-term deflections and ensuring that they are smaller than the acceptable values stipulated in the relevant design code. Alternatively, a simple indirect procedure may be adopted whereby the span/depth ratio of the flexural member is kept below a specified safe value. Crack widths may also be controlled directly by calculation or indirectly by satisfying various requirements concerning the location of tensile reinforcement in the members.

Design for Durability. Until recently, durability has received little attention as a design requirement for reinforced concrete structures. However, newly constructed concrete structures have suffered from poor durability in many countries in the past few years, and DURABILITY has become a matter of serious concern both to researchers and practitioners. Code design procedures are increasingly imposing indirect durability requirements in the form of minimum quality levels for the concrete (Ref. 2).

See also CONCRETE MIX PROPORTIONING; CONCRETE STRUCTURES; DETAILING OF REINFORCED CONCRETE STRUCTURES; LIMIT STATES DESIGN; REINFORCED CONCRETE SLABS; REINFORCEMENT FOR CONCRETE; and SAFETY FACTORS IN STRUCTURAL DESIGN.

References

1. *The Structural Use of Concrete*. CP110:Part 1:1972. British Standards Institution, London, 1972.
2. *Unified Concrete Structures Code*. Draft Australian Standard, Document DR83046, Standards Association of Australia, Sydney, 1983.
3. *Building Code Requirements for Reinforced Concrete*. ACI 318-83. American Concrete Institute, Detroit, 1983.
4. R. F. WARNER, B. V. RANGAN, and A. S. HALL: *Reinforced Concrete*, 2nd Ed. Pitman, Melbourne, 1982.

R.F.W

REINFORCED CONCRETE SLABS for floors and roofs may be of cast in situ construction or formed from PRECAST CONCRETE members generally acting compositely with cast in situ concrete.

Types. A site-cast reinforced concrete floor or roof slab may be supported on columns without or with beams.

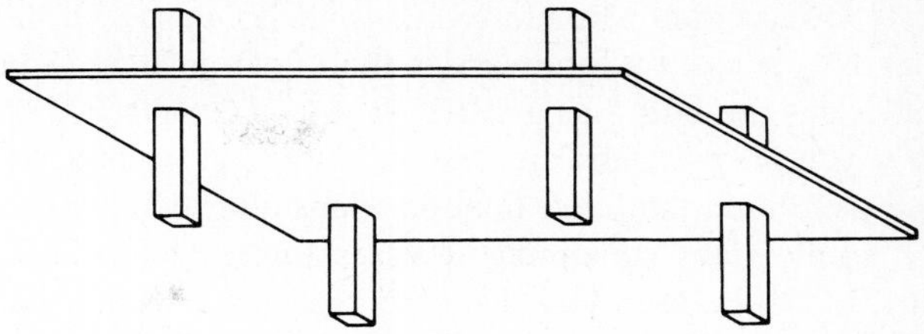

Figure 198 Flat plate.

The *flat plate* (Fig. 198) with uniform thickness is the simplest form of beamless slab construction. The *flat slab* (Fig. 199) has capitals at the tops of the columns and/or drop panels, which are thickened areas of slab surrounding each column, in order to provide the additional slab stiffness and shear strength necessary for larger spans and heavier live loads.

The *two-way slab* (Fig. 200) is supported on beams in two directions on the column lines. The beams may be replaced by walls. If the span between the column lines is large, secondary beams may also be present between the column lines to reduce the spans of the slabs. If the span of the slab in one direction is more than twice that in the other direction, the greater part of the slab loading will be carried in the direction of the short span, and the slab can be regarded as a *one-way slab*. One-way slab action also occurs if the beams or walls are only present in one direction. Beams supporting two-way and one-way slabs are normally of reinforced concrete acting monolithically with the slab, but can be of structural steel (see COMPOSITE CONSTRUCTION).

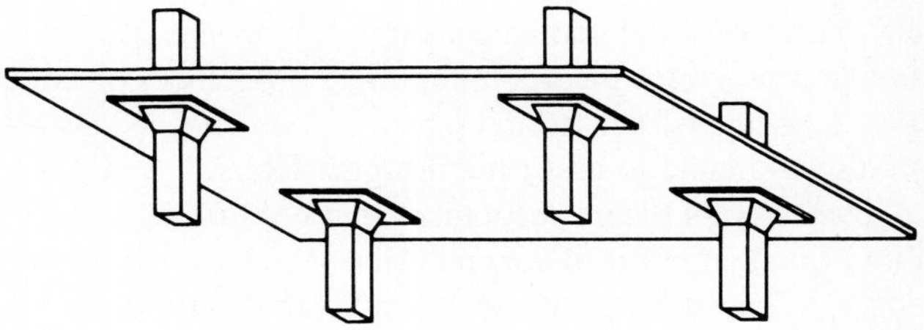

Figure 199 Flat slab.

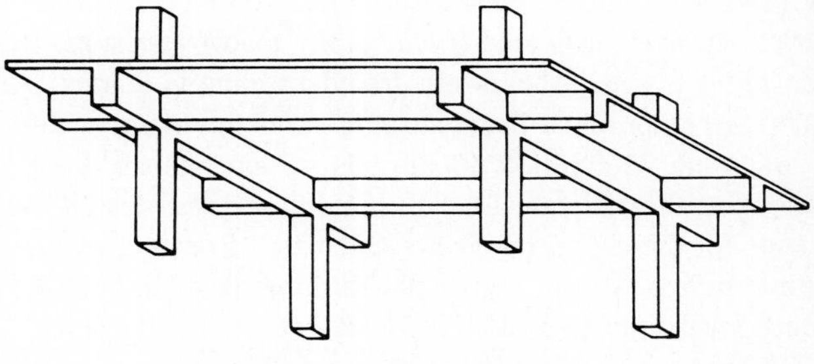

Figure 200 Two-way slab.

Construction. Reinforced concrete floor and roof slabs may be site-cast on formwork in their final position. Some flat plates are built using LIFT-SLAB CONSTRUCTION in which all the slabs are cast in a stack at ground level and then lifted into their final positions after being cured.

Waffle slabs containing recesses formed in the lower surface can be used to reduce the dead load. Waffle slabs can also be used as a means of increasing the effective depth of a slab without an accompanying increase in dead load. Hollow clay tiles or other fillers can also be used in place of recesses as a further variation of a waffle slab.

The choice of slab type depends on the magnitude of the live load, the spans, the cost of materials and labor during construction, and local preference. Two-way slab floors are more labor intensive during construction, but use less reinforcing steel due to the large effective depths of the beams, compared with beamless floors with the same spans between columns and live loads.

Design Procedures. Reinforced concrete slabs are designed so that the slab sections have adequate concrete depth and reinforcing steel to satisfy (1) the strength requirements for flexure, torsion, and shear at the ultimate load, and (2) the limits on deflections, cracking, and vibrations at the service load. Floors are also expected to act as diaphragms, distributing the horizontal forces due to wind or earthquakes to the vertical elements (walls and/or columns) of the structure.

The various approaches available for determining the distributions of moments and shears in reinforced concrete slabs are elastic theory, limit design, and detailed code procedures, as described next. Any one of these approaches can be used as a design method.

Elastic theory solutions for slabs, involving the solution of a fourth-order differential equation, can be obtained using computer programs incorporating finite differences or finite element methods. Charts and tables of moments and shears are also available to assist the designer (Refs. 1 and 2).

Limit design theory recognizes that because of the ductility of most reinforced concrete slab sections a substantial redistribution of moments and shears can occur before the ultimate load is reached. A commonly used approach is Hillerborg's strip method (Refs. 1 and 3) in which the slab is replaced by two systems of strips running in two directions at right angles, which share the external loading. The strips are not considered to carry any load by torsion, and the moments and shears are found by simple statics. An alternative approach is Johansen's yield line theory (Refs. 1 and 4) in which the critical collapse mechanism that develops at the ultimate load is sought, and the design moments for the slab sections are calculated from the required strengths along the lines of plastic hinging of the collapse mechanism (see also LIMIT DESIGN OF STEEL STRUCTURES).

Detailed code procedures are based on theoretical considerations supplemented by experimental data. The moment coefficients for two-way slabs by the long-standing method due to Westergaard is one such approach. Two methods are currently given in detail in the building code of the American Concrete Institute and have been adopted in the codes of many other countries (Ref. 1). The first method, known as the *direct design method*, is for reasonably regular slab floors. The design static moment is found from equilibrium requirements and is distributed between the negative and positive moment regions according to set rules. The second method, known as the *equivalent frame method*, is for a more general range of slab floors. A three-dimensional building structure is considered to be made up of equivalent frames on column lines taken longitudinally and transversely through the building, which are analyzed to determine the design moments.

See also FOLDED PLATES; LIMIT STATES DESIGN; and REINFORCED CONCRETE DESIGN.

References

1. R. PARK and W. L. GAMBLE: *Reinforced Concrete Slabs*. Wiley, New York, 1980.
2. S. TIMOSHENKO and S. WOINONSKY-KRIEGER: *Theory of Plates and Shells*. McGraw-Hill, New York, 1959.
3. A. HILLERBORG: *Strip Method of Design*. Cement and Concrete Association, London, 1975.
4. K. W. JOHANSEN. *Yield Line Theory*. Cement and Concrete Association, London, 1962.

R.P.

REINFORCEMENT FOR CONCRETE is provided by embedding deformed steel bars or welded wire fabric (Ref. 1) within freshly made concrete at the time of casting. The purpose of reinforcement is to provide additional strength for concrete where it is needed. Thus the steel provides all the tensile strength where concrete is in tension, as in beams and slabs, it supplements the compressive strength of concrete in columns and walls, and it provides extra shear strength over and above that of concrete in beams.

Manufacture. Bars are normally hot-rolled from new STEEL billets produced by traditional means from basic oxygen or electric furnaces or from billets produced by the continuous casting method. In the United States a minimal amount is rerolled from standard section T rails (called rail steel) or from axle steel for cars and locomotive tenders (axle steel). In many parts of the world, cold-worked (twisted) bars are the major type of high-strength reinforcement, while in others a quench-and-tempered steel or an alloy steel is prominent.

Sizes. Reinforcing bars, often called *rebars*, are generally available in sizes from 10 through 56 mm (nos. 3 through 18) in straight lengths up to 18 m (60 ft). Welded wire fabric (WWF) is manufactured into sheets or rolls from low-carbon wire with diameters from 3.4 to 12.8 mm (0.134 to 0.50 in.). The wires can be either smooth or deformed. The deformation on rebars consists of raised ribs rolled onto the bar, while deformed wire is generally made by indenting the pattern into the surface.

Design Strength. The required strength of reinforcement (Ref. 1) is controlled by the allowable stresses permitted by building codes and standards (Refs. 2, 3, 4). In all codes the limiting value for strength design is the yield stress for hot-rolled bars (see STEEL) or the proof stress for cold-worked bars and hard-drawn wire. The most widely adopted design strengths for both tension and compression are 400 MPa (60 ksi) for rebars and 450 or 485 MPa (65 or 70 ksi) for fabric.

Fabrication. After manufacture, both WWF and rebars must be cut and bent to the configuration of the concrete, allowing for sufficient cover between the finished concrete surface and the embedded steel (see DETAILING). The cover is normally obtained by supporting the reinforcement on chairs or spacers.

Special Features

Bond Strength. The forces in concrete and reinforcement are transferred to and from each other by bond or mechanical anchorage, such as the welded intersections of WWF or the deformations of the rebar.

Splicing of Reinforcement. The force in one piece of rebar or fabric can be transferred to another by overlapping, welding, or mechanical splices. Modern standards (Ref. 5) require rebars to be fully weldable without preheat to ensure that site erection can be both safe and simple.

Standard Fabric Styles. In countries such as the United Kingdom, Germany, Singapore, New Zealand, and Australia (Ref. 6), fabric is predominantly manufactured to a standardized configuration of wire, mesh, and sheet sizes, and this simplifies the design, specification, and detailing of the fabric.

References

1. *Annual Book of ASTM Standards, Section 1, 1984.* American Society for Testing and Materials, Philadelphia, 1984.
2. *Building Code Requirements for Reinforced Concrete, ACI 318-83.* American Concrete Institute, Detroit, 1983.
3. *The Structural Use of Concrete, CP110:Part 1:1972.* British Standards Institution, London, 1972.
4. *Unified Concrete Structures Code, Draft Document DR83046.* Standards Association of Australia, Sydney, 1983.
5. *Steel Reinforcing Bars for Concrete, AS1302-1982.* Standards Association of Australia, Sydney, 1982.
6. *Hard-drawn Steel Wire Reinforcing Fabric, AS1304—1984.* Standards Association of Australia, Sydney, 1984.

B.J.F.

RHEOLOGY is the science of deformation and flow of materials. Knowledge about the deformation of building components in response to dead and live loads and to stresses created by the environment are of cardinal interest to the designer, construction engineer, regulatory authorities, and user.

It is common practice to assume that for solid materials stress is proportional to strain (Hooke's law) and to calculate expected deformations from elastic moduli. It is similarly often assumed that in liquids shear strain is proportional to rate of flow (Newton's law) and that the material is characterized by a single viscosity coefficient. Such practices are often legitimate, but these idealized concepts do not always describe the behavior of a particular material under given operating conditions sufficiently closely. It is the role of rheology to characterize materials more precisely by their constitutive equations, which describe the dependence of stress on past and present deformations.

Important Types of Rheological Behavior

Viscoelasticity. The deformation of many apparently elastic materials increases gradually under a load maintained for a long period. This type of behavior can be stimulated by a combination of elastic and viscous elements (springs and dashpots) coupled in series or in parallel, and it has therefore been labeled viscoelasticity. The responses of viscoelastic materials, typified in Fig. 201, lend themselves to fairly simple theoretical analysis provided they are additive (principle of superposition). Although the behavior depends on temperature, it is often possible to combine the effects of temperature and time and predict their response from a single master curve.

Plasticity. The term plasticity refers to nonrecoverable deformations at stresses above a critical yield value. This includes the permanent deformation of highly stressed ductile metals as well as the flow of materials, such as pastes, that behave like elastic solids at low loads, but flow like liquids at shear stresses above a critical threshold (Fig. 202, curve a).

Nonlinear Flow. Many liquids fail to follow Newton's law (Fig. 202, curve b), and measurements of apparent viscosity over a range of strain rates are necessary to identify non-Newtonians and to permit reliable flow predictions.

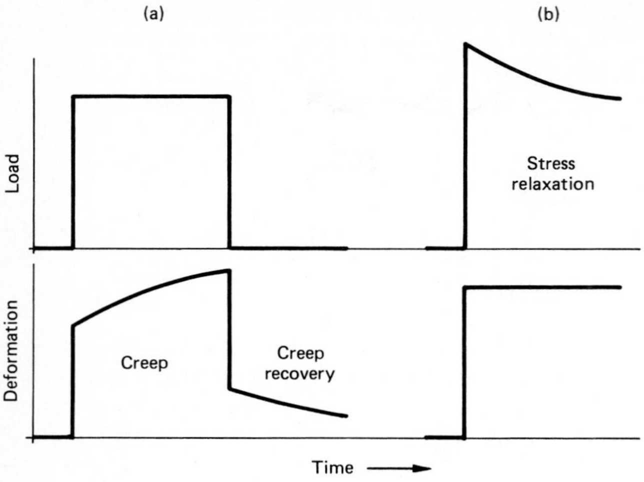

Figure 201 Examples of viscoelastic behavior. (a) Creep at constant load. (b) Stress relaxation at constant deformation.

Among non-Newtonian types of flow the following are of special interest: *dilatant liquids* have a higher apparent viscosity at higher strain rates (Fig. 202, curve c); this property is often associated with a volume increase caused by shear, hence the name. *Pseudoplastic liquids* have lower apparent viscosity at higher strain rates (Fig. 202, curve d); these ratios are independent of the previous flow. In *thixotropic materials* the apparent viscosity is reduced by the shearing process but reverts to its original value when the material is at rest; this differs from pseudoplasticity by its dependence on prior treatment, but many authors deliberately ignore the distinction and use the term thixotropy to include pseudoplastic behavior.

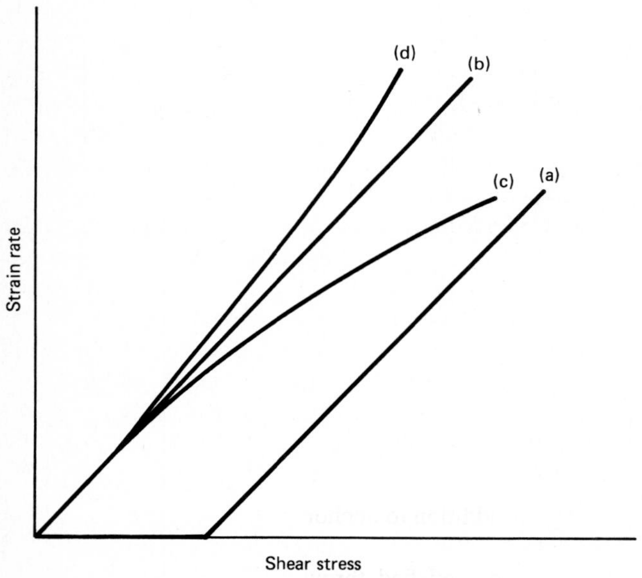

Figure 202 Flow of materials. (a) Plastic flow. (b) Newtonian flow. (c) Dilatancy. (d) Pseudoplasticity.

Rheological Behavior Relevant to Building. For structural materials, the designer and regulatory authority are concerned with the largest deformation likely to arise under service conditions rather than with detailed knowledge of the deformation process. It is common practice to treat viscoelastic materials such as concrete, wood, or plastics as if they were elastic, but to reduce elastic moduli for longer applications of design load, thus allowing for the effect of creep.

To simplify the criteria for acceptance and classification of materials with desired rheological properties, standard tests specify strict, often arbitrary, conditions yielding results that provide a useful basis for comparison and an indication of the general response of the material. It must, however, be remembered that samples giving the same reading in the standard test may differ widely in their responses to other conditions. Soils are classified, among other ways, by consolidation tests and tests for liquid and plastic limit. Fresh concrete is required to pass a slump test. Bitumen is subjected to penetration and viscosity tests.

Control of rheological properties is critical in the application of materials such as sealants, adhesives, and paints. These materials generally exhibit plastic nonlinear flow before curing. Paints are manufactured to have rheological properties suited to the intended method of application by brush, roller, or spray gun. The requirements of good leveling and minimum sag must be reconciled.

References

1. M. REINER (ed.): *Building Materials, Their Elasticity and Inelasticity.* North-Holland, Amsterdam, 1954.
2. F. R. EIRICH (ed.): *Rheology, Theory and Applications,* 5 Volumes. Academic Press, New York, 1956 to 1969.
3. R. HOUWINK and H. K. DECKER (eds.): *Elasticity, Plasticity and Structure of Matter.* Cambridge University Press, Cambridge and New York, 1971.
4. G. H. TATTERSALL and P. F. G. BANFILL: *The Rheology of Fresh Concrete.* Pitman, London, 1983.
5. R. I. TANNER: *Engineering Rheology.* Oxford University Press, Oxford, 1985.

P.U.A.G.

ROOF STRUCTURES. The technology of constructing structural roofs developed along with the establishment of organized settlement. Materials available at that time imposed restrictions on both span and type of construction. For example, stone or other masonry roofs were constructed in the form of either corbeled vaults or true MASONRY ARCHES capable of covering only a limited SPAN.

The introduction of volcanic cementing materials such as ANCIENT ROMAN CONCRETE permitted construction of masonry DOMES, such as that covering the Pantheon in Rome, which spans in excess of 40 m (130 ft), or vaults spanning some 25 m (80 ft). Although there is evidence of ancient Roman structures incorporating crossed diagonals, the earliest surviving use of some form of timber trusses dates from the Middle Ages, as exemplified by the Hammer Truss (Fig. 203), which uses exclusively dovetailed joints. Even trusses built during the Renaissance rarely employ functional web diagonals but rather king-post-type bracing for the purpose of stiffening the rafters (Fig. 204). It is only in the 18th century that timber-trussed roofs using metal bolts and straps for connections permitted economical roofing over large spans. The introduction of rolled steel at the end of the 19th century revolutionized the construction of buildings in general and roofs in particular.

The **Most Common Types of Modern Roof Structures** for low-rise buildings may be classified under the following headings:

Beam and Column is a simple and inexpensive system

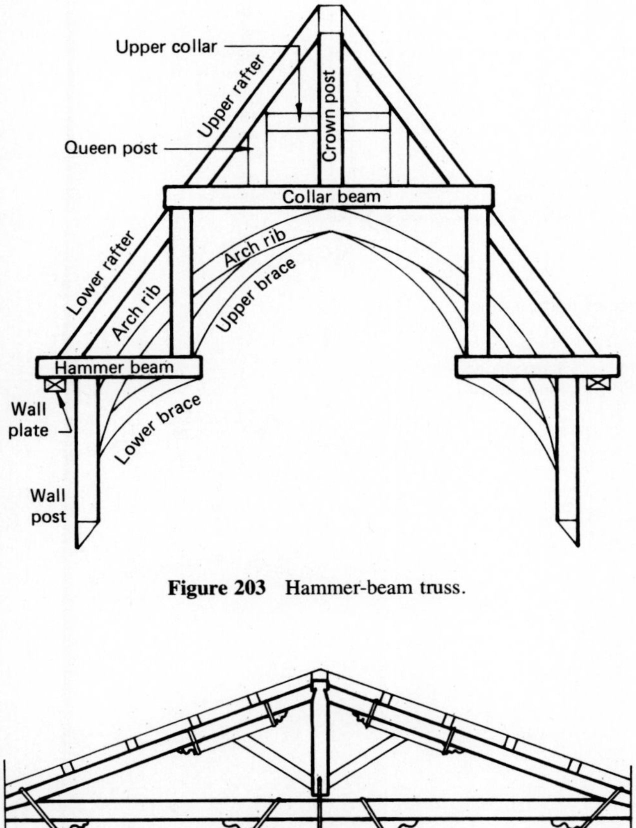

Figure 203 Hammer-beam truss.

Figure 204 King-post truss.

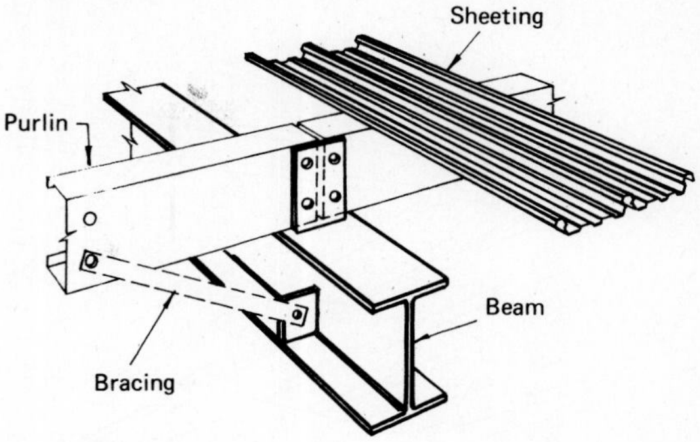

Figure 205 Typical connection of steel beam to steel purlin.

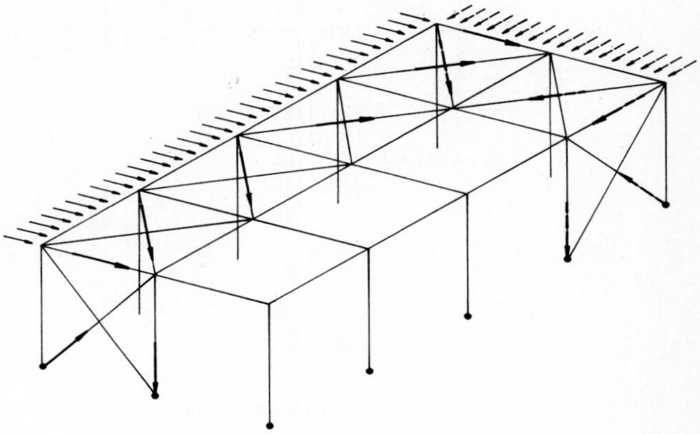

Figure 206 Wind-load path in steel frame.

and, for areas not subject to snow loading, appropriate for spans of up to 12 m (40 ft). It employs beams simply supported at each end by columns, at a maximum of 10 m (30 ft) centers, and supporting the roof sheathing through purlins (Fig. 205). The system requires bracing for wind forces in two directions in the form of diagonal steel brace or masonry panels (Fig. 206). In addition, columns must be anchored to foundation or slabs to resist uplift forces due to wind action. The system is most appropriate for flat or near flat roofs.

Steel Frames, by providing for moment continuity through junction of column and beam (Fig. 207), present a more efficient utilization of materials. Since this permits profiling of the frame and hence distributing of the material to where it is more fully utilized, it has also an inherent ability to resist lateral wind loads (Fig. 208). It is therefore possible to provide column-free space over distances of 30 to 40 m (100 to 130 ft) without an excessive cost penalty. The steel-frame system requires wind bracing in one direction only, in addition to anchorage against uplift forces and lateral thrusts at ground level. The anchorage requirements can usually be satisfied through appropriate foundations, but in soft soils may require the provision of ties engaging neighboring columns. The system is used in most industrial

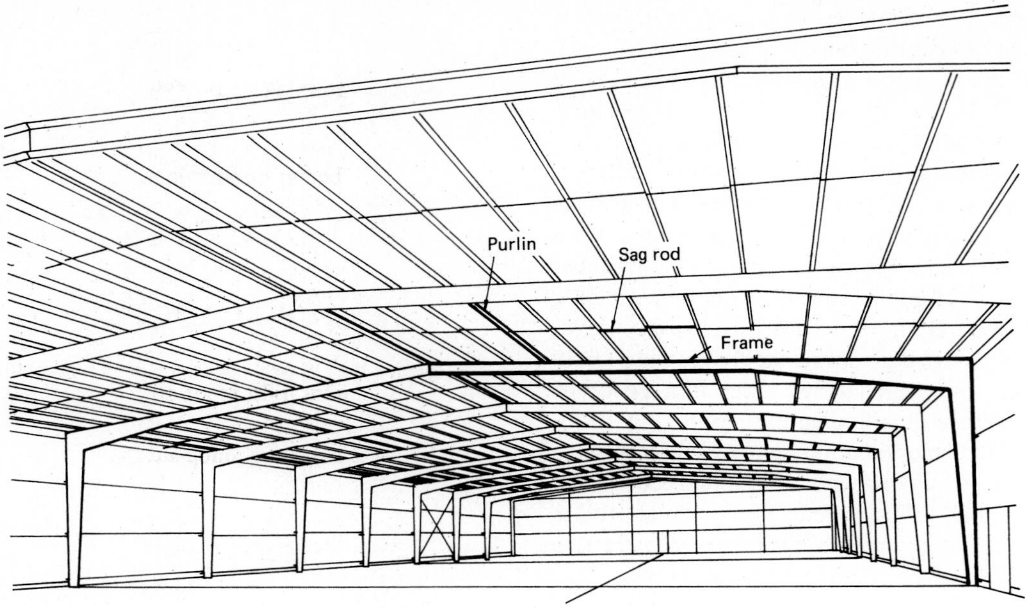

Figure 207 Steel portal frame.

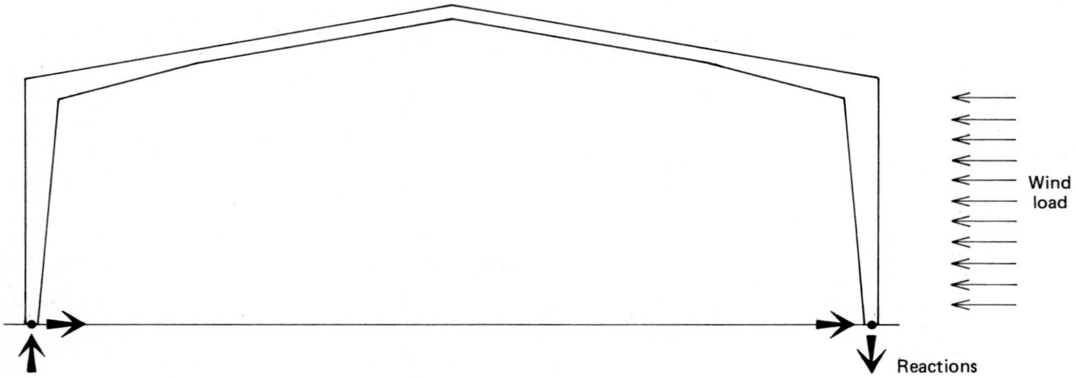

Figure 208 Wind resistance of portal frame.

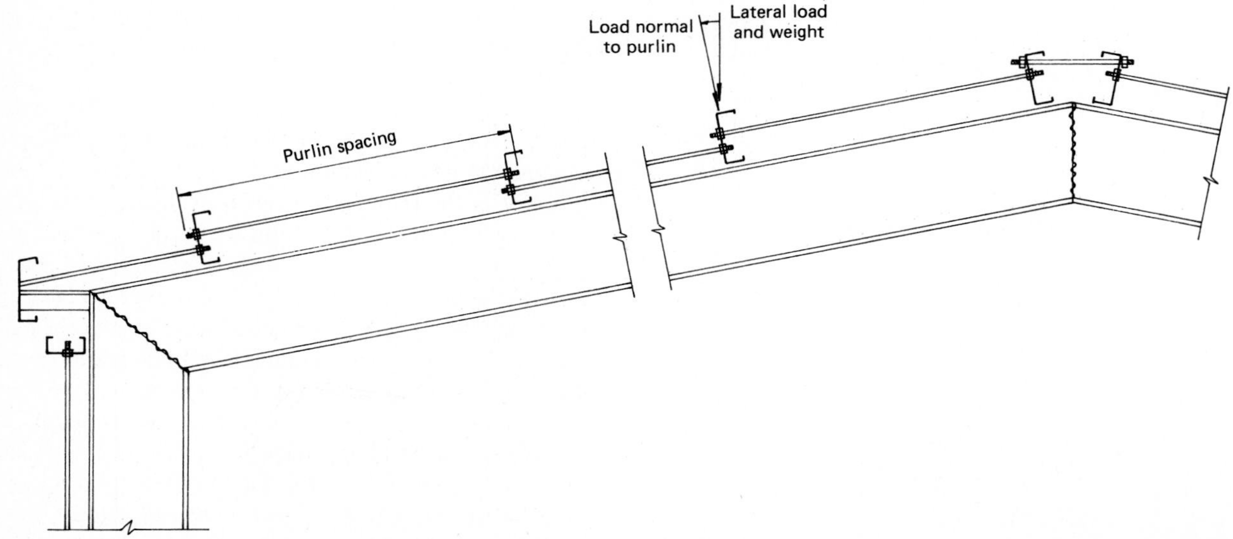

Figure 209 Typical detail for sag rod.

building and usually results in sloping roofs (5° to 30°). Sag rods are required to prevent purlins from downslope bending (Fig. 209). While the weight of steel required varies from 20 to 25 kg/m^2 (4 to 5 lb/ft^2) of roof area, the cost per ton of a fabricated structure is some 40% higher than that required for the simple beam and column system, which requires somewhere on the order of 15 to 20 kg/m^2 (3 to 4 lb/ft^2).

Steel or Timber Truss structures offer better utilization of material by separating the opposing tensile and compressive forces through a configuration of vertical and diagonal web members. The distribution of primary loads is usually calculated by resolution of external forces and in this respect these structures are statically determinate (Fig. 210). Modern trusses, however, are usually fully welded, thus introducing an element of indeterminacy that leads to some enhancement of strength and stiffness. TIMBER trusses that normally employ timber compression members are often reinforced by steel tensile members. Their structural integrity relies extensively on the capacity of the joints to transfer loads. STEEL roof trusses can be designed to span large column-free spaces, and it is not unusual to span over 50 m (170 ft). The most common span-to-depth ratio is on the order of 8 to 12, and steel trusses are usually very rigid and do not suffer from excessive deformations. The truss systems can accommodate flat or sloping roofs and provide some resistance to lateral forces. The building usually requires bracing against wind forces acting normal to the span. Relatively lightweight steel trusses, which are usually employed for the direct support of roof decks between walls or main structural members, are known in the industry as open web joists (Fig. 211). Such joists can assume a variety of shapes depending on the particular application (Fig. 212).

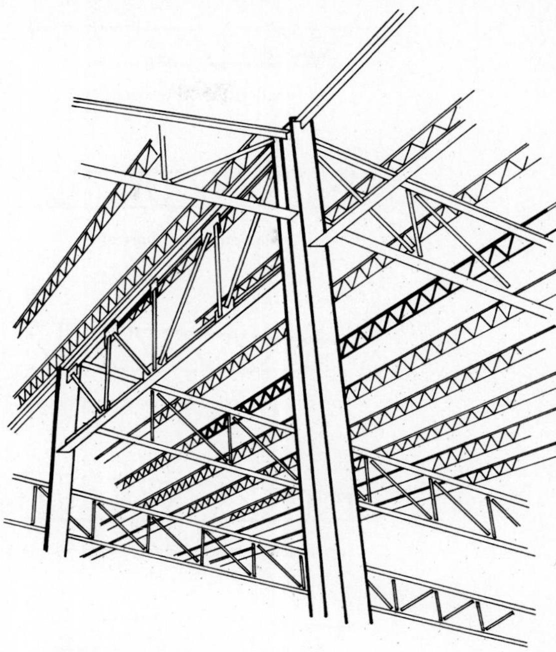

Figure 211 Steel structure with open-web joists.

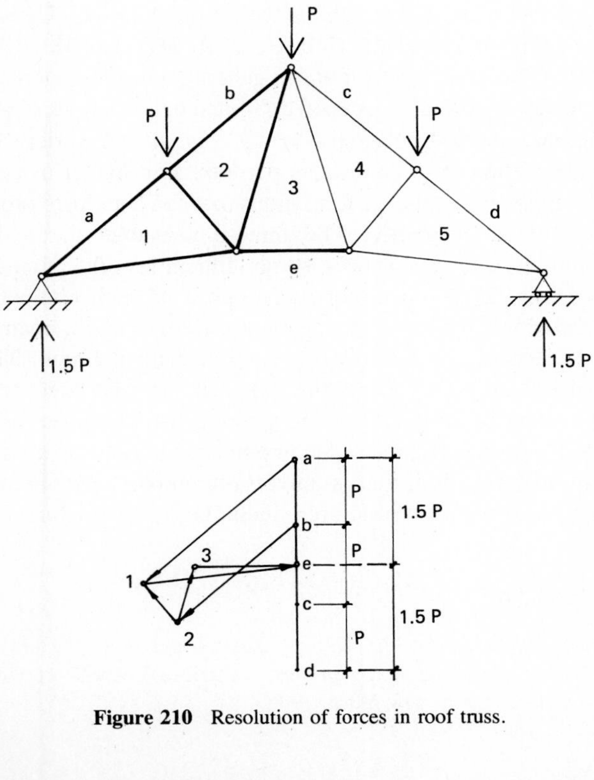

Figure 210 Resolution of forces in roof truss.

Concrete Slabs are rarely used in the construction of low-rise roofs except in special circumstances such as where a fire-resisting structure is required or where the roof is to be subjected to pedestrian or vehicular traffic. Such roofs are usually in the form of flat slab, flat plate, or slab and beam construction.

Other Structural Types. The problem of enclosing larger column-free spaces requires more sophisticated forms of structure such as SPACE FRAMES, FOLDED PLATES, or concrete SHELL STRUCTURES. The cost of constructing such structures is much greater than for the more conventional types described previously; this is particularly true in countries with high labor costs. For example, whereas a conventional steel framed roof costs some \$25 to \$50 per square meter (\$2.50 to \$5.00 per square foot), the cost of a shell roof could escalate (depending on span) to \$200 per square meter (\$20 per square foot). Where buildings require very large unobstructed spans, CABLE AND SUSPENSION STRUCTURES or TENSILE ROOFS are considered suitable. In special circumstances PNEUMATIC STRUCTURES are used.

Design Criteria. In ancient times, structures were mainly designed by intuition and experience, without benefit of science, an appreciation of load and the resolution forces, and/or calculation of stresses. The development of the structural analytical methods of the 18th and 19th cen-

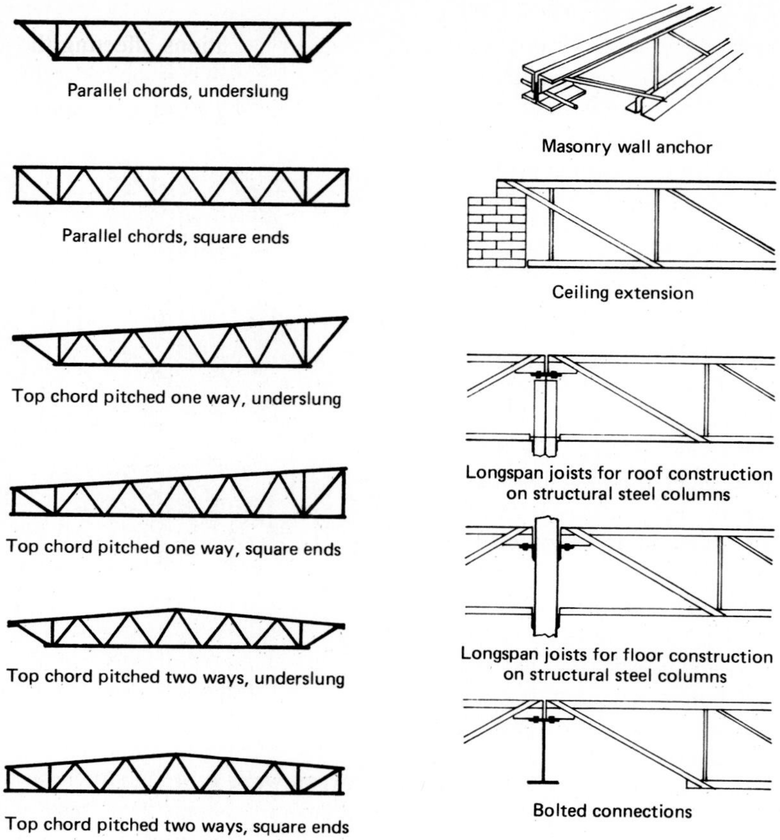

Figure 212 Various types of open-web joist.

turies together with the invention of inexpensive processes of producing steel revolutionized structural engineering in general and structural roofs in particular.

The 20th century saw the introduction of rules and regulations in the form of building codes that prescribe the loads to be applied, the strength of the materials, and the appropriate factors of safety. Modern loading codes specify the design live loads, including SNOW LOADS. Design for WIND LOAD has received considerable attention in the last decade. Statistical concepts, such as a return period of storms, permit a more rational approach to the problems of wind. Modern codes of practice recognize the dynamic effect of wind on buildings, and also that their response cannot be simply treated as a quasi-static effect. The recent trend toward LIMIT STATES DESIGN provides a better understanding of the actual performance of a structure and therefore constitutes a more sophisticated approach to the conventional concepts of SAFETY FACTORS. At the present time limit state design is still in its embryo stages, and universal acceptance of this approach will necessitate extensive further studies.

Structural Analysis. The complexity of structural analysis and the inadequacy of previously existing analytical tools imposes inhibitions and restrictions on the design of modern structures. Therefore, in the precomputer era designers preferred for the sake of expediency to opt for simple STATICALLY DETERMINATE STRUCTURES. The universal use of COMPUTER METHODS FOR STRUCTURAL ANALYSIS in general, and finite-element methods in particular, opened unlimited opportunities to use not only more advanced systems, but also a variety of shapes. The use of computers permits, as a matter of routine, the prediction of deformations that can anticipate problems of serviceability. The three-dimensional nature of concrete structures causes some difficulties in computer modeling. In practice, for the purpose of such computer calculations, the three-dimensional system of slabs, beams, and columns is reduced to a two-dimensional frame. The torsional stiffness of beams and slabs is translated into equivalent bending stiffness in a somewhat imprecise fashion. More recently, the minicomputer has introduced a standard analytical tool that is universally employed in the design of all types of structures, including roof structures.

References

1. *Specification for the Design, Fabrication and Erection of Structural Steel for Buildings.* American Institute of Steel Construction, New York, 1978.

2. *Manual of Steel Construction, Part 1—Dimensions and Properties*, 8th ed. American Institute of Steel Construction, New York, 1980.
3. *Plastic Design in Steel—A Guide and Commentary, ASCE Manual of Engineering Practice No. 41*, 2nd ed. American Society of Civil Engineers, New York, 1971.
4. *1969 Specification for the Use of Structural Steel in Building—B.S.499*. British Standards Institution, London, 1969.
5. CONSTRUCTIONAL STEEL RESEARCH AND DEVELOPMENT ORGANISATION: *Steel Designer's Manual*, 4th ed. Crosby Lockwood Staples, London, 1972.
6. M. R. HORNE: *Plastic Theory of Structures*, 2nd ed. Pergamon Press, Oxford and Elmsford, N.Y., 1979.
7. OMER W. BLODGETT: *Design of Welded Structures*. Lincoln Arc Welding Foundation, New York, 1976.
8. HENRY J. COWAN: *The Masterbuilders*. Wiley, New York, 1977.

A.W.

ROOF SURFACES, the exterior cladding or weathering layer of roof systems, are of many types. For flat roofs, a continuous watertight membrane of bituminous, plastic, or rubber material is used, which is sometimes covered with an additional surfacing, such as gravel, for protection against weather and traffic (see FLAT ROOFS). For steeply sloping roofs, the cladding may consist of small, overlapping, watershedding units of slate, clay, concrete, wood, asphalt felt, or metal. Larger overlapping sheets of metal, asbestos-cement, or plastic, corrugated or troughed to make them more rigid and to give them spanning ability, are also commonly used.

Wood Shingles were widely used in the Middle Ages or earlier in Europe, where oak was the preferred wood, and in North America by the early settlers, who usually made them from local wood. Western red cedar, the best-quality wood for shingles and shakes, has recently regained popularity in the western parts of North America for residential buildings. Treating wood shingles with chemical fire retardants has reduced the hazard from fire, which was one of their main drawbacks.

Wood shingles are sawn on both sides into slightly tapered units, whereas shakes are hand and machine split and may be sawn on one side. Wood shingles come in 400, 450, and 600 mm (16, 18, and 24 in.) lengths, and shakes in 450 and 600 mm (18 and 24 in.) lengths. They are usually laid in horizontal overlapping courses on either open or closed sheathing so that something less than a third of each course is exposed to assure triple coverage. In wet climates, shakes and shingles should be laid on inclines of 1 in 2 (26°) or higher for rapid runoff and drying, but they are sometimes unwisely laid on slopes as low as 1 in 3 (18°). Conditions affecting the service life of wood shingles or shakes vary considerably, and life expectancy may be as little as 15 or more than 30 years.

Asphalt Shingles are the most popular roof surfacing for residential buildings in North America, where they originated late in the 19th century, and are used to a lesser extent in other parts of the world. The current annual production in North America is enough to cover about 5 million homes. The base material is asphalt-saturated cellulose fiber felt or glass fiber mat coated with asphalt and surfaced with colored mineral granules. A wide variety of styles, shapes, colors, and textures is available, but the most common standard shingles are produced in 915 mm × 305 mm (36 in. × 12 in.) strips. Two 127-mm (5-in.) slots are cut at right angles to one side to divide the strip into three equal tabs, making what is called a double-coverage square-butt strip shingle. These shingles are laid fully supported over closed sheathing, sometimes with a felt underlay as well. Weather exposure is 127 mm (5 in.) on inclines of 1 in 3 (18°) or higher. For lower slopes to a minimum of 1 in 6 (9.5°), a continuous two-ply asphalt and felt underlay or triple coverage with full cementing is required.

Asphalt shingles are easy to install at moderate cost and have a service life of 15 to 25 years with little maintenance. In manufacture the material is produced as a continuous sheet material 915 mm (36 in.) wide, which is cut into strip shingles. The sheet material may also be used in roll form for application with lapping cement to form a continuous roofing membrane.

Metal Roofing may be fully supported on closed sheathing or self-supporting on purlins. The fully supported roofing may be of aluminum, copper, lead, zinc, or terne-alloy-coated steel. Available in sheet and strip form, it is usually applied with standing seam or other special joints, with fastenings at the joints that allow thermal movement. Some materials, terne (80% lead and 20% tin)-coated steel in particular, are available in small units to simulate tiles or shingles.

Steel and aluminum, suitably protected by galvanizing, terne coating, enameling, or organic film laminates, are available in a large variety of corrugated or troughed profiles. They are now produced in lengths to 15 m (50 ft), which eliminates the need for end joints. Mechanical equipment has been developed to automatically form and seal the side joints during application and is used for both self-supporting and fully supported metal roof surfacing. Most mechanically formed joints incorporate concealed fasteners, enabling slopes as low as 1 in 100 (0.6°) to be used successfully. A steeper slope, as for other simply overlapped water-shedding roofing, is required to prevent rain penetration where there are exposed fasteners and nonsealed end and side laps. Exposed fasteners usually consist of special drive screws with weathertight washers.

Other Self-Supporting Roofing. Asbestos-cement sheets with corrugated and troughed profiles similar to those for metals have been widely used to cover industrial buildings for many years. Although embrittlement of the material with age does not affect weathertightness, it reduces resistance to impact and makes walking on such roofs hazardous.

Glass-fiber-reinforced polyester resin plastic, acrylic resin sheets, and glass with wire reinforcing are available in profiles similar to those for metal and asbestos-cement and can be used in panels or strips to provide translucent room lighting. There may be a loss of light transmission with age for the plastic sheets.

References

1. JOHN A. WATSON: *Roofing Systems: Materials and Application.* Reston Publishing Co., Reston, Va., 1979.
2. A. J. ELDER (ed.): *A. J. Handbook of Building Enclosure.* Architectural Press, London, 1974.
3. *Principles of Modern Building II:* DSIRO Building Research Station, Her Majesty's Stationery Office, London, 1961.

M.C.B.

S

SAFETY FACTORS IN STRUCTURAL DESIGN have been used throughout all the history of building construction. Structures are normally designed to be able to support more load than the load due to the normal use of the building. The safety factor implies that there are inherent uncertainties in the strength of construction materials, the design assumptions, and the magnitudes of the loads. When a proper safety factor is used, the user of the structure can, with a high degree of certainty, expect that it will perform its intended function for its intended life span. The safety factor thus accounts for the small but finite probability that the strength of the member may be considerably less and that the load may be considerably more than the magnitudes used in the design calculations. Absolute safety cannot be guaranteed by the factor of safety; it is only a device by which the probability of structural failure is minimized to a level acceptable to society. It cannot protect the structure against failure due to gross errors of judgment or calculation or negligence, misuse, or sabotage.

Throughout most of the history of building, construction safety factors evolved by the process of trial and error. However, by the second half of the 19th century building design was no longer a purely artisan activity, but also involved the use of materials test data, climatic observations, and the theory of the mechanics of solids, specifically the theory of elasticity and the theory of elastic stability. The concept of *allowable stress design*, which evolved from this rationalization of the design process and which was extensively used in most structural design specifications until the 1950s, utilized the following notion of the factor of safety:

$$\text{factor of safety} = \frac{\text{limit state stress}}{\text{allowable stress}}$$

The *allowable stress* used by the designers to compare with the stress computed by elastic theory for the maximum expected *working loads* is thus a suitable fraction, say $\frac{3}{5}$ or $\frac{1}{2}$ or $\frac{1}{3}$, of a limit state stress (e.g., crushing stress of stone, fracture stress of cast iron, yield stress of steel, buckling stress of a column). As experience with a certain type of construction increased, the factors of safety went down. For compact structural steel beams in the United States, for example, the factor of safety against yield was 2.0 in 1890, 1.83 in 1923, 1.65 in 1963, and 1.52 in 1983.

Beginning in the 1930s, but increasingly in the 1950s, it was realized that allowable stress design, with its single factor of safety for a given limit state and its assumption of linear behavior, was not rational enough or uniformly economical. Consequently, the concept of LIMIT STATES DESIGN evolved, and it is now the basis of most of the structural specifications in the world.

Modern structural specifications explicitly utilize statistical data (see STATISTICS) and probability theory to develop such load factors that all structures in a given class of structures have the same degree of reliability. The notion of probability and uncertainty was always recognized by designers, but only implicitly. In a series of papers in the 1950s, Freudenthal formulated the mathematical and conceptual basis of probability-based design, and simplifications and further extensions now give the designers or the structural specification writers a broad range of simple or sophisticated methods for the assessment of the reliability and safety of structures. Structural design specifications mandate minimum levels of reliability and in most cases prescribe minimum load factors.

References

1. A. PUGSLEY: *The Safety of Structures*. Edward Arnold, London, 1966.

2. A. M. FREUDENTHAL: Safety and probability of structural failure. *Transactions of the American Society of Civil Engineers*, Vol. 121 (1956), pp. 1337–1375.

T.V.G.

SANITARY APPLIANCES contribute greatly to human health and comfort. To the individual, they are the most tangible evidence of the sanitary revolution that commenced in the latter half of the 19th century. There are two groups: *soil appliances*, which receive human body wastes, that is, water closets, urinals, and slop sinks; and *ablutionary* or *waste appliances*, that is, baths, showers, wash basins, bidets, and, less commonly, scullery sinks, pantry sinks, and drip sinks. All appliances are fitted with a trapped outlet, which is connected directly to the drain in the case of soil appliances. Waste appliances, however, are connected above the water of a water-seal trapped gully, except in the case of one-pipe and single-stack systems (see DRAINAGE INSTALLATIONS).

Although authorities differ in their detailed specifications, all require sanitary appliances to be made of hard, smooth, nonabsorbent, and incorrodible material. Vitreous china is the most common material, except for baths, for which porcelain-enameled cast iron is generally used. The general features of the main types of appliances are discussed next.

Water Closets (WCs) are of two main types, according to the physical position of the user: the western type is used in a sitting position (Fig. 213a), and the eastern type is designed for the squatting position (Fig. 213d). All are water flushed.

Water Closets Sets comprise the WC pan and flushing apparatus, with connecting pipework. The water closet pan of typical western design is cast in one piece complete with flushing rim, outlet trap (either of P or S pattern), and pedestal. There are two main types: the washdown (Fig. 213a) and the siphonic (Fig. 213c). Flushing is effected by the rapid discharge of 9 to 13 liters ($2\frac{1}{2}$ to $3\frac{1}{2}$ US gallons) of water from a cistern mounted either above head height (2.1 m or 7 ft to the underside), as in Fig. 213a, or low down to the pan (Fig. 213b). In multistory commercial and office buildings, flush valves fed from a header tank are often used. Wall-mounted WC pans are also a recent innovation in commercial buildings.

The eastern water closet pan (Fig. 213d), designed for people accustomed to the squatting position, is provided with squatting plates for accuracy in use and is set into the floor.

Urinals, common in men's toilets in public and commercial buildings, are of three types: stall, slab, and basin (Fig. 214). Although the stall type provides greater privacy,

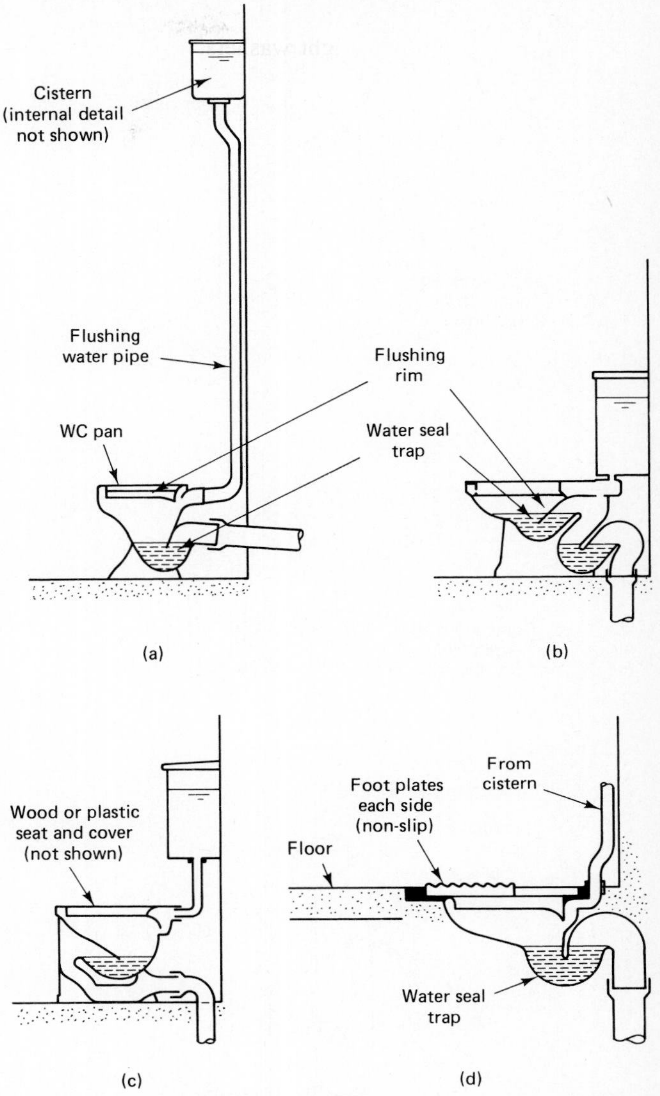

Figure 213 Water closets (WCs). (a) Washdown WC with high-level cistern. (b) Double-trap WC with close-coupled cistern. (c) Single-trap siphonic WC with close-coupled cistern. (d) Eastern-type WC.

the slab type is the more common, especially when made of stainless steel; this presents a large flat, impervious surface free of nooks and corners, which facilitates cleaning. Flushing is either by automatic cistern, connected to a sparge pipe with multiple outlets for stall and slab types, or by individual cisterns or flush valves. Outlets from urinals are 50 to 75 mm (2 to 3 in.) in diameter, covered with a grating, and connected through a water seal trap to the soil drainage pipe.

Slop Sinks are used mainly in hospitals and institutions for the disposal of bedroom slops and wastewater. They are fitted with a trapped outlet of similar size to that of WC pans and are connected to the soil drainage. They may be

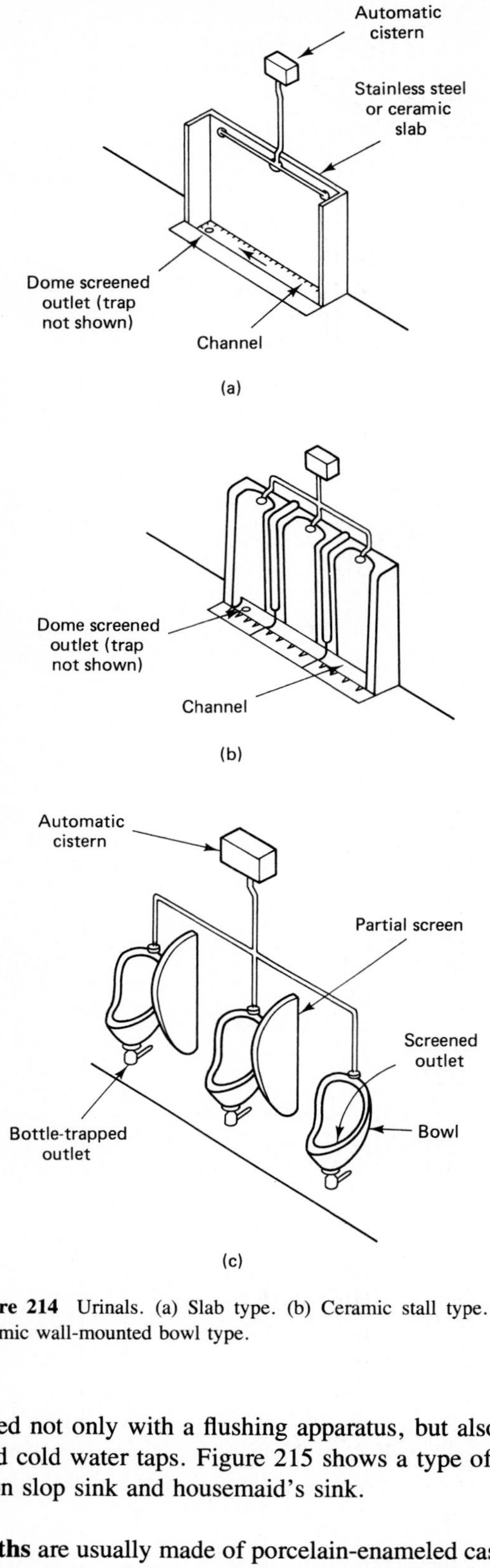

Figure 214 Urinals. (a) Slab type. (b) Ceramic stall type. (c) Ceramic wall-mounted bowl type.

provided not only with a flushing apparatus, but also with hot and cold water taps. Figure 215 shows a type of combination slop sink and housemaid's sink.

Baths are usually made of porcelain-enameled cast iron to a length of about 1.7 m (5 ft 6 in.), although a variety of sizes, colors, shapes, and materials is now available (Fig. 216). They have a restricted outlet size of 38 mm

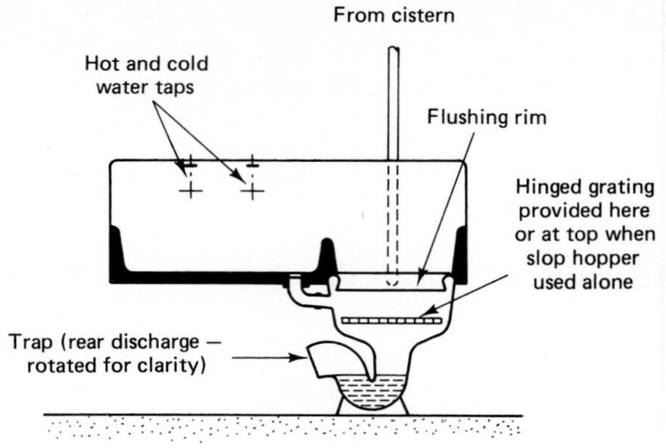

Figure 215 Combined housemaid's sink and slop sink.

($1\frac{1}{2}$ in.) to limit their peak discharge. Each is connected to the waste pipe, often above the water seal of a floor waste trap (see DRAINAGE INSTALLATIONS).

Shower Baths (or simply showers) may consist either of simple forms built in concrete, tiled, and isolated from the general floor level by a kerb, or of more elaborate forms precast from materials similar to baths, finished with non-slip bottoms, and provided with a similar type of outlet plug. The trapped shower outlet also is commonly connected above the water seal of a floor waste or disconnector trap.

Handbasins, available in a wide variety of shapes, sizes, and colors, are fitted with 32-mm ($1\frac{1}{4}$-in.) trapped outlets. These are connected to the waste pipe system in similar fashion to baths and showers.

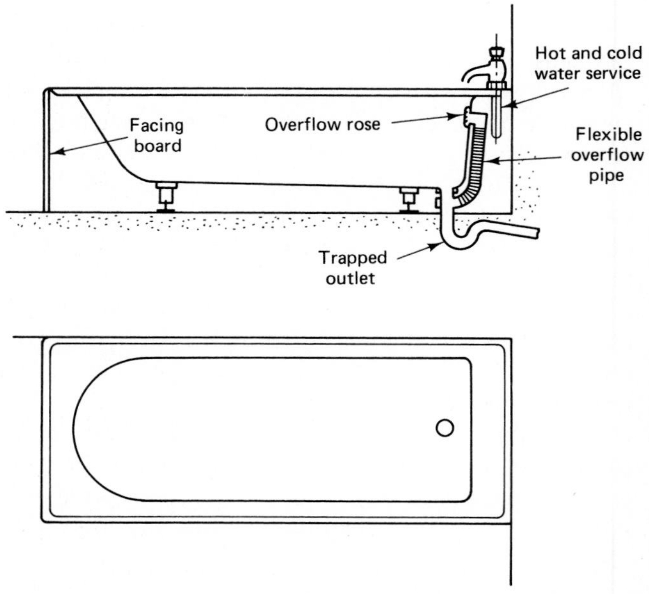

Figure 216 Bath with overflow.

Bidets, widely used in Europe for cleaning perineal areas of the body, are less common in the English-speaking world, except in female wards of hospitals. The bidet has a trapped outlet, hot and cold water connections, and a cleansing spray.

References

1. F. HALL: *Building Services and Equipment*, Volume 1. Longman, London, 1976.
2. F. G. GOODWIN and J. DOWNING: *Domestic Sanitation.* Estates Gazette, London, 1971.
3. T. A. THOMPSON: *A Guide to Sanitary Engineering Services*. Macdonald and Evans, London, 1972.

P.J.B.

SANITATION FOR DEVELOPING COUNTRIES, as for developed countries, aims to provide the means of disposing of refuse and sewerage so as to maintain healthy environmental conditions. This is achieved by controlling the biological oxygen demand (BOD) of effluents, controlling the breeding and access of disease-carrying insects to excreta, and preventing pathogens in excreta from contaminating food or water. The BOD of effluents must be controlled to protect the ecology of streams, rivers, and oceans into which the effluent discharges.

Pathogens that can cause disease in humans are *bacteria*, responsible for gastrointestinal infections such as dysentery, diarrhea, cholera, and typhoid; *viruses*, responsible for viral infections such as hepatitis, poliomyelitis, and dengue fever; *protozoa*, responsible for diseases such as sleeping sickness and amoebic dysentery; and *helminth* (worm) infestations such as roundworm, hookworm, and tapeworm.

Control of such infections can be achieved by effective sanitation, control of disease-carrying insects, and education to improve personal and domestic hygiene.

Effective Disposal of refuse (trash, garbage, and offal) is usually achieved by incineration or burial in land fill. The methods used for disposal of sewerage (excreta and contents of household drains) in developing countries are usually less expensive and less complex to construct and maintain than the fully piped sewers to primary, secondary, and tertiary treatment plants found in developed countries. Less expensive excreta disposal methods can be classified into three groups.

1. Dry on-site treatment: dry pit latrines and composting latrines.
2. Dry off-site treatment: bucket or bulk cartage latrines with landfill disposal.
3. Wet on-site treatment: wet pit latrines, aqua privies, septic tanks, and biogas plants.

All these disposal methods rely on the reduction of BOD and pathogens by aerobic bacteria (where free oxygen is available) and/or anaerobic bacteria (where free oxygen is excluded).

Composting Latrines (Fig. 217) are of two types, the single-vault continuous operation type such as Multrum and Clivus designs, and alternating twin-vault batch systems such as the WHO Vietnamese design. Continuous-operation types utilize aerobic bacteria to act on excreta and

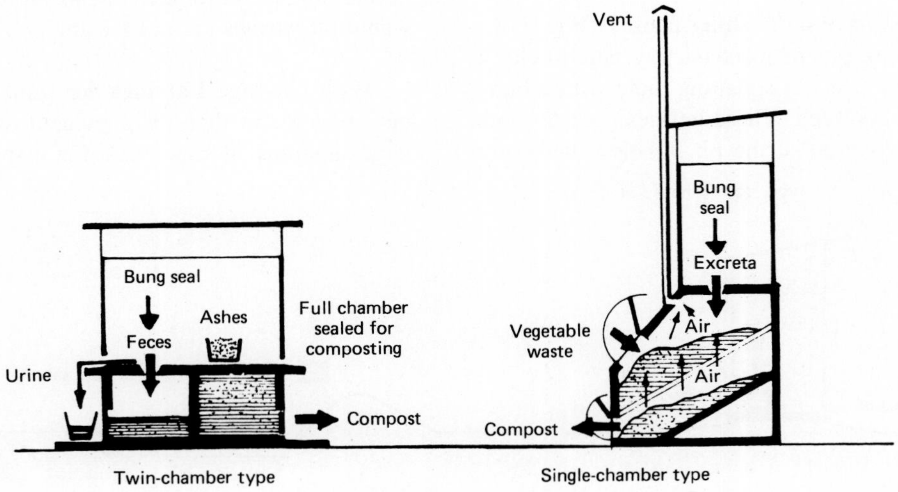

Figure 217 Composting latrines.

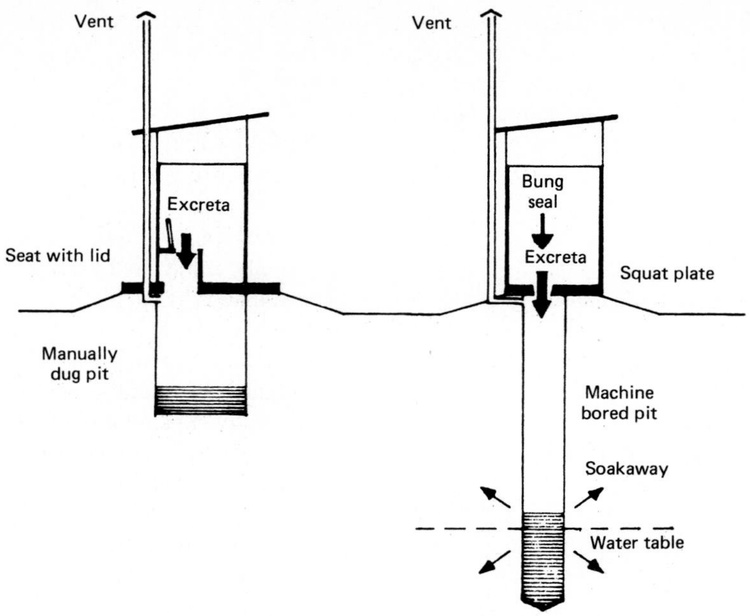

Figure 218 Dry (nonflush) pit latrines.

vegetable wastes suspended on a rack above the floor of the ventilated vault. Urine is evaporated off or drained away. As the mixture decomposes, it falls through the rack and is removed for use as fertilizer.

In the alternating twin-vault type, one vault at a time receives feces. Urine is drained away in a separate surface channel. Feces are covered with loose earth, ashes, or sawdust to reduce odors. When the vault is nearly full, it is sealed with lime mortar and left for a few months to compost by anaerobic bacterial action. Contents are then removed and used for fertilizer. During this time the other vault is used as the latrine. Both types work best in warm climates and with little or no urine loading.

Dry Pit Latrines have no flushing facility (Fig. 218). They are manually dug pits or mechanically bored holes a few meters deep over which a squatting plate with a bung seal or seat with lid is placed. These latrines operate more efficiently when the bottom of the pit is below the water table, which allows excreta to be decomposed by anaerobic bacteria below water level and to soak away into the surrounding ground. Gases generated, such as methane, are vented through a tall vent pipe. When pits are dry, a combination of anaerobic and aerobic decomposition takes place. When a pit is almost full, the surface cover is removed and the top of the pit filled with a mixture of lime and earth. A new pit is then dug a suitable distance from any well or other underground water source.

Bucket Latrine (Fig. 219) collection service and transport can be a heavy ongoing expense. A large suitable disposal site is needed to allow for rapid trench burial. Bucket latrines are useful for settlements on rocky or swampy soils where pit latrines are not feasible.

Bulk Cartage Latrines are similar to bucket latrines, but are used in densely populated areas where there are large amounts of night soil for disposal. Collection and

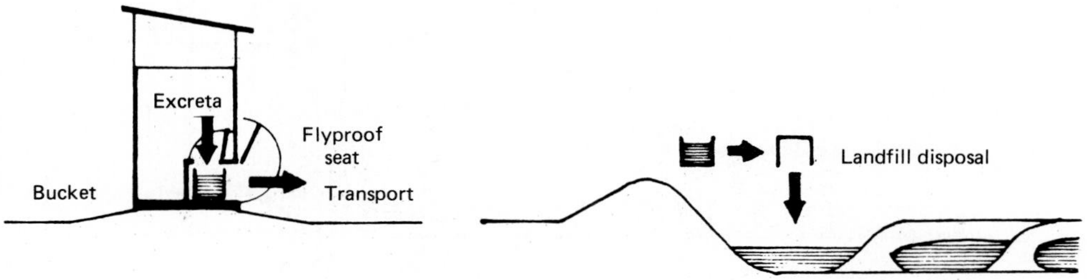

Figure 219 Bucket latrine.

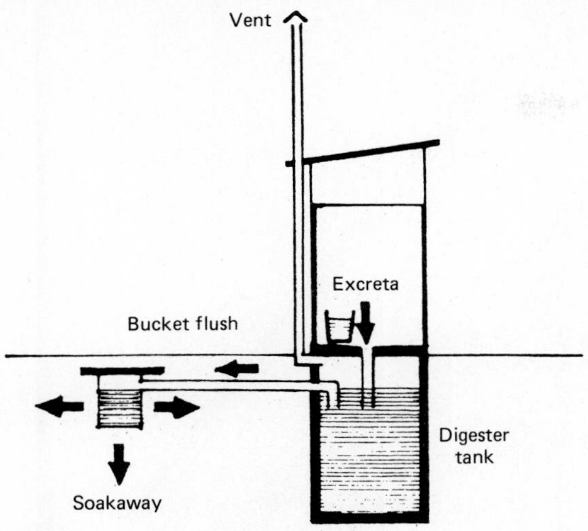

Figure 220 Aqua privy.

transportation can be rationalized by centrally located community latrines and rail car transportation to the disposal site.

Aqua Privies (Fig. 220) are simplified septic tanks with a single chamber and without a full flush pan. Where bucket-flushed squat plates are used, excreta enters the tank through a short pipe that penetrates below the surface of the liquid in the tank to minimize odors. Alternately, excreta may enter through a low-volume, water-seal, bucket-flushed floor trap set in the squat plate. Decomposition is by anaerobic bacteria below water level in a permanent tank, which periodically requires desludging. Gases generated in this process of decomposition are vented through a tall vent pipe. Excess effluent from the tank is drained to soak-away pits.

Septic Tanks can be either single or double chamber. They are generally used with full cistern flush water seal pans. Single-chamber designs use anaerobic digestion; in double-chamber designs the second chamber is ventilated and uses aerobic bacteria for digestion. The permanent tanks need desludging periodically. Excess effluent is flushed into soak-away pits.

Wet Pit Latrines are bucket-flushed, water-seal, floor-pan latrines with a soak-away pit in porous soil. Digestion of excreta is by anaerobic bacteria below water level. The lower section of the pit is lined to retain water when the pit does not reach the water table. Gases from the digestion are vented through a tall pipe.

Biogas (Gobar Gas) Digestors (Fig. 221) operate similarly to a single-chamber anaerobic septic tank, but provision is made to trap the gas, which is largely methane, given off during digestion. The methane gas can be used

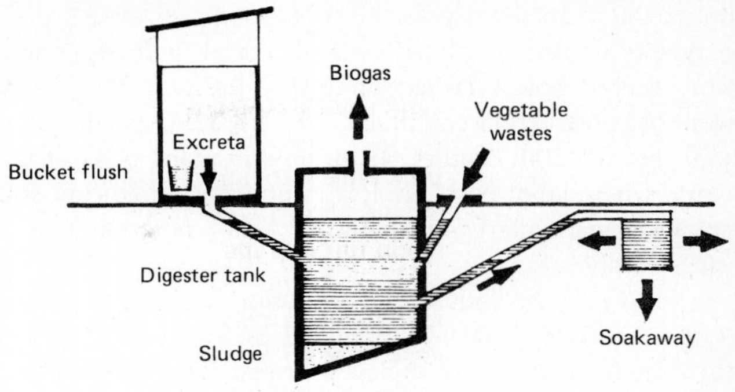

Figure 221 Biogas digester.

as fuel for cooking and lighting buildings. For efficient gas production, the contents of the digester tank should have a carbon:nitrogen ratio of approximately 30:1. Vegetable wastes are usually added to the excrement to raise the carbon content in the tank. Excess effluent from the tank is often drained into ponds, where algae is grown as feed for domestic animals such as ducks. The digestor tank requires desludging periodically.

The choice of excreta disposal method is determined by local ground conditions, rainfall, water table, water supply, ground temperature range, and social, cultural, and religious influences within the community.

See also DRAINAGE INSTALLATIONS; REFUSE DISPOSAL; SANITARY APPLIANCES; and SEWAGE DISPOSAL.

References

1. A. PACEY (ed.): *Sanitation in Developing Countries*. Wiley, New York and Chichester, England, 1978.
2. R. FEACHEM et al. (eds.): *Water, Wastes and Health in Hot Climates*. Wiley, New York, 1977.
3. W. RYBCZYNSKI et al. (eds.): *Low-Cost Technology Options for Sanitation: A State-of-the-Art Review and Annotated Biliography*. International Development Research Centre, Ottawa, 1978.

R.M.A.

SCAFFOLDING is a general term for any temporary open structural framework used for elevated access or support purposes. The name derives from an old French word *escaffaut*, meaning a platform for witnessing dramatic events.

Historically, it can be reasonably assumed that scaffolding is almost as old as construction itself, although earliest records are probably those of early Chinese building construction (Ref. 1). The type of scaffolding used at that time was constructed from a framework of vertical and horizontal timber poles lashed together at working levels. A type of bricklayer's trestle was also used, with legs made from timber poles, lashed in scissor fashion. This basic form of construction continued to be used universally until well into the 20th century. Even now in many parts of the world where labor costs are low, wooden scaffolds are still common, particularly in Southeast Asia where bamboo is used widely.

Following the Industrial Revolution, sawn timber was used extensively, and in the United States by 1915 nailed rectangular timbers had replaced lashed round poles (Ref. 2). Steel tubular scaffolding first appeared in England in 1909 following the invention of special fittings that were used to couple lengths of $1\frac{1}{2}$ in. (38 mm) bore waterpipe (Ref. 3). Various forms of prefabricated frame systems were produced later, the most popular coming from the United States, comprising compact welded tubular steel frames connected with quick-assembly scissor bracing. More recently, there have been a number of developments of modular designs using individual members of set lengths of tube, with components welded to ends of horizontals engaged in receiving pockets on the vertical members (Figs. 222 and 223).

Most scaffold systems are intended for general use, but some are designed specifically for access or support work. Access systems are comparatively light, making use of thin-walled high-tensile steel tube or, as in the case of mobile maintenance towers, aluminum tube. Support scaffolding or shoring is usually required for FORMWORK support work, and greater mass and more structural rigidity are required to carry the very high loads encountered. Rectangular and triangular frames are available for the quick-assembly systems of up to 10 tons capacity per leg, but for loads in excess of this figure it is necessary to use the bolted-component systems, which have capacities of up to 50 tons per leg.

Figure 222 Steel tube and fittings scaffold.

Figure 223 Steel tube system scaffold.

References

1. JOSEPH NEEDHAM: *Science and Civilization in China*, Vol. 4. Cambridge University Press, Cambridge, England, and New York, 1971.
2. RONALD E. BRAND: *Formwork and Access Scaffolds in Tubular Steel.* McGraw-Hill, London and New York, 1975.
3. *This Is SGB*. Publication by Scaffolding Great Britain Ltd., London, 1983.

J.B.D.

SEALANTS AND JOINTING form a very important part of the weatherproofing of buildings.

Building Wall Joints that remain weatherproof with little or no maintenance can be achieved by the use of drained joints or pressure equalization. Experience has shown that in joints subject to movement, long-term durable weatherproofing is not possible by means of a sealant, no matter how good, if it is expected in a single external seal to prevent both wind and water penetration through the joint. Greater care in joint detailing and sealant selection is necessary at the design stage.

Reference 1 discusses these matters and the three prerequisites for rain penetration through a wall: water on the outdoor face, openings to permit its passage, and forces to drive or draw it inward. If any one of the three is eliminated, penetration cannot occur.

The **Drained Joint** eliminates or controls the forces contributing to rain penetration (Ref. 2). The requirements of an effective drained joint are as follows:

1. A *rain screen* at the outdoor face.
2. An *air seal* at the indoor face.
3. A *drainage* system between the screen and the seal that drains and equalizes the pressure in the drainage cavity to that outdoors.

A typical drained joint is illustrated diagrammatically in Fig. 224.

Sealants in Drained Joints are used as air seals. They can also be used as rain screens in place of the vented baffles in vertical joints, provided there are also open horizontal joints with upstands and continuous air seals.

When a sealant is used as a single external seal, it must be weather resistant and retain a 100% seal. In a drained-joint system, the sealant is not only sheltered from the weather but, being an air seal only, performs its function even if a small amount of leakage occurs. However, since a sealant is required to do so for very long periods, particularly if access to the indoor side of the joint is difficult, sealant selection must be done with care, taking into consideration the requirements for movement and durability in

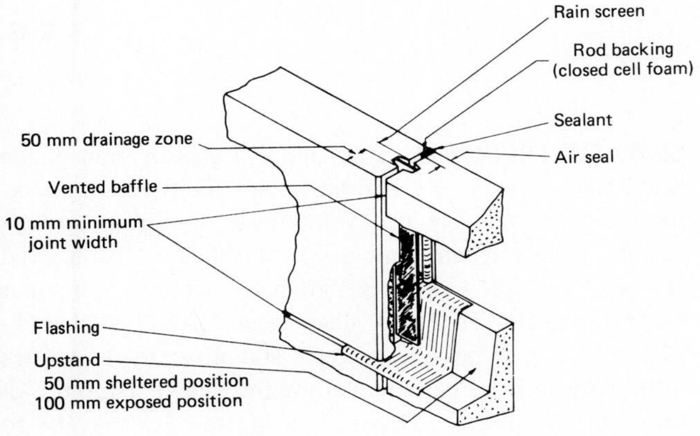

Figure 224 Drained joint in concrete panel construction, which shows the junction between a vertical and horizontal joint.

its position. These influence the choice of the sealant from the large range of bulk sealants, preformed sealants, and gaskets available. Where building tolerances are small and surfaces smooth, for example, in aluminum window glazing, preformed sealants offer considerable advantages in handling and placing.

The **Sealant Types** normally used in buildings can be divided into two types (Ref. 1):

1. *Bulk sealants* are normally applied with a gun, although some can be poured, and some are in the form of preformed tapes.
2. *Preformed sealants* include the *gasket seals*, which depend on compression of the gasket to maintain a weathertight joint and do not require adhesion to the joint boundaries.

Information on the properties and details of the use of the various sealants is best obtained from the manufacturers' literature and Refs. 3 and 4.

Measuring the Performance of Wall Elements is very desirable to check the adequacy of the design and demonstrate to all parties concerned that it is likely to be satisfactory. Checking is especially valuable for designs incorporating drained joints, and equipment (Ref. 5) has been developed for this purpose.

See also CURTAIN WALLS; FENESTRATION; and GLAZING.

References

1. N. G. BROWN: Sealants and jointing. In F. J. Bromilow (ed.), *Building Maintainability and Efficiency Research and Practice*. CSIRO Division of Building Research, Melbourne, Australia, 1982.
2. N. G. BROWN and E. R. BALLANTYNE: Watertight or weatherproof application of drained joint principles. *Building Forum*, Vol. 5 (1973), pp. 2–8.
3. A. DAMUSIS: *Sealants*. Van Nostrand Reinhold, New York, 1967.
4. J. P. COOK: *Construction Sealants and Adhesives*. Wiley, New York, 1970.
5. N. G. BROWN and E. R. BALLANTYNE: *The SIROWET Rig for Testing the Weatherproofness of Building Facades*. CSIRO Division of Building Research, Melbourne, Australia, 1975.

N.G.B.

SECURITY IN BUILDINGS protects property and assets. The alternatives are to transfer the risk to insurance companies or to accept it.

Guard Security covers guard and patrol services. In the course of their duties guards and patrols detect criminal activity and apprehend offenders, but deterrence is their strength. For problems such as stock loss, pilferage, and industrial sabotage, investigation services can be used.

Physical Security involves the protection of an asset through the erection of physical barriers, such as locks, bars, and grilles. Safes range from small business and home safes to treasury safes. The type depends on the value of the risk; however, the objective is to delay the safecracker, as he does not have unlimited time to complete his task. Records and files can be protected by the use of fire-resistant cabinets and data cabinets (which protect computer programs). These physical barriers protect against fire *and* theft.

Electronic Security is a major growth industry. There are four main areas where the intruder is vulnerable: the perimeter, the break-in point, the interior, and the target itself.

The *perimeter* can be protected by photoelectric beams, or vibrator/inertia sensors on fences can warn if someone is trying to climb over or cut through the barrier. Microwave, electric field, and the new buried line (leak cable) sensors detect intrusion and can inform security control. Alternatively, CCTV (closed circuit television) or strong lighting may be most appropriate to a particular security need.

Potential *break-in points* can be protected by glass break detectors that create an alarm when subjected to the frequency of breaking glass. Circuits on doors and windows, designed to break in the event of a break in, are most effective. Examples of these are reed switches and window tape.

The *interior* can be monitored with ultrasonic, microwave, photoelectric, passive infrared, or CCTV equipment. In the unlikely event of an intruder arriving at his or her *final destination*, seismic and break-glass detectors can be built into safes, filing cabinets, and display cases.

The components then form the alarm system through their connection to an alarm panel. The alarm system may be local; upon receiving an alarm condition an audible signal is sounded. The system may be hooked to a telephone dialer that sends a prerecorded message to any one of a number of other premises. For high security, the system may be hooked by a direct line to a central station attended 24 hours each day, which constantly monitors the alarm.

Access control monitors and reports on the movements of people, records who they are and when they arrived and left, which sections of the premises they visited, and then warns of unauthorized attempts to enter high-security areas. This is done through coded cards and card readers positioned on doors, parking gates, elevators, and machinery. A plant can be monitored 24 hours a day through a central station.

Company Policy has a security function. Important information can be lost through speeches, reports, conversations, and disgrunted employees. Documents can be protected partially by safes and filing cabinets; but security policy also involves reducing negligence through employee awareness. Internal theft can be reduced significantly by stock control and effective access control to goods. Effective theft control also requires a knowledge of employee problems and discontent.

Computer Security combines various types of security to reduce theft, fraud, or manipulation of computer hardware and software. The risk is reduced by the protection of disks and tapes by access control to computer facilities through locks, gates, or guards and by microprocessor-based access control. However, computer security also involves strict software controls, such as passwords built into programs and records kept of all duplication processes. Employee use of computer facilities should be recorded, and backup copies of programs should be made and stored at different locations. All obsolete programs and computer printouts should be destroyed.

See also FIRE AND SMOKE ALARMS.

References

1. *The Science of Security*. Wormald Security Group, Sydney, 1980.
2. S. STROUSS: *Security Problems in a Modern Society*. Butterworth, London, 1980.
3. J. M. CORROLL: *Computer Security*. Security World Publishing Co., Los Angeles, 1977.
4. C. A. SENNEWALD: *Effective Security Management*. Security World Publishing Co., Los Angeles, 1978.

R.E.G.

SEWAGE DISPOSAL is essential to preservation of the health and amenity of buildings and their environment. Sewage from residential developments comprises *sullage* (wastes from kitchens, laundries, washbasins, baths, and showers) and *soil wastes* (contaminated wastes from water closets, urinals, and slop sinks) (see SANITARY FITTINGS). The increasing amount and diversity of wastes from industrial and commercial premises have led to use of the term *wastewater*. Disposal of wastewater may be to public sewers, streams, or other surface waters or onto the land.

Discharge to Municipal Sewers is the simplest disposal method for domestic and other wastewaters (see DRAINAGE INSTALLATIONS). Industrial wastewaters often require pretreatment before they can be permitted to enter municipal sewers. Where the discharge point on the property is either too low or too remote for discharge to gravitate, it is often feasible to pump to the sewer.

Modern cities usually have *separate* foul sewers for collecting wastewater. Stormwater from roofs, roads, and paved areas is then collected in a larger storm drainage system. Even so it is not possible to completely prevent stormwater from entering separate foul sewers, and wet weather flow usually controls their design capacity. Many older cities have *combined* sewer systems, in which both wastewater and stormwater are collected in one large system. This avoids the need to keep stormwater separate from the wastewater, but makes treatment and pollution control much more difficult.

Treatment. Discharge to surface waters, groundwaters, or onto land requires that the wastewater be treated to a suitable standard, which depends on the locality, the beneficial uses of the receiving area, and the volume and characteristics of the wastewater. A common inland municipal treatment system (Fig. 225) comprises primary, secondary, and tertiary stages with separate treatment of sludges and other by-products. In primary (physical) treatment, coarse objects are screened out, fine, heavy grit is removed, and much of the fecal matter is settled out. Secondary treatment converts the nonsettlable organic material into biological solids that settle out in secondary sedimentation. Tertiary treatment performs the two functions of further polishing, by filtering out most of the remaining suspended matter, and disinfection, usually by chlorination. Screenings and grit removed in the process are usually buried. The organic sludges from primary and secondary sedimentation are stabilized by anaerobic (or aerobic) digestion and lagooning and finally dewatered and disposed to land.

By this means over 95% of the primary pollutants, biochemical oxygen demand and suspended solids, as well as enteric bacteria, are removed. Additional treatment is necessary if the plant nutrients nitrogen and phosphorus are to be removed.

For small communities, isolated residences, and industrial establishments requiring on-site disposal, simplified treatment systems such as package and factory-built modular treatment units are often used. Such systems usually comprise screens, some form of aerobic biological stage with secondary sedimentation, disinfection, and aerobic digestion of the biological sludge (Fig. 226).

Where less sophisticated treatment is required and soils are suitable for subsurface disposal, septic tanks are an economical means of disposal (see SANITATION IN DEVELOPING COUNTRIES).

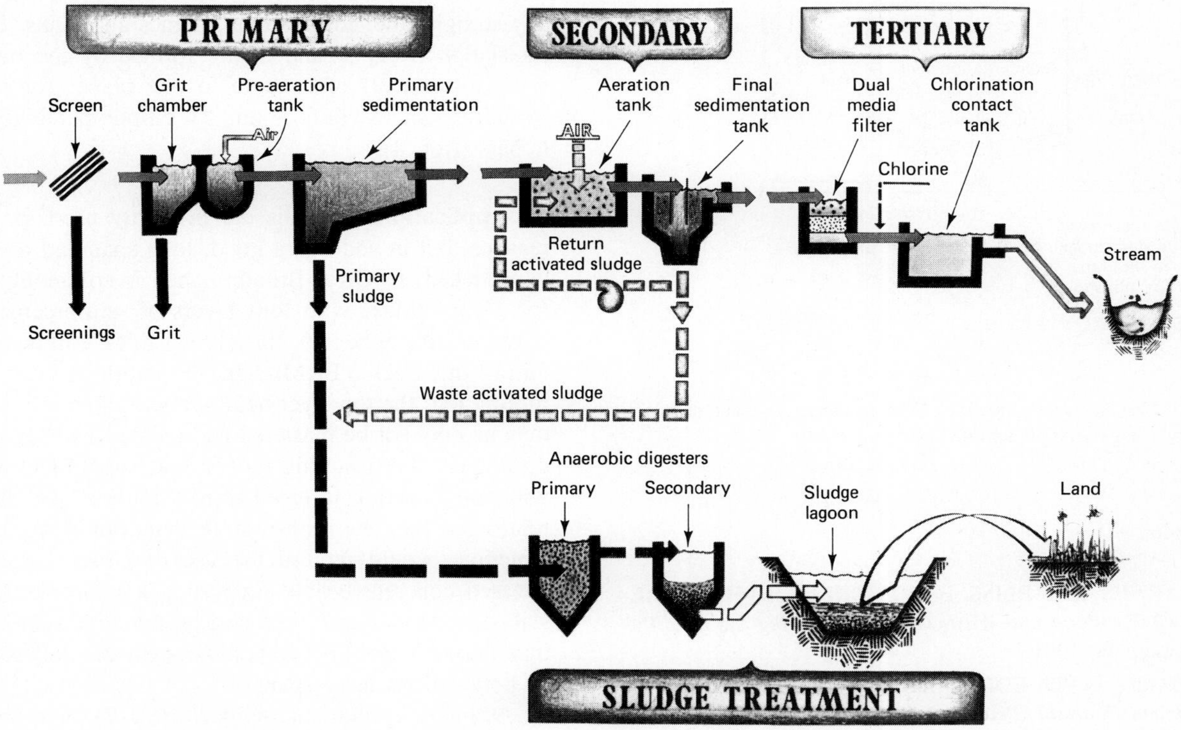

Figure 225 Wastewater treatment plant based on the activated sludge process.

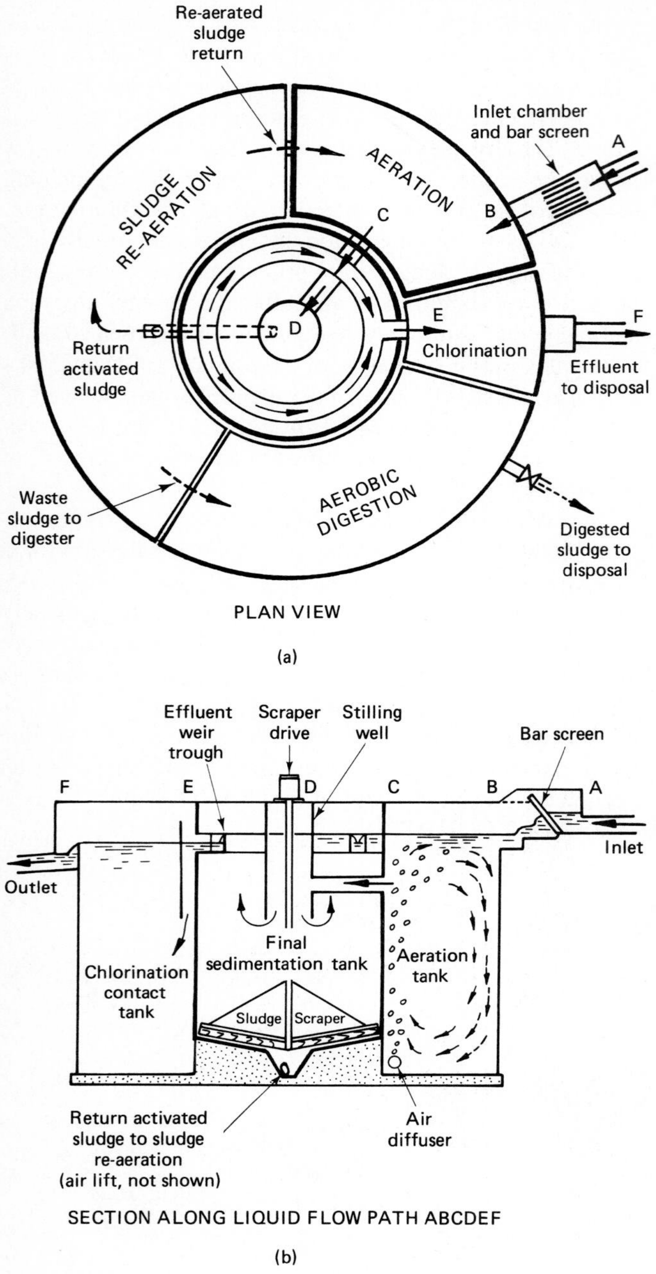

Figure 226 Schematic drawing of package treatment plant based on a two-stage aeration activated sludge system.

References

1. D. BARNES, P. J. BLISS, B. W. GOULD, and H. R. VALLENTINE: *Water and Wastewater Engineering Systems*. Pitman, London, 1981.
2. METCALF AND EDDY, INC: *Wastewater Engineering Treatment Disposal, Reuse*. McGraw-Hill, New York, 1979.

P.J.B.

SHELL STRUCTURES are thin, curved slabs of reinforced concrete. Because of their curvature they can, for the same thickness, span farther than REINFORCED CONCRETE SLABS or FOLDED PLATES.

Designation of Shells. Shells may be classified from various points of view (Ref. 1); the most important are (1) shells of single or double curvature and (2) shells of translation or rotation.

Gaussian curvature is generally used for classification (1) (Ref. 3). Curvature is the reciprocal of the radius of curvature for two-dimensional structures. For three-dimensional shells, the Gaussian curvature at any point is the product of the principal curvatures (Ref. 3), that is, the curvatures on two planes at right angles to one another.

Shells of positive Gaussian curvature (Refs. 1 and 2) are called *synclastic shells*; the surface curves away from a tangent plane at any point on the surface and lies completely on one side of this plane. Spherical domes and elliptic paraboloids are of this type. Shells of negative Gaussian curvature (Refs. 1 and 2) are called *anticlastic shells*. They are formed by two families of curves, each curved in opposite directions. Hyperbolic paraboloids (hypars for short), conoidal shells (conoids), and hyperbolas of revolution are of this type (Refs. 3 and 4). Shells of zero Gaussian curvature are singly curved shells (Refs. 1 and 2); cylinders and cones are of this type.

Classification (2) is also widely used. Shells of rotation (Refs. 1 and 2) are formed by the rotation of a curve about any straight line, for example, domes and tanks. Shells of translation (Refs. 1 and 2) are formed by the translation (i.e., movement) of a curve in one plane, for example, cylindrical shells (Refs. 4 and 5), elliptic paraboloids, and hypars (Ref. 4).

Applications. Shells are primarily used for esthetic reasons; but in addition a good, long span and robust roof is produced. In Great Britain a shell is commonly 65 mm (or $2\frac{1}{2}$ in.) thick, with four layers of reinforcement. It is covered with either 25 mm (1 in.) thick corkboard or 50 mm (2 in.) thick VERMICULITE and three-layer built-up roofing felt, the top layer being mineral finished. The soffit may or may not be plastered and painted with emulsion or distemper. The resulting roof is thus superior to one built with roof sheeting fastened to roof purlins. An alternative reinforced concrete roof of large span, not using shell construction, would require the use of closely spaced prestressed concrete beams supporting a reinforced concrete slab. Its overall depth is much less and its weight is greater than that of a shell; hence both strength and deflection limit its potential for large spans.

Ordinary cylindrical shells (Fig. 227) have been used where large reinforced concrete spans were required for factories and bus stations. They have also been used for

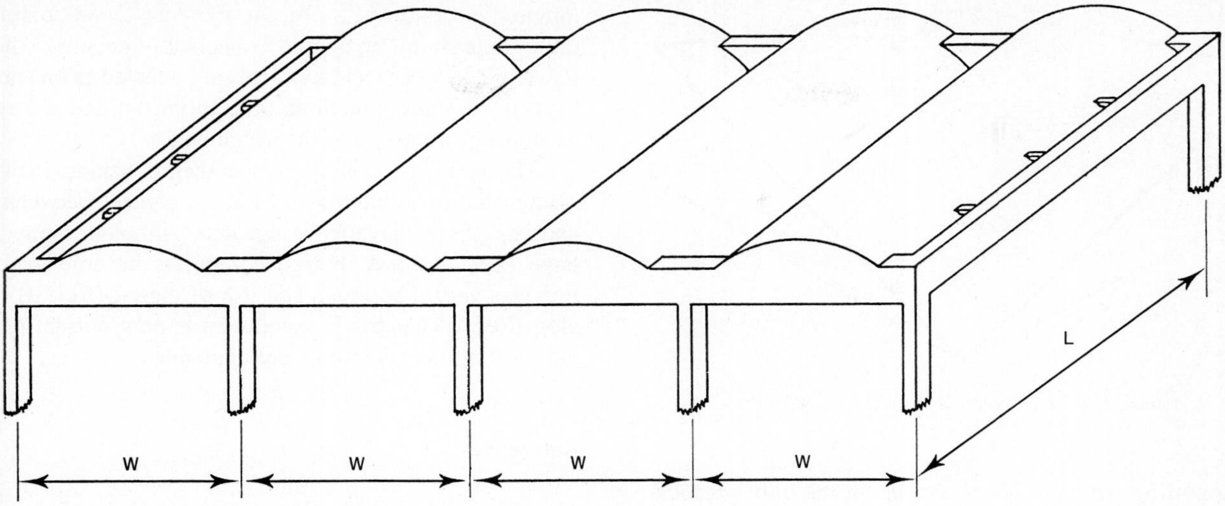

Figure 227 Cylindrical shells (barrel vaults). W = width of shells; L span of shells.

offices, schools, and colleges, even though shorter spans would have been satisfactory.

North-light (south-light in the southern hemisphere) cylindrical shells give good natural lighting. They have been built over colleges, schools, waterworks, and factories.

Conoids (Fig. 228) give a good distribution of light. They have been used as an alternative to north-light cylindrical shells. They are often more attractive, use less materials, and are thus lighter in weight; but they are more expensive for formwork and steel fixing and slightly more expensive for concreting.

DOMES have been used extensively over circular tanks for sewage, water, and chemical waste. Domes, not necessarily spherical, have been used over many prestigious buildings, such as exhibition halls, temples, mosques, and synagogues. Before World War II they were used for major libraries, but the dome shape creates ACOUSTIC problems.

Cylindrical shells are economical because the formwork is straight in one direction and conforms to the simplest curve, the circle, in the other direction.

Hypars (Fig. 229) have the advantages of doubly curved shells and yet can be formed with straight boards. Even though the straight boards need to be tapered because of the doubly curved geometry, they are possibly the easiest of the doubly curved shells to construct. They have been used for many esthetically attractive structures as roofs over factories, garages, markets, churches, and exhibition halls.

Economics. Many different types of construction can be used for roofs. If they are compared only on first cost, the type of shell satisfactory in Great Britain would never be less expensive than competing forms of construction. Corrugated sheeting with a fiberboard lining for insulation would be much cheaper than a 65 mm ($2\frac{1}{2}$ in.) thick shell. It would also be lighter in weight, helping the economy of

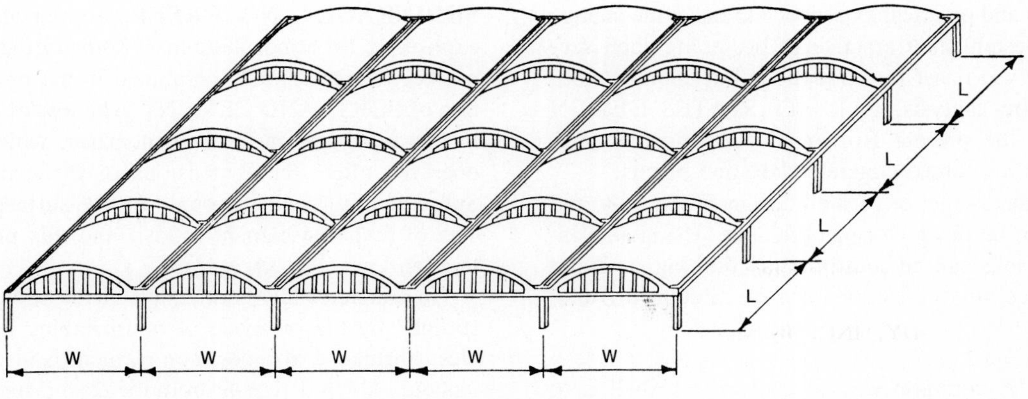

Figure 228 Conoid shells. W = Width of shells; L = span of shells.

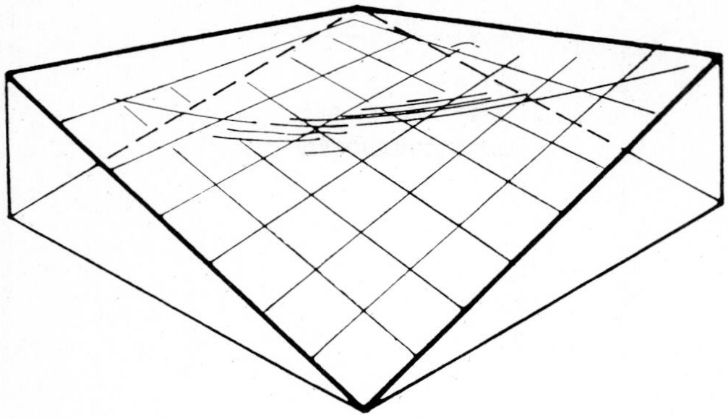

Figure 229 Hyperbolic paraboloid (hypar) shell.

the supporting structure. First cost is not the only practical consideration; the life of the roof and its maintenance cost are also important.

The difficulty of generalizing about British and American costs becomes worse internationally as it is complicated by a different family of less expensive shells being used in some countries. These are not in total construction as watertight (maintenance is effected when necessary) and as resistant to heat transmission, fire, and localized forces and minor knocks from, for example, maintenance personnel. These shells are often much thinner than the 65 mm ($2\frac{1}{2}$ in.) minimum thickness generally used in Great Britain. Practice in the United States is similar to that in Great Britain. In several countries hypars have been constructed only 30 mm ($1\frac{1}{4}$ in.) thick and without finishes such as roofing felt, thermal insulation, or plaster.

References 6 and 7 give examples of the costs of shells in the United States. In many underdeveloped countries, labor is less expensive relative to materials than in the more advanced countries. This point is made in Ref. 8 with regard to India.

Design. To design a shell, it is first necessary to assess its dimensions from past experience. It is then analyzed elastically for internal forces and moments. There is insufficient research and practical experience to make the change from the long-established practice of designing shell sections elastically, to resist forces and bending moments obtained by elastic analysis, to LIMIT STATES DESIGN conforming to the present British and American recommendations for reinforced concrete slabs and beams.

The established practice of shell design (Refs. 3, 4, and 5) has proved to be satisfactory and reliable. Using an elastic analysis, cracks can be controlled by limiting steel and concrete stresses, and deflections can be calculated (Refs. 3, 4, and 5).

Analysis. A common way of analyzing a shell is to consider an infinitesimally small element of it with all the longitudinal and shear forces, moments, and torsions in three dimensions that could conceivably occur on the element. These forces and moments are equated to one another by general static equations of equilibrium and are related to strains, stresses, and displacements.

The boundary conditions are then considered; the displacements of points on the line of contact between shell and edge beam have to be equated. Although many shells have been designed before computers became available, this was very laborious. The use of digital (Ref. 4) or analog (Ref. 5) electronic computers is now normal practice when computing facilities are available.

References

1. J. N. BHARUCHA; Proposed classification of shell roofs. *Indian Concrete Journal*, Vol. 33, No. 12 (Dec. 1959), pp. 419–421.
2. D. P. BILLINGTON: *Thin Shell Concrete Structures*. McGraw-Hill, New York, 1965.
3. C. B. WILBY and M. M. NAQVI: *Reinforced Concrete Conoidal Shell Roofs*. Cement and Concrete Association, Eyre and Spottiswoode, London, 1973.
4. C. B. WILBY and I. KHWAJA: *Concrete Shell Roofs*. Applied Science Publishers, London, 1977.
5. C. B. WILBY: *Design Graphs for Concrete Shell Roofs*. Applied Science Publishers, London, 1980.
6. C. S. WHITNEY: Economics, in *Conference on Thin Concrete Shells*. Massachusetts Institute of Technology, Cambridge, Mass., 1954, pp. 22–24.
7. C. S. WHITNEY: Cost of long-span concrete shell roofs. *Proceedings of the American Concrete Institute*, Vol. 21, No. 10 (June 1950), pp. 765–776.
8. B. K. CHATTERJEE: *Theory and Design of Concrete Shells*. Edward Arnold, London, 1971.

C.B.W.

SHRINKAGE AND CREEP. Hardened CONCRETE exposed to the atmosphere loses some of the water that has not become chemically combined in the process of hydration of PORTLAND CEMENT. The loss of free water held in the capillaries in the cement paste, which occurs first, does not affect the dimensions of the concrete element. Subsequently, water adsorbed on the surfaces of the particles of hydrated cement is lost, and this produces a contraction known as *shrinkage*.

In concrete, this contraction of the cement paste is restrained by the presence of nonshrinking aggregate particles. Shrinkage of a concrete element is also restrained by its core, which is remote from the drying surface and therefore unable to dry out. It follows that shrinkage of small laboratory specimens is larger than in the case of large sized elements.

For typical concrete mixes, shrinkage of laboratory-sized specimens is between 300×10^{-6} and 900×10^{-6} (on a linear basis) when the relative humidity of the air is 50%. About one-half of the shrinkage occurs in 1 month and most of it in 9 months, except in massive elements.

Shrinkage occurs in concrete regardless of whether or not it is subjected to load. When a sustained load acts, there is a gradual increase in strain with time. This is called *creep*. Creep is assumed to be independent of shrinkage, so the two strains can be considered to be additive, but it is affected by the process of drying: creep of drying concrete is significantly greater than creep of concrete in hygral equilibrium with the air. If follows that creep of large elements (where the core is unable to dry) is smaller than in the case of laboratory specimens.

It can be seen that creep is related to the movement of adsorbed water in hydrated cement paste, but is not a simple consequence of evaporation of water; indeed, even concrete immersed in water, provided it is subjected to a sustained stress, undergoes creep. As in the case of shrinkage, aggregate restrains the creep of hydrated cement paste.

Creep takes place over many years, but does so at a decreasing rate, as shown in Fig. 230. Typically, creep is two to four times the elastic deformation.

The structural consequences of shrinkage and creep in concrete structures can be large. Restrained shrinkage can cause cracking unless appropriate reinforcement is provided. Creep rarely leads to failure (except in large concrete masses subject to a thermal cycle), but it increases the deflection of horizontal members and causes loss of PRESTRESS.

Shrinkage and creep occur also in TIMBER. Shrinkage is caused by drying of the timber and creep by a sustained load. The phenomena are similar to those occurring in concrete, but they cause fewer problems if timber is seasoned, that is, dried, before use and if joints in timber floors permit some temperature and moisture movement.

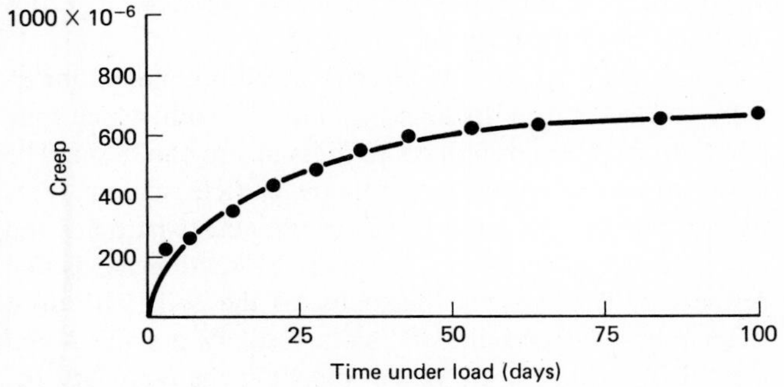

Figure 230 Development of creep with time for a concrete specimen (water/cement ratio, 0.57; aggregate/cement ratio, 6.7; strength of concrete, 35 MPa; applied stress, 12 MPa).

See also RHEOLOGY.

References

1. A. M. NEVILLE: *Properties of Concrete*. Pitman, London, 1981.
2. A. M. NEVILLE, W. H. DILGER, and J. J. BROOKS: *Creep of Plain and Structural Concrete*. Construction Press, London, 1983.

A.N.

SMOKE AND TOXIC GASES cause the majority of deaths in fires. *Smoke* is defined as "the airborne solid and liquid particulates and gases evolved when a material undergoes pyrolysis or combustion." The term *toxic gases* is often used to refer to the components in the smoke that adversely affect body functions.

The effects of smoke are several: (1) the particulates reduce visibility and thus impede a person's escape from fire; (2) certain chemical species in the fire gases, even at very low concentrations, may irritate the eyes, skin, and respiratory tract; and (3) inhalation of acutely toxic gases may cause narcosis and poisoning, resulting in incapacitation and death.

Smoke and toxic gases are not confined to the immediate vicinity of the fire, particularly in a building with mechanized ventilation. Soon after their formation, they move along the ceilings of rooms and corridors and to upper floors of tall buildings. Upon cooling, they become more evenly distributed, increasing the chance that a victim may be exposed. Thus danger extends to areas remote from the fire source.

Practically all fires, no matter what material is burning, produce smoke and toxic gases. The number and size of particles produced determine the light obscuration, which, with the composition and concentrations of the toxic gases, changes with the material and the conditions of burning. As a general rule, any burning material containing carbon, such as wood, emits carbon monoxide and carbon dioxide, among many other gases. A material containing nitrogen as well as carbon, such as wool or nylon, also gives off hydrogen cyanide and oxides of nitrogen.

Building design can reduce the perils presented by smoke and toxic gases. Products of combustion may be confined to an area isolated from the rest of the building. Pressurized corridors and stairwells provide smoke-free zones and thus facilitate escape from fire. High ceilings, as in atria, may allow accommodation of a large volume of smoke and hot gases. Roof vents and high-capacity fans may extract the smoke.

Selection of materials based on their fire properties is another possible control measure. Among these properties are the propensities to generate smoke and toxic gases. It should be noted that consideration of these in isolation from other fire properties may not always be appropriate.

The smoke-producing propensity of a material is usually determined on the basis of light obscuration or the weight of the smoke produced. Fire gas toxicity could be evaluated by two methods. In the first, the toxic gases are chemically analyzed and the peril is estimated from the known toxicity of each component. Alternatively, the integrated toxic effect of fire gases on test animals is observed.

See also EVACUATION OF BUILDINGS IN CASE OF FIRE and FIRE AND SMOKE DETECTORS.

References

1. C. F. CULLIS and M. M. HIRSCHLER: *The Combustion of Organic Polymers*. Clarendon Press, Oxford, 1981.
2. H. L. KAPLAN, A. F. GRAND, and G. E. HARTZELL: *Combustion Toxicology: Principles and Test Methods*. Technomic Publishing Co., Lancaster, Pa., 1983.
3. *Fire Protection Handbook*, 15th ed. National Fire Protection Association (U.S.A.), Boston, 1983.

K.S. and F.R.S.C.

SNOW LOADS

Snow Accumulation. Snow is a major consideration in the design of roofs in cold climates (Ref. 1). Its distribution depends on the size, shape, roughness, color, and heat loss of the roof, the location of the building in relation to other structures, trees, and hills, and on the weather, which varies from winter to winter and with location, elevation, and exposure to sun and wind.

The wind has a large effect on the distribution of snow on roofs. In its absence, snow accumulates uniformly and is redistributed only by drainage of meltwater and by sliding or creep. Wind moves snow in three ways: (1) falling snow is carried along in the flow before landing, (2) snow particles are rolled and bounced along the roof surfaces (saltation), and (3) snow is whipped up from the roofs to great heights (turbulent mixing). Snow accumulates in drifts at changes in roof levels, at parapets, in roof valleys, on the lee side of peaked or arched roofs, and at other obstructions that retard wind speed below that sustaining transport. Designers therefore prudently consider (1) uniformly distributed loads (reduced in exposed locations), (2) drift loads, and (3) other unbalanced loads due to sliding and the natural irregularity in the snow cover.

Design Standards. In building codes (Refs. 2, 3, and 4) roof loads are obtained by multiplying statistically derived ground snow loads by a series of factors accounting for roof shape, slope, and roughness, heat loss, exposure to wind, and the specific gravity of roof snow, which varies from about 0.2 to 0.45, depending on the climate. Code and standards writing bodies use their experience and judgment to evaluate data from measurements on roofs (Refs. 1 and 5) and to derive these factors for common roof shapes.

For unusual or complex roofs, some codes allow novel approaches, such as analysis of hourly or daily climatic data and wind tunnel or water flume modeling of snow drifting, to aid in selecting design snow loads (Ref. 6).

References

1. D. A. TAYLOR: Roof Snow Loads in Canada. *Canadian Journal of Civil Engineering*, Vol. 7, No. 1 (1980), pp. 1–18.
2. *Minimum Design Loads for Buildings and Other Structures*. American National Standards ANSI A58.1—1982. American National Standards Institute, New York, 100 p.
3. *Bases for Design of Structures—Determination of Snow Loads on Roofs*, ISO 4355—1981(E). International Organization for Standardization, 1981.
4. *Snow Loads*. Commentary H of the Supplement to the National Building Code of Canada, NRCC 23178. National Research Council Canada, Ottawa, 1985.
5. M. J. O'ROURKE, R. REDFIELD, and P. VON BRADSKY: Uniform Snow Loads on Structures. *Journal of the Structural Division, Proceedings of the American Society of Civil Engineers*, Vol. 108, ST12 (1982), pp. 2781–2798.
6. N. ISYUMOV and A. G. DAVENPORT: A Probabilistic Approach to the Prediction of Snow Loads. *Canadian Journal of Civil Engineering*, Vol. 1, No. 1 (1974), pp. 28–49.

D.A.T.

SOIL CLASSIFICATION consists of dividing soils into groups such that all soils in a particular group have similar characteristics and exhibit similar engineering behavior. A classification system therefore provides a means by which information regarding the significant engineering properties of a soil may be conveyed.

The tests performed to classify a soil are almost invariably carried out on disturbed samples of soil, which may not accurately represent the soil in its in situ condition. The most frequently performed tests are particle size analyses (which may include a sieve analysis for coarse particles and a hydrometer analysis for finer particles) and analyses that define in some way the plasticity of the soil. The most frequently used measures of soil plasticity are the *liquid limit* (LL), which is the water content at the boundary between the liquid and plastic states of the soil, and the *plastic limit* (PL), which is the water content at the boundary between the plastic and semisolid states of the soil (see also

RHEOLOGY). Soils that are highly plastic generally have a relatively high liquid limit (> 50%) and a relatively high *plasticity index:* PI = LL − PL.

There exist a number of soil classification systems, some of which depend only on particle size and some of which require specification both of particle size and plasticity characteristics. The most widely used system for soil mechanics and foundation engineering applications is the *Unified Soil Classification System* (Ref. 1) in which a soil is described by two letters:

Primary letter	*Secondary letter*
G: gravel	W: well graded
S: sand	P: poorly graded
M: silt	M: nonplastic fines
C: clay	C: plastic fines
O: organic	L: low plasticity
Pt: peat	H: high plasticity

Coarse-grained soils (gravel, sands) are classified by means of their grading curves (which describe the particle size distribution characteristics), while finer-grained soils (silts, clays, organic soils) require that both the grading and plasticity characteristics be determined.

From the viewpoint of foundation performance, the soils most likely to cause problems are silts, clays, or organic soils of high plasticity (MH, CH, and OH classifications) and peat (Pt). A comprehensive summary of the engineering properties of the various soil groups is available in Ref. 2.

It is frequently useful to extend the classification of a soil by describing its geological origin, using such terms as residual soils and transported soils. The latter can be further subclassified into alluvial, marine, glacial, aeolian, and lacustrine soils.

References

1. A. A. WAGNER: *The Use of the Unified Soil Classification System by the Bureau of Reclamation.* Proceedings of the 4th International Conference on Soil Mechanics and Foundation Engineering, London, Vol. 1, Butterworths, London, 1957.
2. A. CASAGRANDE: Classification and Identification of Soils. *Transactions of the American Society of Civil Engineers*, Vol. 113 (1948), p. 901.

H.G.P.

SOIL MECHANICS may be defined as the application of the laws of mechanics and fluid flow to engineering problems relating to soils. Soils here are considered to be the sediments and deposits of solid particles produced by the mechanical and chemical disintegration of rocks. The major facets of soil mechanics are as follows:

1. Theoretical soil mechanics, which involves the development of theories of behavior of soils under stress, often based on extremely simplifying assumptions.
2. Experimental soil mechanics, which is the investigation of the physical properties of real soils.
3. Applied soil mechanics, which is the application of theoretical, experimental, and empirical knowledge of soil behavior to the solution of practical problems.

History. Little attempt appears to have been made to solve soil problems on a scientific basis until toward the end of the 18th century, when the French scientist Coulomb published a law of failure for soils. Almost one century later, the British engineer Rankine developed his theory of earth pressure, and these two classical theories still form the basis of practical methods of computing earth pressures.

The problem of slope stability was first explored by Alexandre Collin in 1846 and subsequently by the Swedish engineer Fellenius in 1911, and the latter's analysis remains one of the more widely used analysis methods. The modern era of soil mechanics began with the work of Karl Terzaghi, who published the book *Erdbaumechanik* in 1925. Terzaghi developed the principle of effective stress in soils as the cornerstone of soil mechanics and formalized the analysis of many soil engineering problems. His textbook *Theoretical Soil Mechanics* (Ref. 1), had an enormous influence on succeeding generations of engineers.

In 1936, the first International Conference on Soil Mechanics and Foundation Engineering was held in Boston, and following World War II these international conferences have been held every four years or so. Four-yearly regional conferences have also been established in six regions of the world. The development of powerful numerical techniques and sophisticated testing procedures, in conjunction with the use of computers, has led to a major expansion of knowledge and analytical capability, but the heterogeneous and time-dependent nature of real soils still poses major problems in the modeling of soil behavior and the application of soil mechanics theories to practical problems.

Principle of Effective Stress. The principle of effective stress has proved vital to the solution of practical problems in soil mechanics and the understanding of the influence of pore water on the deformation and strength of soils. In a soil whose pores are wholly filled with water (i.e., a saturated soil), the effective stress in any direction, σ', is defined as the difference between the total stress in that direction, σ, and the pressure in the pore water (the pore pressure) u; that is, $\sigma' = \sigma - u$. The effective stress represents the stress carried by the soil particles or the *soil skeleton* and has sometimes been termed the *intergranular stress*, although this description is not strictly accurate. In partially saturated soils whose voids contain air as well as water, the principle of effective stress is modified to allow for the pressure in the pore air as well as the pore water pressure (Ref. 3).

The deformability and strength of a soil are dependent on the effective stress, and the principle of effective stress can be stated in terms of the following two hypotheses (Ref. 2):

1. Deformation or strain within an element of a soil mass can only arise if there is a change in effective stress in the element.
2. The shear strength of a soil on any plane depends on the effective stress acting on that plane:

$$\tau_f = c' + \sigma' \tan \phi'$$

where τ_f = shear strength
σ' = effective stress on the plane considered
c' = cohesion
ϕ' = angle of shearing resistance

c' and ϕ' are often termed the shear strength parameters of a soil and may be determined from appropriate field or laboratory tests on the soil. These parameters depend largely on the soil type, but will also depend on the conditions of loading and the previous stress history of the soil.

Components of Deformation of Soils. The principle of effective stress has enabled the time-dependent nature of deformations in clay soils to be understood and the components of deformation to be isolated. If a foundation on a clay soil is loaded relatively rapidly, the total stress applied to the foundation is distributed through the soil mass. At any point in the soil mass, the change in total stress causes an excess pressure to be developed in the pore water. The increase in pore pressure usually lies between the increases in vertical and horizontal stresses; hence the vertical effective stress increases and the horizontal effective stress decreases, leading to distortion of the soil without volume change. The integrated effect of these deformations gives rise to a deformation of the foundation, termed the *immediate* deformation. With the passage of time, the excess pore pressures gradually reduce to zero (at a rate depending largely on the permeability of the soil); consequently, the increased total stress gradually transfers to the soil skeleton. There is therefore a gradual (time-dependent) additional deformation due to this increase in effective stress, the *consolidation* deformation. The final deformation ρ_f is the sum of the immediate deformation ρ_i and the consolidation deformation ρ_c:

$$\rho_t = \rho_i + \rho_c$$

Figure 231 illustrates the problem of a vertically loaded foundation and the progress of settlement with time subsequent to loading. In many soils, a further time-dependent component of deformation, due to soil creep, may also occur, but this is often small relative to the immediate and consolidation settlement components.

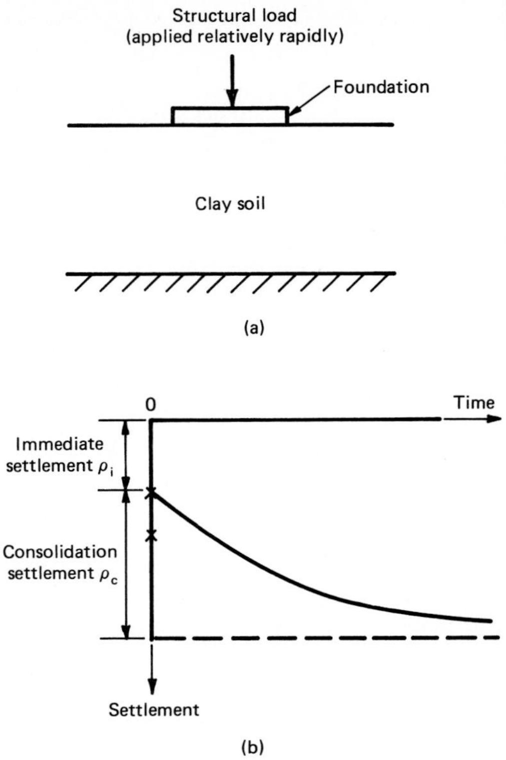

Figure 231 Time-dependent deformation. (a) Loaded foundation on clay soil. (b) Time–settlement relationship for foundation.

Application of Soil Mechanics. The principles of soil mechanics are generally used to solve problems in one or more of the following categories:

1. Stability of soils or FOUNDATIONS; such problems require a knowledge of the shear strength properties of soils.
2. Deformations of soil masses or foundations; these problems require a knowledge of the compressibility of soils.
3. Flow of fluids through soil; such problems require a knowledge of the permeability of soils.

Within the first category (stability problems) are those involving the bearing capacity of foundations (the load applied to a foundation that will just cause it to collapse or fail), the stability of slopes (e.g., the determination of the height or inclination of a slope to cause failure of the slope), the stability of excavations, and the earth pressures against retaining walls. Among deformation problems are the determination of the vertical movement (settlement) of foundations under load, the settlement and horizontal movement

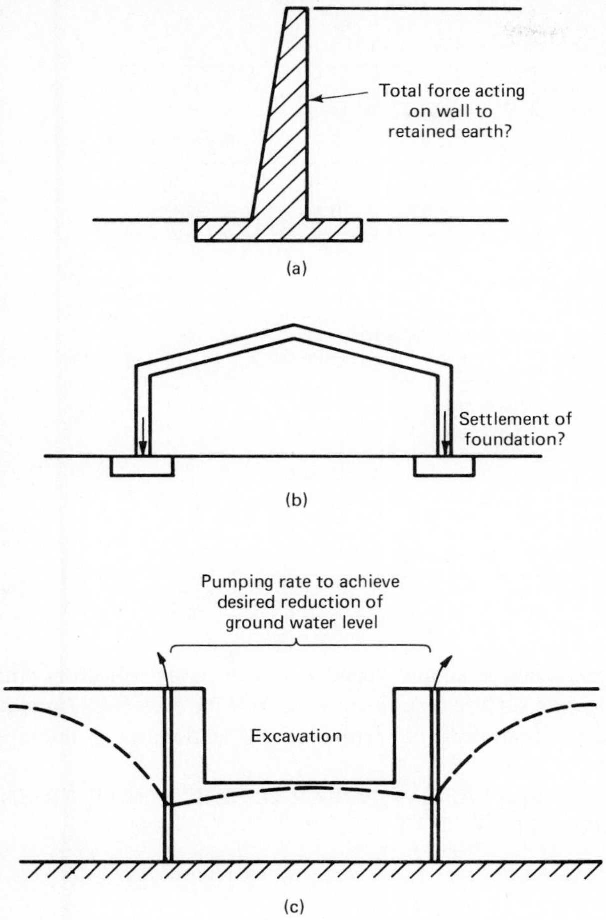

Figure 232 Soil mechanics problems. (a) Stability problem: retaining wall. (b) Deformation problem: settlement of building. (c) Flow problem: dewatering around excavation.

below an embankment on soft clay, and the horizontal movements around an excavation. Fluid flow problems include the determination of the quantity of seepage through and beneath an earth dam and the pumping requirements for a dewatering system in order to lower the groundwater tables in the vicinity of an excavation. Some of these problems are illustrated in Fig. 232.

References

1. K. TERZAGHI: *Theoretical Soil Mechanics*. Wiley, New York, 1943.
2. N. E. SIMONS and B. K. MENZIES: *A Short Course in Foundation Engineering*. Newnes-Butterworths, London, 1977.
3. A. W. BISHOP, I. ALPAN, G. E. BLIGHT, and I. B. DONALD: *Factors Controlling the Strength of Partially Saturated Cohesive Soils*. Proceedings of the Conference on Shear Strength of Cohesive Soils, Boulder, Colo., American Society of Civil Engineers, New York, 1960.

H.G.P.

SOLAR COLLECTORS FOR SOLAR THERMAL SYSTEMS. All collectors designed to produce a thermal (or heat) output from a solar radiation input consist of components with specific functions. The primary element, called the *absorber* and generally made of metal, traps the maximum amount of sunlight incident by ensuring that the surface finish has a high solar absorptance. It uses a matt black-paint surface or one of the many solar selective absorbing surfaces.

The second essential component of any collector is a well-insulated box structure designed to minimize the heat loss from the hot absorber element. These boxes can later be coupled together to form a continuous roof structure on a building, or they can be simply laid on to an existing roof structure.

As all collectors have to allow solar radiation into the absorber, any cover element to minimize top heat loss from the collector must be transparent to radiation in the solar spectrum from 0.3 to 3.0 μm; but it should preferably be totally opaque to long-wave (or infrared) radiation beyond 3.0 μm. A cover allows use to be made of the greenhouse effect observable in horicultural glasshouses. A wide range of specially designed low-iron glasses is available to give high solar transmittance and zero infrared transmittance. In addition, many transparent plastics can be used for covers. Caution should be exercised in the choice of the plastic, as many suffer badly from ultraviolet radiation degradation, both optically and mechanically; however, they are lighter than glass (Table 1).

The final essential element of any collector is the provision of fluid-flow passages in good thermal contact with the absorber, to extract the heat from the collector with a fluid such as water, water/glycol mixture, heat transfer oil, refrigerant fluid, or air (Table 2).

Collectors can be classified according to the degree of concentration of the incoming solar radiation. The most common *flat-plate collectors* (Fig. 233) have a concentration ratio of unity, and any collectors used in conjunction with domestic systems are likely to have a concentration ratio of less than 10/1. This low concentration ratio means that the collectors do not need to be continually tracked, but may have to undergo daily, monthly, or seasonal adjustments to optimize their thermal performance.

Collectors of higher concentration ratio, such as those with parabolic reflectors used for higher-temperature applications, are not particularly useful for space heating or cooling applications and so are not likely to be used in dwellings. Such collectors require continuous and precise tracking of the sun to attain their optimal efficiency.

TABLE 1
Some Properties of Plastics and Glass Used in Solar Collectors

Property	Glass	Polymethyl Methacrylate (PMM)	Teflon FEP	Polyvinyl Fluoride (PVF)	Tedlar 20	Mylar PET	Polycarbonate
Specific gravity	2.50	1.19	2.15	1.40	1.38	1.39	1.20
Solar transmittance	0.86	0.90	0.98	0.94	0.92	0.85	0.88

TABLE 2
Some Properties of Heat Transfer Fluids at Atmospheric Pressure

Property	Water	Air	50% Propylene Glycol/ Water	Freon 12
Freezing point (°C)	0	—	−33	−155
Boiling point (°C)	100	—	110	−30
Specific heat (kJ/kg°C)	4.18	1.01	3.55	0.90

To minimize the heat loss from the absorber, it is possible to surround this element with a vacuum sheath, and this has led to the development of evacuated *tabular collectors* (Fig. 234). Such collectors, because of their low loss coefficient, have the ability to retain their operational effectiveness at higher temperatures and are likely to prove effective in conjunction with absorption chillers of the water/lithium bromide or ammonia/water type.

Collector characteristics are normally expressed in the form of collection efficiency versus operational temperature; it can be seen from Fig. 235 that the selection of a particular collector depends on the predicted load temperature. A simple unglazed flat-plate collector may be adequate for heating a low-temperature swimming pool or to supply energy at a low temperature to solar-assisted heat pump systems; but the higher temperature required for industrial processes demands more sophisticated collectors. These may be single-glazed low-iron cover collectors with selective surface absorbers, vacuum-tube collectors, or simple concentrating collectors of the stationary or tracking type.

See also PASSIVE SOLAR DESIGN and SOLAR ENERGY: ITS ACTIVE USE IN BUILDINGS.

References

1. J. A. DUFFIE and W. A. BECKMAN. *Solar Engineering of Thermal Processes*. Wiley, New York, 1980.
2. W. W. S. CHARTERS, Chapters 6 and 7 in A. A. M. SAYIGH (ed.): *Solar Energy Engineering*. Academic Press, New York, 1977.
3. W. W. S. CHARTERS and B. WINDOW: Solar collector

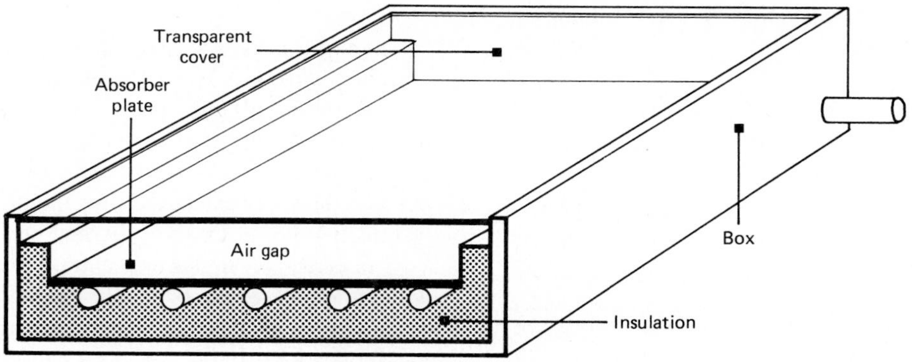

Figure 233 Schematic diagram of a standard fin and tube flat-plate collector with one transparent cover and insulation behind the absorber plate.

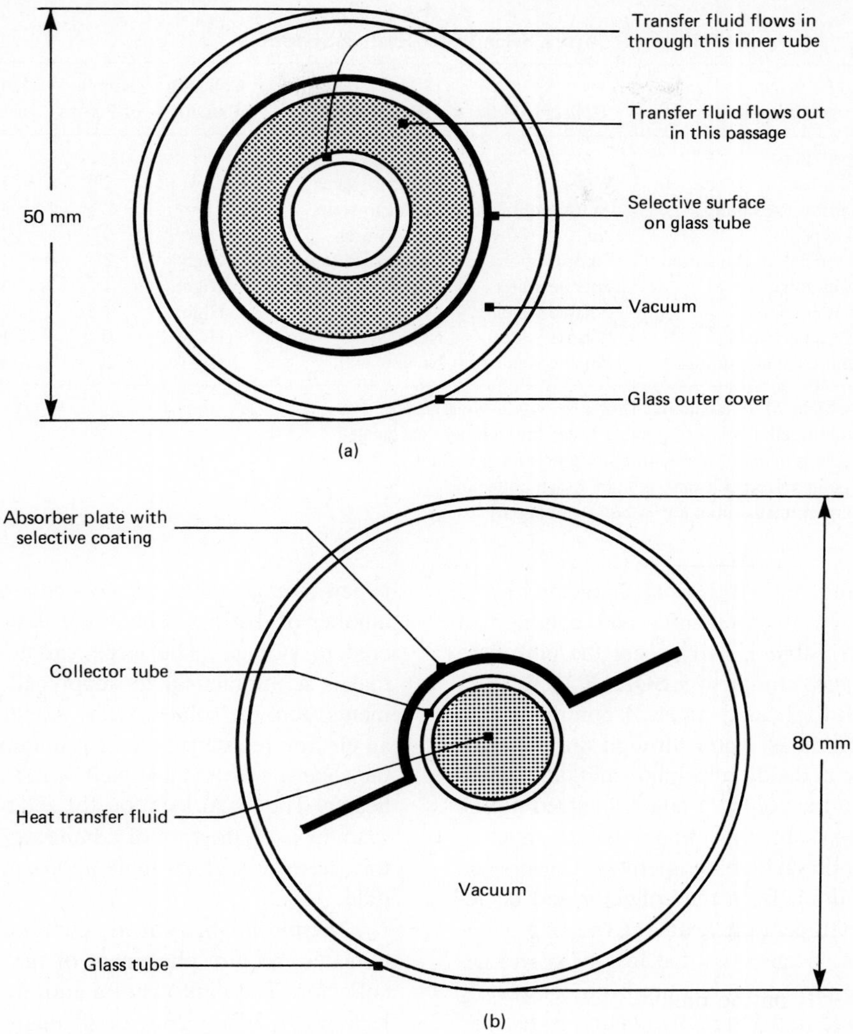

Figure 234 Categories of evacuated tube collectors. (a) Evacuated tubular collector using concentric glass tubes. (b) Evacuated tubular collector with absorber surface within single glass tube.

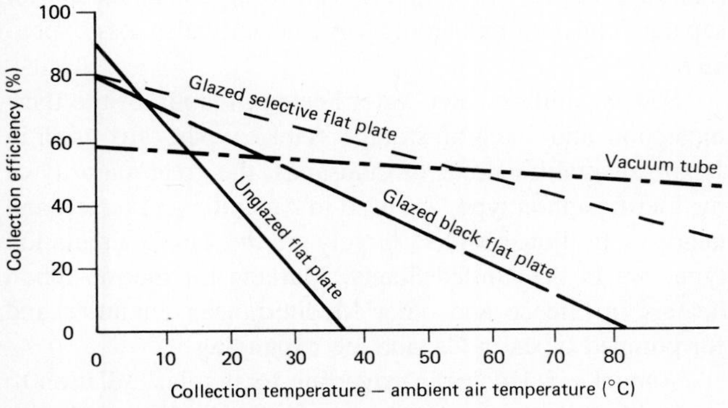

Figure 235 Typical collector efficiency curves.

design and testing. *Search: Science, Technology and Society (Sydney)*, Vol. 9, No. 4 (1978), pp. 123–219.

W.W.S.C.

SOLAR ENERGY: ITS ACTIVE USE IN BUILDINGS. Solar energy can be used in buildings for water heating, space heating, and space cooling. Active systems for water heating and space heating are in common use, but solar cooling is limited to installations for experimental purposes.

TABLE 1
Active Solar Water-Heating Systems

System	Collector Fluid	Heat Exchanger	Collector Pressure	Number of Pumps	Extent of Use[a]
Pumped Types					
Drain-back	Water	Water to water	Low	2[b]	H
Nonfreezing liquid	Antifreeze or oil	Liquid to water	Low	2[b]	M
Air type	Air	Air to water	Low	2[b]	L
Vaporization–condensation	Fluorocarbon	Condensate to water	High	2[b,d]	L
Drain-down	Potable water	None	High[c]	1	M
Recirculation	Potable water	None	High[e]	1	L
Thermosiphon type	Potable water	None	High[e]	0	H
Integral collector storage	Potable water	None	High[e]	0	L

[a]H = high, M = moderate, L = low, world-wide basis.
[b]Assuming circulation of potable water through heat exchanger.
[c]Assuming fluorocarbon with boiling point below 30°C.
[d]No pumps needed if tank is higher than collector.
[e]Assuming mains pressure in hot-water supply.

Solar Water Heating. The principal characteristics of solar water heaters used in residential and commercial buildings are shown in Table 1. All except the tank type (integral collector storage) comprise a SOLAR COLLECTOR and a storage tank for heated water. A control system and a pump for circulation of water through the collector are also required except in the thermosiphon and tank types.

Freeze protection to the collector and associated piping must be added to the system in climates where freezing may occur. In the drain-down type, temperature sensors and valves allow water to drain from the collector and be replaced by air when the temperature nears the freezing point. Although the system is designed so that loss of power activates drainage, freeze protection has not been completely reliable. In regions where freezing rarely occurs, water may be recirculated instead of draining the collector in cold weather.

There are three other types of systems. In one, solar heat is collected in a nonfreezing aqueous solution of ethylene glycol or propylene glycol or in a mineral or synthetic oil, and transferred to water in a heat exchanger. In another, air is heated in the collector and circulated through an air-to-water exchanger. In the third type, a liquid with a low boiling point, such as a fluorocarbon, vaporizes in the collector and heats water by condensing in a coil usually immersed in the water storage tank.

High heat transfer effectiveness and low maintenance are achieved in the widely used drain-back systems. Water is circulated in a closed loop through the collector, a small reservoir, heat exchanger, and pump. Water from the mains is circulated from a storage tank through a heat exchanger coil in the reservoir. When the pump stops, the collector drains into the partially filled reservoir as air replaces the water.

Most *pumped types of solar water heater* are controlled by actuating the pump when the temperature difference between storage tank and collector outlet exceeds a preset number of degrees. Solar heat is then collected and delivered to storage. The large variation in solar availability makes it impractical to supply all the hot-water requirements from the solar system. Auxiliary heat is provided by an electric resistance element in either the uppor portion of the solar hot water tank itself or in a separate gas-fired water heater. The usual location for solar collectors in hot-water systems is on the roof of a building, tilted toward the equator, preferably at an angle approximately equal to the latitude.

Thermosiphon systems, commonly used in nonfreezing climates, require placement of the storage tank above the collector. The tank may be entirely separate from the collector, as in Fig. 236, or in many packaged systems the tank is attached to the collector frame at its upper edge. The lower density of heated water in the collector tubes causes it to flow upward into the top of the tank. Water returns to the lower edge of the collector from the bottom of the tank.

The *integral collector storage (ICS)* water heater comprises some type of black tank in an insulated box with a transparent cover. Although less efficient than a heater with separate collector and storage, it is usually also less expensive.

Several million solar water heaters, mainly of the thermosiphon and integral storage (tank) types, are used in Japan; several hundreds of thousands, the great majority of the thermosiphon type, are used in Australia and Israel; and nearly a million heaters, largely of the pump-circulation type, are in the United States. Markets for thermosiphon heaters in Greece and other Mediterranean countries and for pumped types in Canada are expanding.

Annual efficiencies in coverting total solar radiation to useful heat in the hot-water supply are typically 30% in well-designed and well-operated systems. In sunny cli-

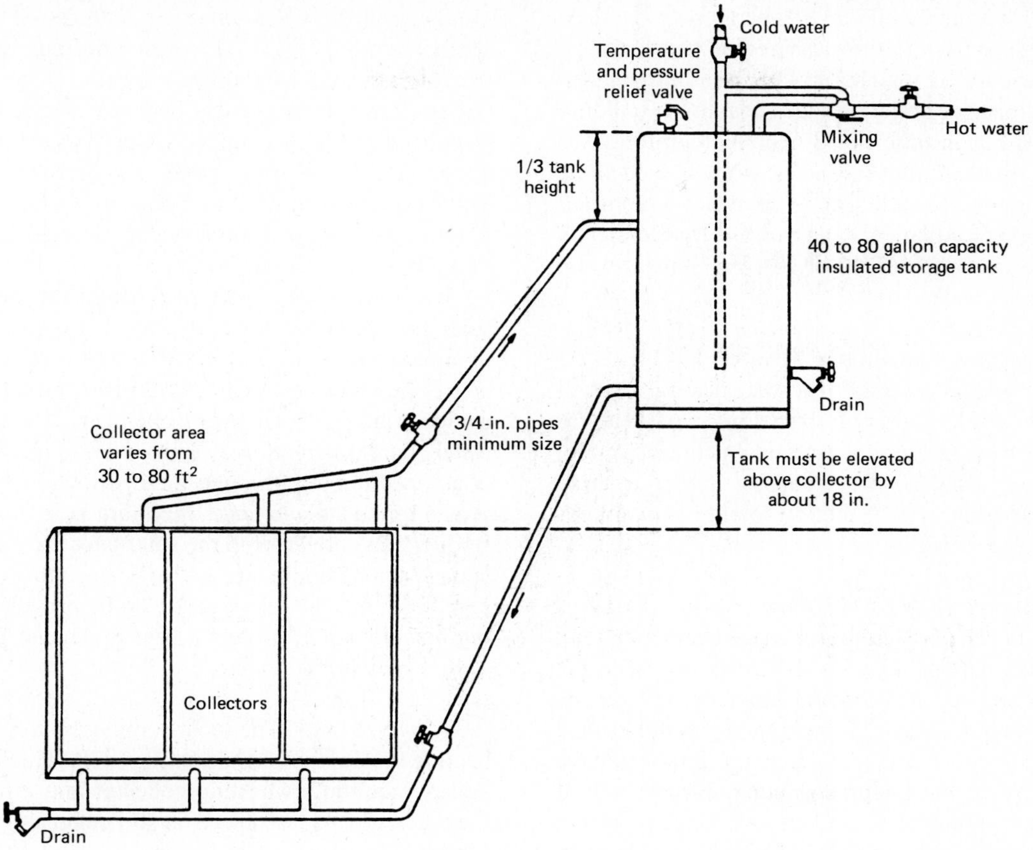

Figure 236 Direct-heating, thermosiphon-circulating solar water heater.

mates, annual heat deliveries of 2000 MJ/m^2 (200,000 Btu/ft^2) of collector can be obtained. A solar water heater with an efficient 4-m^2 (40-ft^2) collector can provide an average daily heat output 25 MJ (25,000 Btu or 7 kWh), equivalent to about 200 liters (50 US gallons) at 40°C (122°F), assuming a supply temperature of 20°C (68°F).

The cost of solar water heaters varies widely with the system type, size, and manufacturer and with the marketing method. A complete system of the pumped-circulation type with 4 to 6 m^2 (40 to 60 ft^2) of solar collectors costs the final user in the United States $3000 to $5000. The installed cost of thermosiphon heaters with 2- to 3-m^2 (20- to 30-ft^2) collectors is as low as $1000 in countries where freeze protection is not necessary.

Solar Space Heating. There are two main types of active solar space heating system; one uses a liquid as a heat collection medium, and the other is based on solar-heated air. Solar space-heating systems differ from hot-water systems in two important respects. Because of higher heat demand, solar space-heating systems are generally three to ten times larger than water heaters. Second, in addition to seasonal variability in solar energy, little demand exists in summer, whereas peak demand occurs on the coldest winter days. Designers rarely provide more than 60% of the annual heat requirements from solar energy, because a higher fraction would require a larger system poorly utilized in mild weather. However, the fact that solar space-heating systems are substantially larger than solar hot-water heaters produces some economy of scale; total installed cost is appreciably less per kilowatt of heat delivery capacity than for water heaters.

Even lower cost of delivered heat can be achieved by incorporating water heating with the solar space-heating system. A heat-exchanger coil, a pump, and a solar hot-water tank are the only additions required for utilization of the solar hardware for water heating during periods of low space heat demand, which range from 4 to 8 months of the year. With collection of efficiency about the same as for solar water heaters, the annual useful solar delivery for space heating and hot water in a moderately sized system supplying not more than 50% of the annual total heat requirements approaches 500 kWh/m^2 (50 kWh/ft^2) in climates where winters are sunny.

In most of the United States and in other relatively cold climates in the industrialized world, space heating is considered a necessity, and building temperatures are automatically controlled. Auxiliary heating equipment is therefore generally used in combination with the solar supply. Some type of furnace or boiler provides heat to the building when solar energy is insufficient to maintain comfort. Automatic controls first call for solar heat, which is used alone when its temperature is sufficient to maintain a constant temperature in the building. Auxiliary heat serves either as an alternative to the solar supply or as a supplement when the solar source is inadequate.

Most of the demand for space heating occurs at night when the atmospheric temperature is at its lowest level. Storage of solar heat is essential if a solar heating fraction higher than one-third is desired; but storage of heat for periods longer than one day is uneconomical and rarely attempted. Storage may not be necessary if only a small fraction of the demand is to be met by solar or if the building is not occupied at night.

Liquid Systems. Figure 237 shows the principal features of a drain-back solar system for space heating and hot water. Water circulating in a closed loop is heated in the collector and stored in an insulated tank, usually on the ground floor or in a basement. Temperature sensors located near the collector outlet and in or near the bottom of the storage tank activate the circulating pump whenever heat can be collected. When the incident solar energy is not sufficient for a temperature rise to occur, the pump stops running and the collector drains into the storage tank. In another type of liquid solar space-heating system, a nonfreezing aqueous solution of ethylene glycol is circulated through the collector and a heat exchanger; heat is transferred and stored in a tank of water. Typical operating conditions in a liquid solar space-heating system involve collector flow rates of 40 liters/hour m^2 (1 U.S. gallon/hour ft^2) of collector and a water storage volume of 80 liters/m^2 (2 gallons/ft^2) of collector.

In *forced warm-air systems*, heat is transferred from the solar storage to the circulating air in a conventional water-to-air exchanger, usually of the finned-tube type. Auxiliary heat is provided in a heater or furnace through which air is passed from the solar heat exchanger. If a hydronic (hot-water) heat-distribution system is used, solar-heated water is pumped directly through the distribution system and returned to the storage tank. Auxiliary heat from a hot water "boiler" in parallel with the solar tank is called upon when storage temperatures are not high enough to meet the demand. Service hot water may also be provided by circulating potable water through a heat-exchanger coil immersed in the solar storage tank.

Air Systems. The most commonly used air-type solar heating system is sketched in Fig. 238. The principal components are the air heating collector and a pebble bed for heat storage. Air is heated in the collector, typically at a rate of 0.6 m^3/m^2 (2 ft^3/ft^2) from an inlet temperature of

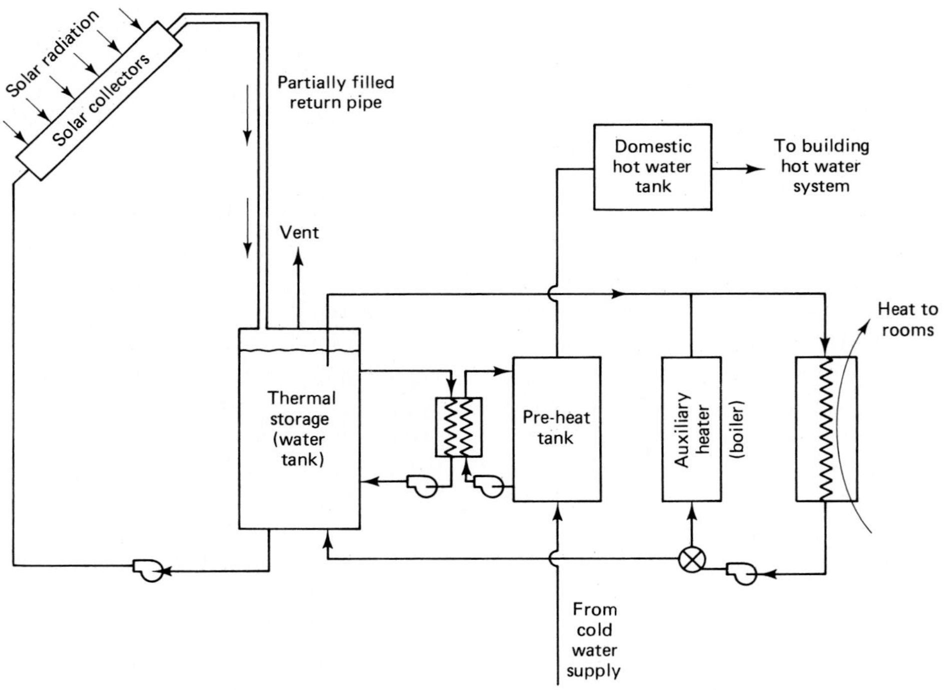

Figure 237 Schematic diagram of a single-liquid (drain-back) space-heating and water-heating system (gravity return).

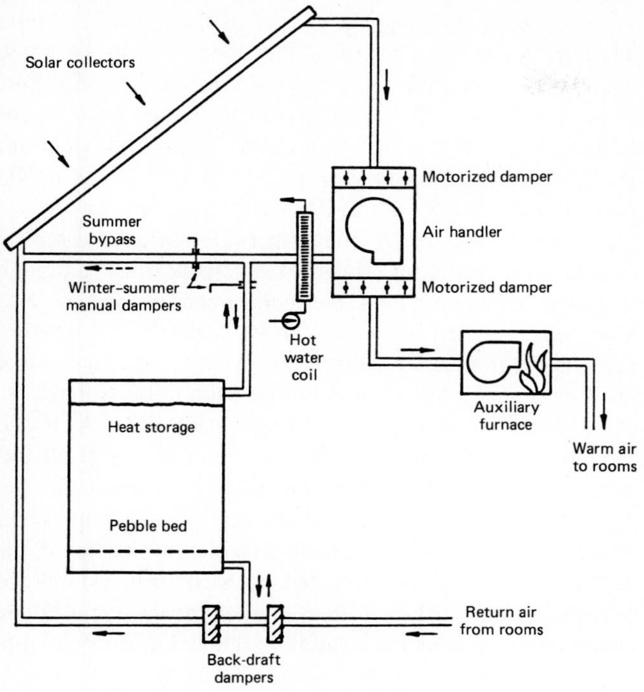

Figure 238 Two-blower air-heating solar system.

20°C (68°F) to a temperature of 50 to 60°C (122° to 140°F) during the middle of the day in clear weather. When space heating is not called for, the heated air is supplied to one end of the pebble bed; air is withdrawn from the other end and returned to the collector. Heat is stored in the pebble bed in temperature-stratified zones; the higher temperatures are near the point of entry of air flowing from the collector.

When space heating is needed at night and during cloudy periods, the position of the dampers is altered, and the circulating fan in the heat-distribution system draws air through the pebble bed in a direction opposite to that employed during the heat-collection cycle. The room air is heated by contact with hot pebbles and supplied to the distribution system at a temperature nearly equal to that at the end of the pebble bed from which the air is delivered.

A pebble bed of sufficient mass and volume for storage of one day's collector heat output should have 250 kg of rock per square meter of collector (50 lb/ft^2). Rock of reasonably uniform size, 20 to 30 mm ($\frac{3}{4}$ – $1\frac{1}{4}$ in.), is used. A collector of 30 m^2 (300 ft^2) would therefore involve the use of about 7.5 metric tons (8 U.S. tons) of rock having a volume of about 4.5 m^3 (160 ft^3). Temperature stratification in the pebble bed results in the return of air to the collector at favorably low temperatures during most of the year. Air from either the rooms or the cooler end of the pebble bed, both usually at about 20°C (68°F), is supplied to the collector, assuring maximum efficiency of collection. The delivery of air is also favorably affected by the storage stratification, as the maximum temperature is always available. Auxiliary heat, when needed, can be conveniently supplied by a conventional forced-warm-air furnace, in which the distribution fan is also located. Fuel is used only when the control system detects that solar heat from the collectors or from storage is inadequate.

Service hot water can also be provided in an air system by use of a conventional heat exchanger in the duct, which brings solar-heated air to the pebble-bed storage unit. Potable water is circulated from a tank through the coil when solar heat is available and when water can be heated.

Daytime Solar Heaters. There has been an increasing market, particularly in the United States, for solar collectors, primarily of the air heating type, that provide solar heating only when the sun is shining. No storage is used in these installations. Because the daytime heat requirements of a residence in sunny weather are usually a small fraction of the total daily requirements, these systems are small. For a typical 150 m^2 (1500 ft^2) residence in a relatively cold climate, 2 to 5 m^2 (20 to 50 ft^2) of solar collectors can effectively supply heated air to one or two rooms or to the distribution system without storage. This is a relatively low cost supply of solar heat. The system comprises a small duct supplying air from the rooms to the collector, a duct and small fan to deliver the heated air to the rooms, and a simple controller to operate the fan when solar heat is available and when heat is needed. Only in mild weather and in the summer is such a system idle. Its low cost justifies its use in climates where reasonably good solar radiation and long, cold winters prevail. The concept can also be effectively used in large commercial and industrial buildings, which are not occupied at night and therefore require heat mainly during daylight hours.

See also PASSIVE SOLAR DESIGN and SOLAR COLLECTORS FOR SOLAR THERMAL SYSTEMS.

References

1. W. C. DICKINSON and P. N. CHEREMISINOFF (eds.): *Solar Energy Technology Handbook.* Marcel Dekker, New York and Basel, 1980.
2. J. F. KREIDER and F. KREITH (eds.): *Solar Energy Handbook.* McGraw-Hill, New York, 1981.
3. J. A. DUFFIE and W. A. BECKMAN: *Solar Engineering of Thermal Processes.* Wiley, New York, 1980.

O.G.L.

SOUND is a sensation perceived when the eardrum and its associated structures are made to vibrate by rapid fluctuations of air pressure in the external part of the ear. Human hearing extends over a frequency range from about 20 to

20,000 Hz (20 kHz), although this upper limit decreases with increasing age. In its most sensitive region (1 to 3 kHz), a normal ear can detect pure tones with sound pressure amplitudes as low as 20 μPa, and this pressure is used as the standard level (0 dB) on the sound pressure level (SPL) scale. Human hearing is much less sensitive below 100 Hz and above 10 kHz. Sounds with amplitudes greater than about 20 Pa (120 dB) are painful and damaging to hearing. (See ACOUSTICAL TERMINOLOGY AND MEASUREMENT; SOUND-LEVEL METER; and NOISE).

Any object vibrating in a fluid such as air generates sound waves by causing, alternately, small compressions and rarefactions of the fluid in front of its surface. These waves travel with a velocity v that is characteristic of the particular fluid, although v generally increases with increasing temperature ($v = 340 \text{ m s}^{-1}$ in air at 20°C). Such waves can also travel through solids but can be heard as sound only after the vibrating surface of the solid excites the neighboring air. If the frequency of a pure tone is f, then its wavelength λ in a medium with sound velocity v is $\lambda = v/f$. Wavelengths of audible sounds in air range from about 20 m down to about 2 cm. Vibrating objects excite sound waves efficiently only if their size is not too small compared with the sound wavelength involved.

Pure tones (e.g., from sine-wave oscillators) have only a single frequency. Complex tones with regularly repeating pressure patterns (wave forms) such as single steady notes on musical instruments have many components, the frequencies of which are all integer multiples of the fundamental (lowest) frequency. Such components are said to be *harmonics* of the fundamental, which is itself the first harmonic of the series. In percussive instruments, like bells, the component frequencies are not usually harmonics of the fundamental and are referred to as *overtones* or *upper partials*. The particular patterns of relative intensities of the harmonics or overtones (i.e., the *frequency spectrum*) give musical instruments their characteristic sound qualities.

Musical instruments consist of primary vibrating systems (strings, reeds, etc.) coupled to *resonant systems* (pipes, cavities, etc.) that vibrate most readily at particular frequencies or to large light structures (violin body, piano soundboard) that enhance sound radiation to the air. The *resonances* of these systems influence the harmonic spectrum and thus the sound quality (see HELMHOLTZ RESONATOR).

The human voice is produced by air forced past vocal folds in the throat, which vibrate at around 150 Hz in adult male speech and 300 Hz in adult female speech. This frequency is varied in singing. The puffs of air released into the throat have high harmonic content with audible energy up to over 3 kHz. Resonances in the vocal tract emphasize bands of harmonics in three separate frequency ranges (*formants*), the location of which is determined by the positions of lips, mouth, and tongue, to give distinct vowel sounds. The formant frequencies of human speech lie in the range of most acute hearing (300 Hz to 3 kHz). Telephones produce very intelligible speech although they transmit essentially only this range. For high-quality music reproduction, the full audible frequency range is required (see SOUND REINFORCEMENT SYSTEMS).

Sounds produced by impact between hard solid bodies contain vibration components of all frequencies (although in instruments like gongs, particular musical notes are emphasized). Impact vibrations can be transmitted with very little loss through long distances in rigid structures (like steel or stone buildings) and produce sound by radiation to the air (see SOUND INSULATION AND ISOLATION). Rough continuous noises also have components of all frequencies, although particular bands may predominate.

Sound waves spread spherically away from small sources and decrease in intensity with the square of the distance. They can be reflected by hard surfaces and absorbed by soft surfaces. Because sound wavelengths are relatively long, sound is generally scattered diffusely except by large smooth surfaces (see ACOUSTICS OF AUDITORIA and SOUND ABSORPTION).

References

1. L. E. KINSLER, A. R. FREY, A. B. COPPENS, and J. V. SANDERS. *Fundamentals of Acoustics*. Wiley, New York, 1982.
2. T. D. ROSSING: *The Science of Sound*. Addison-Wesley, Reading, Mass., 1982.
3. A. H. BENADE: *Fundamentals of Musical Acoustics*. Oxford University Press, New York, 1976.
4. L. L. BERANEK: *Acoustics*. McGraw-Hill, New York, 1954.
5. P. M. MORSE: *Vibration and Sound*. McGraw-Hill, New York, 1948 (reprinted by the Acoustical Society of America, 1982).

N.H.F.

SOUND ABSORPTION (ATTENUATION OF SOUND IN ROOMS) is the process by which sound energy is removed from a sound field. If sound is introduced into an enclosure, a sound field is established dependent on the rates of supply and absorption. If the source produces sound of constant power, the resulting sound field reaches steady-state levels when the sound energy supplied is just equal to the energy absorbed. If the source is abruptly shut off thereafter, the sound field decays with time until it has been completely absorbed.

The persistence of the sound field in an enclosure, after the source of sound has been shut off, is known as *rever-*

beration, and the sound field remaining during reverberation is known as the *reverberant* sound field. The reverberation decay time, defined as the time required for a sound field to decay to one-millionth of its initial value, is accepted as the single most important descriptor to define the quality of a room for all acoustical purposes; it is intimately tied to the absorption of sound in a room.

Sound absorption may occur during propagation, by dissipation of the sound energy into heat, and during reflection at any boundary either by dissipation at the boundary or by transmission through it without return. Thus sound absorption is associated with sound-energy dissipation within the acoustic field and with transmission out of the field. Generally, propagation loss is negligible compared to other losses except in large rooms and at high audio and ultrasonic frequencies when it becomes important.

Sound absorption at a boundary depends both on properties of the boundary and properties of the incident sound field. As the reverberant field generally dominates the direct field of the source at the boundaries of an enclosure, the incident field can only be described in a statistical sense, and even then the description is difficult. These considerations suggest that the dependence of absorption at a boundary on properties of the sound field be removed by assuming that the sound is incident from all directions with equal probability and intensity. The absorption at the boundary then becomes solely a property of the boundary construction.

In the case of a boundary comprised of a surface of the order of a wavelength or more in extent, a statistical absorption coefficient may be defined as the fraction of sound energy, incident equally from all directions, that is absorbed. The total absorption is assumed proportional to the total surface area. In contrast, if the boundary is comprised of a discrete object less than the order of a wavelength in extent, due to various possible effects, its absorption is not generally proportional to its surface area. However, in either case the absorber is characterized by an equivalent area of total absorption. One square meter (or one square foot) of equivalent totally absorptive surface is given the name 1 sabin.

Implicit in the concept of a statistical absorption coefficient is the assumption that events at any location on a surface are independent of events occurring at all other locations on the surface. When this assumption is true, the surface is said to be *locally reactive*. This concept is interconnected with that of statistical absorption; one implies the other. The concept of locally reactive is an analytic convenience, appropriate for the description of porous acoustic tiles and similar structures. It is not appropriate in the description of edge-mounted panels and even heavy walls in the frequency range of lowest-order resonance and again at higher frequencies in the range of coincidence.

In the case of the lowest-order resonance, a disturbance anywhere causes the whole surface to vibrate, supported at the edges like the diaphragm of a drum. At coincidence the speed of propagation of flexural waves in the surface, which is proportional to the square root of the frequency, matches the speed of propagation of sound waves so that at coincidence, and higher frequencies, the surface is again driven in resonant response to the incident sound. The surface then becomes a very effective absorber of sound, either through dissipation or transmission.

The analytic description of steady-state and especially of reverberant decay is extremely complicated. No completely satisfactory analysis exists to describe reverberation decay unambiguously, even in the practical case of the reverberant room used for the measurement of the sound absorption of absorptive devices. Thus absorption coefficients greater than 1 are occasionally encountered; these are generally truncated and reported as 1, suggesting total absorption. Theoretical considerations, supported by experiment, indicate an upper bound of about 0.93 for all uniformly flat, porous surfaces.

See also HELMHOLTZ RESONATORS and SOUND INSULATION AND ISOLATION.

References

1. HEINRICH KUTTRUFF: *Room Acoustics*, 2nd ed. Applied Science Publishers, London, 1979.
2. LEO L. BERANEK (ed.): *Noise and Vibration Control*. McGraw-Hill, New York, 1971.
3. P. M. MORSE and R. H. BOLT: Sound waves in rooms. *Reviews of Modern Physics*, Vol. 16 (1944), pp. 69–150.

D.A.B.

SOUND INSULATION AND SOUND ISOLATION concern methods for preventing intrusion of airborne noise into an inhabited space in a building, either from a neighboring space (apartment, office, schoolroom, or mechanical room) or from the outdoors (street or air traffic, playgrounds, etc.) (Ref. 1).

In the developed countries, minimum noise-control requirements are set forth in building codes. Two conditions must be met for them to be effective: the code requirements must be *relevant* to their goal and they must be adequately enforced. Usually, neither condition is met.

Insulation versus Isolation. In considering acoustical privacy, it is essential to distinguish between the *sound insulation properties of the partitions* (wall and floor–ceiling constructions) and the *sound isolation actually achieved between rooms* (Refs. 2 and 3).

The distinction is clear-cut in standard test procedures.

In the method for measuring transmission loss, the only route for sound to travel from the source room to the receiving room is the path directly through the test specimen; all other paths are deliberately blocked. The transmission loss is a property of the partition: the higher the number, the better the insulation provided by the partition (Ref. 4).

The test method for *noise reduction*, a measure of the sound isolation between two rooms, makes no attempt to focus on the party wall, but evaluates the effect of all the sound paths that connect the two rooms. A loud noise is generated in the source room, and the sound levels in the source and receiving rooms are compared. Their difference represents the acoustical isolation between the two rooms, no matter how the noise gets from one to the other (Ref. 5). The codes do not make this distinction; they specify only the kind of partition required, either by stating the minimum acceptable transmission loss or by describing partitions whose construction is ''deemed to satisfy'' the requirements.

The achievement of good sound isolation between rooms requires that the party walls (or floors) be massive and impermeable to air. It is also possible to achieve good isolation with walls of less weight, but then the construction must be more complicated, using resilient elements, staggered studs, separate plates, and sound absorptive material in the wall cavities (Ref. 1).

Flanking Transmission. Choosing the correct sound *insulation* is only one step along the way to achieving the final goal of adequate sound *isolation*. An equally important step is to ensure that the partitions are correctly constructed and assembled without sound leaks between or through them. Unfortunately, the choice of suitable partition construction is often spoiled by sound traveling from one room to another by *flanking paths* that have nothing to do with the partition in question.

Building codes can do little to cope with the problem of flanking transmission. It must be solved by responsible on-site inspection during the construction of the building, and preferably with a requirement for performance tests to demonstrate adequate sound isolation between apartments when the building is completed (Ref. 2).

Acoustical Privacy (Ref. 3). In addition, in order that the achieved isolation between apartments actually leads to the desired acoustical *privacy* for the occupants, some account must be taken of the life styles of the building inhabitants and the quietness of the neighborhood: it is easier to achieve privacy for well-behaved than for rowdy tenants and easier in an urban neighborhood than in the quiet of the suburbs.

The loudness of the noises in the source room, the amount of sound absorptive material in the source room, and the expectations for privacy on the part of the occupants can be combined into a privacy index (PI). One value (PI = 70) can be chosen for defining a minimum grade of sound privacy, and a higher value (PI = 82) for a better grade.

The other two factors of importance to acoustical privacy are (1) the sound isolation achieved between the two rooms in question in the finished building, expressed in terms of the A-weighted noise reduction (NR_A) between the source and receiving rooms, and (2) the typical background noise level in the receiving room, expressed in terms of the A-weighted sound level, L_A (see SOUND-LEVEL METER). If the sum of these two numbers equals or exceeds the privacy index, the desired privacy between the two rooms will be achieved.

For example, if we wish to achieve the better grade of privacy (PI = 82) and if the background noise in the building is measured (or estimated) to be L_A = 35 dBA, then the partitions must be chosen with a high enough intrinsic *sound insulation* (transmission loss or sound transmission class, STC) that, when finally built, the construction will achieve *sound isolation* between adjacent apartments corresponding to a noise reduction (NR) of 47 dBA. Then we will have $L_A + NR_A = 35 + 47 = 82$.

Note that if the typical background noise in the building were 5 dB quieter, this condition would have to be compensated for by planning for 5 dB *more* sound isolation between rooms in order to achieve the same degree of acoustical privacy.

Exterior Walls. An analogous approach is taken to providing isolation from outdoor noises, except that here we must take account of the fact that the weakest sound path through the facade of a building is the windows and the leaks around them. Tight-fitting gaskets and/or double windows can significantly improve isolation from outdoor noises. If such measures are not taken, the sound entering through the windows will dominate the indoor noise environment and it will hardly matter what construction is selected for the facade walls (Ref. 1).

The same construction considerations that improve the sound insulation of a facade also improve its thermal insulation. For this reason, it is often easier to evaluate the thermal insulation for a building using acoustical tests than with the customary (but awkward) tests of thermal energy flow.

Sound Treatment. A final word on classification may be helpful here: people frequently speak loosely of ''sound treatment,'' usually having in mind (erroneously) the application of acoustical tile to help prevent noise intrusions from the neighbors. In fact, there are two kinds of sound treatment: one concerns the installation of sound-absorptive materials (like acoustical tile) in a room to prevent the noise *due to sources within the room* from building up to disturbing levels. To the extent that the treatment is effective for this purpose, it also slightly reduces the intrusion of the noise into adjacent rooms.

The other kind of sound treatment is the subject of the present article: the provision of heavy, impervious party walls to minimize the transmission of noise from one room to another. The addition of porous acoustical tile to the party wall is almost entirely without effect in improving the transmission loss of that partition. After all, one would not try to make a raincoat out of a diaper.

See also ACOUSTICAL REQUIREMENTS OF BUILDINGS; IMPACT AND VIBRATION ISOLATION; NOISE; and SOUND ABSORPTION.

References

1. JOHN H. CALLENDER (ed.): *Time-Saver Standards for Architectural Design Data*. McGraw-Hill, New York, 1982; see the chapter on *Acoustics* by R. B. NEWMAN.
2. THEODORE J. SHULTZ: How noise creeps past the building codes. *Noise Control Engineering*, Vol. 1, No. 1 (Summer 1973), pp. 4–14.
3. THEODORE J. SCHULTZ: A-level differences for noise control in buildings. *Noise Control Engineering*, Vol. 1, No. 2 (Autumn 1973), pp. 90–97.
4. *Airborne Sound Transmission Loss of Building Parritions, Laboratory Measurement of—ASTM E 90-81*. American Society for Testing and Materials, Philadelphia, 1981.
5. *Airborne Sound Insulation in Buildings, Measurement of—ASTM E 336-77*. American Society for Testing and Materials, Philadelphia, 1977.

The corresponding international standards for Refs. 4 and 5 are *Acoustics—Measurements of Sound Insulation in Buildings and of Building Elements—ISO R 140. Part III, Laboratory Measurements of Airborne Sound Insulation of Building Elements* and *Part IV, Field Measurements of Airborne Sound Insulation between Rooms*. International Organization for Standardization, Geneva, 1978.

Note: The distinction between sound insulation and sound isolation, which is emphasized in this article, is not carefully preserved in international practice.

T.J.S.

SOUND-LEVEL METER. Portable sound-level meters were developed in the 1930s when miniature vacuum tubes and reliable microphones became available. They have several uses, in particular noise source identification on machinery and measurement of noise levels in buildings and communities. A small hand-held sound-level meter with a piezoelectric microphone is shown in Fig. 239 and its block diagram in Fig. 240.

The two sound attenuators are interconnected and operated by a single range switch to give 10 dB steps in attenuation (see ACOUSTIC TERMINOLOGY AND MEASUREMENT). The indicating meter has a range of 20 dB; its zero position is governed by the attenuator setting. The meter circuit must include a rectifier, and most are designed to measure the root-mean-square (rms) value; the internationally standardized *fast* and *slow* meter dampings are provided in this circuit.

Weighting Networks. The meter has A-, B-, and C-weighting networks, constructed from resistance–capacitance circuits, and it can be operated with any of these three networks. They are designed for measurements of sound-pressure levels below 55 dB (A-weighting), 55 to 85 dB (B-weighting), and above 85 dB (C-weighting). The A-weighting network is the one most commonly used, because it seems to give the best correlation with human response to noise. Some sound-level meters have octave or one-third octave attachments (see ACOUSTIC TERMINOLOGY), and others have provision for digital meter readouts.

Sound-level meters are built to meet various standards or recommendations (Refs. 1, 2, and 3). The American standard (Ref. 1) specifies four types of sound-level meters, for *precision* (type 1), *general-purpose* (type 2), *survey* (type 3), and *special-purpose* (type S) measurements. The requirements for precision and accuracy are greatest for type 1 and least for type 3. The choice of sound-level meter depends on the precision and accuracy required, the other capabilities of the instrument, and its purchase price (which is normally highest for a type 1 sound-level meter).

Measurement Problems. Sound-level meters have several measurement problems, in particular undesired sensitivity to vibration, humidity, temperature, magnetic and electrostatic fields, and wind. Sound-level meters should not be used in high winds (above about 20 km/h or 12 mph); otherwise, the wind noise can mask the noise to be measured unless the latter is very intense. Windscreens are available to reduce wind noise problems.

To minimize reflections caused by the operator, the sound-level meter should be held at arm's length and pointed at the noise; some sound-level meters are designed to be orientated at right angles to the noise and instructions for orientation are supplied by the manufacturer. Alternatively, a separate microphone can be used with an extension cord between the microphone and the meter.

References

1. *Specification for Sound Level Meters—ANSI S1.4-1971*. American National Standards Institute, New York, 1971.

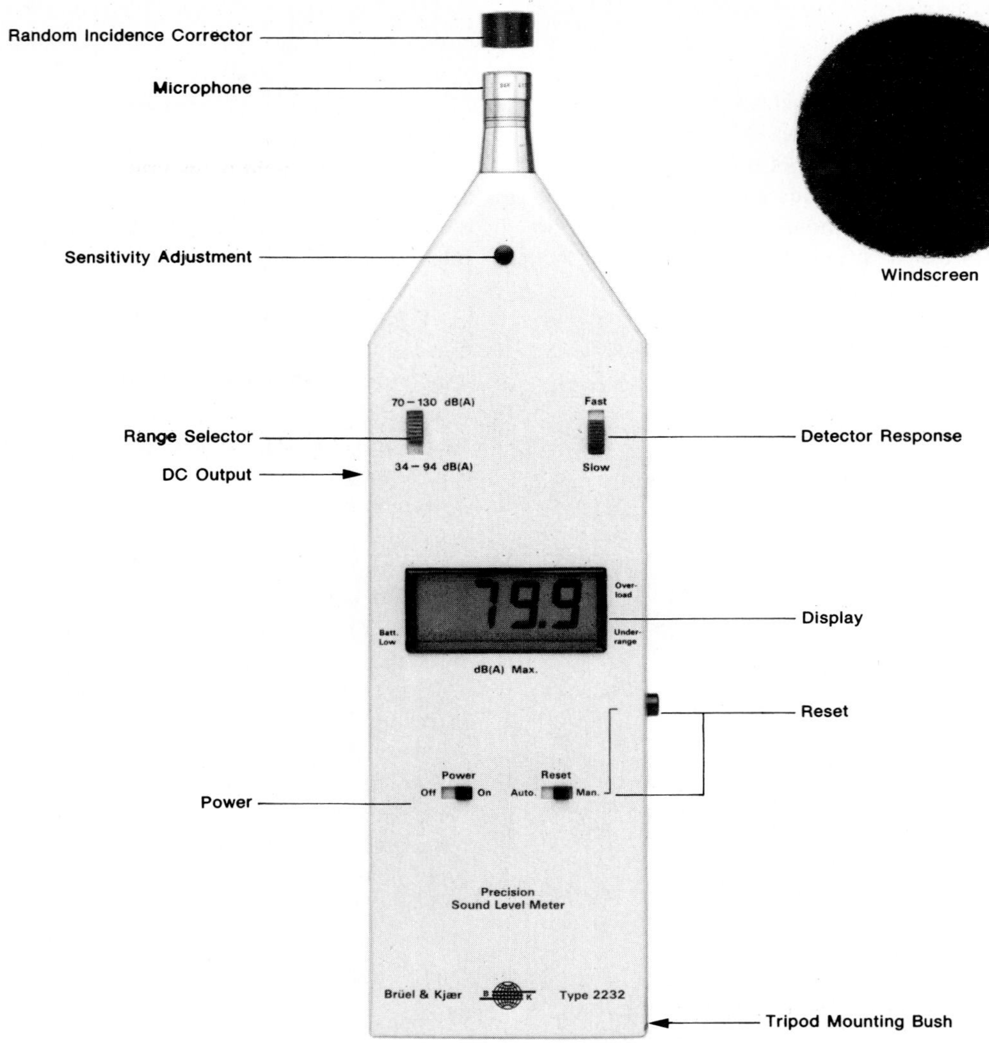

Figure 239 Simple sound-level meter made by Brüel and Kjaer. (By courtesy of the manufacturer.)

2. *Precision Sound Level Meters—IEC No. 179.* International Electrotechnical Commission, Paris, 1965.
3. *Recommendations for Sound Level Meters—IEC No. 123.* International Electrotechnical Commission, Paris, 1961.
4. M. J. CROCKER and A. J. PRICE: *Noise and Noise Control,* Vol. 1. CRC Press, Boca Raton, Fla., 1975, pp. 154–162.

M.J.C.

SOUND-REINFORCEMENT SYSTEMS comprise one or more *microphones, amplifiers,* and *loudspeakers* whose purpose, respectively, is to collect, amplify, and distribute sound from one or more sources to hearers distributed over a large area, either in the open air or inside a building. Microphones are electroacoustic transducers that convert sound into electrical energy for amplification; loudspeakers then reconvert it into sound.

Acoustical Environment. A modern sound-reinforcement system should collect and distribute sound with such overall fidelity that performers and hearers alike are barely aware of its operation. The acoustical environment may range from a simple outdoor situation, free from adjacent reflecting surfaces, to a cathedral, in which multiple reflections of sound may occur from surrounding walls and large stone pillars located in the listening area. The ideal loudspeaker system is a single source so arranged in height and directional properties that sound is distributed uniformly over the audience, avoiding multipath transmissions with arrival-time differences greater than 20 ms. When the audience area is so large that uniform sound distribution cannot be achieved from a single loudspeaker system, addi-

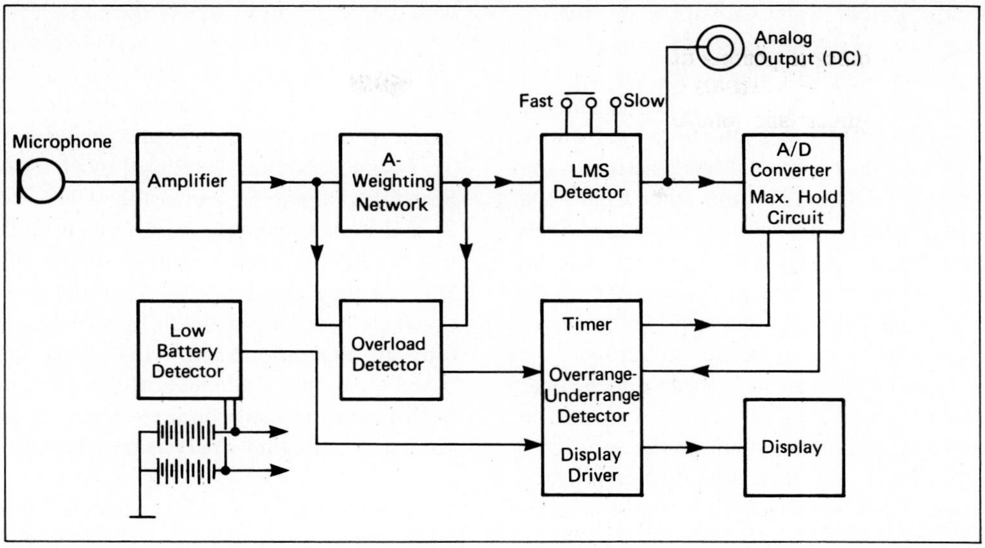

Figure 240 Block diagram of the sound-level meter shown in Figure 239. (By courtesy of Brüel and Kjaer.)

tional systems must be provided, each fed with a signal delayed by the time of travel of sound between the two systems; digital delay equipment is commercially available.

If a properly installed sound system fails to achieve satisfactory sound distribution in an auditorium because of undesirable reflections, surface treatment and, possibly, structural modification will be needed to improve the natural acoustics of the building (Refs. 1 and 2).

Electroacoustical Feedback occurs when sound from the loudspeaker(s) is fed back, either directly or after reflection, into the microphone (Ref. 3). The tendency toward feedback is increased by the use of more than one live microphone and by the presence of strong eigentones or natural room modes. These are standing waves, as occur in organ pipes, caused by reflections between hard parallel surfaces or by multiple reflections around the various internal surfaces of the building. Feedback tendency is reduced by the proper choice, adjustment, and placement of microphones.

The **Linear Array,** or column loudspeaker, enables sound to be directed into the audience and away from the ceiling, whence it would otherwise be reflected back as reverberant sound. This helps to increase the ratio of direct to reverberant sound in the listening area, particularly in buildings with very high ceilings. A useful by-product of the directional property of columns is the existence of a region of negligibly low intensity along an extension of the major axis. Location of the microphone in this region greatly reduces the tendency toward acoustic feedback. When the column is located high above the microphone, the amplified sound reaches the listener later than the natural sound. Because of the Haas effect (Ref. 4) the hearers gain the impression that the amplified sound is emanating directly from the live source.

Speech Intelligibility. For good speech intelligibility, the amplifier system, often taken for granted, should be of high quality and have negligible transient distortion. Also, in reverberant buildings, the permissible *rate* and *dynamic range* of speech delivery, for good intelligibility, are limited by the reverberation characteristics of the building with the sound system in operation and finally by the ability of the speaker to interact sympathetically with the whole system in a manner appropriate to its acoustical environment.

See also ACOUSTICS OF AUDITORIUMS.

References

1. V. O. KNUDSEN and C. M. HARRIS: *Acoustical Designing in Architecture.* American Institute of Physics, for the Acoustical Society of America, New York, 1980.
2. J. E. BENSON: Acoustics and the community. *Proceedings of the Institution of Radio and Electronics Engineers of Australia.*, Vol. 25 (July 1964), p. 464.
3. J. E. BENSON and D. F. CRAIG: A feedback-mode analyser/suppressor unit for auditorium sound-system stabilisation. *Proceedings of the Institution of Radio and Electronics Engineers of Australia*, Vol. 30 (March 1969), p. 76.
4. P. H. PARKIN: The application of the Haas effect to speech reinforcement systems. *Acustica*, Vol. 4 (1954), p. 87.

J.E.B.

SPACE FRAMES. ROOF STRUCTURES are usually divided into two groups: (1) two-dimensional or plane systems and (2) three-dimensional or space systems. A roof truss is a typical example of a plane system consisting of all members lying in one plane and interconnected into a stable framework able to resist external loads acting in the plane of the structure. A DOME, on the other hand, is a typical example of a three-dimensional structure. Such structures, usually referred to as space structures, can be subdivided into (1) skeleton (or braced) frameworks, often called space frames, (2) SHELL and stressed-skin systems, and (3) suspended (CABLE or membrane) structures.

A space frame can be defined as a three-dimensional assembly of structural elements resisting loads that can be applied at any point, inclined at any angle to the surface of the structure and acting in any direction. Skeleton space structures consisting of pin-connected bars are sometimes called space trusses, whereas the term space frame usually applies to space structures with rigid joints. In practice, it is virtually impossible to produced pin-connected joints, and therefore a space frame denotes any stable three-dimensional framework of members.

Interest in space frames is growing constantly, and the large number of such structures built all over the world shows clearly that, through prefabrication and industrialization, these systems often compete very successfully with more conventional structures, at the same time providing architects with more impressive forms. The great publicity received by space frames during the last decade is largely due to the realization of their structural advantages, their inherent lightness combined with remarkable stiffness. The acceptance of various types of space frames is the result of research and development work done during several decades by progressive designers, architects, and engineers, who already many years ago appreciated the huge potential of such systems.

The early and highly original development work of Mengeringhausen, Buckminster Fuller, Tsuboi, Matsushita, du Château, Richter, Castaño, and Fentiman provides typical examples of progressive designers who have left their imprint on the architectural acceptance and widespread use of space frames (Ref. 1).

As a rule, space frames are built from simple, prefabricated units, which are often of standard size and shape. Such modular units, mass-produced in the workshop, can be easily and rapidly assembled on site by semiskilled labor.

Calculations. The introduction of the electronic computer and the development of several versatile connectors are, no doubt, the two main reasons for the present rapid acceptance of space frames. With modern high-capacity computers it is possible to analyze even very complex space frames with much greater accuracy than ever before and with a marked reduction in the time involved. Recent developments of algebraic representation and processing of structural configurations drastically reduce the amount of time required for the tedious and time-consuming preparation of data for the computer analysis of space frames.

Formex algebra, developed by H. Nooshin at the Space Structures Research Centre of the University of Surrey (Ref. 2) is a typical example of a modern and very convenient tool for algebraic representation of any type of space frame, whether they are double-layer grids, braced barrel vaults, domes, towers, or skeleton hyperbolic paraboloidal structures of any description. This newly developed process, based on simple mathematical descriptions of the interconnected pattern of structural members, is especially suited to automated structural analysis based on the stiffness method.

Double-Layer Grids. Special developments have taken place within the last decade in the field of double-layer grids. These space frames consist of two plane grids forming the top and bottom layers parallel to each other and interconnected by vertical and diagonal members.

The great interest expressed in double-layer grids is reflected in the large number of commercial systems developed all over the world. The most successful have proved to be MERO, Unibat, Space Decks, Triodetic, Diamond Truss, and Nodus. They have been discussed in detail elsewhere (Ref. 3). Originally, space frames were used to cover large span roofs; today they are also used for floor construction in multistory buildings (Fig. 241). A typical example of the great impact that space frames have made on contemporary architecture is provided by a vast space frame for the Convention and Exposition Center in Manhattan, New York City, USA. This huge building stretching 360 meters (1200 ft) along Manhattan's 11th and 12th Avenues and costing over $200 million, is the biggest space frame structure in the whole of the United States. Two-way double-layer steel grid has been chosen for the roofs, floors, and enclosing vertical walls.

The two large hangars erected at Heathrow Airport, London, and covering an overall area of 25,000 m^2 (270,000 ft^2) are an example of diagonal double-layer grids. In each hangar the space frames provide an uninterrupted span of 138 m (453 ft) (Fig. 242). The recently completed (1982) hangar in Singapore, covered also by a diagonal grid, has an even larger clear span of 200 m (656 ft).

Three large exhibition halls covered with space frames are especially well-known.

1. The first is the aluminum double-layer two-way grid at São Paulo, Brazil, covering an area of 260 m × 260 m (850 ft × 850 ft) and supported by only 25 columns (Fig. 243).
2. The second example is the huge complex of exhibition halls in Düsseldorf, West Germany, covering an area

Figure 241 Example of the use of double-layer grids in multistory buildings.

Figure 242 External view of the diagonal double-layer grid covering one of the BOAC hangars at Heathrow, London Airport.

of 156,000 (1,700,000 ft^2). Thirty-two pavilions, each measuring 30 m × 30 m (100 ft × 100 ft), and resting on four corner points only, have been built in Oktaplatte, and 82 pavilions in the MERO system. All of them are examples of the orthogonal two-way double-layer grid.

3. The third and more recent example is the National Exhibition Centre in Birmingham, which is, so far, the largest single project to have utilized the Nodus system. It comprises a total of 93 identical space frame grids, all in tubular steel, each 27.9 m × 27.9 m (92 ft × 92 ft).

In addition, there is another hall at the same Centre, one of the largest clear span buildings in the world, 108 m (355 ft) long and 90 m (295 ft) wide. Four latticed box girders span along and across the hall supporting nine infill Nodus space frame grids. The total area of some 100,000 m^2 (1,000,000 ft^2) covers exhibition space (Fig. 244). More than 6000 t of structural steelwork was used. Simplicity of site assembly and speed of steelwork erection were important factors in the completion of this major project to schedule.

Construction. It is normal practice for double-layer space frames to be constructed on the ground and then lifted into position. This permits easy ground level installation of

Figure 243 Lifting of the aluminum space frame for the São Paulo Exhibition Center.

services and reduces safety hazards. When this is not possible due to restricted site, the space frame can be erected on site supported on scaffolding until self-supported. This technique has been used recently during the construction of the Gatwick Railway Concourse, U.K. (Fig. 245), where a tubular steel space frame of the Nodus type allowed an in-situ piece-small assembly, thus eliminating disruption and allowing the railway station to operate normally throughout the building program.

Space frames were a particularly attractive proposition for various large social and leisure complexes erected during the last decade in various countries in the Middle East. Sky canopy for the Dubai Municipality, built in the Nodus space frames, attracted a great deal of admiration on account of the exquisite beauty of the structure. Five large exhibition buildings for the Sudan Exhibition and Fairs Corporation at Khartoum again illustrate the use of Nodus two-way rectangular, double-layer tubular steel grids.

A very recent example is the Al Diafa Exhibition Centre, a 50.8 m × 102.1 m (167 ft × 335 ft) steel space frame structure set in Riyadh, Saudi Arabia, built in the Space Decks system. This rather unusual structure, an example of a suspended roof design, consists of two double-layer grids, 50 m × 50 m (164 ft × 164 ft), each supported by one central column protruding above the roof to provide fixed points for eight hangars. Although space frames are used principally in the construction of horizontal double-layer grids, many designers that adapted them to a variety of other applications. This includes vertical, inclined, and multilayer grids (Ref. 4). These applications are favored by modern designers as they are particularly effective in the covering of exhibition areas, shopping centers, and similar buildings where the space frames are exposed or viewed through glazing.

Vaults and Domes. Significant developments have taken place in the construction of some other types of space frames, especially braced barrel vaults and braced domes.

Figure 244 Space frames covering the National Exhibition Centre in Birmingham, England. (By courtesy of the British Steel Corporation.)

Figure 245 Internal view of the multilayer space frame covering the Gatwick Railway Concourse, England. (By courtesy of the British Steel Corporation.)

The earlier work of Buckminster Fuller on geodesic domes proved to be instrumental in reviving the interest of architects and engineers in braced skeleton domes. From numerous structures of this type built in various parts of the world, one must mention the Al Ain aluminum done built in 1982 in the MERO system. It is a double-layer tubular dome having a diameter of 70 m (230 ft) and a height of 25 m (82 ft). The total weight of the structural framework is only 77.5 tons, the total number of tubular modular bars is 7078, and the number of MERO joints is 1889. The upper layer is a regular three-way grid; the inner layer is a skeleton grid consisting of bars arranged in hexagons interconnected with triangles. The Al Ain dome is a splendid example of what high technology can achieve in prefabricated space frames. It is a part of the social, cultural, and sports center built recently in the United Arab Emirates. In addition to the dome, open grandstands, various pavilions, and huge entrance canopies are all covered with space frames, an excellent example of the architectural freedom of form achieved through the use of modern space frame concept.

See also COMPUTER METHODS OF STRUCTURAL ANALYSIS; POLYHEDRA; AND STEEL STRUCTURES.

References

1. Z. S. MAKOWSKI: Recent trends and developments in prefabricated space structures. *Proceedings of the Pacific Symposium on Space Frames. International Association for Shell Structures.* Tokyo and Kyoto, 1971, pp. 33–45.
2. H. NOOSHIN: Algebraic representation and processing of structural configurations. *Computers and Structures*, Vol. 5 (1975), pp. 119–130.
3. Z. S. MAKOWSKI (ed.): *Analysis, Design and Construction of Double-Layer Grids.* Applied Science Publishers, London, 1981.
4. U. K. BUNI, P. L. DISNEY, and Z. S. MAKOWSKI: *Multi-Layer Space Frames.* Constructional Steelwork Research and Development Association, London, 1980.

Z.S.M.

SPAN. Some of the ancient civilizations produced buildings of prodigious size; the Great Pyramid at Gizeh held the record for height from 2580 B.C. to 1876 A.D., more than 44 centuries. However, its height was a by-product of its size. Great horizontal and vertical spans did not exist prior to Ancient Rome.

The **Longest Interior Span** (see Table 1) in Ancient Rome was the Pantheon, still standing in modern Rome. It

TABLE 1
Interior Spans

Building	Structure	Place	Year A.D.	Span m	Span ft
Pantheon	Concrete dome	Rome	123	44	143
S. Maria del Fiore	Masonry dome	Florence	1434	42	138
Train Shed, New Street Station	Iron trusses	Birmingham England	1854	64	211
Royal Albert Hall	Iron dome	London	1871	56	185
Exhibition Hall	Iron dome	Vienna	1873	110	360
Galeries des Machines	Iron portals	Paris	1889	111	365
C.N.I.T. Exhibition Hall	Reinforced concrete dome	Paris	1958	206	676
Louisiana Superdome (longest span, 1984)	Steel dome	New Orleans	1975	207	680

TABLE 2
Bridge Spans

Bridge	Structure	Place	Year A.D.	Span m	Span ft
Trajan's Bridge	Masonry arch	Alcantara, Portugal	105	30	98
Trajan's Bridge	Timber truss	River Danube below Iron Gates	104	52	170
Rhine Bridge	Timber truss	Schaffhausen	1758	122	400
Menai Bridge	Suspension	Menai Straits, Wales	1826	177	580
Ohio River Bridge	Suspension	Wheeling, W. Va.	1849	308	1010
Forth Bridge	Steel cantilever	Edinburgh, Scotland	1889	521	1710
George Washington Bridge	Suspension	New York City	1931	1067	3500
Humber Bridge (longest span, 1984)	Suspension	River Humber, N.E. England	1980	1410	4626

TABLE 3
Tall Buildings and Towers Accessible to People

Building	Structure	Place	Year	Height m	Height ft
Great Pyramid	Masonry	Gizeh, Egypt	2580 B.C.	147	481
Colosseum	Masonry	Rome	82 A.D.	48	158
Torre del Mangia, Town Hall	Brick	Sienna, Italy	1348	102	334
Cathedral spire	Masonry	Strasbourg, France	1439	142	465
Cathedral spire	Masonry and iron	Rouen, France	1876	148	485
Cathedral spire	Masonry	Cologne, Germany	1880	156	513
Washington Monument	Masonry, with iron frame for stairs and elevators	Washington, D.C.	1884	169	555
Eiffel Tower	Iron	Paris	1889	301 (original height)	986
Chrysler Building (77 stories)	Steel frame	New York City	1930	319	1046
Empire State Building (102 stories)	Steel frame	New York City	1931	381	1250
TV tower	Concrete	Moscow	1967	385 (original height)	1263
Sears Tower (tallest building, 1984; 110 stories)	Steel frame	Chicago	1974	442	1450
CN Tower (communications and observation tower)	Concrete	Toronto	1975	446	1464

was approached, but not surpassed, by the Duomo of Florence (S. Maria del Fiore) in 1434, and it retained the record until 1854.

The **Longest-Spanning Bridge** (see Table 2) surviving from Ancient Roman times is in Portugal; but there is convincing evidence that during the same century a much bigger span was built over the Danube. No longer-spanning bridge was erected until the 18th century. During the 19th century the longest bridge span, which had previously been of approximately the same size as the longest interior span in a building, increased rapidly so that Table 2 merely gives some important advances.

The **Tallest Structure** (see Table 3) until 1876 was the Great Pyramid at Gizeh. The tallest building surviving from Ancient Rome is the Colosseum. The tallest surviving medieval church spire is that of Strasbourg Cathedral. The tallest surviving medieval secular tower is the Torre del Mangia of the Sienna Town Hall. Table 3 gives details of these structures and of the tall structures that set records since 1876. Television and radio masts not accessible to the public are excluded.

References

1. J. H. STEPHENS (ed.): *The Guiness Book of Structures.* Guiness Superlatives, London, 1976.
2. H. J. COWAN: *The Master Builders.* Wiley, New York, 1977; *Science and Building.* Wiley, New York, 1978.

H.J.C.

SPRINKLER SYSTEMS are automatic devices combining valves activated by heat from a fire with orifices and deflector plates so arranged as to distribute water on the fire. A sprinkler head may be operated by a fusible element (which is more reliable) or a glass bulb filled with liquid, and it may be (1) upright, (2) pendant, or (3) sidewall. The sprinkler heads are attached at precalculated intervals to pipes suspended from the ceiling and have a separate permanent water supply with its own set of control and alarm valves. In the event of a fire reaching a predetermined temperature, the sprinkler mechanism collapses and releases the water held in the pipework. This strikes the sprinkler deflector plate, which converts the stream of water into

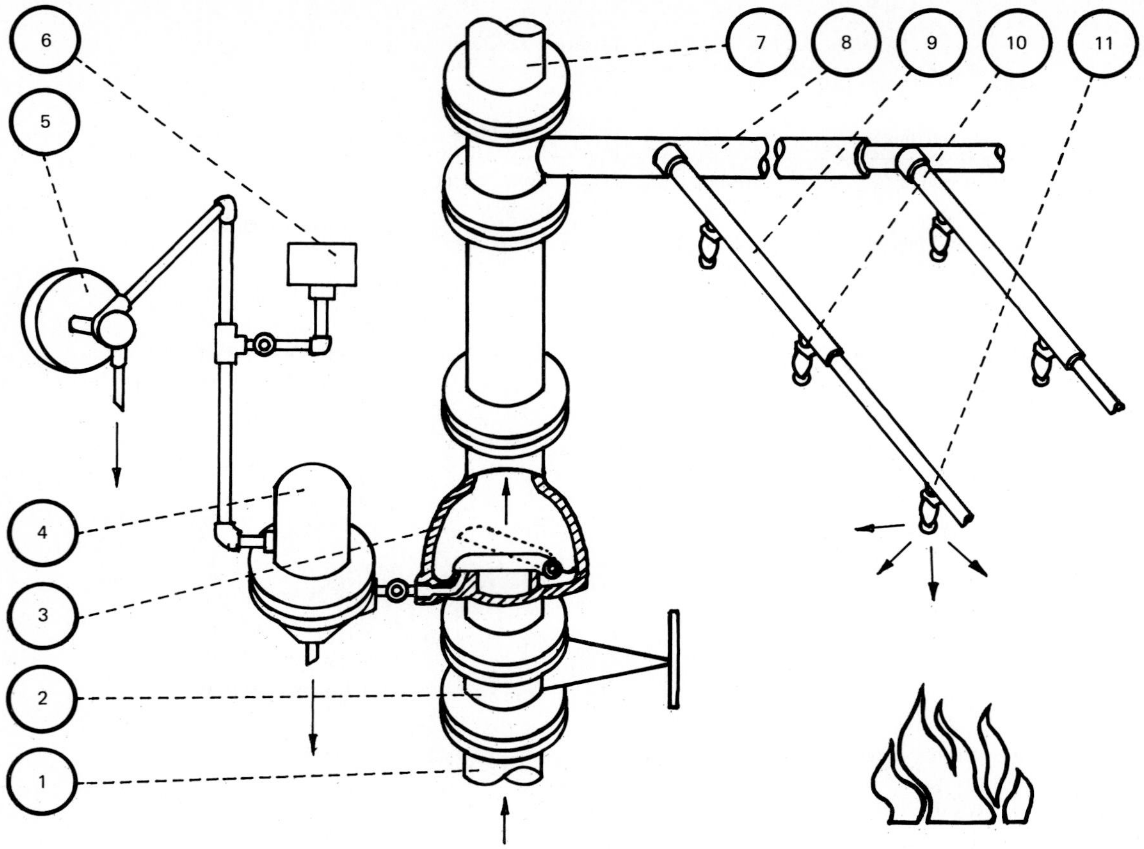

Figure 246 Elements of a sprinkler system. (1) Water supply. (2) Main stop valve. (3) Sprinkler alarm valve [cut away to show how water is released to operate the alarm devices (5) and (6), when the sprinkler head (11) operates]. (4) Retard chamber to eliminate false alarms due to surges in water main pressure. (5) Local water motor alarm bell. (6) Pressure switch for starting pumps, calling the fire department, tripping the air-conditioning system, and closing the fire doors. (7) Main riser. (8) Distributing main. (9) Range pipe with sprinklers. (10) Sprinkler heads. (11) Only the sprinklers in the fire zone operate.

droplets that extinguish the fire. The demand for water caused by the operation of the sprinkler head automatically activates an alarm valve, an alarm bell, and a signal to the fire department. It also starts the pumps and shuts down the air-conditioning plant (Fig. 246).

In freezing conditions, air-filled pipes with special alarm valves are used. When sprinklers open, the air is released and water is immediately supplied to the pipes.

The minimum volume of water per unit area to be provided over the expected area of the fire is specified in the relevant building codes. These codes are based on many years of experience with fires.

The three standard sizes of sprinklers are 10 mm ($\frac{3}{8}$ in.), 15 mm ($\frac{1}{2}$ in.), and 20 mm ($\frac{3}{4}$ in.) for light, ordinary, and high hazards.

Automatic sprinklers are also available for specialized applications of foam, dry chemicals, and vaporizing liquids. These are used where water would be dangerous.

For safety to life and property, nothing has so far surpassed a properly designed and installed sprinkler system.

See also FIRE AND SMOKE ALARMS and FIRE-FIGHTING EQUIPMENT.

References

1. *Fire Protection Handbook*, 13th ed. National Fire Protection Association, Quincy, Mass., 1969. *Section 16.* Fixed Fire Protection Systems—Sprinklers, pp. 1–119.
2. NORMAN J. THOMPSON: *Fire Behaviour and Sprinklers.* National Fire Protection Association, Quincy, Mass., 1964.

A.L.

STAINLESS STEEL is an alloy of iron and chromium or iron, chromium, and nickel; the vital element is chromium.

Research has shown that the corrosion resistance of alloys containing a minimum of 12% chromium is due to a dense, hard, impervious, invisible film of chromium-rich oxide that forms on the surface. This film is self-healing when broken and re-forms readily when the metal surface is exposed to air.

Grades. There are many types of stainless steel, each developed for a specific purpose and differing in chemical composition and resistance to corrosion. The American Iron and Steel Institute (AISI) classification for stainless steels lists more than 40 different grades. These are grouped as follows:

1. The 200 series contains chromium, manganese, and nickel. These steels are metastable and contain austenite, a solid solution of γ-iron (see METALLURGY and STEEL). They are essentially nonmagnetic.
2. The 300 series contains 18% to 25% chromium and 8% to 20% nickel; other elements may be added for special purposes. These steels are austenitic and essentially nonmagnetic.
3. The 400 series (hardenable) is generally chromium–iron alloys, although other elements may be added for special purposes. These steels, when hardened, are mainly martensite, a solid solution of carbon in α-iron. They are magnetic and are hardened by heat treatment.
4. The 400 series (nonhardenable) is generally chromium–iron alloys, although other elements may be added for special purposes. These steels are mainly ferrite, a solid solution based on α-iron. They are magnetic and are not hardenable.

Of the many grades of stainless steel produced, only four are of primary interest to architects, because they offer the most economical combination of beauty, strength, and resistance to weathering and corrosion. These four grades are AISI types 301, 304, 316, and 430. T301 can be temper rolled, that is, cold rolled to give much higher strength with a slight reduction in corrosion resistance. T304 is the most widely used austenitic stainless steel for exterior corrosion-resisting applications. T316 is used where greater corrosion resistance is required in marine and polluted industrial environments. T430 has a lower corrosion resistance than T304 and is generally used for internal trim.

Manufacture. Stainless steels are usually made in an electric arc furnace and refined by an argon/oxygen blast. This is called the AOD process. The molten stainless steel is cast into ingots. When cold, these ingots are dressed all over, reheated, and hot rolled into bar or strip. The hot-rolled strip is annealed, descaled, and then cold rolled into thinner strip. This strip is finished to 2D, 2B, BA, No. 3, No. 4, or other surface finishes and supplied as coil or sheet.

In addition to sheets, strips, and bars, stainless steel is made in tubes, plates, structural sections, and castings. The greatest tonnages used in architecture are in sheet and strip in a variety of finishes.

Applications. Stainless steel is widely used for CURTAIN WALLS, cladding, decorative panels, windows, flashing and gutters, fittings and SANITARY APPLIANCES. The Chrysler building in New York was clad with T304 stainless steel in 1929, and the cladding is still in excellent condition. Since then many large buildings with stainless steel curtain walls have been erected in the United States and Canada. Interesting examples are the Toronto City Hall and the Union Carbide Building, New York City.

Stainless steel has been widely used as a cladding material for roofs and walls. A good example is the Maritime Services Board Control Tower in Sydney (Fig. 247). Stainless steel is also widely used for gutters, both for appear-

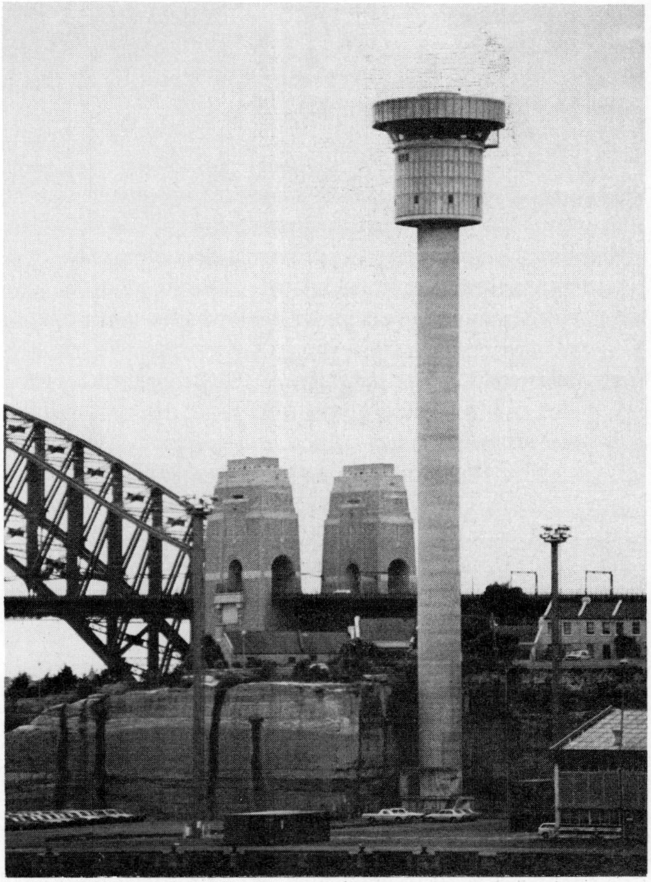

Figure 247 Control Tower of the Maritime Services Board, Sydney, Australia, clad with T304 stainless steel with 2B finish (designed by Sabemo, fabricated by Cabeng Rendell).

ance and long life. The latter is increasingly important due to the escalation of labor costs.

Stainless steel sinks, laundry tubs, and urinals are now regarded as standard equipment in modern buildings; an extensive range of sizes and shapes is available from stock. Stainless steel is specified for kitchens in hospitals, cafeterias, and clubs, where its ease of maintenance, hygienic quality, and "as new" appearance after years of wear have made it the universal choice.

Protection. As with many other building materials, it is advisable to protect stainless steel during erection in order to prevent splashing it with cement and plaster, which, although they do not attack stainless steel, may be difficult to remove. Strippable plastic coatings are often used for protection, but many are affected by strong sunlight, leading to removal difficulties. The manufacturers' advice should be sought, and all surfaces thoroughly cleaned after the building is completed. Together with regular maintenance, this ensures that the surfaces retain their attractive appearance.

Steel scrapers and steel wool should not be used, as they leave rusty deposits on stainless steel. Cleaning materials should not contain any iron oxides, chlorides, or hypochlorites, as these produce pitting. A detergent solution, followed by thorough rinsing with water, is all that is necessary for regular cleaning.

References

1. J. H. CALLENDER, H. A. JANDLE, R. W. McLAUGHLIN, and J. C. RITCHIE: *Curtain Walls of Stainless Steel.* Committee of Stainless Steel Producers, American Iron and Steel Institute, New York, 1955.
2. *Architects' Stainless Steel Library.* International Nickel Company, New York, 1970.
3. *Steel Products Manual: Stainless and Heat Resisting Steels.* American Iron and Steel Institute, New York, 1963.
4. C. B. ROLFE: *Stainless Steel in Building.* Commonwealth Steel Company, Sydney, 1975.

C.B.R.

STAIRS. The design of stairways is constrained by considerations of comfort and safety. Building codes, especially those covering safety from fire, may dictate the number and location of stairways in a building and specify their design and construction in detail.

Proportions of Steps. See Fig. 248. Many traditional formulas relate the going to the rise or the tread length to the rise; most are based on only coarse anthropometric assumptions, and there are no scientific grounds for adhering

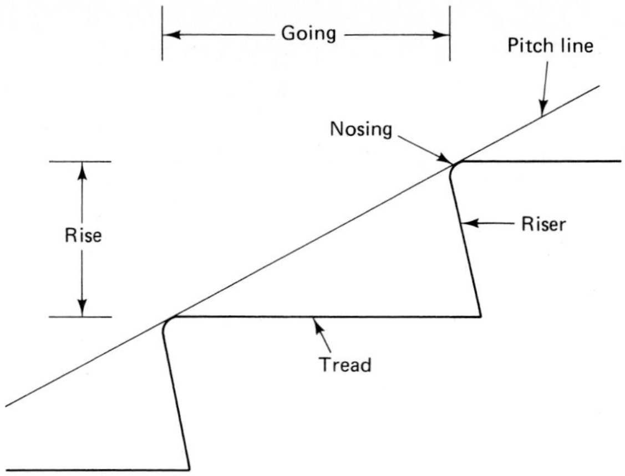

Figure 248 Definition of terms.

exactly to any particular rule. The concurrence of modern practice is that the sum of the going and twice the rise should be between 550 mm ($21\frac{1}{2}$ in.) and 700 mm ($27\frac{1}{2}$ in.), dimensions that accord with the central range of length of an adult's pace. Stairs within private dwelling may be as steep as 210 mm ($8\frac{1}{4}$ in.) rise to 240 mm ($9\frac{1}{2}$ in.) going; in public buildings the pitch should be shallower and may be as low as 100 mm (4 in.) rise with 360 mm ($14\frac{1}{4}$ in.) going. Laboratory studies of energy expenditure on stairs show that, for a given story height, the total energy used grows less as stairs increase in steepness but that the rate of energy expenditure increases. Statistical analyses of accidents show that more occur when descending stairs than when people are ascending and that small treads are associated with a high rate of accidents.

Walking Speeds and Traffic Flow Capacity. Average walking speeds on stairs tend to be lower than the speeds adopted along level corridors, but most people climb short flights more quickly than the speed that minimizes energy expenditure. Individual persons tend to walk down steps faster than upward, but, in dense crowds, the converse can be true. Table 1 lists approximate mean walking speeds and design capacities of stairs for various categories of user. In fire safety codes, the recommendations given are frequently equivalent with a flow capacity of 80 persons/minute/meter width, but this is a greater rate than is normally acceptable in everyday use.

Stairs for Fire Escape. Some stairways may be used in an emergency by a crowd in panic; in almost all countries the design of escape stairs is regulated by mandatory codes. The requirements differ in detail, between national codes, but the following are typical:

1. The stairways must be built within an enclosed shaft of specified fire resistance.

TABLE 1
Approximate Mean Speeds of Movement up Stairways, with Corresponding Flow Capacities

	Free Flow: Mean Plan Density 0.6 P/m²* or Less		Full Design Capacity, One-way Flow: Plan Density 2 P/m²*	
	Speed along Slope (m/s)	Limit of Stair Capacity with Free Flow (P/min/m width)	Speed along Slope (m/s)	Stair Capacity (P/min/m width)
Young and middle-aged men	0.9	27	0.6	60
Young and middle-aged women	0.7	21	0.6	60
Elderly people, family groups	0.5	15	0.4	40

*P/m^2 = persons per square meter

2. The design must be such that a person following the stair is led without impedance to an exit from the building; the stair must not continue without a break down into a basement; all doors into the stairway, except the final escape door, should open inward but without causing obstruction to persons already on the stair.
3. There must be not less than a minimum number of steps (often 3), and not more than a maximum number (often 16) in any flight; the steps must be all of specified uniform dimensions.
4. Adequate handrails must be provided, including central rails to divide broad stairways.

The width of escape stairs is normally determined from the number of occupants of upper floors of the building, using tables or formulas given in the code.

See also EVACUATION OF BUILDINGS IN CASE OF FIRE and HUMAN BEHAVIOR IN FIRES.

Stairs and the Disabled. Although ramps should be provided for wheelchair users, many ambulant disabled prefer short flights of steps. The preferred going is 250 mm (10 in.); the rise should not exceed 190 mm ($7\frac{1}{2}$ in.) and preferably should be less than 170 mm ($6\frac{1}{2}$ in.). Handrails should be provided on both sides at about 850 mm (33 in.) above the line of the nosings; they should run continuously at landings and extend at least 300 mm (12 in.) beyond the top of the stairs. Open risers should be avoided, and splayed risers with slightly rounded nosings are better than undercut square nosings. (See also BARRIER-FREE DESIGN).

References

1. PETER TREGENZA: *The Design of Interior Circulation.* Van Nostrand Reinhold, New York, 1976.
2. SELWYN GOLDSMITH: *Designing for the Disabled*, 3rd ed. Royal Institute of British Architects, London, 1976.

P.R.T.

STATICALLY DETERMINATE STRUCTURES can be designed directly, choosing whatever form satisfies the functional requirements, because the strength of their elements (members) may be determined uniquely using elementary statics and the maximum safe stresses for the material of construction. The equations of statics employed specify that the structure as a whole and all its parts shall be in equilibrium. The design of such structures was developed extensively after 1850 due to the railroad boom, which created a need for heavy-duty iron and masonry bridges and large buildings with iron roof structures. However, some time earlier MASONRY ARCHES were designed as statically determinate structures, relating to the conditions of equilibrium for each voussoir; Brunel used that method for his masonry railroad bridges. Other typical examples of elementary statically determinate structures are the single beam resting freely on supports at each end, the cantilever, the triangular truss (Fig. 249), and the three-pin arch (Fig. 250). SPACE FRAMES also may be statically determinate.

Graphic statics, based on the triangle of forces, is well suited to the design of statically determinate structures. It was developed extensively in the 19th century to avoid tedious manual calculations of strength, with the likelihood of error. Later simple scale MODELS were recognized as complementary to graphics in the design office. More recently, COMPUTER METHODS have revolutionized the design process, with facilities to examine and OPTIMIZE alternative designs automatically for function and safety. Those methods are especially useful for the ELASTIC ANALYSIS OF STATICALLY INDETERMINATE STRUCTURES, which require the use of equations of compatibility of elastic strains in addition to those of statics.

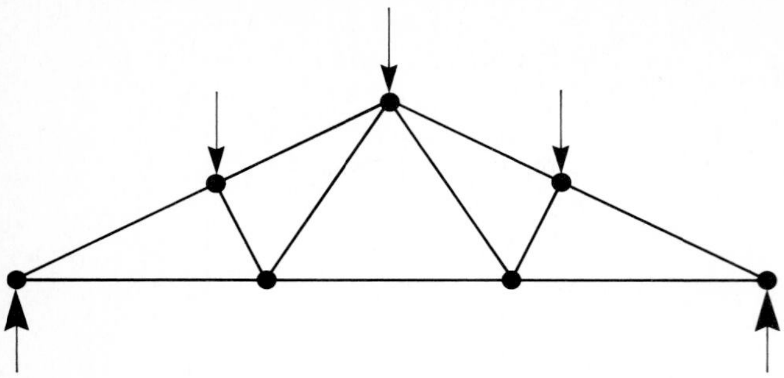

Figure 249 Triangulated truss with simple (pinned) joints. Triangulated trusses with simple joints are always statically determinate. If one or more members are removed, the structure becomes a mechanism. If one or more members are added or if any of the simple joints are made rigid, the structure becomes statically indeterminate.

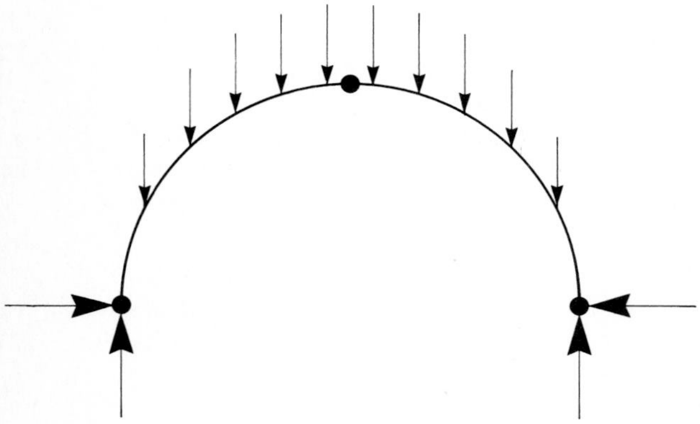

Figure 250 Three-pinned arch. The structure collapses if another pin joint is added (see MASONRY ARCHES). The arch becomes statically indeterminate if one or more pin joints are removed.

It is not always easy to distinguish between the two categories of structure; STEEL STRUCTURES are more likely to be statically determinate than reinforced CONCRETE STRUCTURES; but a detail such as the nature of the joints (simple joints or rigid joints) between members (see JOINTS FOR STEEL STRUCTURES) may introduce indeterminacy. The reason for choosing indeterminacy might be a saving in structural material or better resistance to WIND LOADS or EARTHQUAKES.

A homely illustration of the difference between the two categories is provided by three- and four-legged stools on a rough floor. Statics is sufficient to find the compressive force in each leg of the three-legged stool, but is insufficient with regard to the four-legged stool, which has one leg more than is needed by statics for equilibrium. Also, the former is stable on uneven surfaces by virtue of its statical determinacy, while the four-legged stool may wobble.

References

1. R. C. COATES, M. G. KONG, and F. K. KONG: *Structural Analysis*. Nelson, London, 1980.
2. W. MORGAN: *The Elements of Structure*. Pitman, London, 1977.

T.M.C.

STATISTICS is concerned with the collection, presentation and interpretation of numerical data. The distribution of numbers in a set $\{x_i: 1 \leq i \leq n\}$ may be described by a function of the numbers that is called a *statistic*. For example, the arithmetic *mean*

$$\bar{x} = \frac{1}{n} \sum_{i=1}^{n} x_i$$

indicates where the set of numbers is centered, and the *median* is the middle number of the set. Each of these statistics could be described as an average, and an author using this term must explain which statistic he or she is using. The sample standard deviation S for a set of n numbers $x_1, \ldots, x_n$ is defined by

$$S = \sqrt{\frac{1}{n-1} \sum_{i=1}^{n} (x_i - \bar{x})^2}$$

where $\bar{x}$ = arithmetic mean

It is a measure of the spread of the set of numbers.

Probability. The interpretation of statistics requires a mathematical model to describe the behavior of random phenomena; this is the *mathematical theory of probability*. Its axioms are based on the empirically observed property of *statistical regularity*. In an idealized *random experiment* it is impossible to predict the outcome of a single trial, and the result of one trial is independent of the results of other trials. A variable whose value depends on a random experiment is a *random variable*. The distribution of a continuous random variable, X, is described by its *probability density function*, $f(x)$. The area under the graph of this function $f(x)$ between $x = a$ and $x = b$ is equal to the relative likelihood that X takes a value between a and b (Fig. 251). All values of this function are taken nonnegative, and the total area under the graph is 1.

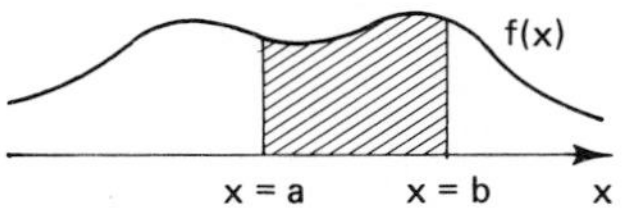

Figure 251 Probability density function.

Normal Variable. The most important random variable has probability density function

$$f(x) = \frac{1}{\sqrt{2\pi}\sigma} e^{-(x-\mu/\sigma)^2}$$

where π = circular constant 3.14 . . .
e = base of the natural logarithm

The parameters μ and σ are, respectively, the mean and the standard deviation of the variable. The graph of $f(x)$ is symmetrical about $x = \mu$ and has a bell shape (Fig. 252). The variable is called the *normal variable* (also known as the Gaussian variable) and is denoted by $N(\mu, \sigma^2)$. Many naturally occurring variables, such as the height or weight of a population of people, can be modeled accurately by this variable. However, its particular usefulness is due to a theoretical result, called the *central limit theorem*, that establishes that the sum of identically distributed independent random variables will be approximately normally distributed provided that the number of variables is sufficiently large, and regardless of the shape of the individual distribution.

Sampling. To obtain information about various properties of a population of numbers when it is inconvenient or impractical to take a census of the whole population, a sample of values $\{x_i: 1 \leqq i \leqq n\}$ may be taken to infer from it properties of the whole population. The sample chosen should be representative of the whole population. Ideally, each member of the population should be equally likely to be included in the sample; but in practice this is often diffcult to achieve, and sampling techniques for a given population are continually modified by experience. Probability theory models how particular sample statistics should be distributed (in many cases normally) and hence provides an indication of just how accurately a sample statistic is likely to estimate a particular population parameter.

Hypothesis Testing. To test a hypothesis about a population, one can choose an appropriate statistic and by sampling (or by performing an experiment) obtain a sample value for the statistic. *Probability theory* enables one to work out the theoretical distribution of the test statistic assuming that the hypothesis is true. In particular, one can work out the probability of obtaining an experimental result as or more extreme than that actually observed. The hypothesis is rejected if this probability is less than that specified, say 0.05 or 0.01. The figure selected as the criterion for rejection of the test is called its *level of significance*. Common tests include the chi-square test for goodness of fit, the (Student's) *t*-test for comparing population means, and the *F*-test for comparing population variances.

See also QUALITY CONTROL OF MATERIALS.

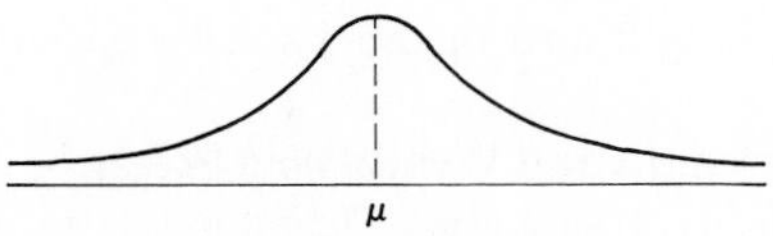

Figure 252 Normal variable.

References

1. I. GUTMAN and S. S. WILKS: *Introductory Engineering Statistics*. Wiley, New York, 1965.
2. WILLIAM FELLER: *Introduction to Probability Theory and Its Applications*, Vol. 1. Wiley, New York, 1968.
3. J. J. MORONEY: *Facts from Figures*. Penguin Books, Harmondsworth, England, and New York, 1974.
4. DARREL HUFF: *How to Lie with Statistics*. Penguin Books, Harmondsworth, England, and New York, 1977.

R.G.Cr.

STEEL is an alloy of IRON and carbon and various other alloying elements such as silicon, manganese, chromium, nickel, tungsten, and molybdenum. As used today in building it is over 98% iron, less than 1% carbon, and about 1% manganese, with small quantities of various other elements.

Manufacture. Iron ore is any mineral that contains an adequate proportion of iron oxide. Iron is extracted from it and refined, and chemicals are added to achieve the particular grade of steel desired. The first process is the production of pig iron from the iron ore in a blast furnace, a massive furnace well over 35 m (115 ft) in height, surrounded by ancillary buildings and ore yards. The ore, coke, and limestone are introduced into the top of the furnace, and a blast of heated air enters at the bottom. The products are molton pig iron, slag, and gas; all may be used in other manufacturing processes. Normally, the pig iron is used immediately in its molten state to make steel.

Pig iron can also be made in electric furnaces or from steel scrap. It is expected that the direct reduction of iron ore using natural gas will become feasible; this would remove the need for the expensive and huge blast furnace. The manufacture of pig iron results not only in a purification of the iron, but also in the addition of carbon and many other elements. The next process is the oxidation of carbon and other elements to produce the chemistry of the alloy specified for the grade of steel. Until about 20 years ago, this was done in an open-hearth furnace. Today the oxidation process normally uses a basic oxygen furnace. Electric furnaces are also used.

Oxygen is introduced into the furnace of molten pig iron, steel scrap, and slag-forming fluxes. The carbon is transformed into the gases carbon monoxide and carbon dioxide, and impurities such as manganese, silicon, sulfur, and phosphorus are eliminated in the slag, which floats to the surface and is removed. The molten steel is led into a ladle and, upon solidification, becomes a steel ingot.

For use in buildings, steel is produced as steel plate, steel shapes, or sheet steel. To produce steel plate and steel shapes, the steel ingot is reheated and then hot rolled to the desired geometry in a mill. The process is designed to produce the desired grain size, elongation, and other mechanical properties of the final product. After the rolling, the plate or shape is cooled on a cooling bed and, if necessary, straightened to specification tolerances. Sheet steel, used for roofing or siding, may be either hot rolled or else hot rolled and then cold rolled; the surface of cold-rolled steel must be treated either by galvanizing or painting to protect it against CORROSION.

Properties. Steel combines strength with ductility at relatively low cost, and it is a very versatile building material. Its metallurgical and mechanical properties depend particularly on the amount of carbon. They are also affected greatly by temperature, by speed of testing, and by the shape and size of the test specimen. For this reason, standardized tests have been devised, such as those of the American Society for Testing and Materials (ASTM), that specify size of specimen, temperature, and speed of testing (see TESTING OF MATERIALS).

The mechanical properties are obtained from the results of a stress–strain test. Figure 253a shows the behavior of ASTM A36 steel, a standard American structural steel. Figure 253b shows the initial portion of the curve, which contains the range of the service or working stresses. The yield strength, σ_y, defines the onset of yielding in the material, which remains essentially constant for a relatively large strain; this ability of the steel to undergo large plastic deformations without fracture is called its *ductility*. It is the ductility of steel that makes it such a safe material and allows its analysis and design to be exact (see STEEL STRUCTURES and LIMIT DESIGN).

The chemical composition has an important effect on the mechanical properties (see METALLURGY). Carbon increases the strength and hardness, but adversely affects ductility and weldability. Manganese is important for strength, ductility, and heat-treating processes. Sulfur and phosphorus tend to make steel brittle. Chromium, nickel, and tungsten increase both strength and ductility.

Whereas yield strength and ductility are the most important mechanical properties of steel, a number of its other mechanical properties are significant:

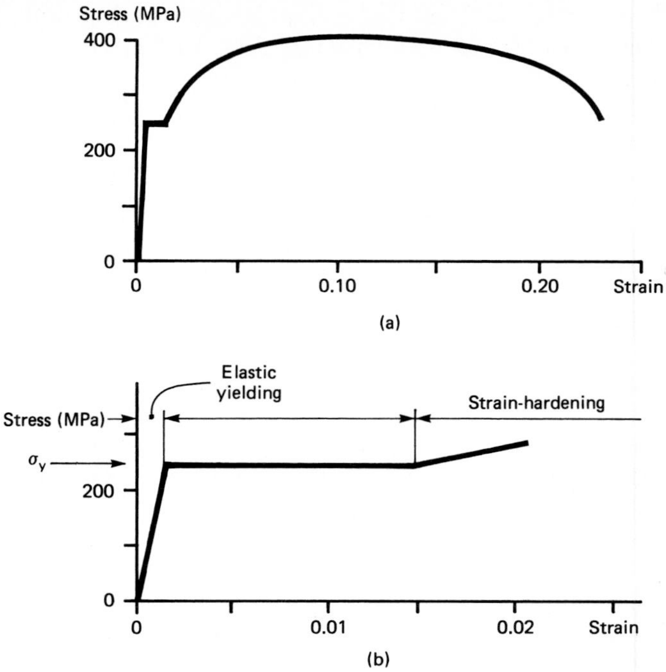

Figure 253 Typical stress–strain relationship for structural carbon steel in accordance with ASTM specification A 36 (Ref. 3).

Tensile strength: the maximum strength observed in a test on a tension specimen, defined as the maximum axial load divided by the original cross-sectional area.

Modulus of elasticity, E: the ratio of stress to strain (see MEASUREMENT OF STRAIN) in the elastic range (Fig. 253). This is the same for all structural steels, equal to 200,000 MPa or 30,000 ksi at room temperature. The modulus of elasticity determines the deflections that a structure can undergo. Since it is essentially a constant, deflections are not reduced by the use of very high strength steels; indeed, deflections are increased because of smaller cross-sectional areas made possible by the use of high-strength steel. This explains the limited use of such steels.

Weldability: the ability of the steel to be welded without affecting its mechanical properties (see JOINTS FOR STEEL STRUCTURES).

Fatigue strength: the ability to withstand repeated dynamic loads.

Fracture toughness: a measure of the resistance to brittle fracture, which is an instantaneous CRACK propagation due to a flaw or other stress concentration.

High-Strength and Weathering Steels. Steel is available in a variety of strengths. This ranges from mild steels (*structural carbon steels* such as ASTM A36 with a yield of approximately 250 MPa or 36 ksi) and the high-tensile steels (*high-strength low-alloy steels* such as ASTM A572 of approximately 350-MPa or 50-ksi yield), to the very high strength steels (the *quenched* and *tempered alloy steels* such as ASTM A514 of approximately 700-MPa or 100-ksi yield.) Special-purpose steels of even higher yield points (1000 MPa or 145 ksi) are available, although these combine increased price with decreased ductility.

The strength of the steel depends in some part on the size of the cross section used, and therefore not all cross-sectional shapes are available for the higher-strength steels. Since the reduction of the cross section from the ingot during the hot-rolling process contributes greatly to the strength through the hot-working of the grain structure, heavy shapes are not generally available for high-strength steels since their size precludes extensive rolling. Heavy shapes may obtain higher strengths through a control of the chemical composition, that is, higher carbon, or through the use of heat-treatment techniques.

High-strength low-alloy steels have been in wide use since COR-TEN was introduced in 1933, and they offer higher strength, good weldability, and, moreover, better corrosion resistance than structural carbon steels. Corrosion resistance becomes important with higher-strength steels because the smaller cross sections have less thickness to resist rusting over the life of the structure. "Weathering steels" have been introduced in the past two decades. These steels are left unpainted and allowed to oxidize; they eventually develop a pleasing patina whose color depends on the chemistry of the steel and the pollution of the atmosphere. This oxidation (or rust) forms a protective coating that prevents further oxidation. Weathering steels are available over a wide range of yield strengths, set out in specification ASTM A588 (see also STAINLESS STEEL).

References

1. L. TALL (ed.): *Structural Steel Design*, 2nd ed. Wiley, New York, 1974.
2. H. E. McGANNON (ed.): *The Making, Shaping, and Testing of Steel*, 8th ed. U.S. Steel Corp., Pittsburgh, Pa., 1974.
3. *The 1984-Annual Book of ASTM Standards* (in 66 parts). American Society for Testing and Materials, Philadelphia, 1984.

L.T.

STEEL STRUCTURES have high strength and stiffness, and excel in long-span buildings such as public assembly buildings and industrial warehouses (Fig. 254). They have many other applications, ranging from the frames of TALL

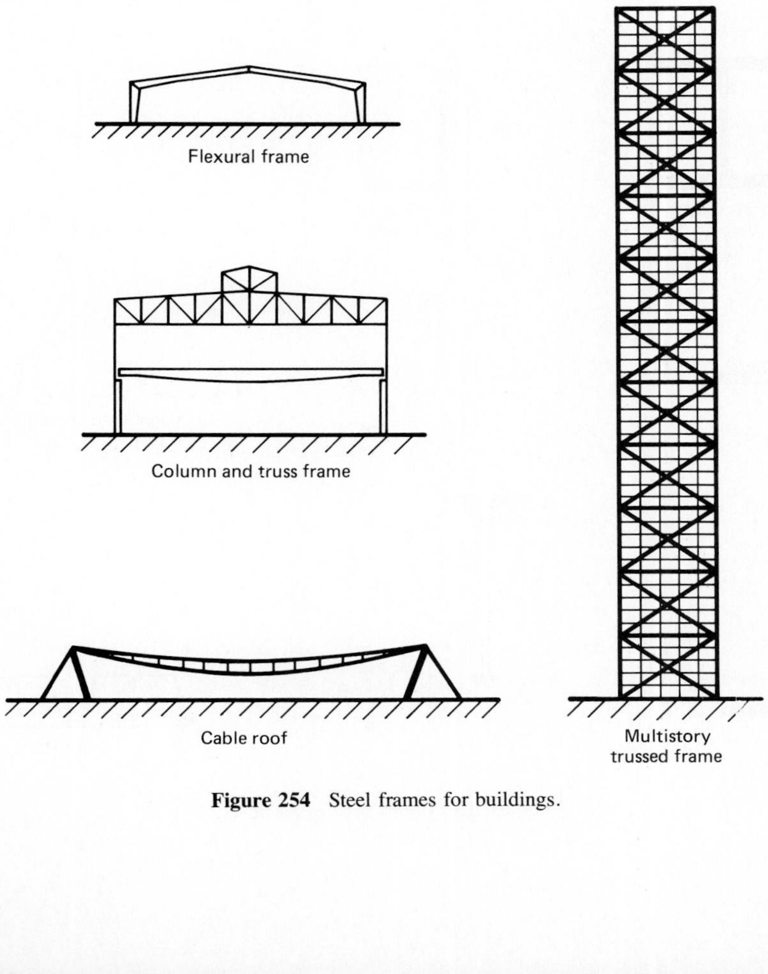

Figure 254 Steel frames for buildings.

BUILDINGS to special uses in single-story domestic buildings. Steel sheeting is widely used as cladding for roofs and walls and for flooring systems. In many uses, the sheeting acts in conjunction with other structural elements to form a stressed skin structure. In COMPOSITE CONSTRUCTION, steel members are combined with reinforced concrete elements such as slabs.

Structural Forms. CABLE AND SUSPENSION STRUCTURES or SPACE FRAMES are often used for very long spans, with triangulated trusses for moderate spans and flexural frames for shorter spans. In flexural frames, horizontal beams transfer the loads by bending actions to vertical columns, which transmit them by axial compression to the FOUNDATIONS. The loads on trusses are transmitted by tension and compression actions in the members. Steel members are thin walled for efficiency and may be hot rolled, fabricated from plates by cutting and WELDING, or cold formed (Fig. 255). The structure is formed by connecting the members at JOINTS by welding or bolting.

The **Strength** of a steel structure depends principally on its resistances to yielding and BUCKLING. Other failure modes include brittle fracture, which is rare in buildings made with modern steels, and fatigue cracking, which is only a consideration in structures such as crane runways that are subject to many load repetitions.

Yielding governs the strengths of tension members and of other members that are short or braced against lateral deflection and that have comparatively thick plate elements. When the force intensity (stress) becomes high enough, yielding takes place; the STEEL flows plastically, and large permanent deflections may occur.

Buckling governs the strengths of slender members or plate elements in compression. The action is one of lateral deflection (or twisting) perpendicular to the action of the applied loads. The buckling resistance depends on the slenderness of the member. For a thin plate element, this is related to its width/thickness ratio and for a long member to its length/depth ratio. The strengths of members of intermediate slenderness also depend on yielding. Frame buckling may affect the strength of a complete structure, as in the case of tall unbraced multistory buildings, in which lateral displacements due to wind are increased by the action of heavy gravity loads.

The **Stiffness** of a steel structure is measured by its resistance to deflection. Slender members may deflect excessively under load or may give rise to vibration problems.

Design. In the design of a steel structure, the loads that act on it are first assessed, and then the structure is analyzed (see COMPUTER METHODS OF STRUCTURAL ANALYSIS) to determine the forces transmitted by its members. These are usually classified as tension or

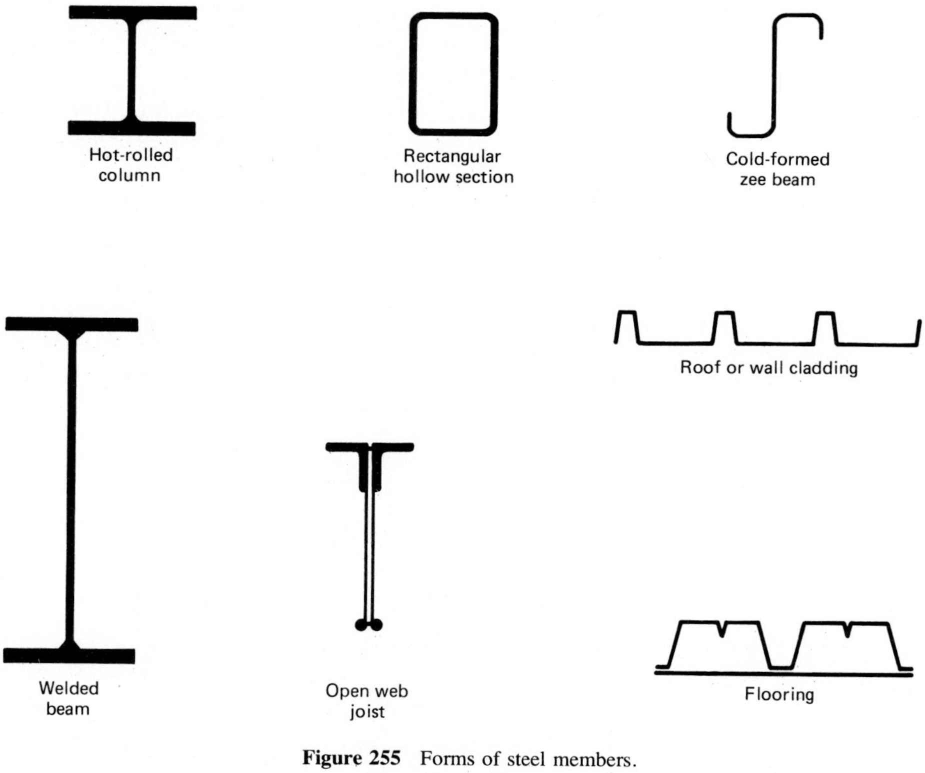

Figure 255 Forms of steel members.

compression members, as beams, or as beam-columns (which transmit both transverse and axial forces).

In the ELASTIC DESIGN of *hot-rolled members*, structural adequacy is assessed at working load level by comparing the calculated stresses with permissible stresses obtained from yielding or buckling considerations by using factors of safety, which are often close to 1.7. In the *plastic design* (see LIMIT DESIGN OF STEEL STRUCTURES) of stocky flexural structures that have sufficient bracing to prevent buckling, the structure is assessed by comparing increased loads (usually by a factor of 1.7) with the loads that would cause flexural collapse. More logical LIMIT STATES DESIGN methods are now being introduced, and these will lead to more consistent designs. The design methods for *cold-formed members* account for the reduced effectiveness of the thin walls, but allow advantage to be taken of the increased yield strength caused by cold working.

Fabrication and Erection. Steel structures are fabricated in workshops by connecting together the plates and members to form segments. This allows the use of sophisticated cutting, folding, bending, drilling, and welding equipment, and in modern workshops an assembly-line technique is often used to advantage. The segments are then transported onto sites and erected using a variety of cranes and the like to position them before their final connection by welding or bolting. Steel structures are usually provided with FIRE PROTECTION and treated to ensure CORROSION PREVENTION.

References

1. N. S. TRAHAIR: *The Behaviour and Design of Steel Structures*. Chapman and Hall, London, 1977.
2. *Specification for the Design, Fabrication, and Erection of Structural Steel for Buildings*. American Institute of Steel Construction, New York, 1975.
3. *Specification for the Design of Light-Gage Cold-Formed Steel Structural Members*. American Iron and Steel Institute, Washington, D.C., 1977.
4. E. H. GAYLORD and C. N. GAYLORD: *Structural Engineering Handbook*, 2nd ed. McGraw-Hill, New York, 1979.

N.S.T.

STRUCTURAL MODEL ANALYSIS is used for the analysis of structures or structural members of complex shape for which exact mathematical solutions are not readily available. A full-scale test is normally too expensive, and an accurately scaled MODEL is made as small as the technologies of model making and measurement permit. The strains and/or deformations due to the loads are measured (see MEASUREMENT OF STRAIN and TESTING OF MATERIALS). The model may also be tested to destruction to determine the ultimate load.

Model analysis is normally undertaken in conjunction with a theoretical analysis based on a simplified version of the structure or structural member. The result of a model test is not as precise as a theoretical analysis, but the dimensions of the structure and its support conditions can be accurately copied. Thus the theoretical analysis of the simplified structure complements model analysis. The increasing versatility of COMPUTER-BASED STRUCTURAL DESIGN has gradually reduced the scope for structural model analysis.

History. Models have been used as an aid to structural design at least as far back as the Renaissance. We know of models for the dome of the Florence cathedral, for St. Peter's in Rome, and for many churches. However, it seems likely that these models were not used quantitatively until well into the 20th century and that their purpose was purely visual, to assist the architect in understanding the structure.

In the late 18th century the elastic theory made rapid progress. It was used more and more for structural design during the next few decades, and empirical rules gave way to structural calculations. The growing sophistication of structural theory was largely responsible for the increasing complexity of engineered structures in the 20th century. This encouraged the search for methods that obviated the need for laborious calculations. Measurements on scale models provided one possible solution, and the first attempt to use them for structural design was made by G. E. Beggs at Princeton University in 1922.

In the 1930s, imaginative concrete structures of complex shape came into use, particularly in Southern Europe, for which a precise mathematical analysis was impossible. P. L. Nervi in Italy and E. Torroja in Spain used scale models to aid their design. During World War II, miniature electric resistance strain gauges were developed for tests on aircraft structures. These could be attached to models with ease, they were more accurate than previously available instruments, and the results could be recorded automatically (see MEASUREMENT OF STRAIN). This led to rapid development in the use of model analysis for architectural structures during the 1950s and 1960s. During the 1970s there has been a decline due to the development of COMPUTER-BASED STRUCTURAL DESIGN, which makes it possible to model mathematically many structural members that could previously be modeled only physically.

Dimensional Theory. The π-theorem, enunciated by E. Buckingham in 1914, established the number of dimensionless ratios (πs) required for dimensional similarity between two physical phenomena, for example, a structure and its reduced-scale model. If the two phenomena are determined by r quantities that can be expressed in terms of n primary dimensions (mass, length, time, etc.), the number of πs is $r - n$.

These πs are easily satisfied by elastic models under static loads, for which there are only five quantities reducible to two basic dimensions. However, when models are tested to destruction or when models are subjected to dynamic loads due to earthquakes, it may be impossible to satisfy all πs, and the designer must use his or her judgment to determine which of the ratios have a significant influence on the measurements and which may be neglected.

Principal Methods Employed. The oldest method was to model the elastic behavior of the entire structure under static loads (see ELASTIC DESIGN). Scale ratios are normally between 1:100 and 1:30. Most models are made from a plastic material, such as Plexiglas. Figure 256 shows the construction of a model, and Fig. 257 shows the completed model under test.

A model made of a dissimilar material can be tested only within the elastic range, and it therefore gives information only on the elastic behavior of the structure. For reinforced CONCRETE STRUCTURES, in particular, it is sometimes desirable to make a model that can be tested to destruction. It is necessary to model not only the dimensions of the structure and the loads acting on it, but also the physical properties of the structural materials. Thus the steel reinforcement must be modeled by steel wires with similar properties, which often means that it must be specially made. The concrete must also be modeled with microconcrete in which the aggregate is suitably scaled. Evidently, there is a limit to the scale reduction possible, and tests reported lie within the range of 1:15 and 1:3. Thus the cost is much higher than for an elastic model.

Figure 257 Dynamic test on a 1:40 model of the Parque Central Buildings in Caracas, Venezuela, at the ISMES Laboratory in Italy to test its response to earthquakes. (Photograph by courtesy of Professor G. Oberti.)

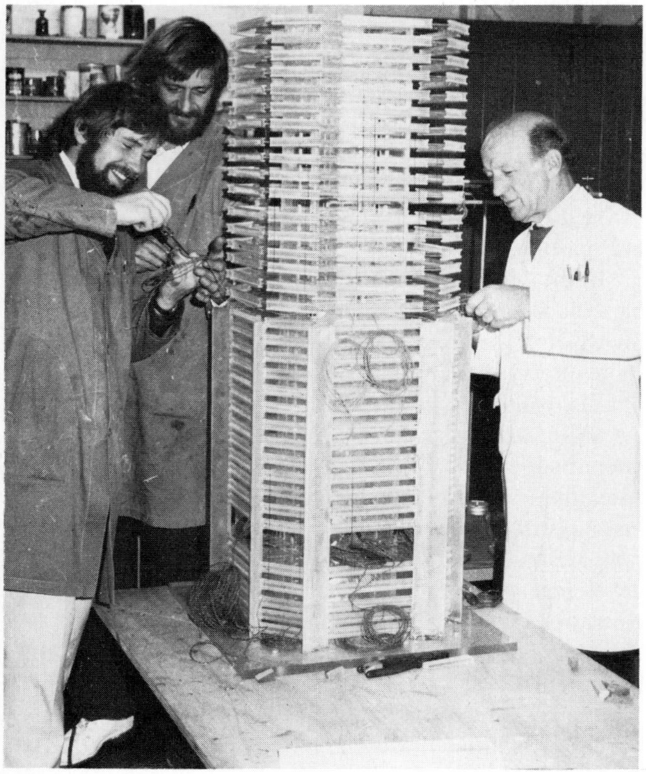

Figure 256 Construction at the University of Sydney of an elastic model for the MLC Tower [a 64-level reinforced concrete building in Sydney, 250 m (800 ft) high] from Plexiglas to a scale of 1:95. The wires lead to 80 electric resistance strain gauges, which record strains due to wind loads.

A static load test is adequate for strain analysis due to gravity loads and also, in most cases, due to wind loads. In regions subject to EARTHQUAKES, however, the dynamic response of the structure may be required. Dynamic analysis introduces several additional variables, and so far all model tests of architectural structures have been confined to the elastic range.

A fourth type of model employs PHOTOELASTICITY. This is a property of certain transparent materials to break up incident light into two components polarized in the directions of the principal stresses. Two polarizing filters or prisms are placed on each side of the model at right angles to one another so that no light is transmitted when the model is unloaded. When the model is loaded, fringes appear that connect points of equal difference between the two principal stresses. The stresses can be calculated by counting a number of fringes. This method is particularly useful for measuring stress concentrations in reentrant corners or to gauge the effect of opening or sudden changes of cross section in walls. It is thus a method of analyzing a part of the structure, rather than the whole structure.

References

1. E. FUMAGALLI: *Statical and Geomechanical Models*. Springer, Vienna and New York 1973.
2. H. J. COWAN et al.: *Models in Architecture*. Elsevier, London, 1968.
3. W. B. PREECE and J. D. DAVIES: *Models for Structural Concrete*. CR Books, London, 1964.
4. J. W. DALLY and W. F. RILEY: *Experimental Stress Analysis*. McGraw-Hill, New York, 1978.

H.J.C.

SUBSURFACE EXPLORATION defines the ground conditions beneath a building (and sometimes adjacent to the building if, for example, a steep unstable cliff or some other important feature occurs next to it) to allow the design of a safe FOUNDATION. Many buildings have failed because ground problems were not discovered due to inadequate subsurface studies or the omission of any such studies. Subsurface exploration is normally planned and undertaken by a geotechnical engineer using the disciplines of geology and SOIL MECHANICS.

Planning of a subsurface exploration program may vary from a simple judgment based on experience to a significant study in its own right. In all cases, however, an assessment must be made of the probable ground conditions in conjunction with the foundation requirements of the structure to set the detail objectives of the investigation. For example, at a residual clay site with shallow rock, the objectives for a single-story dwelling would be to determine the swell–shrink behavior of the clay. If a multistory building was planned for the same site, the objectives would be different and would include ease of excavation for basements and the capacity of the rock to support high concentrated loads.

Once these objectives are set, a suitable means of subsurface access must be selected, such as drilling or test pits, and a testing program designed to allow the determination of appropriate soil and rock properties. Boreholes and test pits should be located to cover specific features such as watercourses and changes in slope, rather than a strict grid. There are no set rules for the number or spacing, although 30 m (100 ft) is a common maximum spacing.

The depth of exploration must be sufficient to reach beyond the depth affected by the building and, where feasible, should extend to rock. A depth of two to three times the footing width below foundation level is a reasonable guide for stronger, uniform soils and widely spaced footing;

TABLE 1
Methods of Subsurface Access

Method	Material	Depths Commonly Achieved	Notes
Spiral flight auger	Soil	30 m (100 ft)	Good where holes stay open. Commonly, 100 to 200 mm (4 to 8 in.) diameter holes to allow tube sampling.
Wash boring	Soil		Logging more difficult and frequent sampling essential for high-quality sampling. Hole should be water or preferably mud filled.
Rock coring	Rock	No practical limitation	Triple tube core barrel at least NMLC size desirable.
Unshored test pits with excavator	Soil or weak rock	6 m (20 ft) generally	Only used in stable ground. Allows very detailed logging of hole. Good-quality undisturbed samples can be obtained by hand trimming.

but deeper depths are necessary in weak soils, where the soil weakens with depth or where footings are close enough to interact.

Subsurface Access is necessary to allow examination of subsurface materials and to obtain samples for testing. The reduced level of each hole or pit should be measured and the depths and descriptions of the soils and rocks encountered logged and reported by a qualified person using an acceptable standard (see SOIL CLASSIFICATION and Refs. 1 and 2).

Methods of subsurface access are summarized in Table 1. The most commonly used forms of drilling in soil are spiral-flight auger drilling, where holes will stay open, and wash boring with casing or in ''mud'' filled holes, where support is required. In both cases samples are recovered for identification and testing of the soils. In rock, diamond core drilling is most commonly used to recover continuous cores of the rock 50 mm (2 in.) or more in diameter.

Disturbed samples dug from test pits or recovered from auger flights are useful for soil identification, but not for laboratory strength or compressibility testing. For these tests, undisturbed samples are required and are obtained by hand trimming in test pits or by thrusting into the soil steel tubes with specially prepared cutting edges. Samples in some form are usually recovered at least every 1 to 2 m (3 to 6 ft) and whenever a change of soil is apparent. In critical situations continuous samples are recovered.

TABLE 2
Site Testing

Test	Description	Comment
Standard penetration test	*N* value is number of blows of 63.5-kg (140-lb) weight falling 455 mm (18 in.) to drive a standard 5 cm (2 in.) diameter split sampler 305 mm (12 in.) (ASTM D1586-67 1976).	Produces a semidisturbed sample. Approximate correlations available with strength and deformation properties.
Cone penetration test	A 1000-mm^2 (1.5-in.^{2}) cone is pushed into the soil and the resistance measured.	Approximate correlations available with strength and deformation properties.
Vane shear test	A cruciform vane is pushed into the soil and the torque to twist the vane is measured.	Used for shear strength testing in soft clays.
Plate bearing test	A plate resting on the soil is loaded and the deflection measured.	Measures immediate compressibility and strength of soil. Only tests a depth of about the plate diameter.
Pressuremeter tests	A cylinder is expanded against the borehole sides, and pressure and radial deflection are measured.	Good for deformation and strength testing. Self-boring instruments in soils avoid sample disturbance.
Permeability tests	Water pumped into or out of hole.	Only reliable means of soil permeability measurement.
Geophysical methods	Measures seismic velocity, resistivity, or other properties.	Useful for broad-scale studies and planning of borehole locations. Needs specialist interpretation.

Testing of the soil or rock may be undertaken on the building site (Table 2) or in the laboratory on undisturbed samples.

Reporting of subsurface exploration should include a full description of the methods used, the location and reduced level of boring and pits, detail logs of the materials encountered, and all test results. A geological interpretation of site conditions to define the occurrence of the various soil and rock types and an assessment of their properties that are relevant to the project is also required.

References

1. JOSEPH BOWLES: *Foundation Analysis and Design*, 2nd ed. McGraw-Hill, Kogakusha, Tokyo, 1977.
2. Report of the Task Committee on Subsurface Investigation for Design and Construction of Foundations of Buildings. *Journal of the Soil Mechanics and Foundation Division, American Society of Civil Engineers*, Vol. 98, Nos. SM5, 6, 7, and 8 (May, June, July, and August 1972), pp. 481–492, 557–578, 749–766, 771–786.
3. *Code of Practice for Site Investigations BS 5930*. British Standards Institution, London, 1981.
4. *SAA Site Investigation Code AS1726*. Standards Association of Australia, Sydney, 1981.
5. M. C. ERVIC (ed.): *In situ Testing for Geotechnical Investigations*. A. A. Balkema, Rotterdam, The Netherlands, 1983.
6. COMMISSION ON RECOMMENDATIONS ON SITE INVESTIGATION TECHNIQUES: *Recommendations on Site Investigation Techniques*. International Society for Rock Mechanics, Lisbon, July 1975.

C.P.T.

The **SUN** is the source of most energy used in buildings. Firewood is produced directly by photosynthesis, the conversion by green plants of water and carbon dioxide into organic compounds. This chemical reaction requires energy, which is provided by solar radiation. Coal, natural gas, and oil are derived from plants that grew millions of years ago. SOLAR ENERGY can be generated with SOLAR COLLECTORS or PHOTOVOLTAIC CELLS.

Hydroelectric power is an indirect form of solar energy, for the sun evaporates the water that forms the clouds that eventually fall as rain to fill the reservoirs. Wind is partly produced by solar radiation and partly by the rotation of the earth. Most important of all, the energy expended by people and animals is replenished by eating plants or meat. Except for fossil fuels, these forms of energy are renewable.

Solar Spectrum. The effective temperature of the sun is about 5900 K (kelvins measured from absolute zero, equal to about 5600°C or about 10,000°F). The sun emits energy approximately as does a black body (i.e., a body with a full absorptive power and no reflecting power), and the energy reaches the earth in that form. The distribution of radiation outside the earth's atmosphere is shown in the upper curve in Fig. 258. The mean value of this radiation is 1395 W/m^2 (442 Btu/ft^2 · h), and this is called the *solar constant*. Some solar radiation is absorbed by the earth's atmosphere, notably by ozone, water vapor, and carbon dioxide. The lower curve in Fig. 258 shows the distribution to be expected on a sunny day at sea level.

The solar spectrum ranges approximately from 300 to 3000 nm (nanometers, or meters × 10^{-9}). The spectrum of visible light ranges from 390 nm (violet) to 760 nm (red).

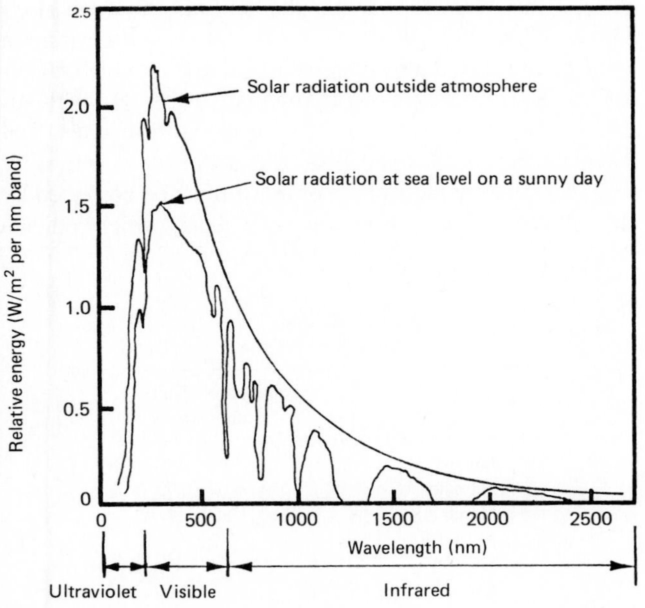

Figure 258 Variation of solar radiation with the wavelength of the radiation. The upper curve shows the solar radiation outside the earth's atmosphere and the lower curve the radiation received at sea level on a sunny day. Constituents of the atmosphere, such as ozone, carbon dioxide, and water vapor, absorb radiation at particular wavelengths. (Courtesy of Van Nostrand Reinhold.)

At one end beyond the visible spectrum is the ultraviolet range, and on the other side the infrared range. The infrared solar band is six times as wide as the visible spectrum, but the amount of radiation received from it is relatively low. The ultraviolet solar band is narrow. However, it is of particular importance when solar radiation is converted to electricity by PHOTOVOLTAIC CELLS, because only radiation with a wavelength of less than 1000 nm is effective when silicon cells are used.

Solar Radiation. The amount of solar radiation varies with the altitude of the sun and height above sea level. The sun never reaches an altitude of 90° outside the tropics, so the intensity of the solar radiation varies with latitude. Solar radiation is further reduced by cloud cover, air pollution, and water vapor. The greatest annual total solar radiation (about 2500 kW · h/m^2 or 800,000 Btu/ft^2) is received in the Sahara Desert, Saudi Arabia, and Central Australia.

The *total solar radiation* received at the earth's surface consists of the direct solar radiation, received directly from the sun, and the diffuse solar radiation. The sun provides illumination through diffused sunlight, even when the sun is hidden behind clouds. Similarly, the sun provides diffused thermal radiation at the earth's surface.

Solar radiation is measured with a *solarimeter*. The simplest type measures the thermal radiation with a thermopile, that is, a number of thermocouples (see TEMPERATURE MEASUREMENT) connected together. Photographic exposure meters cannot be employed because their photovoltaic cells respond only to visible light, not to infrared radiation. Thermopiles measure the temperature due to the total solar radiation, but the diffuse solar radiation can be obtained by eliminating the direct solar radiation with a small sunshade. The direct solar radiation is then the difference between the total and the diffuse radiation.

From the radiant temperature recorded by the thermopile, the *irradiance* (or flow of solar energy) can be determined in W/m^2 or Btu/ft^2 · h. Alternatively, the *irradiation* (or total energy over a period of time, such as an hour or a day) can be measured in kW · h/m^2, J/m^2, or Btu/ft^2.

Data are available for some, but not all, weather stations in North America and in other parts of the world from the U.S. Weather Bureau. Both total and diffuse solar radiation are available separately for some of these; for others, only the total solar radiation is recorded.

Solar radiation measurements are expensive, and in the absence of data the total irradiation can be estimated from the solar constant and the hours of sunshine, measured with a *sunshine recorder*. This is a simple instrument consisting of a spherical magnifying glass and a paper chart. While other recording instruments require a clock mechanism, this is not needed for a sunshine recorder because of the sun's motion through the sky.

An alternative and more accurate method requires the

observation of the type of cloud in accordance with a standard specification (e.g., stratus cloud, cirrus cloud) and the amount of cloud cover in tenths. From these data a ratio called the *cloud cover modifier* is determined by means of a table. The approximate solar radiation actually received on a cloudy day is then taken as the product of the clear radiation and the cloud cover modifier.

Annual and Daily Movements of the Sun. The earth rotates once each day about an axis connecting the north and south poles, and it revolves around the sun once each year. Until the 17th century it was generally believed that the earth was stationary and the sun moved around it; for the purpose of solar design it is convenient to retain this concept and to consider the apparent path in the sky described by the sun each day. This differs from day to day, because the earth's orbit around the sun is not circular, but elliptical, and because the earth's equator is inclined to the plane of this orbit at an angle of 23° 26′ (Fig. 259).

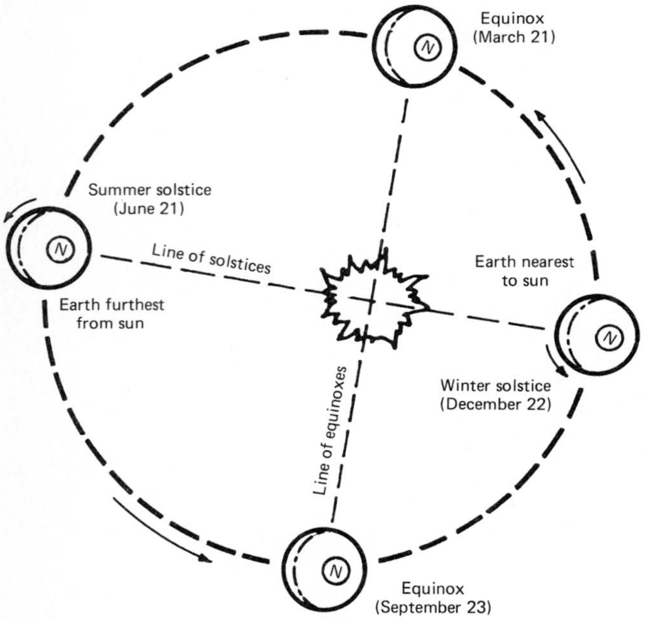

Figure 259 Earth's orbit around the sun and seasons in the Northern Hemisphere. The earth rotates once daily about an axis joining the North and South Poles. The axis of rotation is inclined to the plane of the earth's orbit around the sun at an angle of 23°26′. On the equinoxes (March 21 and September 23, when the day and night are of equal length) the line joining the centers of the earth and the sun is in the plane of the earth's equator, shown in the equinox positions. On the solstices (June 21 and December 22, respectively, the longest and the shortest days of the year), the line joining the centers of the earth and the sun is perpendicular to a line joining the earth's Arctic and Antarctic polar circles, shown in the solstice positions. The earth is farthest from the sun in early July and nearest to the sun in early January. (Courtesy of Van Nostrand Reinhold.)

The rotation of the earth alone would cause the sun to rise daily in the east at the equator, reach an altitude of 90° at noon, and set in the west. At other locations, say with latitude λ, the highest point of the sun's movement would be $90° - \lambda$. The inclination of the earth's equator to the plane of the earth's orbit causes the seasons. The two planes intersect on March 21 and September 23. On these dates the length of the day and the length of the night are equal everywhere on the earth's surface, and these days are therefore called the *equinoxes*. The greatest difference between the length of the day and of the night occur on the solstices, which are on June 21 and December 22. In the Northern Hemisphere the winter solstice is in December, and in the Southern Hemisphere it is in June.

Between March and June the sun appears to move northward as far as the *Tropic of Cancer*, 23° 26′ north of the equator. It then appears to move south as far as the *Tropic of Capricorn*, 23° 26′ south of the equator. The tropical zone is the region between 23°N and 23°S, where the sun appears directly overhead at some time, and may shine on both the south and the north face of a building. Outside the tropics a building facing north and south has one face permanently in the shade, except for brief periods in the morning and the evening during summer; however, the angle of the sun shining on the south face (in the Northern Hemisphere) gets lower as one moves away from the equator, so that SUN SHADING must project more from the facade.

Time. The ellipticity of the sun's orbit causes a variation in the distance between the sun and the earth (Fig. 259) by about 2%, which is not significant, and a variation in the earth's actual movement around the sun, which ranges from $-14\frac{1}{2}$ minutes to $+15\frac{1}{2}$ minutes (Table 1). Watches and clocks are designed to operate at uniform rates, and they indicate *mean solar time*. To obtain *apparent solar time* indicated by the sun's motion, it must be corrected by the *equation of time*. Similarly, sun dials, which indicate

TABLE 1
Equation of Time

Date	Equation of Time (minutes)
January 15	$+9$
February 15	$+14\frac{1}{2}$
March 15	$+9\frac{1}{2}$
April 15	$+0\frac{1}{2}$
May 15	-4
June 15	0
July 15	$+5\frac{1}{2}$
August 15	$+4\frac{1}{2}$
September 15	$-4\frac{1}{2}$
October 15	-14
November 15	$-15\frac{1}{2}$
December 15	-5

apparent solar time, must be corrected by the equation of time to give the mean time shown by a watch:

equation of time = apparent solar time − mean solar time

Furthermore, watch time must be corrected to local time. In 1884 it was agreed that the world's standard time should be that corresponding to the longitude of Greenwich Observatory near London. The time throughout the world differs generally by a multiple of 1 hour from that in Greenwich. The world's circumference is divided into 360° of longitude and 24 hourly time zones, so 1 hour corresponds to 15° of longitude.

Thus Central Standard Time used in Chicago is 6 hours behind Greenwich Standard Time, but the longitude of Chicago is 87° 37′ W. If the time indicated by a watch in Chicago on November 5 is 3 hours 45 minutes, we need to correct the watch time by $(90 - 87.6) \times \frac{60}{15} = +10$ minutes to obtain local time, and by the equation of time ($= -15$ minutes) to obtain local apparent time. Thus the local apparent time is 3 h 45 m + 10 m − 15 m = 3 h 40 m.

References

1. FARINGTON DANIELS: *Direct Use of the Sun's Energy*. Yale University Press, New Haven, Conn., 1964.
2. T. A. MARKUS and E. N. MORRIS: *Buildings, Climate and Energy*. Pitman, London, 1980.
3. A. BANNISTER and A. RAYMOND: *Surveying*. 4th ed., Pitman, London, 1977.

H.J.C.

SUN SHADING AND SUN-CONTROL DEVICES. Sun-shading devices are mainly used in conjunction with the transparent parts of the building envelope to dispose of solar radiation, partly by instantaneous reflection and partly by absorption and subsequent reradiation, convection, and conduction. They should reflect most of the solar energy outdoors rather than indoors, and internal shading devices,

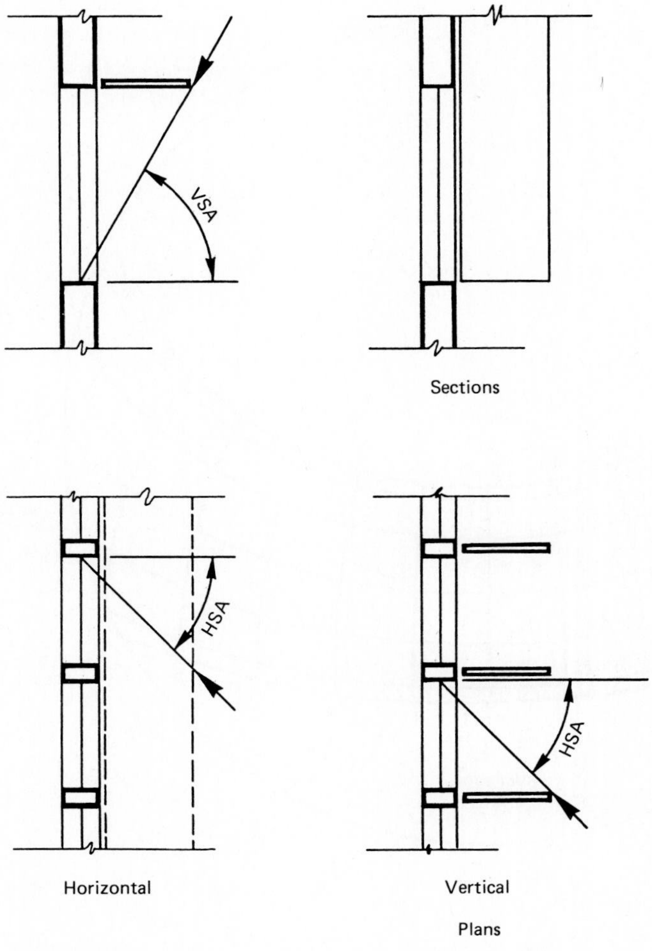

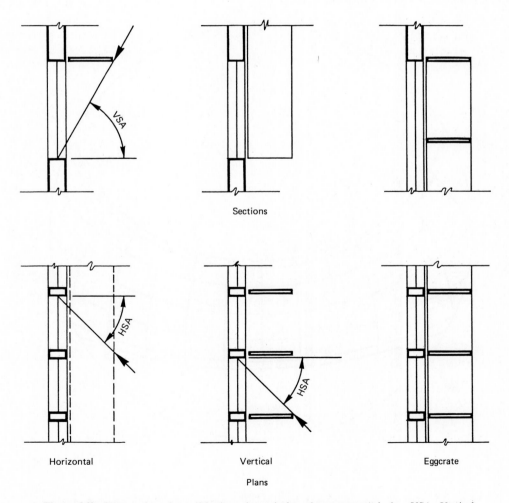

Figure 260 Plans and sections of horizontal, vertical, and eggcrate sunshades. VSA: Vertical shadow angle (profile angle). HSA: Horizontal shadow angle (wall solar azimuth).

like draperies, venetian blinds, and roller shades are not as effective as external shading devices.

External sun-shading devices fall into three main **categories:** horizontal, vertical, and eggcrate (Fig. 260).

To **design** a sun-shading device, the dates and times during which shading is required must be defined, for example the "overheated" period during which the outdoor dry bulb temperature rises above a certain level (Ref. 1), usually 22°C (71°C). During this period, the extreme positions of the sun relative to the area to be shaded are determined, either from universal sun charts (Refs. 2 and 3), which are good for all latitudes; from a local sun chart for the latitude of the building in question, as shown in Fig. 261 (Refs. 4 and 5); from the local shadow chart shown in Fig. 262 (Ref. 3); from tables (Ref. 7); or by calculation (Ref. 8), usually with a computer program. All mathematical formulas and most of the charts use solar time, and the relation between it and the time indicated by clocks is given by the equation of time (Refs. 3, 5, and 8). Transparent overlays are frequently used with local sun charts to give the vertical shadow angle (profile angle) and the horizontal shadow angle (wall solar azimuth). These determine the shape and dimensions of shading devices. The final outline of the shading devices depends on esthetics, structure, cost, and other practical considerations. The use of a shadow template extract from the local shadow chart, rather than the sun chart, cuts down and simplifies the design process (Ref. 6).

The outline of a sun-shading device for a window is a function of the solar positions during the overheated period at that location, and thus a function of the latitude of the place, its climate, and the particular orientation the window is facing. The projection of horizontal shadowing devices on south facades (in the northern hemisphere) is of reasonable size; but the projection for east and west facades needs to be much larger. Openings facing east or west should be kept as small as possible. North facades are self-shading.

Louvers, whether horizontal or vertical, have the advantage of requiring only small projections; however, they

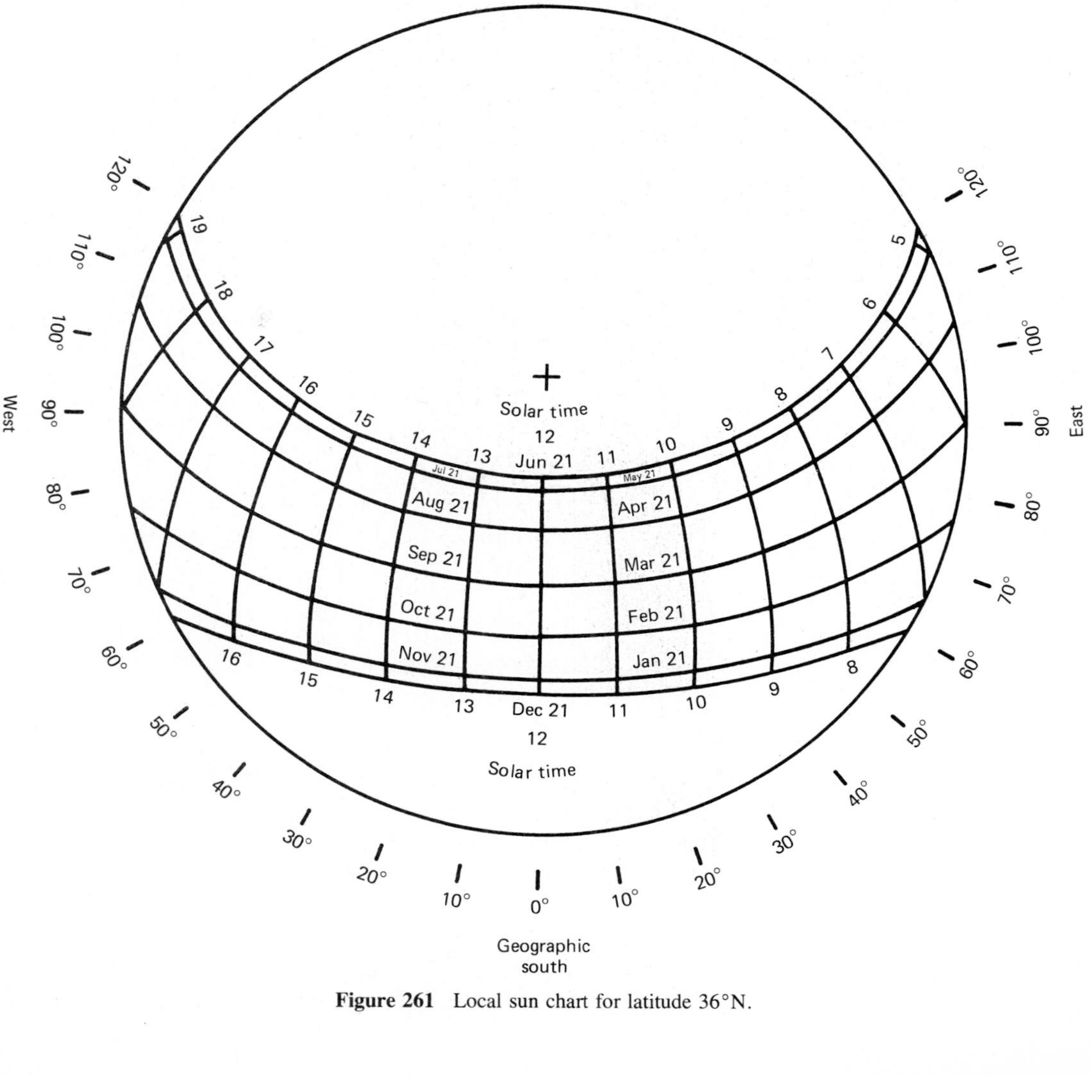

Figure 261 Local sun chart for latitude 36°N.

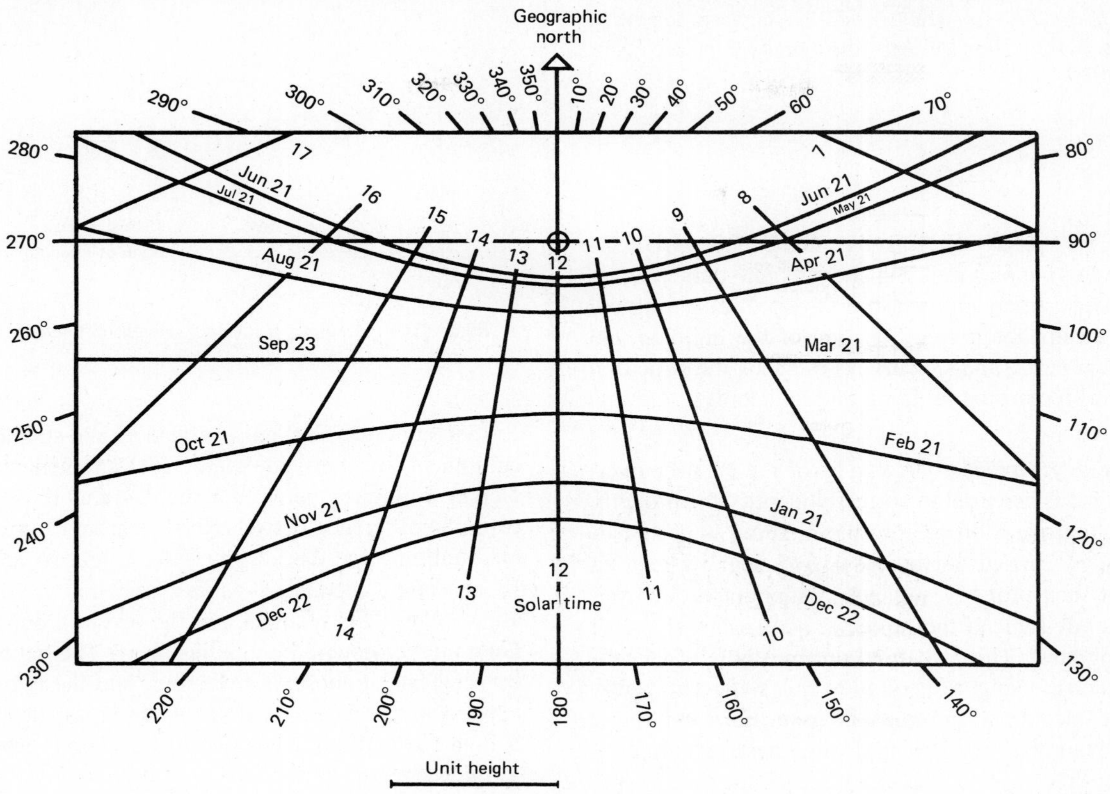

Figure 262 Local shadow chart for latitude 36°N.

obstruct daylight and the view to the outside. Louvered metal screens, composed of a series of mini louvers, can give complete shading while only partially blocking the view.

Trees, shrubs, and vines in conjunction with pergolas are useful as sun-shading devices for low-rise buildings, and transpiration helps to cool the air in the vicinity. Deciduous trees offer shading during the overheated period, and allow the sun to penetrate around their leafless branches during the cool period. The shadow template (Ref. 6) can be used to arrive at the geometrical outline of the trees needed to give shade in a particular situation by projecting the template on vertical planes along the proposed lines of trees.

Limitations. The path of the sun in the sky is the same on days equidistant from a solstice, for example, on April 21 and August 21 (Fig. 262). Consequently, sun shading designed to shade a window on August 21 gives the same shading on April 21; but whereas shading in a particular location is needed in August when the weather is hot, insolation is needed in April when it is cool. This can be achieved by the use of movable sun shading, such as canvas or aluminum awnings, roller shutters, or side hung shutters, the latter being a traditional means of shading in the Mediterranean and the Middle East.

See also FENESTRATION; PASSIVE SOLAR DESIGN; and SUN.

References

1. A. OLGYAY and V. OLGYAY: *Solar Control and Shading Devices.* Princeton University Press, Princeton, N.J., 1976.
2. C. RAMSEY and H. SLEEPER: *Architectural Graphic Standards.* Wiley, New York, 1966.
3. A. MONEM SALEH: The Design of Sun Shading Devices, in H. J. COWAN, *Solar Energy Applications in the Design of Buildings.* Applied Science Publishers, London, 1980.
4. *Sun Angle Calculator.* Libbey-Owens-Ford, Toledo, Ohio, 1975.
5. R. PHILLIPS: *Sunshine and Shade in Australia.* Bulletin 8, Experimental Building Station, Department of Housing and Construction, Australian Government Publishing Service, Canberra, 1975.
6. A. MONEM SALEH: The Shadow Template. *Solar Energy,* Vol. 28, No. 3 (1982), pp. 239–256.
7. J. CALLENDER: *Time-Saver Standards.* McGraw-Hill, New York, 1974.

8. *ASHRAE Handbook of Fundamentals.* American Society of Heating, Refrigeration and Air-Conditioning Engineers, New York, 1976.

A.M.S.

SURVEYING AND SETTING OUT OF BUILDINGS. Surveying on a building site falls into three phases: (1) preliminary surveys to measure the site, (2) setting out surveys to ensure the correct location of the building and its component parts, and (3) surveys to check the structural stability or to record the building and its details.

Preliminary Surveys. A survey of the legal boundaries of the land is essential to correctly identify the property and obtain its dimensions before the building is designed. This should be carried out by a surveyor legally licensed to carry out such property surveys. A topographic or contour plan is advisable if the project covers a large area or is on sloping land. This is carried out by stadia or tacheometric methods, using a theodolite and staff. On gently sloping land a level and staff must be used. For a very large project the plan might be prepared using aerial photogrammetry (Ref. 1).

Setting-Out Surveys: Horizontal. Methods adopted for setting out positions of corners and other construction details depend on the complexity of the building and the precision required. Precision is specified, preferably, as a standard deviation, and the surveyor works to one order of precision higher. Horizontal positions may be set out by offsets (Fig. 263, point R) or by radiation (point S). For distance measurement a steel tape or band is used. Depending on the precision required, corrections for tape standardization, temperature, slope, and sag may be appropriate. For minor measurements a cloth or fiber-glass tape may be used. The theodolite measures the horizontal projection of angles and vertical angles. It should be kept in good adjustment and the angles measured in face-left and face-right positions (Refs. 1 and 2).

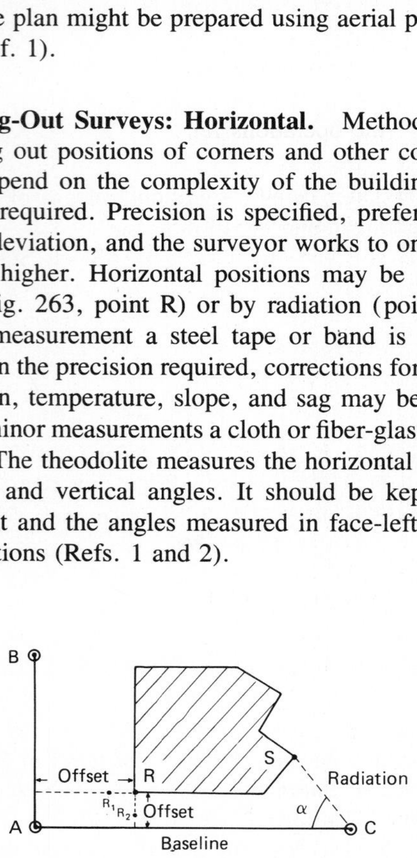

Figure 263 Methods of setting out. R: set out by offsets from AB, AC. Corner recovery points R_1, R_2. S: set out by radiation from C, distance CS, angle α.

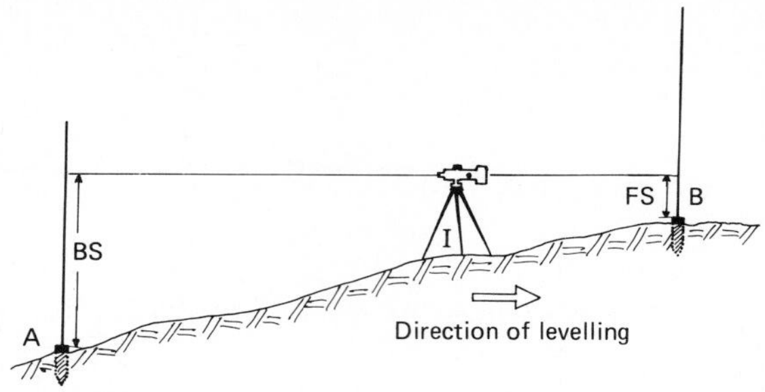

Figure 264 Principle of leveling. Height difference: AB = BS − FS.

Setting Out: Vertical. The level and staff are standard equipment for vertical height measurement (Fig. 264). Height difference between points A and B is determined from the difference between backsight and foresight. Since on a building site the lengths of backsight IA and foresight IB are often unequal, it is essential that the level be in adjustment. This requires that the leveling bubble and the horizontal crosswire be in adjustment. The bubble is tested by centering it and then reversing, and the two-peg test is applied to check the position of the crosshair (Ref. 1). In a high-rise building, heights at the various levels are determined by a steel tape suspended vertically, but the level is used for relative heights on an individual level.

The verticality of a high-rise building and some components, for example, the guiderails in an elevator (lift) shaft, are determined by a suspended plumb bob. Where this is impracticable, steep sights with a theodolite may be used, taking care to level it precisely and to measure on both left and right faces. Alternatively, special-purpose optical plumbing instruments are available.

In setting out for construction, the approach should be systematic, the measurements should be made carefully, and particular care should be taken to ensure the stability of marks placed. As in all surveys, two general principles apply: (1) the work should be "from the whole to the part," to avoid accumulation of error, and (2) the survey should be designed so that every point is subjected to an independent checking measurement.

Deformation Surveys. Deformation surveys measure the effects on the structure of its own weight and loading and of thermal and wind effects. The survey measurements, often interpreted in conjunction with the engineer's stress and strain measurements, provide a sensitive indication of the soundness and stability of the structure. Measurements include small changes in horizontal and vertical position, changes in dimension and tilts. They are specialized and call for the utmost precision, with techniques including high-precision leveling, angle measurements with geodetic theodolites and distances measured with invar bands, electronic distance meters, or laser interferometers (Ref. 3).

Recording Existing Buildings. Surveys to record buildings and their details may be required for archival purposes or for restoration. Conventional measurements suffice for drawing plans, but measurement of detailed facades soon becomes tedious or impossible. Here architectural photogrammetry is an appropriate method. The photographs, taken using special metric cameras, can be stored in the archives, where they are available for measurement in the photogrammetric plotter if ever they are needed for restoration or other purposes (Ref. 4).

References

1. A. BANNISTER and S. RAYMOND: *Surveying*. Pitman, London, 1973.
2. L. M. HAYWARD and S. P. COLLINS: *Survey Practice on Construction Sites*. Pitman, London, 1975.
3. F. KOBOLD. Measurement of displacement and deformation by geodetic methods. *Journal of the Surveying and Mapping Division, American Society of Civil Engineers*, Vol. 87 (July 1961), pp. 37–66.
4. *Handbook of Non-Topographic Photogrammetry*. American Society of Photogrammetry, Falls Church, Va., 1980.

P.V.A.-L.

SYSTEM BUILDING involves the use of a group of components and subsystems and a set of rules and conventions, which are combined in a great variety of configurations to provide a number of solutions to any given building problem. In this way Babylonian brick house construction, the American log cabin, the igloo and the Bedouin tent are building systems. Each uses a limited palette of materials, assembled within a well-understood set of rules, to make a seemingly endless set of variations on a theme.

In recent years, as builders have tried to make more effective use of the many new materials, products, tools, and techniques of modern industry, a number of new systems have been developed for use on contemporary building problems. Most involve the use of manufactured products, components, and equipment, joining them with new materials and techniques and designing connections, dimensional standards, and tolerances to simplify changes, additions, and replacements.

The best systems are those developed with a larger view than the construction of the building. Particular programatic requirements, industry, climate, transport, and market conditions are considered, as well as issues of adaptability, maintenance, and operating costs. Modern system building is conceived as part of the larger system within which buildings are constructed and used. Thus, prefabricated, factory-built and "package" buildings are not necessarily systems building; nor can all prefabricated buildings be considered systems buildings. Many manufacturers take advantage of industrial processes and components without seeking the flexibility, variability, and operational advantages of a system of coordinated components. On the other hand, some very simple technologies, created to aid owner–builders in developing countries, are systems.

Closed and Open Systems. Building systems can be classified as closed or open systems. Closed systems use a specific set, or kit, of parts. These are often manufactured specifically for use in the system, and, while they can be combined in many permutations and combinations, they are limited to what is produced in the system. Open systems may have some specially made components, but they also draw on products from outside the system. Thus, an open system sets up rules for dimensions and interface configuration that allow the use of many different products: windows, wall panels, or air-conditioning units are some examples. Closed systems tend to focus on permutations and combinations of a limited number of parts. As a system becomes more open, the rules become more and more important, and the specific products become less important. To use an analogy from a parallel field, closed systems focus on hardware, open systems on software.

The value of a building system lies in the characteristics of its elements, the number of elements compatible within the system, the operations required to join them, and the versatility of the joinery. Obviously, the development of a series of national and international open systems would be of great value to manufacturers, designers, builders, and users. Over the years, a number of these have emerged. At the same time, many of the closed systems of the 1960s and 1970s have been abandoned or opened. Increasingly sophisticated standards, set by trade associations, manufacturers, and governments, have done a great deal to help this trend toward larger, more open systems. The capital cost of developing an up-to-date kit of parts and of maintaining the necessary inventory for a useful system makes it increasingly difficult for closed systems to compete in general construction. Their most effective use is in special-use buildings. Increasingly today, major office buildings, large-scale residential complexes, hospitals, and shopping facilities are built with open systems. Closed systems are finding a market for warehouses, motels, and small office buildings.

Structure. Although a building's structure is only one of a number of subsystems that form the building system, and in cost and complexity it is rarely the most significant of the subsystems, the nature of the building industry makes structure central to any building system. The structure sets the pace for the building: enclosing, supporting or providing stability for mechanical equipment, enclosure, finishes,

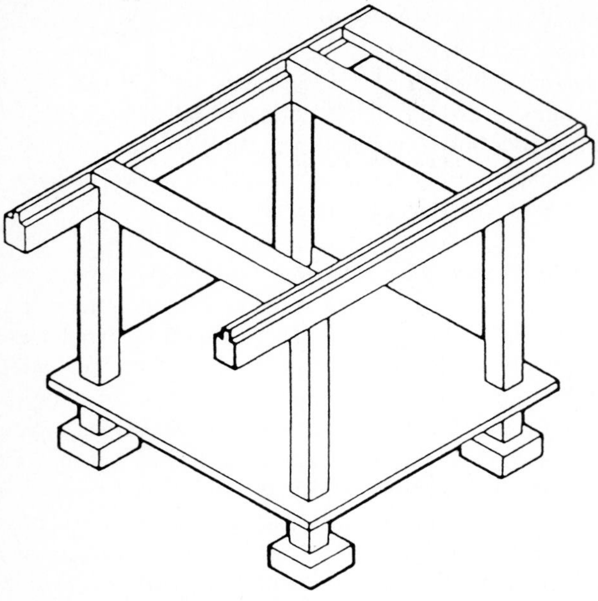

Figure 265 Skeleton (post and beam) system.

and furnishings. For this reason, the classification of building systems by structural characteristics is also most useful.

In general, all systems can be considered as one-directional, two-directional, three-directional, that is, as linear, planar, or volumetric systems. In more familiar terms, they can be thought of as skeletons (Fig. 265), slabs (Fig. 266), and boxes (Fig. 267). These approaches cannot be ranked on an absolute scale of quality or characterized as generally better or worse. Each, however, has characteristics that make it more or less appropriate to specific situations. As we move from skeleton to box systems, we tend to trade off flexibility for the advantages of increased factory production.

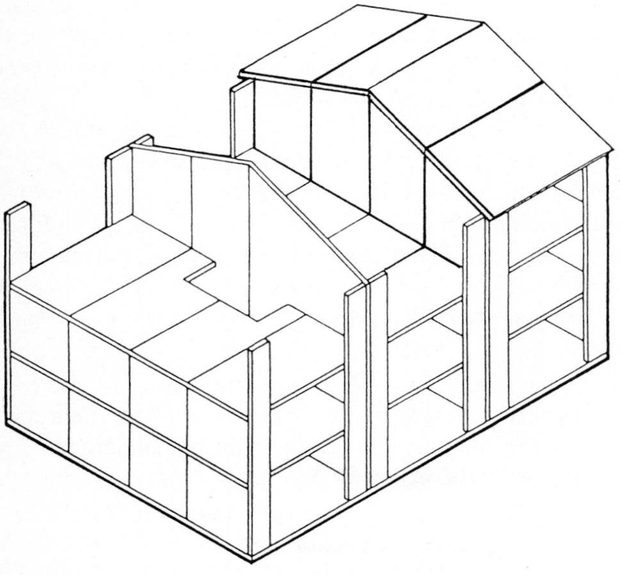

Figure 266 Slab system.

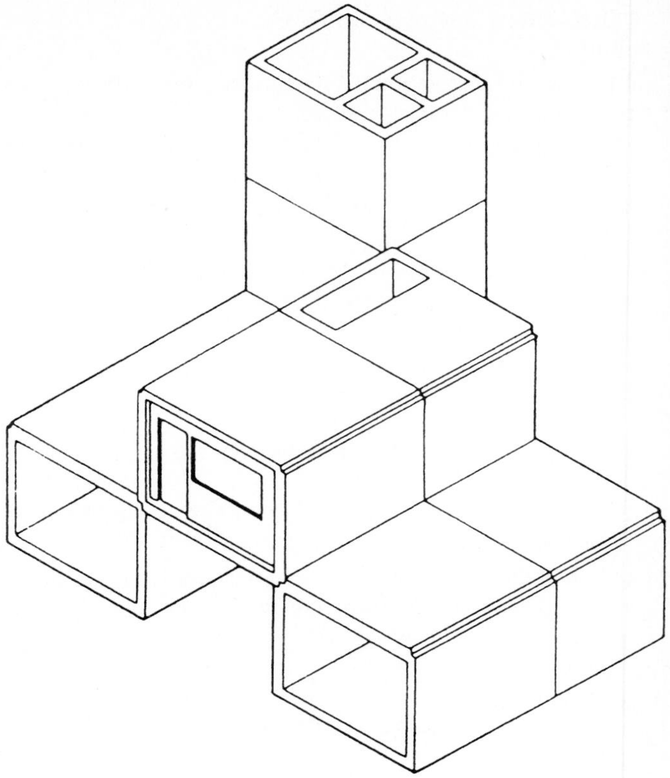

Figure 267 Box system.

A sectional, or box, unit arrives at the site with little more than utility hookups and site work to be done. Thus it can take great advantage of the economies of the factory. But the box is a quite limited planning unit; dimensions are fixed by transport requirements and plan variation by the limited number of permutations and combinations of the few box types. Finish and detail design are set at the factory for the full range of production.

At the opposite end of the scale, the skeleton allows much more planning freedom and individual input, but it is at the expense of the economies of factory assembly. These trade-offs and the choice of system, then, are related to building type (a hotel can use a box more easily than an apartment house), program (those planning housing for large families with low income will value the planning freedom of a skeleton, while senior citizen housing may not suffer from the restriction of the box), and the expectations of the user.

See also DIMENSIONAL COORDINATION; MOBILE/MANUFACTURED HOMES; and PREFABRICATION: ORIGINS AND DEVELOPMENT.

References

1. R. BENDER: *A Crack in the Rear-View Mirror.* Van Nostrand Reinhold, New York, 1973.
2. J. R. BOICE: *A History and Evaluation of SCSD, Building Systems.* Information Clearing House, Menlo Park, Calif., 1968.
3. *The Project and the School.* Education Facilities Laboratories, New York, 1967.
4. R. BENDER and J. PARMAN: A Framework for Industrialization, in S. DAVIS (ed.), *The Form of Housing.* Van Nostrand Reinhold, New York, 1977, pp. 173–189.
5. R. BENDER and J. PARMAN: The Factory without Walls: Industrialization in Residential Construction, in R. MONTGOMERY and D. R. MANDELKER (eds.), *Housing in America: Problems and Perspectives*, 2nd ed. Bobbs-Merrill, Indianapolis, Ind., 1979, pp. 113–115.
6. B. J. SULLIVAN: *Industrialization in the Building Industry.* Van Nostrand Reinhold, New York, 1980.

R.B.

T

TALL BUILDINGS. In most countries the large cities are getting larger. Cities of 5 million in 1980 will grow to 20 million by the year 2000. A dominant physical factor is the tall building. As cities have grown, so have the tall buildings. Starting in 1885 with the ten-story Home Insurance Building (Fig. 268), generally recognized as the first skyscraper, within 100 years their height had increased by a factor of more than ten to the 110-story, 443-m (1455-ft) Sears Tower (Fig. 269).

Topical Scope. When we think of a tall building, our first reaction is that it *looks* tall. It has an environmental impact, but it must fit into the community. It must "operate." This requires mechanical systems; heating, ventilating, and air conditioning; a structural framework; and a working internal space arrangement. It must provide protection from the elements and it must support the normal and accidental loads. Finally, it must be safe and economical. Figure 270 systemizes and further defines these four features of the overall tall building system and sets the stage for our further consideration.

Functional Criteria of Planning and Environmental Systems. A building is tall not because of its height, but because its design or operation is influenced by "tallness." In some places five stories is tall because an elevator is required. In large cities, ten stories or more is tall because special fire-fighting apparatus is needed. For office buildings, the need for tallness is a matter of economics and agglomeration. As countries become industrialized and service oriented, tall buildings are more and more required to conduct business in urban centers. For residential buildings the reasons are more complex. They include encroachment on agricultural land, cost of energy, and delivery of urban services.

The earliest recorded multistory buildings, that is, structures designed for human occupancy as distinct from towers, were probably those of the Minoan civilization on the island of Crete, about 1900 B.C. So the idea of multistory life and work is not new. But it was not until 1885 that building took a leap upward in the ten-story Home Insur-

Figure 268 William LeBaron Jenney's Home Insurance Company Building, Chicago, considered to be the first skyscraper. Built in 1885 and demolished in 1931. Photograph prior to a two-story addition of 1891. (By courtesy of Chicago Historical Society.)

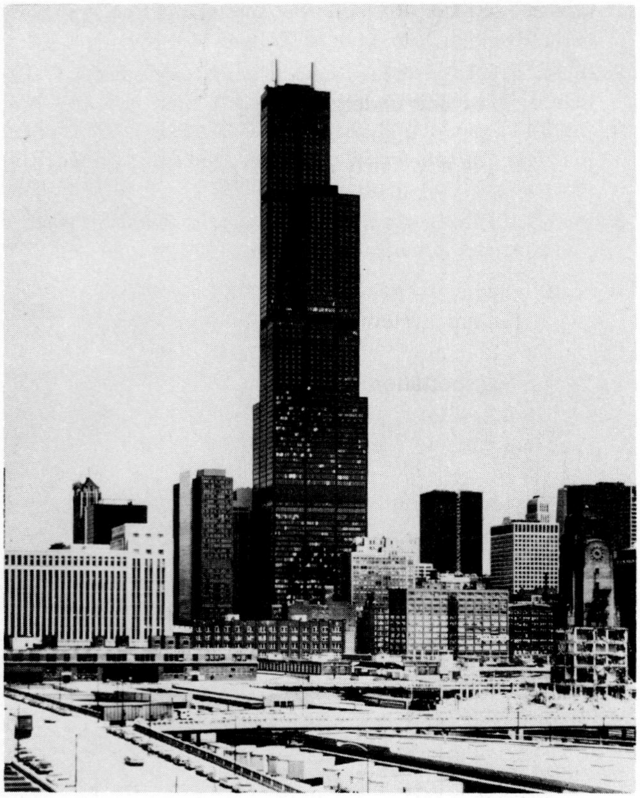

Figure 269 Sears Tower, Chicago, the world's tallest building. It is an office building of structural steel. (By courtesy of Ezra Stoller.)

ance building (later increased to 12). Major technological breakthroughs that made the tall building possible included the elevator with automatic brake, a method of fireproofing iron, and the development of steel.

Social, Cultural, Political, and Economic Considerations. In some parts of the world there is no discernible reaction to the tall building. It is accepted as the normal way of life. In other places the negative reaction has been strong. The most successful tall buildings are those that take into account the social, cultural, political, and economic factors (Ref. 1).

Past functional failures of tall buildings are largely due to incomplete planning. The control of a system as large as a city is difficult, and even when controls are instituted, too often we fail to look at the *total* system. We tend to examine separately problems of pollution, transportation, and housing, forgetting that the city is one organism.

Each city should develop a deliberate urban plan with citizen input and investor input, looking into the functional base of operation of the city and taking into particular account the high-rise. There has been enough experience with office buildings, so it is fairly easy for the architect to design one that works properly, no matter what the culture.

Functional Systems

Utilization (commercial, residential)	Parking
Ecological	Ownership, financing
Site	Operation
Esthetics	Maintenance
Space cognition (signing)	Management
Access and evacuation	Building services
Infiltration protection	Communication
Environmental	Security
Transportation	Fire protection
Energy efficiency	Urban services

Physical Systems

Foundation	Architectural (cladding, walls and partitions, floors, ceilings)
Structural framework	Fittings and furnishings
Mechanical and service systems	Contents (equipment, stores, people)
Electrical	
Utilities	

Loading Systems

Gravity
Temperature
Earthquake
Wind
Fire
Accidental loading
Water and snow

Building Implementation Systems

Need
Planning
Design
Construction
Operation
Demolition

Figure 270 Tall building systems.

For residential tall buildings the situation is entirely different. One of the best investments a developer of a major project can make is to include a social scientist in the design team to provide input on the following:

1. The amenities that should be included.
2. The provisions for emergency service.
3. How often the likely tenants will expect rehabilitation.

In the United States, high-rise apartments work well for the affluent, singles, couples without children, and the elderly, but they are not the first choice of families with small children. For this latter group, extra play space, interior day-care facilities, and outdoor facilities become prime considerations if there is no alternative to the high-rise environment. In cultures with extended families the high-rise solution could indeed be difficult.

Sociopolitical factors go beyond the individual reaction, and citizen participation will continue to increase (Refs. 2

and 3). There will be new emphasis on the local neighborhood. The actual input of the user may be direct, but most often it will be through the social scientist who can better articulate user norms and values. We will see more social scientists as members of the design team (Ref. 4). In cities such as New York and Tokyo the cost of the land alone is more than what entire skyscrapers cost in the 1920s, even when inflation is taken into account. This requires a market analysis, but even more an analysis of the social costs and benefits.

Architecture and the Urban Environment. The architect's first concern is the client, and then the regulations and zoning. However, the needs of the end user must also be considered. Esthetics and style are obvious concerns. Although for some designers environmental factors were always important, since the 1970s their consideration has been mandatory. Access to the building from public space is another prime architectural consideration.

Recent developments in architectural treatment are multiuse (Fig. 271). The 100-story John Hancock Center and the 76-story Water Tower Place in Chicago, where the world's tallest steel and concrete multiuse buildings stand side by side, contain plazas or other public facilities at street level, and interior atriums. Cities in very hot or very cold environments exploit underground space for shopping with interconnections to high-rise centers (Houston and Montreal). Zoning is now fairly common, but except for height limits, it is a rarity to find a city with an urban plan that considers the tall building; San Francisco is one of these exceptions.

Figure 271 Water Tower Place (left) is the world's tallest reinforced concrete building. The John Hancock Center (right), 100 stories tall, is the world's tallest *multiuse* steel building.

A phenomenon in countries with well-developed freeway or interstate systems are the tall buildings, usually hotels, that develop at the interchanges. These are natural nodes and transportation interfaces, and planning authorities would do well to anticipate growth in the form of minicities at such points.

Regional planning usually has no constituency. Who controls the ''spaces in between''? Is the prime agricultural land adjacent to cities to be gradually absorbed by ''development''? Is the alternative more high-rise in town?

A building of modest height can be added to the urban scene without undue concern. But in the case of a major tall building the prior consideration of urban services is crucial. Chief among these is transportation and associated parking. In the World Trade Center in New York, for example, probably more research was done on transportation (pedestrian, auto, bus, rail, water, and air) than on any other factor.

The average automobile in the United States spends over 90% of its time parked. Decisions have to be made in concert with the regulations and public authorities as to whether parking will be provided by the central business district, in separate structures, on parking lots, or within the building itself.

Operation and Maintenance. Once the building is completed, its success or failure depends more than anything else on building management.

The **Physical Systems** provide for the activity that goes on within and around the building (Ref. 5).

The *Structural System* is the major scheme for supporting load and providing stiffness. The major classifications of structure are framed structures, bearing wall, core, and tube. They are made up of columns, beams, connections, walls (both gravity and shear), and floors. Concrete, masonry, and steel are the principal building materials that are used in high-rise construction.

The *Electrical System* covers lighting, power distribution, and emergency power. (A more comprehensive term, *services*, incorporates these and, in addition, some of the functions that the systems themselves drive or control, such an energy conservation and computer control facilities for building management and security.)

Vertical Transportation refers primarily to ELEVATORS but also includes ESCALATORS (when the lower floors of tall buildings are department stores), as well as mail and equipment conveyors.

The *Envelope* consists of the cladding and the roof, which houses the elevator deck house and lightning arrestors; some provide communications antennas, observation areas, and helicopter pads.

In bearing-wall structures the structure itself is the skin, coupled with the windows. The same is true of some tubular structures. Core and frame systems use curtain walls that hang on the exterior frame and can be glass, masonry, or metal, as long as they give proper fireproofing.

PARTITIONS provide for space separation, influence climatology, are a factor in fire protection, and surround the staircases.

The *Ceiling* is more than the underside of the floor. It involves an architectural treatment and a provision of services such as light and heat. The interaction between architecture and structure and between structure and mechanical/electrical systems can spell the difference between success and failure in a high-rise project.

The *Foundations* support the whole, transferring load from the structure to the soil by individual footings, spread footings, or mats. These can be supported by PILES or caissons. In the case of mat foundations, they could be of the compensated or floating type in which the weight of the excavated soil is equal to the weight of the building.

Loading. A building must resist the loads (Ref. 6). *Gravity loads* are common to all buildings, both dead loads (weight of structure and fixed equipment) and live loads (movable loads). In the case of tall buildings, it is important not to overlook construction loads, which can be the greatest gravity load the structure will ever carry. Live-load reduction factors are another feature unique to tall buildings and are permitted because of the improbability that all live loads will be present on all floors at the same time.

Temperature Effects in columns are also unique to tall buildings. Unless insulated, the exterior columns will expand and contract with changes in temperature, whereas the interior columns remain at room temperature. Either the design must accommodate the effect through strengthening or provision of hinging is needed so that the differential movement does not produce cracking. In addition EARTHQUAKE RESISTANT DESIGN, WIND LOADS, and FIRE-RESISTING CONSTRUCTION must be considered.

A modern fire-safe design code provides for the following in addition to the traditional requirements: SPRINKLER SYSTEMS in hotels and offices; ventilation systems that will automatically shut down in the presence of smoke; an emergency power system to operate elevators, sprinklers, and emergency lights; and an automatic device to return elevators to the ground floor in case of fire. More advanced systems provide warning.

Recent estimates suggest that the likelihood of death in a high-rise fire is about one-half as great as that of being struck by lightning. Even so, it is important that regular inspections be carried out, that there be fire marshalls and fire drills, and that the occupants proceed in accordance with instructions in the event of the inevitable fire accident. See also EVACUATION OF BUILDINGS IN CASE OF FIRE and HUMAN BEHAVIOR IN FIRES.

The **Structural Design** aims to provide a structural system to meet functional requirements to support load, and to satisfy economical requirements consistent with needed safety. Although many factors are similar in the design of steel, concrete, and masonry buildings, there are enough differences to justify their separate treatment (Refs. 7 and 8). These are covered in the following articles: COMPOSITE CONSTRUCTION OF STEEL AND REINFORCED CONCRETE; COMPUTER METHODS IN STRUCTURAL ANALYSIS; CONCRETE STRUCTURES; ELASTIC DESIGN OF STATICALLY INDETERMINATE STRUCTURES; FOUNDATIONS FOR BUILDINGS; LIMIT DESIGN OF STEEL STRUCTURES; LIMIT STATES DESIGN; LOAD BALANCING; PRESTRESSED CONCRETE; REINFORCED CONCRETE SLABS; SAFETY FACTORS IN STRUCTURAL DESIGN; STEEL STRUCTURES; STRUCTURAL MODEL ANALYSIS; and VIBRATIONS IN TALL BUILDINGS.

Building Services. The provision of *building services* and the efficiency with which they are designed can make the difference between success and failure in the functioning of a tall building. AIR CONDITIONING AND HEATING are of primary importance. The same ducts, usually in the ceiling, can serve both functions.

Practice in the 1960s and 1970s was energy intensive, but the emphasis in recent years has reduced energy consumption levels from 1500 $MJ/m^2/year$ (150,000 $Btu/ft^2/year$) to as low as 300 $MJ/m^2/year$ (30,000 $Btu/ft^2/year$). This has been accomplished by exploiting windows that open, SUN SHADING, the application of PASSIVE SOLAR DESIGN, and the modification of mechanical systems to enhance energy efficiency.

After temperature comfort, we need light. Reliance is now more on DAYLIGHT, with controls to shut off artificial lights when outside intensity is adequate. Designs in the 1980s also capitalize on the demonstrated adequacy of lower LIGHTING LEVELS, although there is current controversy on this matter. Recent designs have emphasized desk lighting instead of the standard ceiling-mounted fixtures.

WATER SUPPLY and SEWAGE DISPOSAL are the next essentials. In tall buildings the plumbing systems handle both. Because of the volumes involved, stormwater runoff must be provided at the base of the building and at each plaza.

REFUSE DISPOSAL is frequently handled by chutes, with collection at the basement. Cleaning services in some high-rise installations feature centralized vacuum source with hallway and room receptacles.

Recent advances in computers now make possible the full integration of control systems so that fire safety and SECURITY as well as all of the power-driven systems come under intelligent command. Buildings so equipped may be called "smart" or "intelligent" buildings.

References

1. COUNCIL ON TALL BUILDINGS, COMMITTEE 37: Social effects of the environment, Chapter PC-3 of Vol. PC, *Monograph on Planning and Design of Tall Buildings*. American Society of Civil Engineers, New York, 1981, pp. 169–256.
2. J. BURNS: *Anthropods, New Design Futures*. Praeger, New York, 1972.
3. Planning and environmental criteria for tall buildings, Vol. PC of the *Monograph on Planning and Design of Tall Buildings*. American Society of Civil Engineers, New York, 1981.
4. J. E. REIZENSTEIN: Linking social research and design. *Journal of Architectural Research*, Vol. 4, No. 3 (Dec. 1975), p. 26.
5. Tall building systems and concepts, Vol. SC of the *Monograph on Planning and Design of Tall Buildings*. American Society of Civil Engineers, New York, 1980.
6. Tall building criteria and loading, Vol. CL of the *Monograph on Planning and Design of Tall Buildings*. American Society of Civil Engineers, New York, 1980.
7. Structural design of tall concrete and masonry buildings, Vol. CB of the *Monograph on Planning and Design of Tall Buildings*. American Society of Civil Engineers, New York, 1978.
8. Structural design of tall steel buildings, Vol. SB of the *Monograph on Planning and Design of Tall Buildings*. American Society of Civil Engineers, New York, 1979.

L.S.B.

TELEPHONE SYSTEMS involve the terminal instrument, the switching or interconnection system, and the transmission path between them both and other systems. Residential and business facilities in a building are generally connected to a network authority service that provides the transmission path from one end terminal or switching system to another.

Terminal Instrument. The Integrated Service Digital Network (ISDN) is a universal circuit switching network (Fig. 272). It is currently being expanded throughout the world under international regulation, and it permits the communication of voice, text, data, and images between subscriber terminal devices connected to it. This places telecommunications on a new and significantly broader base than that previously provided by the simple telephone service. The subscriber to a regional, national, or international network can now access another subscriber with only a single directory number; once the connection is made, the user may then use voice, text, data or image.

Modern manufacturers of private branch exchange (PBX) switching equipment already provide systems that can merge voice, text, and data communications; they can also process and store information in many forms (Ref. 1). A typical terminal for use within ISDN provides access to several services, for example, voice and text communications or data/text communications in blocks of varying capacity. ISDN services offer the user a wide range of communication options with increased convenience and quality of reproduction. Modern commercial accommodation should be designed to be compatible with the provision of all these facilities.

Change is a characteristic of modern office accommodation; up to 50% of installed telephones and communication terminals may have to be modified or moved to a different location over a period of 12 months. This presents problems because of the variety of cable types required to connect terminal equipment; furthermore, many terminals require mains power in addition to the necessary communication links (Ref. 2).

Cable Entry. Telephone company lines or cables are fed from the service reticulation network into a building by one of three principal means:

1. *Underground conduit entrance:* cables encased in buried conduit ducts from an authority manhole or tunnel and extended to the property line or building alignment.
2. *Buried entrance:* cables buried in a trench and entering a building via a chase or penetration through the building wall or basement.
3. *Aerial entrance:* cables that provide service from an aboveground pole to a point on the building wall or external facade.

External cables are usually fed from the building entrance point to a central terminating frame; from this, interconnection to the internal wiring is arranged.

Internal Cable Reticulation is required to distribute physical line connections from the central terminating frame to the various terminal devices in an orderly, flexible, and economical manner. Local building regulatory codes may limit the choice or define design requirements for such facilities. Vertical cables in a multistory building should be distributed through predesigned riser shafts, conduits, or interfloor penetrations to cross-connection points for each floor level (or significant area of occupancy).

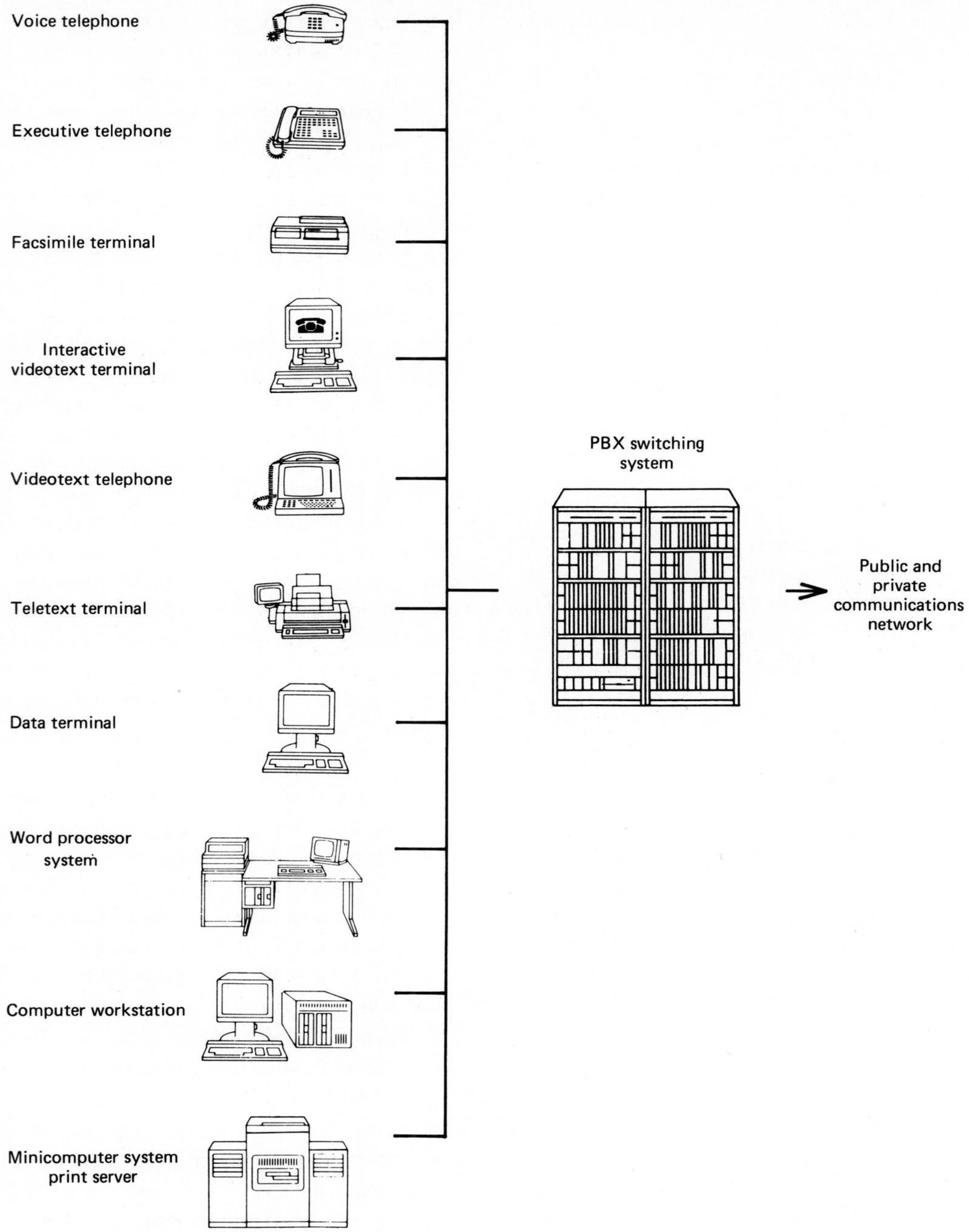

Figure 272 ISDN telecommunications services.

Lateral or area reticulation provides the final connection to telephone instruments and workstation terminals from the intermediate distribution points. It utilizes interfloor ducting, ceiling or raised-floor voids, or a combination of perimeter/partition ducts. Its design requires research into the present and future communication requirements of the area to be served to avoid expensive and disruptive alterations later (Ref. 3).

The ideal reticulation system provides maximum flexibility for plug-in connection as close as possible to the desired location. A modern design for lateral distribution by perimeter, partition, or interfloor ducting is shown in Fig. 273.

Accommodation for Switching Equipment. Although the physical size of modern PBX switching equipment is gradually being reduced, it is often necessary to provide a separate room for the central switching equipment

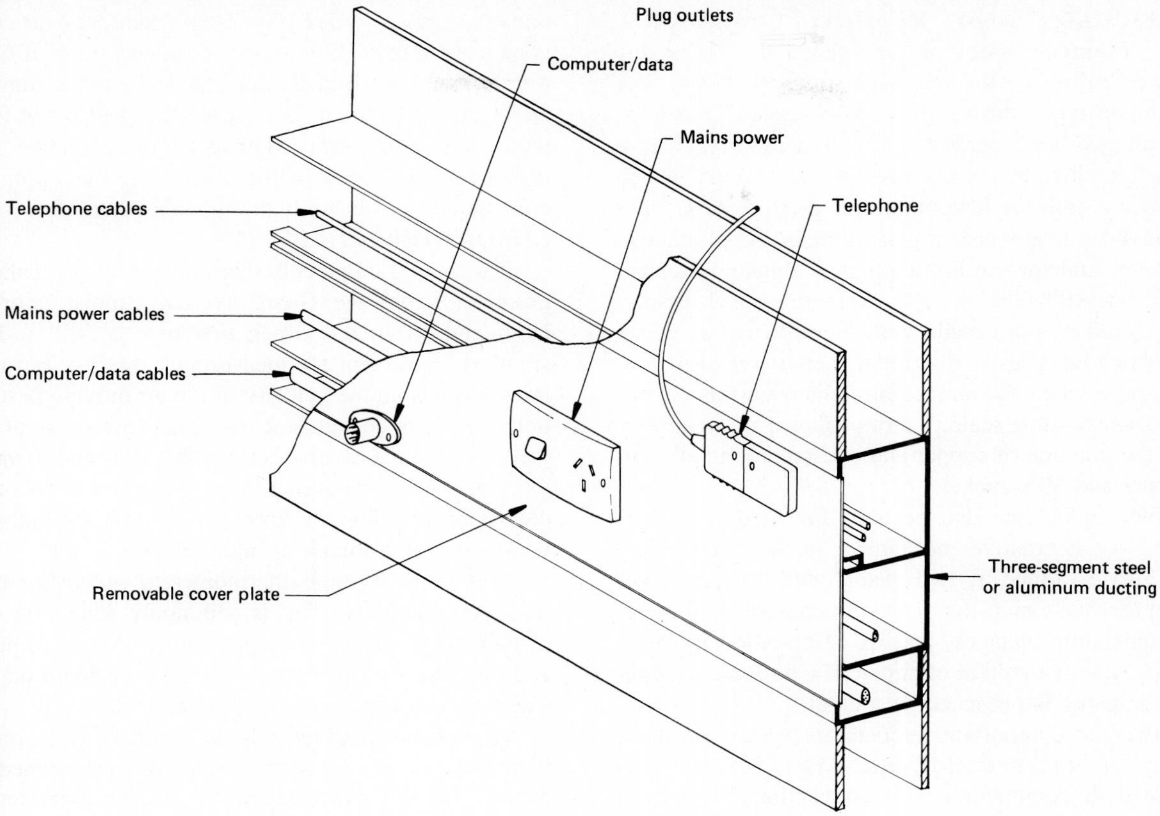

Figure 273 Lateral reticulation ducting system.

and, where necessary, the transmission interface equipment. Major considerations for the design of switching equipment accommodation include the following:

1. Cable access to the main terminating frame and building reticulation facilities.
2. Commercial power and backup facilities (if required).
3. Environmental conditioning.
4. Security and maintenance service access.
5. Floor space and floor loading requirements.

Recommendations for detailed design requirements should be sought from the equipment manufacturer and the network authority. Where the service includes real-time online data communications, commercial power backup facilities may extend to the provision of an *uninterruptible power supply (UPS)* system. This ensures against the disruption of service from transients, surges, and unpredictable breaks in commercial power supplies, which could cause the loss of important data in the transmission.

Security of Communications. Measures to prevent unauthorized access to communication lines and terminals are becoming an essential part of modern design. Advice from a specialist communications consultant or consultant services from the network authority should be sought to ensure that the most effective security measures are incorporated in the design of the accommodation provided for the communication system.

References

1. REINHARDT LUX: EMS Communications Systems. *Siemens Telcom Report 5/83*, Vol. 6 (1983), pp. 256–260.
2. WERNER SCHUBERT: *Communications Cables and Transmission Systems*. Siemens Aktiengesellschaft, Berlin, 1976.
3. FRANCIS DUFFY: *The Orbit Study—Information Technology and Office Design*. EOSYS Limited, Slough SL2 3PQ (England), April 1983, Sections 4 and 10.
4. *A Building Planning Guide for Communication Wiring*. International Business Machines Corporation Special Publication, Raleigh, N.C., 1982.
5. *Practical Methods for Electromagnetic Interference Control*. L. M. Ericsson Special Publication, Stockholm, Sweden, 1984.

J.W.S.

TEMPERATURE AND HUMIDITY MEASUREMENT. Thermal comfort is determined mainly by *air temperature*. This is measured with a *thermometer*. The most common type consists of a closed-ended glass tube with an enlarged head, partly filled with mercury. The temperature is measured by the expansion of the mercury, which remains liquid until the temperature drops to −40°C; it is still liquid when heat-resistant glass softens. An alternative thermometer fluid for use in buildings is alcohol, which is a liquid from −80°C to +70°C. Remote-control temperature measurements are made with *thermocouples*, which consist of two electrically dissimilar metals joined to produce a current when the temperature changes; this is converted to a temperature scale. Thermocouples are frequently made by the junction of copper and constantan, an alloy of 60% copper and 40% nickel.

Thermocouples can also be used for *recording thermometers*. An alternative mechanical method employs a *bimetallic strip* formed by two metals with different coefficients of thermal expansion. The strip consequently bends as the temperature changes, and its deformation moves a pen vertically across a piece of paper. The horizontal movement of the paper is produced by a clock.

The two most important temperature scales are those due to Celsius (used in the METRIC SYSTEM) and Fahrenheit (used in conjunction with conventional American units); both men were physicists in the 18th century. Both scales are now calibrated by the freezing point (0°C and 32°F) and the boiling point (100°C and 212°F) of water.

Humidity is also important for THERMAL COMFORT, particularly in hot-humid climates. It is most conveniently measured with a *wet-bulb thermometer;* this is a thermometer whose bulb is covered by a damp wick or piece of cloth dipping into water. Next to the wet-bulb thermometer is mounted an ordinary or *dry-bulb thermometer* that measures the air temperature. When the air is completely saturated with water vapor, the temperature recorded by the wet-bulb thermometer is the same as the dry-bulb temperature; but when, as is commonly the case, the air is unsaturated, the wet-bulb temperature is below the dry-bulb temperature. The drier the air, the more rapidly is water evaporated from the wet bulb, and the more the bulb is cooled. Hence, at any given temperature the difference between the dry-bulb and the wet-bulb thermometers increases as the humidity of the air decreases.

The difference can be translated into either absolute humidity or relative humidity by means of a chart or table. The *absolute humidity* is the mass of water vapor per mass of dry air. As the temperature decreases, the ability of the air to absorb water vapor also decreases; therefore, as the temperature is reduced, some of the water vapor in the air may be condensed into water. This accounts for the condensation on windows on a cold night.

The *relative humidity* is the ratio of water vapor contained in the air to that contained in saturated air. As air is cooled, its relative humidity rises unless some of the moisture is removed from the air. Thus a relative humidity of 80% is quite common in cold weather and is not felt to be particularly unpleasant, whereas the same relative humidity in hot weather would strike most people as oppressive; it corresponds to a much higher *absolute* humidity (see PSYCHROMETRIC CHART).

The wet- and dry-bulb thermometer is also called a hygrometer (from the Greek *hygros*, ''moist'') or a psychrometer (from the Greek *psychros*, ''cold''). It is the simplest instrument for measuring humidity, but its accuracy depends on the velocity of the air moving past the wet bulb. This can be standardized either by the use of a fan to drive air past the thermometer bulbs, as in the *Assmann hygrometer*, or by mounting the two thermometers on a handle, as in the *sling psychrometer*, so that the bulbs can be whirled through the air at high velocity.

Wet- and dry-bulb thermometers cannot be used for *recording* humidity. This is commonly done with a ribbon of hair that changes its length with a change in humidity, and thus causes a pointer to move up or down on a chart. The time element is provided by a clock.

A *globe thermometer* is an ordinary thermometer or thermocouple placed at the center of a blackened copper sphere, 150 mm (6 in.) diameter. If the temperature of the surroundings is higher than the air temperature, the globe thermometer absorbs long-wave radiation (see THERMAL CONDUCTION, CONVECTION, AND RADIATION) and records a temperature higher than the air temperature. Conversely, if the surroundings are cooler than the air, the globe thermometer records a temperature below air temperature. The globe thermometer is also influenced by air movement. The mean radiant temperature can be calculated from the temperature of the globe thermometer, the air temperature, and the velocity of the air by means of a formula or a chart.

See also SUN and THERMAL COMFORT CRITERIA.

References

1. THOMAS BEDFORD: *Basic Principles of Ventilation and Heating*. H. K. Lewis, London, 1964.
2. LLOYD W. TAYLOR: *Physics*. Dover, New York, 1941.

H.J.C.

TENSILE ROOFS comprise constructions in which membranes, cables, or nets are loaded in tension; the majority are either linear or surface structures. In addition to wire rope and bridge strand for cable supported or suspended roofs and nets, a range of fabrics has been used, including cotton, canvas, and coated synthetics; of these, PVC-coated polyester has been the most successful. The experimental application of vinyl-coated fiberglass for the shallow cable-reinforced dome of the U.S. Pavilion at Expo '70 (1970) stimulated the search for more permanent fabric materials. Teflon (fluorocarbon resin) coated fiberglass fabric was first used at La Verne College, California, in 1973, and subsequently for many different types of buildings, the most famous being the Pontiac Silverdome, Michigan, 1975, and the giant Haj Terminal in Jeddah, Saudi Arabia, completed in 1982.

The origins of tensile roofs can be traced to three archetypes: tents, suspension bridges, and balloons and airships. Although tents were the oldest, suspension bridges were at first more influential as models. Balloons and airships antedate air-stabilized or PNEUMATIC STRUCTURES, in which a small pressure differential serves instead of mast or arch supports.

Cable Roofs. The earliest adaptation of suspension bridge technology was made in 1826 by Bederich Schnirch in his quest for fireproof roof construction. The main types of nonprestressed cable roofs resemble specific bridge structures: in a cable-suspended roof the deck is carried directly on the cable in the manner of a CATENARY bridge. Alternatively, flat roofs may be hung from either a suspended cable similar in arrangement to a suspension bridge or by means of inclined cable stays as in a cable-stayed bridge. The Wuppertal Municipal Swimming Pool, 1957, and an aircraft hangar at Kempten, 1957, had heavy cable-suspended roofs, but the best known instance is Dulles Airport Terminal, built outside Washington, D.C., 1962 (Fig. 274). Cable-supported roofs with rigid cantilevered decks supported by inclined cable stays, in either a linear or radial pattern, enable long spans to be achieved economically, for example, in the Pan-American Hangar and Terminal buildings at John F. Kennedy International Airport, New York, 1959.

Cables need not be parallel; Bernard Lafaille produced a dished, cable-suspended roof of radial cables contained within a compression ring at Zagreb in 1936. Double-layer systems of prestressed cables were an obvious next step; the cables in the two layers are interconnected and pretensioned so as to vary the natural frequency of the individual cables (see CABLE AND SUSPENSION STRUCTURES), for example, in the Utica Memorial Auditorium, New York, 1959, and the New York State Pavilion for the New York World's Fair, 1964–1965. Cable trusses such as the Jawerth System employ a similar principle.

Traditional Tensile Roofs. Tents and ancient Roman theater *vela* cotton awnings are among some of the earliest examples of traditional tensile roofs. Among nomads tents evolved from huts, the simplest types, such as the conical tent of northern Eurasia and North America and the trellis tent of the Central Asian steppelands, retained a portable timber frame around which bark, animal skins, or felt was

Figure 274 Dulles International Airport for Washington, D.C., at Chantilly, Va., 1958–1962. Eero Saarinen and Associates, architects. Precast concrete panels were placed between each net of 25-mm-diameter strand cables; these were then preloaded and encased in reinforced concrete ribs to produce a rigid shell.

Figure 275 Medieval parasol-roofed tents are an important instance of traditional tensile architecture. Relief on the Reliquary of St. Charlemagne, Aachen, 1200–1215.

draped. The black tent of goat's hair of North Africa and the Middle East is the only nomad type supported on poles that is comparable to modern tensile roofs, but even here stability is as much a factor of the weight of the tent cloth as of prestressing and shape.

Very large and sumptuously decorated tents, some resembling portable palaces, were constructed in the Near East and Northern Europe after A.D. 1200 (Fig. 275) and their shapes imitated in architecture, especially in the design of roofs, possibly as a way of recalling the Israelite tabernacle of the Old Testament. The habit of imitating tent shapes became so ingrained that it survived well into the 20th century in such notable buildings as Le Corbusier's Chapel at Ronchamp and Kenzo Tange's two stadiums for the Tokyo Olympics in 1964.

Prestressed Surfaces. The Raleigh Arena, North Carolina, 1952, served as a seminal structure by pioneering the classic saddle shape for prestressed cable roofs. It became the prototype for many later structures, such as the Ingalls Hockey Rink at Yale University, 1953–1958, and the Sidney Myer Music Bowl, Melbourne, 1958.

Frei Otto's work represents a departure from the bridge-based engineering development of the time, as he chose the tent as the starting point for his development of modern tensile roofs. He was anticipated in this by a Russian engineer, V. G. Zhukov, who, in 1896, had designed a series of four steel tents for the Exhibition at Nijny-Novgorod. In his early projects Frei Otto concentrated on finding the most efficient minimal surfaces for modest pavilions constructed of prestressed textile fabric; but with the challenge of larger structures, such as the West German Pavilion at Expo '67, Montreal, in 1967, he devised a system of flexible prestressed wire-rope cable nets to which the building envelope was attached. This form of construction was further refined in the much larger stadium roofs for the Munich Olympic Games in 1972 (Fig. 276), which were also significant for the application of COMPUTER METHODS OF STRUCTURAL ANALYSIS. Subsequently, prestressed cable net sports structures similar to those at Munich were built at the University of Jeddah in 1982.

Convertible Roofs. Frei Otto was also fascinated by the ability of tent membranes to be bunched, and this led to the creation of convertible roofs, tent canopies suspended from a system of radiating cables that could be automatically extended or retracted. An early example of a centrally bunching roof was constructed at Bad Hersfeld, 1966–1968 (Fig. 277), and roofs were designed in collaboration with Roger Taillibert for an open-air theater at Cannes, 1965, and a swimming pool at Boulevard Carnot, Paris, in 1966–1967.

As yet, the full potential of tensile roofs is difficult to assess. In addition to making long-span roofs cheaper, the development of a range of new tensile technologies has greatly extended the freedom of the designer to choose a structure that best meets the specific functional and programmatic requirements of each project.

References

1. Convertible Roofs. *Information des Instituts für leichte Flächentragwerke* (IL) No. 5. University of Stuttgart, Stuttgart (German Federal Republic), 1972.
2. C. ROLAND: *Frei Otto: Structures*. Longman, London, 1972.
3. F. OTTO (ed.): *Tensile Structures*, Vol. 1, Pneumatic Structures, 1967; Vol. 2, Cables, Nets and Membranes, 1969. M.I.T. Press, Cambridge, Mass.
4. P. DREW: *Frei Otto, Form and Structure*. Crosby Lockwood Staples, London, 1976.
5. P. DREW: *Tensile Architecture*. Granada, St. Albans, England, 1979.
6. R. D. DENT: *Principles of Pneumatic Architecture*. Architectural Press, London, 1971.

P.D.

TESTING OF MATERIALS FOR STRENGTH, DUCTILITY, AND HARDNESS. The **strength** of ductile building materials, such as steel, aluminum, or the plastics, is usually determined by applying a tensile load. For brittle materials, such as concrete, fired clay products, and other ceramics, a compressive load is applied. Both forms of loading can be achieved by a hydraulically operated testing machine of the type shown in Fig. 278. Such

Figure 276 Principal stadium roofs for the XX Olympic Games, Munich, Germany, 1968–1972. Behnisch and Partners architects, F. Otto, and IL roof consultant. The prestressed cable net was constructed with a 750-mm regular square mesh double net, overlaid with 3-m acrylic panels.

a machine can also be adapted to test materials in bending. Hence, because of the multipurpose nature of the machine, it is known as a *universal* testing machine.

For each material, specimen dimensions are specified by relevant standards. These dimensions have been determined for reasons of reducing statistical variation between individual tests, avoiding stress concentrations (and hence preventing premature failure), and enabling deformation measurements to be made on the specimen.

Standards also specify the rate of application of load because this can influence the measured strength. Some materials, such as plastics, are sensitive to the temperature of the test environment, and hence it must be specified. For materials such as concrete and timber, the moisture con-

Figure 277 Convertible roof over the open-air theater at Bad Hersfeld, West Germany, designed by Frei Otto and Ewald Buber, 1968.

dition of the specimen at test influences the measured strength and hence test conditions are specified.

The term *strength* is defined as the maximum load sustained by the material during test, divided by the original cross-sectional area and, hence, is in fact a failure stress.

During loading in the testing machine, both the specimen and the testing machine undergo deformation. Deformation measured between the cross heads of the testing machine contains components due to both sources. Hence, to measure the deformation of the specimen, it is necessary to place a device on the test specimen itself. Such a device can either be mechanical, electrical, or optical and allows a load–deformation and hence stress–strain curve to be determined (see MEASUREMENT OF STRAIN). Material behavior is characterized by features of the stress–strain curve as well as the failure mode of the specimen.

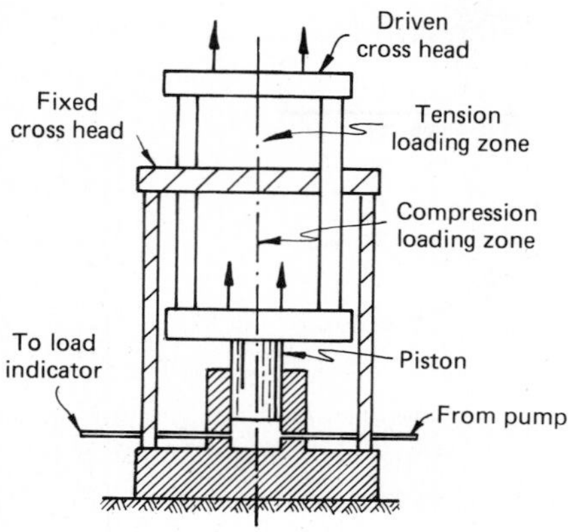

Figure 278 Hydraulically operated universal testing machine.

Ductility. For tension loading, ductile behavior is defined in terms of elongation over a gauge length or reduction in area at the point of fracture. For brittle materials, these definitions are not appropriate, and ductility is sometimes defined in terms of the maximum deflection measured at fracture in a bend test. Conversely, most ductile materials cannot be *fractured* in a bend test. The cold bend test consists of bending a bar around a pin over a wide angle. The angle at which a crack starts is taken as a measure of ductility in bending. Specifications define the required ductility by stating the minimum angle through which the material must bend for a specimen and pin of specified size.

Hardness. Although hardness is an ambiguous term, for building materials it relates to resistance to elastic and plastic deformation. Hardness tests fall into three categories: (1) indentation methods, (2) rebound tests, and (3) scratch tests. All three tests measure deformation characteristics of the material surface.

Hardness tests were the precursor of modern-day nondestructive tests and were designed to provide a rapid method of assessing the strength of materials. For each test, a hardness number is determined and this is correlated with material strength.

Indentation Tests use a hardened steel ball or diamond-tipped indentor, which is forced into the surface of the material. For metals, the Brinell, Rockwell, and Vickers tests are used; each of these tests uses a special indentor and a specified indentation load. The hardness number is related to the depth of the indentation (Rockwell) or its area (Brinell and Vickers). For wood, the Janka test is used. In this test, the depth of penetration is specified and the indentation load is the variable.

Rebound Tests measure the height of rebound of a standard weight when it falls from a specified height onto the surface of the material. The Shore test is used for metals, but it suffers from the disadvantage that it cannot be used on vertical surfaces. The Schmidt hammer, designed for assessing the compressive strength of concrete, uses a spring-loaded mass; it can be used on vertical surfaces, although a correction must be made for the effect of gravity.

The *Moh Scratch Test* is used to estimate the relative hardness of two materials, the harder one being capable of scratching the other, but not vice versa. The Moh scale ranges from 1 (talc) to 10 (diamond); it is essentially used for differentiating between various minerals. The scale is considered too coarse for most building materials.

Correlation of the hardness number, determined from a particular test, with strength, is based on substantial corroborative laboratory testing, and extrapolation beyond the limits of the laboratory testing must be viewed with caution.

Reference

C. W. RICHARDS: *Engineering Materials Science*. Brooks/Cole Publishing Company, Belmont, Calif., 1961.

D.J.C.

THERMAL COMFORT CRITERIA. Since comfort implies only the absence of discomfort, it can be said that a thermal environment will be judged to be comfortable if no sensation of heat or cold is experienced or, if either sensation should be evoked, that sensation is not considered to be unpleasant. Thermal comfort is therefore a subjective judgment peculiar to the person concerned and as such is not susceptible to instrumental measurement or objective verification. Thermal comfort is nevertheless amenable to experimental investigation by the simple device of enquiry, the basis of the well-known "comfort vote" technique. A representative sample of the population to be investigated is exposed to a range of conditions either naturally occurring or artificially produced in a climatic chamber and, after a sufficient period of time has elapsed to permit equilibration to occur, the subjects are questioned as to their thermal sensations, which are then recorded on a standardized numerical scale.

As would be expected from the subjective nature of the response, there is always considerable variation between the subjects and, to a lesser degree, in the response of the same subject from time to time. For example, in one such investigation in which the subjects were all dressed alike and engaged in the same light indoor activity, it was found that whereas 80% of the sample agreed that maximum comfort was experienced when the air temperature was 23°C (73°F), 10% found this temperature too cool for comfort, and the remaining 10% complained that it was too warm.

The maintenance of a constant deep body temperature requires that the metabolic heat produced within the body must be balanced by an equal loss to the environment. The degree of thermal comfort experienced depends on the ease with which this end is achieved. This is determined in part by the prevailing environmental conditions and in part by certain attributes of the individual exposed to these conditions.

There are four environmental determinants of thermal stress: (1) the temperature of the air (usually the dominating factor), (2) the mean radiant temperature of the surroundings, (3) the degree of air movement, and (4) the prevailing humidity. Contrary to popular belief, the last is probably the least important within the range of thermal comfort. There is considerable interaction between all four factors, so the effect of any one may be much modified by the value of one or more of the other three. A familiar example of this is the modification of our perception of air temperature by a change in the degree of air movement. The complexity of these interactions has led to many attempts, with varying degrees of success, to express the combined effect of two or more of the factors in the form of a single index. Of the many indexes proposed, *effective temperature*, or one of its modifications, is the best known and, although not without its shortcomings, probably the most suitable for use within the range of thermal comfort.

Of the individual determinants three are of major importance: (1) the rate of working, (2) the clothing worn, and (3) the degree of acclimatization. The degree of activity and the clothing worn, by their effect on the rate of heat production and heat loss, respectively, profoundly affect whether any given combination of environmental factors is deemed to be comfortable. The rate of energy expenditure is commonly determined by external circumstances, but adjustment of the amount of clothing worn constitutes for the individual the most important means of extending the range of thermal comfort. Acclimatization is a familiar but imperfectly understood physiological process as a consequence of which the ability to regulate body temperature is greatly enhanced. As a result of this enhancement, environmental conditions that would otherwise be regarded as unacceptable may produce only a minor degree of discomfort, which, with habituation, may come to be disregarded.

The effect on the preferred temperature of such physical characteristics as age, sex, race, and body-build has been much investigated with varying results. A major difficulty is to be sure that such differences as have been observed cannot be explained in terms of clothing, activity, or acclimatization. There are also many less tangible factors, such as bodily health, mental attitude, socioeconomic status, and past experience, which seem to influence an individual's judgment of what is thermally comfortable, but which remain as yet inadequately investigated.

References

1. D. A. McINTYRE: *Indoor Climate*. Applied Science Publishers, London, 1980.
2. R. K. MACPHERSON: The assessment of the thermal environment—a review. *British Journal of Industrial Medicine*, Vol. 19 (1962), pp. 151–164.

R.K.Ma.

THERMAL CONDUCTION, CONVECTION, AND RADIATION are the three methods whereby heat can be transferred from the outside to the inside of the building, and from one part of the building to another.

Thermal Conduction is the transfer of heat from one part of a body to another, or from one body to another body with which it is in contact. All substances consist of very small particles that are in continuous motion, which depends on the temperature. The particles on the warmer side of the body move more intensely than those on the colder side, and they transfer some of the energy to the colder particles by colliding with them. As their motion increases, that part of the body becomes warmer. The rate of heat transfer depends on the temperature difference and on the thermal conductivity of the material:

$$Q = \frac{k(T_1 - T_2)}{d}$$

where Q = quantity of heat passing through a unit area of a body (e.g., a floor or wall) in unit time, measured in watts per square meter (W/m^2) or Btu per square foot per hour ($Btu/ft^2 \cdot h$)

k = thermal conductivity of the material, measured in watts per square meter per kelvin ($W/m \cdot K$) or Btu per foot per hour per degree Fahrenheit ($Btu/ft \cdot h \cdot °F$)

T_1 and T_2 = temperatures on the warmer and the cooler side of the body, whose difference is measured in kelvins or degrees Fahrenheit

d = distance between the two sides of the body in meters or feet

All materials, whether solid, liquid, or gaseous, conduct heat; but some are better conductors than others. Thus metals are excellent conductors of heat, while still air is a poor conductor. A porous material that contains small pockets of air thus provides good THERMAL INSULATION.

Convection. In liquids and gases convection is generally a more effective method of heat transfer. The term is used to describe the transfer of heat by mixing one portion of a liquid or a gas with the remainder, which is at a higher or a lower temperature. While conduction cannot be seen, it is possible to make thermal currents due to convection visible, for example, by introducing particles of dye into water that is heated by a burner.

Convection is due to the fact that most liquids and gases change their density as they change their temperature. Generally, they get lighter as they get hotter. Thus, if a pot of water is heated at its base by a burner, the warm particles of water rise to the top and the colder particles sink to the bottom, where they are heated in turn. Water between 0°C and 4°C (32°F and 40°F) is an exception; the density of water increases with cooling until it reaches 4°C, but then decreases again until it freezes. This is an important property for life on earth; if it were otherwise, lakes would freeze solid from the bottom, instead of forming a skin of ice on top.

In buildings convection is of particular importance, because it increases the transfer of heat by air flowing over the surfaces of walls, floors, roofs, and pipes used for heating and cooling. However, air is a poor *conductor* of heat.

The quantity of heat transferred by convection is

$$Q = h(T_1 - T_2)$$

where h = boundary layer heat transfer coefficient, measured in watts per square meter per kelvin ($W/m^2 \cdot K$) or Btu per square foot per hour per degree Fahrenheit ($Btu/ft^2 \cdot h \cdot °F$)

The numerical value of h depends on the nature of the flow, the physical properties and velocity of the fluid, the shape and dimensions of the surface, and temperature. Values for use in buildings are given in Chapter 22 of Ref. 1.

Thermal Radiation is energy in the form of electromagnetic waves, that is, the same type of energy as transmits visible light, television, radio signals, and x-rays. Solar heat energy has a wavelength ranging from 0.3 to 3.0 μm. Energy reradiated from the surfaces of a building has a longer wavelength, from 2.5 to 100 μm.

The quantity of heat radiated is governed by the Stefan–Boltzmann law, first proposed by Joseph Stefan and proved by Ludwig Boltzmann in 1884:

$$Q = \sigma \epsilon T^4$$

where σ = Stefan–Boltzmann constant, which is $5.68 \times 10^{-8}\ W/m^2 \cdot K^4$

ϵ = emissivity of the surface (it ranges from 1.0 to 0 and is about 0.9 for concrete and brick and about 0.2 for bright galvanized steel

T = temperature, measured from absolute zero in kelvins

References

1. *ASHRAE Handbook—1981 Fundamentals Volume*. American Society of Heating, Refrigeration and Air-Conditioning Engineers, Atlanta, Ga., 1981.
2. T. A. MARKUS and E. N. MORRIS: *Building, Climate and Energy*. Pitman, London, 1980.

H.J.C.

THERMAL INSULATION. The purpose of thermal insulation in buildings is twofold:

1. In cold countries it serves to keep the internal temperature at a comfortable level at a minimum cost of fuel expenditure, whatever the level of external ambient temperature.
2. In hot countries it is necessary to reduce the amount of solar ingress into dwellings to reduce the cost of air-conditioning.

The *k*-value of a material (see Table 1) is its thermal conductivity. In modern SI units, this is measured in watts/meter kelvin or W/m K (Fig. 279). Traditional U.S. practice employs as units Btu/foot hour °F or Btu inch/foot^2 hour °F. The relation between these two systems of units is given by

$$1 \text{ Btu/h ft °F} = 1.730\,073 \text{ W/m K}$$

$$1 \text{ Btu in./h ft}^2 \text{ °F} = 0.144\,228 \text{ W/m K}$$

Dry ceramic building materials (brick, concrete, etc.) have a *k*-value that depends almost entirely on the density of the material. It can be worked out from the following regression analysis equations:

$$k = 0.0588d^2 + 0.0726d + 0.05267 \quad (\text{for } d < 1.0)$$

$$k = 0.2899d^2 - 0.360d + 0.2485 \quad (\text{for } d \text{ between 1.0 and 1.8})$$

$$k = 0.436d^2 - 0.784d + 0.527 \quad (\text{for } d > 1.8)$$

where k = thermal conductivity in W/m K
d = specific gravity

If the building material is moist, this value has to be multiplied by a moisture factor (Table 2).

To provide maximum insulation of a building, one should endeavor to employ materials with as low a *k*-value as possible, provided they comply with other criteria such as strength and water resistance.

TABLE 1
Typical *k*-Value of Various Materials

Material	*k*-Value in W/m K
Copper	166
Steel	147
Marble or similar materials	1.3
Glass	0.85
Rubber	0.16
Timber, depending on type, content, and moisture	0.08–0.18
Foamed plastics, mineral wool, and similar materials	0.03–0.04

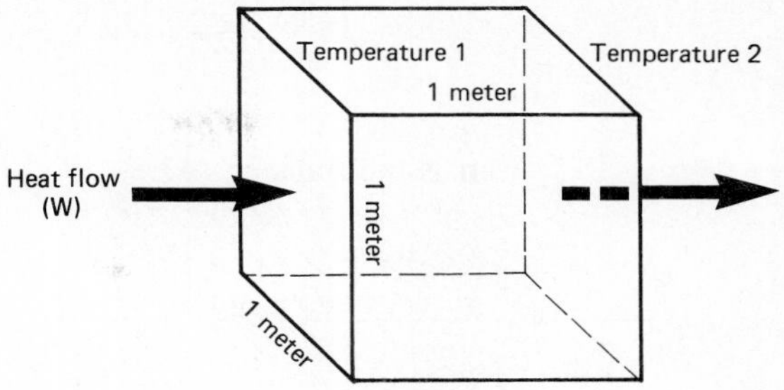

Figure 279 The thermal conductivity of a material is expressed by the ratio of the heat flow per distance per temperature difference (W/m K).

The **U-value** or thermal transmittance of a wall, roof, floor, or window structure is the overall heat transfer coefficient, which is expressed in modern SI units as watts/meter^2 kelvin or W/m^2 K. Traditional U.S. practice employs BTU/h ft^2 °F. The relation between these units is given by

$$1 \text{ BTU/h ft}^2 \text{ °F} = 5.678\,26 \text{ W/m}^2 \text{ K}$$

The *U*-value of a structure is made up of three parts:

1. The conductance of the external air layer, h_0, which depends markedly on wind speed and nature of the surface (Fig. 280).
2. The nature and thicknesses of the various layers from which the wall is constructed, including air spaces if present (Fig. 281).
3. The internal boundary layer thermal conductance, h_i, which is given as 8.5 W/m^2 K for a wall, 5.7 W/m^2 K for a floor, and 11.5 W/m^2 K for a ceiling (assuming that the inside is warmer than the outside). If this is not the case, the values for floor and ceiling are reversed.

The *U*-value of a structure (U) or the thermal resistance (R) can be evaluated from the following equation:

$$\frac{1}{U} = R = \frac{1}{h_0} = \frac{d_1}{k_1} + \frac{d_2}{k_2} + \frac{d_3}{k_3} + \cdots + \frac{1}{h_i}$$

TABLE 2
Moisture Factor for Determining *k*-Value for Moist Building Materials

% Moisture	Moisture Factor
1	1.3
3	1.6
5	1.75
10	2.1
20	2.55

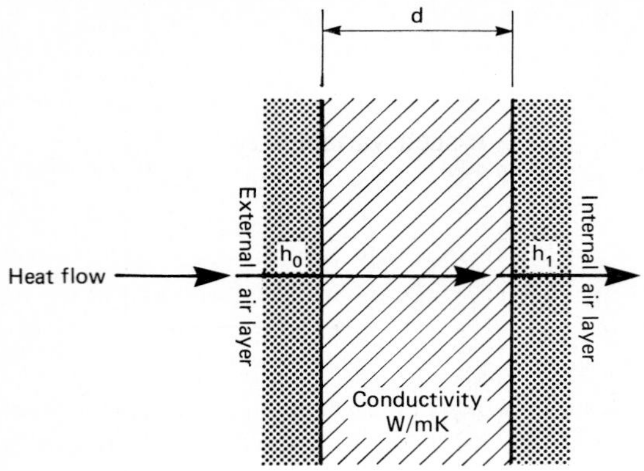

Figure 280 The thermal transmittance (U) of a wall depends on the conductance of the external boundary air layer (h_o) and the internal boundary air layer (h_i).

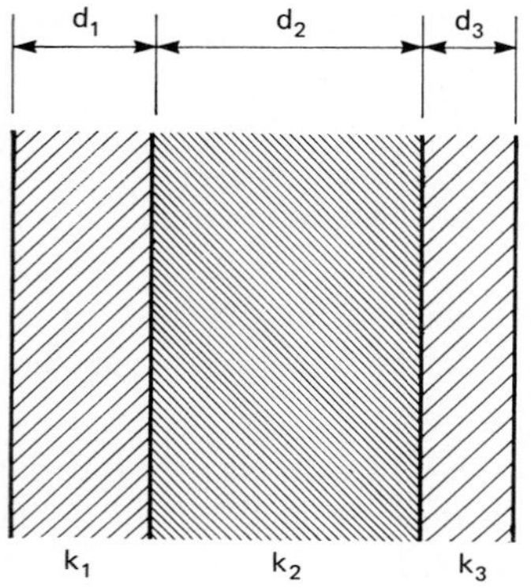

Figure 281 The thermal transmittance also depends on the thickness (d_1, d_2, and d_3) and the thermal conductivities (k_1, k_2, and k_3) of the layers of the wall.

where $k_1, k_2, k_3, \cdots$ = k-values of the different materials of construction of the wall, roof, and so on, in W/m K

$d_1, d_2, d_3, \cdots$ = thicknesses in meters

Air cavities have d/k values of 0.179 K m^2/W when plain and 0.345 K m^2/W when shiny on one or both sides.

R is expressed in K m^2/W and is called the thermal resistance, while U is expressed in W/m^2 K and is called the thermal transmittance. For optimum insulation we aim at as low a U-value as possible, while considering other criteria such as cost, strength, and water resistance.

References

1. R. M. E. DIAMANT: *Thermal and Acoustic Insulation of Buildings*. Iliffe, London, 1965.
2. R. M. E. DIAMANT: *The Internal Environment of Dwellings*. Hutchinson, London, 1971.
3. R. M. E. DIAMANT: *Insulation Deskbook*. Heating and Ventilation Publications, Croydon (Surrey), England, 1977.

R.M.E.D.

TILT-UP CONSTRUCTION is a form of precasting in which concrete walls are cast at the job site horizontally on the floor slab (or a temporary casting slab) and then erected or ''tilted'' into position using a mobile crane (Fig. 282). Once erected, the panels are temporarily braced until they can be joined permanently into one unified structure. The roof acts as a diaphragm, and the end walls serve as shear walls, so lateral loads due to wind or seismic disturbance are transferred to the foundations. Tilt-up construction had its beginnings in the early part of this century. However, its adoption on a large scale began with the advent of the mobile crane.

Reference

G. WEILER: Approximate methods for analysis of tilt-up concrete wall panels. *Concrete International*, Vol. 4, No. 11 (Nov. 1982), pp. 16–22.

K.H.

TIMBER is by volume the major construction material of the 20th century, a position it has always held. Some of its traditional uses have been replaced by newer materials such as steel or concrete; but it still predominates. In all probability this will continue because timber has one feature that makes it unique among structural materials: it is renewable.

Apart from its availability, low energy demand, low cost, and esthetic attraction, other qualities of timber make

Figure 282 Tilt-up construction.

it sought after. Some of these are its high strength to weight ratio, its corrosion resistance, its good insulating properties against heat, sound, and electricity, and its comparative durability; but most of all it is the simplicity with which it can be worked by hand or machine. This attribute enables it to be produced easily in a large range of sizes and shapes, and permits it to be easily fastened with adhesive, nails, screws, bolts, and other types of connectors. This characteristic is valued equally by the supervisor of a large sophisticated production line and the home handyman.

Despite its renewable nature, there is increasing recognition that timber must not be wasted, as it often has been in the past. Its many useful properties have not always been used to their best advantage; but through intensive research and development current practices in most countries are now tending to use it in construction much more efficiently.

One of the most useful methods employed to achieve these aims is through the use of standards. The developed countries and most of the developing ones have standards organizations that arrange for the preparation by experts of documents that permit even the most casual user to quickly become familiar with the latest information on material properties and recommended practices. The timber standards of most countries are usually considered the prime reference documents that are issued by the standards body or form the basis for rules issued by trade associations. They become critical for the designer when, as often happens, they are incorporated into building codes.

Structural Grading Rules make up one set of standards. While it is appreciated that the strength properties of timber vary from species to species, it is not always recognized that it is possible that the same degree or even larger variation may be present within pieces sawn from the same log. Structural grading rules classify pieces of timber on the basis of their mechanical properties into a small number of groups called *stress grades*. From this it can be seen that one stress grade may contain pieces from several species.

Visual grading is the oldest stress-grading method. It is based on the premise that the mechanical properties of structural timber differ from the properties of clear wood because of characteristics that can be seen and judged by eye. Examples of such characteristics are knots, slope of grain, decay, splits, and checks. The best-known reference in this field is the American Society for Testing and Materials (ASTM) Designation D245.

Stress grading may also be carried out by mechanical means, based on an observed relation between modulus of elasticity and bending strength. In this technique the grading device continuously measures the stiffness of a piece of wood as it passes through the machine at a comparatively fast speed.

Once the stress grade is known, timber design codes provide the information necessary to develop the permissible design stresses, by detailing the modifications necessary to cope with such variables as moisture content, duration and direction of loading, form factors, and types of connectors.

In critical structures, good workmanship is essential, as overcutting of notches and poor fixing can weaken an otherwise efficient design. There is consequently an increasing trend toward factory prefabrication.

Like any material, timber has its limitations, which must be recognized. Many of these relate to its biodegradable nature; but these limitations may usually be overcome through attention to construction details, such as keeping the timber dry, or by specifying either an appropriate preservative treatment or a naturally durable species.

With improved understanding and modern technology, timber is capable of holding its place as a major construction material in perpetuity.

See also BALLOON FRAME; LAMINATED TIMBER; PLYWOOD; PRESERVATION OF TIMBER; TIMBER JOINTS; and TIMBER STRUCTURES.

References

1. W. G. KEATING: Utilisation of mixed species through grouping and standards. *Australian Forestry*, Vol. 43 (1981), pp. 233–244.
2. A. J. PANSHIN and C. de ZEEUW: *Textbook of Wood Technology*, 3rd ed. McGraw-Hill, New York, 1970.
3. B. A. RICHARDSON: *Wood in Construction.* Construction Press, Hornby, England, 1976.
4. *Standard Method for Establishing Grades and Related Allowable Properties for Visually Graded Lumber.* ASTM Standard D245-74. American Society for Testing and Materials, Philadelphia, 1982.
5. *Wood Handbook.* Agricultural Handbook No. 72, U.S. Department of Agriculture, Forest Products Laboratory, Washington, D.C., 1974.

W.G.K.

TIMBER JOINTS FOR STRUCTURES AND JOINERY

Structural Timber Joints are made with mechanical fasteners and/or glue.

Nails (Fig. 283) in joints are subject to shear. The strength of the joint depends on the number, spacing, and diameter of the nails and on the specific gravity, moisture content, and thickness of the timber members. Square, grooved or square twisted, ring shanked, and annular ringed

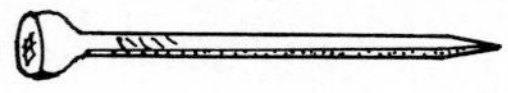

Figure 283 Nail.

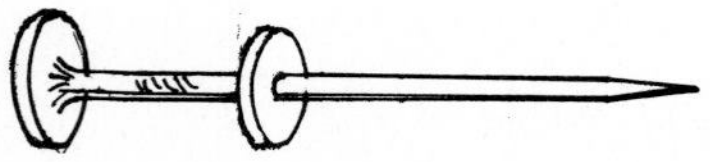

Figure 284 Double-headed nail for temporary joint, used for ease of withdrawal.

nails have higher load capacities than ordinary nails. Ringed (Fig. 284) or double-headed nails are used for temporary joints where ease of withdrawal is important.

Screws (Fig. 285) are used like nails, but provide greater resistance to withdrawal forces than nails. A larger screw thread is needed for particle board.

Bolts (Fig. 286) used in timber structures differ from those used in joints for steel structures by having a larger ratio of length to diameter. The strength of joints depends on the direction of the applied loads relative to the grain of the timber, the diameter of the bolt, and the strength of the timber in compression parallel to the grain.

Bulldog connectors (Fig. 287) are used when the loads at the joint are too large for simple bolts. The toothed plate and the bolt are both in shear.

Split rings (Fig. 288) or *shear plates* (Fig. 289) are needed for high loads. The split ring and its bolt are in single shear. Alternatively, one or more shear plates and their bolts can be used; their shear strength is higher.

Figure 285 Screw.

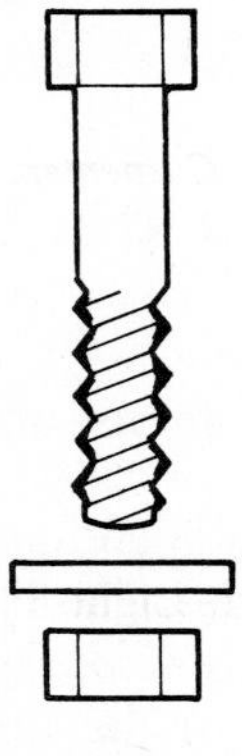

Figure 286 Bolt.

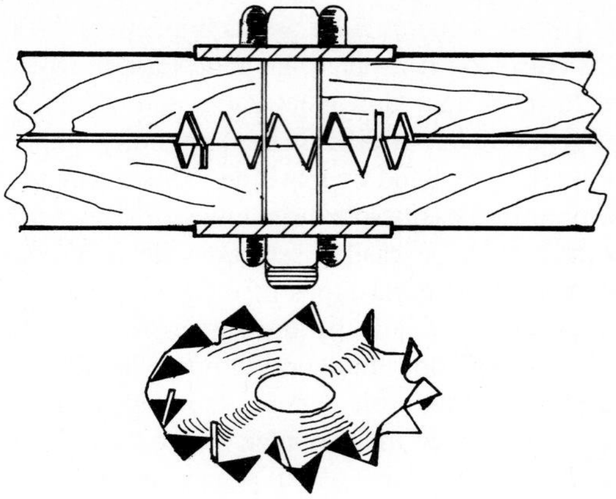

Figure 287 Bulldog connector.

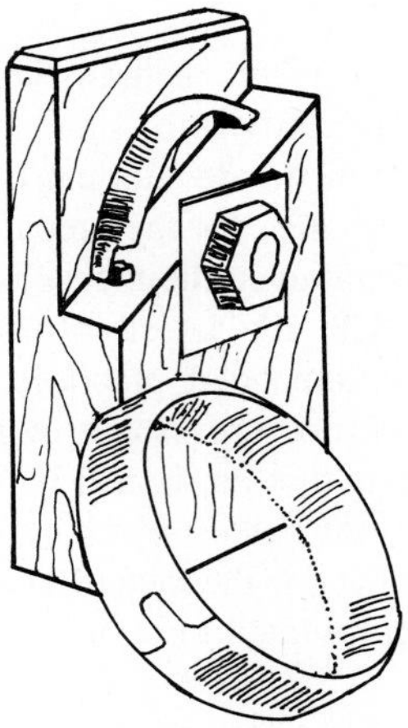

Figure 288 Split-ring connector.

Multitooth plate connectors (Fig. 290) are particularly useful for prefabricated roof trusses. They are fabricated from coil steel 1.2 to 2.0 mm thick. The integral nails are cut and pressed out at right angles to produce teeth at fre-

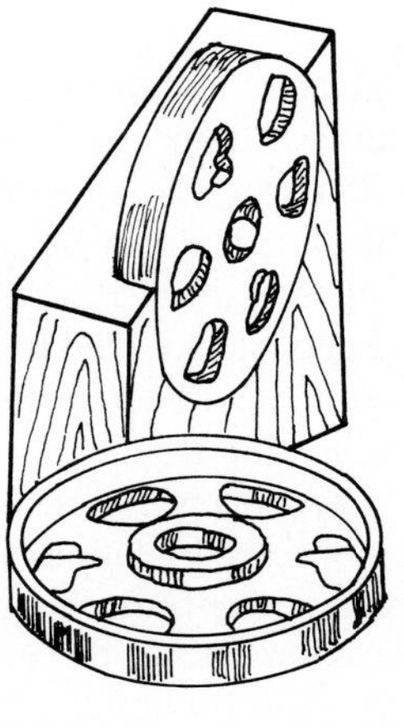

Figure 289 Shear-plate connector.

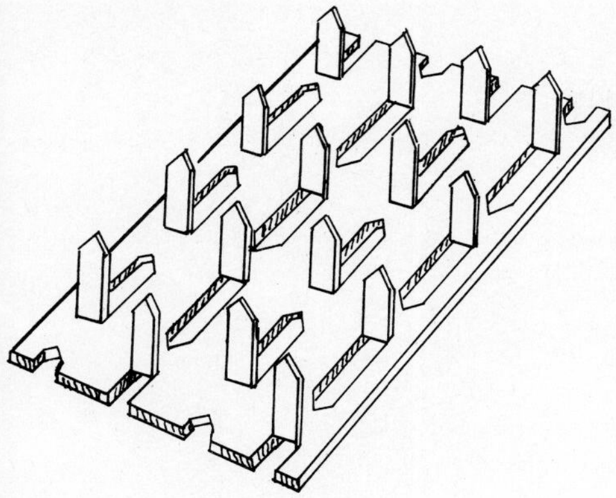

Figure 290 Multitooth plate connector (nailplate).

quent intervals. These are subsequently forced into the timber members by a hydraulic press in a factory under strict supervision.

Glued joints are designed to have the glue subject to shear only. The glue may be applied to one or both faces of the members in the joint, and it should be as thin as possible. The pieces of the joint must be held in close contact during curing so that slip does not occur in the life of the joint.

Joints Used for Joinery are generally glued. They can be divided into those used in intersections of timber pieces and those used for increasing the width of members. The two pieces of a joint for intersections may be in the same plane or at an angle to each other. Butt, halving, housing, dovetail (Fig. 291), bridle, and mortise, and tenon joints (Fig. 292) are members of this group. Joints for widening are needed when sufficiently wide planks of timber are not available. Butt edge, rubbed glue, doweled (Fig. 293), feathered, and slip-tongued joints belong to this group.

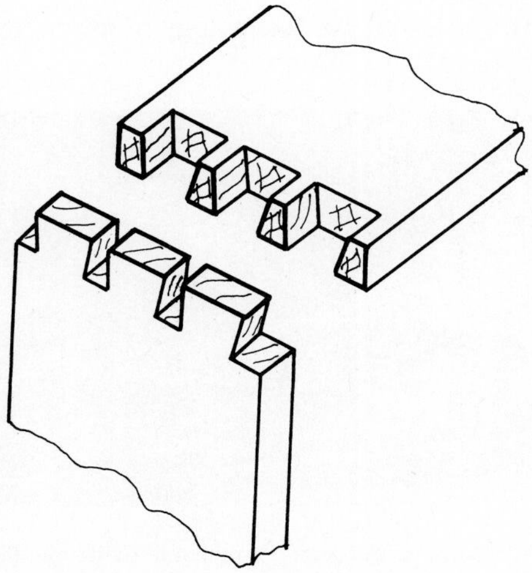

Figure 291 Dovetail joint.

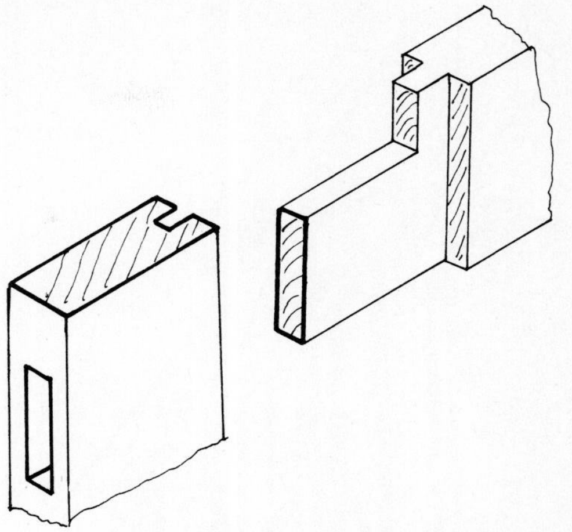

Figure 292 Mortise-and-tenon joint.

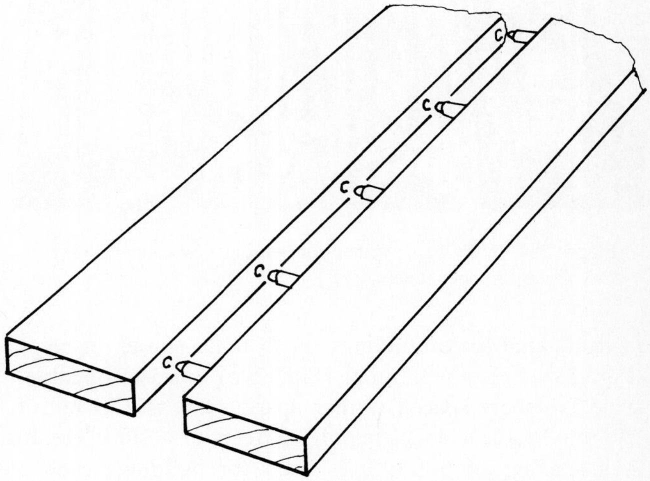

Figure 293 Longitudinal joint for joinery.

References

1. E. C. OZELTON and J. A. BAIRD: *Timber Designers Manual.* Granada, London, 1976.
2. AMERICAN INSTITUTE OF TIMBER CONSTRUCTION: *Timber Construction Manual*, 2nd ed. Wiley, New York, 1976.
3. E. SHARP: *Australian Carpenter*, Metric Edition, Macmillan, Melbourne, 1979.

G.V.

TIMBER STRUCTURES are of two basic types, light framed and heavy framed. Light-framed structures use relatively small-sized members close together, such as 35 mm × 90 mm studs spaced 400 mm apart ($1\frac{1}{2}$ in. × $3\frac{1}{2}$ in. studs spaced 16 in. apart); these are most commonly found

Figure 294 Light frame system for housing. (By courtesy of the U.S. Forest Products Laboratory.)

Figure 296 All-wood foundation, using pressure-treated studs, bearing plates, and plywood. (By courtesy of Barnes Lumber Corporation, Charlottesville, Va.)

in residential-scale buildings up to three stories in height (Fig. 294). Heavy framing (Fig. 295) consists of larger-sized members spaced farther apart, such as 150 mm × 250 mm beams at a spacing of 1.5 m (6 in. × 10 in. beams at a spacing of 5 ft); it is used for buildings such as churches, warehouses, and arenas, although it is sometimes used for housing and offices. Heavy framing tends to be more expensive than light framing, but it is fire retardant due to its mass, while light framing (unless specially treated) is not. Wood can be used to construct any or all parts of a building, such as walls, floors, roofs, and even foundations (Fig. 296). It can be used in a wide variety of forms as round poles (Fig. 297), sawn beams, fabricated trusses, plywood sheets, and glue-laminated girders or arches, assembled in numerous ways by nails, screws, bolts, adhesives, and various special metal connectors. Timber structures, when properly constructed, are particularly effective against earthquakes as they possess an inherent flexibility. However, timber structures in general (except for housing) are not used in contemporary construction to the extent they once were historically, due to the competition from newer materials such as steel and reinforced concrete. In addition, large solid sawn members are difficult to obtain because of the increasing scarcity of large-diameter trees.

However, research and technology in recent years have

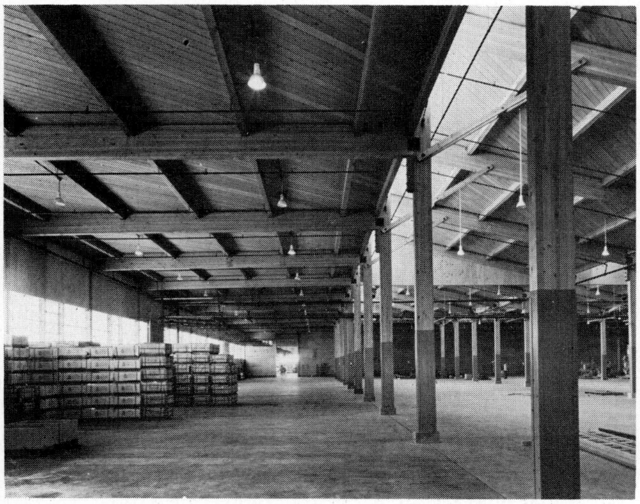

Figure 295 Heavy framing for warehouse, using glue-laminated girders and columns. (By courtesy of the American Institute of Timber Construction.)

Figure 297 Pole construction for a school in California; the system is earthquake and flood resistant. (By courtesy of Koppers Company, U.S.A.)

Figure 298 Heavy framing in a church, using both straight and curved glue-laminated members. (By courtesy of the American Institute of Timber Construction.)

done a great deal to enhance the use of wood in building construction. A large-sized member, instead of being cut from a single log, can be made from a lamination of smaller members, which are then glued together (Fig. 298). Depending on laminate thickness and fiber orientation, these laminates are called by such names as glulam, press-lam, or micro-lam. They can be made in either straight, tapered, or curved shapes of almost any size or length. On occasion, metal rods, plates, or prestressing tendons are incorporated into the laminates. Small particles or chips of wood, formerly considered waste products, can now be bonded by adhesives and reconstituted into sheets, panels, and other moldings of appreciable strength. Plywood, a cross-grained glue-laminated sheet product in use for many years, has taken on new structural applications in the form of stressed-skin panels for roofs, walls, and floors. Many of these panels are modularized and prefabricated (often incorporating wiring, plumbing, and insulation) for rapid field assembly. Plywood, fabricated into box beams, can be used for heavily loaded or long-span girders. Pressure-treated plywood, in combination with treated sawn lumber reinforcement, is often used to construct entire foundation systems for small homes, using no concrete or mortar whatsoever. Engineered wood construction, using light framing principles, is also commonly seen in the construction of modular homes; entire rooms or sections of homes are prefabricated and shipped to the building site in one piece.

Technological developments related to better ways of joining timber members have resulted in an assortment of new types of metal connectors and adhesives. One of the most widely used new metal connectors is the light-gauge steel plate (sometimes called the gang-nail plate) used to join components of trusses by pressing the teeth of these punched plates into the portions of the truss members to be joined. This development has greatly reduced the cost of fabricating light trusses.

Other new developments include automatic electronic grading of wood, new chemical preservatives, and refinements in the theory and design of timber structures, the latter through controlled tests (as at the U.S. Forest Products Laboratory) and sophisticated computer modeling.

See also BALLOON FRAME; LAMINATED TIMBER; PLYWOOD; TIMBER; and TIMBER JOINTS.

References

1. *Wood Structures.* American Society of Civil Engineers, New York, 1975.
2. D. E. BREYER: *Design of Wood Structures.* McGraw-Hill, New York, 1980.
3. *Timber Construction Manual.* American Institute of Timber Construction, Englewood, Colo., 1974.
4. *The Wood Book.* Avery-Phares, Inc., Seattle, Wash., 1981.
5. *A E Concepts in Wood Design.* American Wood Preservers Institute, McLean, Va., a bimonthly magazine.

W.Z.

TOLERANCES. All manufactured articles have some inaccuracies in their dimensions due to human error, or machine or instrument imprecision. To ensure proper fit and functioning, deviations must be limited; this is done by the specification of tolerances. The calculation of tolerances is a necessary prerequisite for effective joint design, especially in DIMENSIONAL COORDINATION.

Tolerances and deviations are normally considered together. Induced deviations from specified design dimensions are caused by imprecise manufacture, layout (setting out), and on-site assembly, and such deviations cannot be predicted accurately. Inherent deviations caused by physical or chemical changes are much more predictable and can be taken into account when the limits of size or position and the required tolerances are determined.

Application. In the building context, several tolerances combine to give the *building tolerance:* the manufacturing tolerance, the layout tolerance, and the erection tolerance. Not all of them are cumulative and neither are they relevant in every case; therefore, they should be considered separately.

The *manufacturing tolerance* represents the space within the limits of which a point, line, or plane of a component must lie after manufacture. The manufacturing tolerance is made up of several other tolerances:

1. *Dimensional tolerance*, which governs the dimensions of a component, such as length, width, height, and thickness.
2. *Angular tolerance*, which determines the relative orientation of straight lines.
3. *Form tolerance*, which limits the form of a line or shape relative to a basic form.

The *layout tolerance* represents the space within the limits of which the points or lines that set out the building or its parts must lie. This tolerance is made up of two separate tolerances:

1. *Position tolerance*, which governs the permissible deviation of a position relative to a specified position.
2. *Orientation tolerance*, which governs orientation parameters such as level, verticality, and angular deviation.

The *erection tolerance* determines the spaces on site within the limits of which the components must be located so that they fit. This tolerance also is made up of a position tolerance and an orientation tolerance.

Specification. Tolerances may be shown in several ways:

1. As the specified difference between an upper and a lower limit, in which case the tolerance is an absolute value, for example, a tolerance of 12 mm ($\frac{1}{2}$ in.).
2. As an equal or unequal positive and/or negative deviation from a specified dimension, for example, 6000 ± 6 mm (20 ft 0 in. ± $\frac{1}{4}$ in.) or 6000 mm + 12 mm, −6 mm (20 ft 0 in. + $\frac{1}{2}$ in., −$\frac{1}{4}$ in.). In special cases, the permissible deviation occurs in one direction only.
3. As a ratio, for example, 1 mm/m or L/1000 ($\frac{1}{4}$ in. per 20 ft 0 in. or L/960).

References

1. *Tolerances for Building—Vocabulary.* International Standard ISO 1803—1973. International Organization for Standardization, Geneva, 1973.
2. *Tolerances for Building—Relationship between the Different Types of Deviations and Tolerances Used for Specifications.* International Standard ISO 4464—1980. International Organization for Standardization, Geneva, 1980.
3. *Tolerances and Fits for Building—The Calculation of Work Sizes and Joint Clearances for Building Components,* DD22:1972. British Standards Institution, London, 1972.

H.J.M.

V

VALUE ENGINEERING (VE) is a technique whereby a multi-disciplined team of design professionals takes a systematic and creative second look at key design decisions for the purpose of assuring owners of optimum return on investment. This VE team identifies areas of action and suggests alternative means of initial and LIFE-CYCLE cost improvement, while maintaining or improving the specified quality and the functional requirements. As a recognized discipline, value engineering has evolved over the past 40 years, starting in the General Electric Company, which was the first manufacturer to organize a number of techniques into a formalized decision-making system; it is now used in all areas of industry and government, including the construction industry.

Value engineering optimizes both building and civil works projects, and it has the following benefits for owners, designers, and contractors:

1. Minimizing initial construction costs by 5% to 15% overall.
2. Rationalizing project efficiency, in both functional and operating terms.
3. Improving the quality of facilities by improving the ratio of net to gross areas and redistributing benefits.
4. Selecting the most appropriate overall life-cycle cost-effective systems with respect to energy, maintenance, staffing, and other ongoing costs.
5. Improving energy efficiency and life-cycle cost by 10% to 20%.

Normally, value engineering techniques are utilized throughout the procurement process and serve as a formalized design review at key points such as the following:

Project programming/feasibility
Concept design
Schematic design
Design development

Construction documentation
Construction bidding

Methodology. The analysis of the function of all elements is the key to value engineering to ensure that the required function of an element is obtained at the lowest total cost for a given level of quality. Value is the relationship between cost and worth of a particular element:

$$\text{value} = \frac{\text{cost}}{\text{worth}}$$

where *cost* = price paid for a specific element
worth = minimum cost to achieve the required function.

This value ratio acts as a guide toward selecting areas of VE study and as a stimulant to creativity and the generation of alternatives.

Study Procedure. The value engineering team at any given point in the procurement cycle may consist of the owner, the original designers, engineers, and architects, and the builder; at times it also involves specialists in process, lighting, geotechnical concerns, and acoustics. The value engineering procedure generally follows three stages, as shown in Fig. 299, and includes the following:

The Pre-study Stage, in which information is gathered for the formal study, including development of cost, energy, space, and other models using a one-page visual display, related to the functional components of the project.

The Study Stage, comprising the following:

1. *The information phase*, which includes review of information concerning the project, the designer briefing, the review of cost and other models, and the analysis of functions. Study areas are then selected based on this information.
2. *The idea phase*, during which creative alternatives are suggested as a means of satisfying the functions established during the previous phase.
3. *The analytical phase*, which is a critical evaluation in both quantitative and qualitative terms. It lists advantages and disadvantages and prepares cost estimates and life-cycle cost analyses.
4. *The recommendation phase*, which produces formal documentation of the suggested ideas for value improvement, including both the original and the proposed designs.

The Post-study Stage includes preparation of a report of recommendations, coordination of final implementation, and follow-up of actual savings achieved.

Applications. The value engineering methodology and tools are appropriate for any project. They produce optimum results in particular elements, as illustrated in Fig. 300. The techniques can also be used to determine key decisions, such as whether to renovate, expand, replace, substitute, or eliminate present facilities. Through the use of value engineering techniques, design professionals can identify the functions to be satisfied and select the most cost-effective solution.

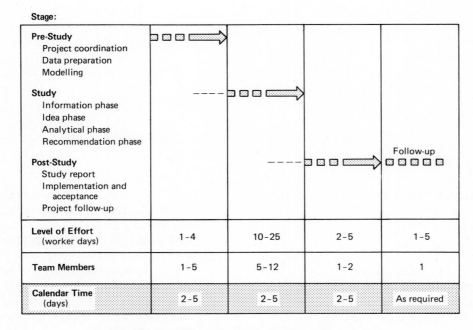

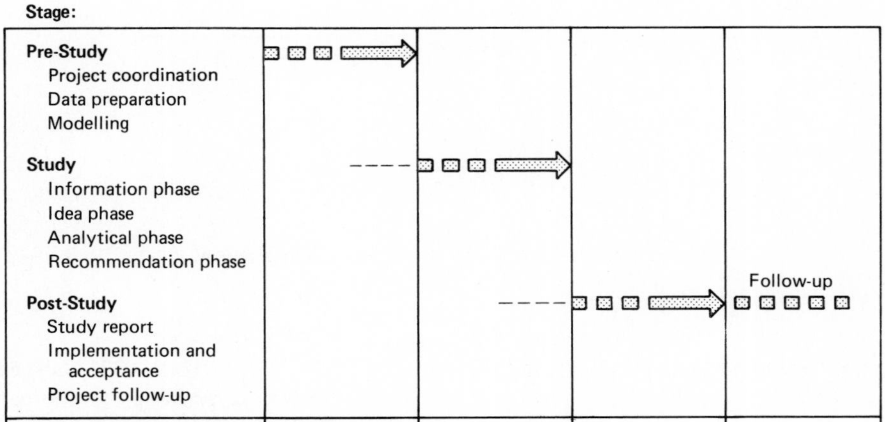

Level of Effort (worker days)	1-4	10-25	2-5	1-5
Team Members	1-5	5-12	1-2	1
Calendar Time (days)	2-5	2-5	2-5	As required

Figure 299 Value engineering study schedule and level of effort.

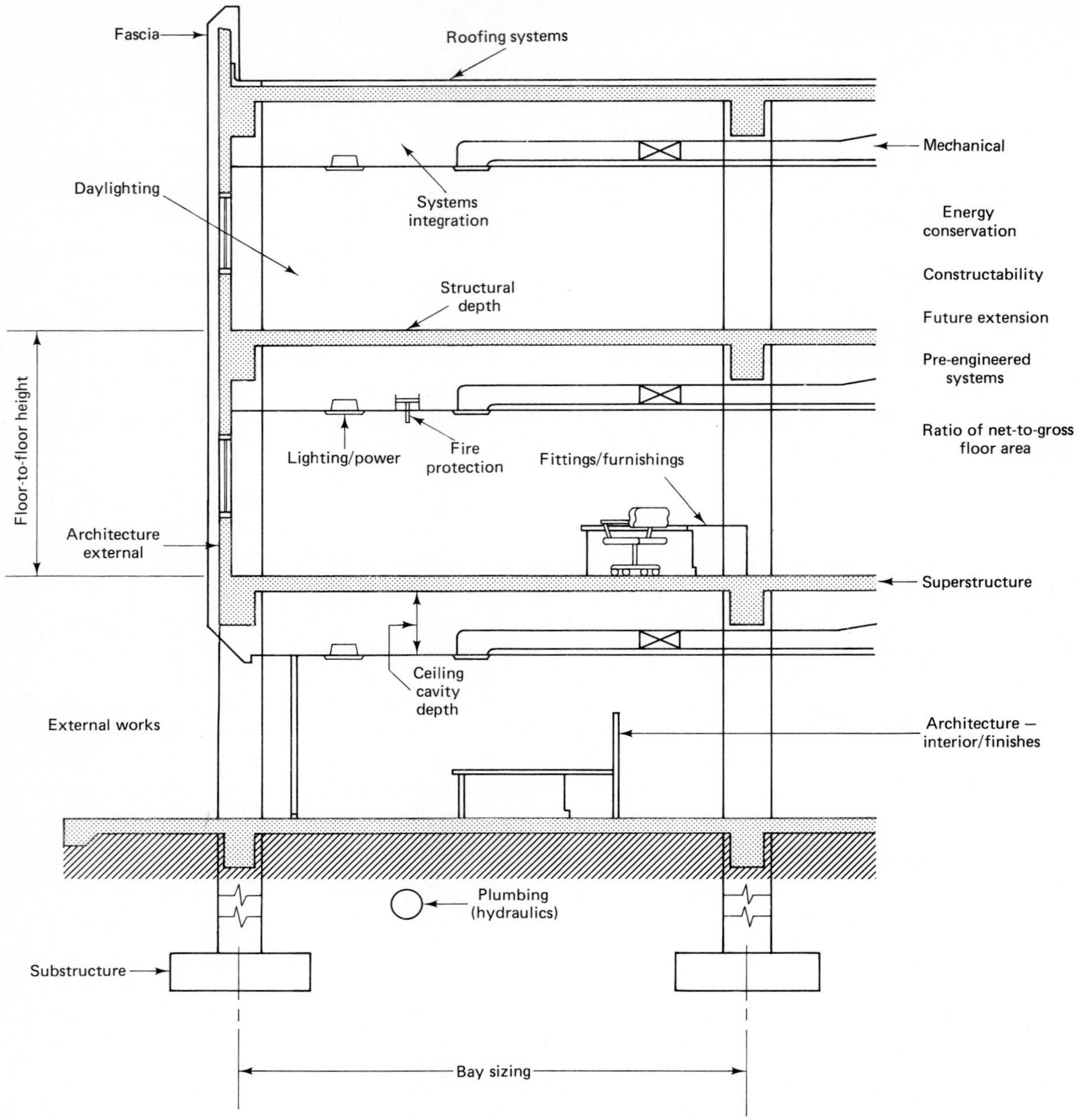

Figure 300 Areas applicable to value engineering review.

See also NETWORK ANALYSIS and OPTIMIZATION.

References

1. ALPHONSE J. DELL'ISOLA: *Value Engineering in the Construction Industry*, 3rd ed. Van Nostrand Reinhold, New York, 1983.
2. ALPHONSE J. DELL'ISOLA and STEPHEN J. KIRK: *Life Cycle Costing for Design Professionals*. McGraw-Hill, New York, 1981.
3. ALPHONSE J. DELL'ISOLA and STEPHEN J. KIRK: *Life Cycle Cost Data*. McGraw-Hill, New York, 1983.
4. *Value Management Handbook*, PBS P 8000,1A. General Services Administration, Washington, D.C., 1978.
5. LAWRENCE D. MILES: *Techniques of Value Analysis and Engineering*, 2nd ed. McGraw-Hill, New York, 1972.

6. *Value Engineering Workbook for Construction Projects*, EPA-43019-76-008. U.S. Environmental Protection Agency, Washington, D.C., 1976.

R.J.D. and S.J.K.

VAPOR BARRIERS are moisture-impervious building materials used to control, contain, and/or direct water vapor in buildings.

Reasons for Vapor Control. Barometric, thermal, and vapor pressure are continually acting on water vapor to equalize interior and exterior pressure by diffusion, possibly causing condensation. CONDENSATION produced when vapor comes in contact with cold surfaces may cause structural and/or health problems. High relative humidity may also result in higher energy use and costs. In addition, museums and libraries often need moisture control to preserve their contents.

Methods of Vapor Control. Excess vapor may be reduced by decreasing its production, increasing building air changes, dehumidifying and/or allowing diffusion into ventilated areas with selective use of vapor barriers.

Materials. Only metal and glass are completely impervious to the diffusion of water vapor. Polyethylene, other plastics, tar-coated paper, some paints, and other liquid applied coatings are used as vapor retarders. Vapor barriers and their adhesives may create a fire hazard.

Ratings. The efficiency of vapor barriers is expressed in *perms*. In metric units, one perm represents a transfer rate of one nanogram of moisture per square meter per second with a vapor pressure difference of one pascal. In customary American units, 1 perm equals 1 grain per square foot per hour per inch mercury. The lower the rating, the more effective the material as a retarder of vapor transmission. The necessary rating is determined by service conditions. Canadian residential standards require ratings of 45 metric perms (0.79 perms in American units), while in the United States ratings of 57 (1.00) are specified.

Installations. Vapor barriers should be installed so that they retard the movement of moisture from warm areas to cold areas. They should be placed as close as possible to the warm surface at a location where they will be kept above the dew point temperature of the humid area. Vapor barriers are also gas barriers and as such prevent the passage of gases such as radon, carbon dioxide, formaldehyde, etc. Adequately ventilated attics usually allow vapor diffusion without the use of a ceiling "gas/vapor barrier." Improper placement of a vapor-impervious material may cause moisture problems.

Air/Vapor Barriers. Frequently, the function of a vapor barrier (to stop vapor diffusion) is combined with the function of an air barrier (to stop the transportation of vapor by air movement) in a composite air/vapor barrier.

References

1. G. O. HANDEGORD: Vapor barriers in home construction. *Canadian Building Digest CBD9*, Ottawa, 1960.
2. K. M. KELLY: Indoor moisture: Effect on structure, comfort, energy consumption and health. *Proceedings of ASHRAE/DOE Conference*; *Envelope of Buildings II*, Atlanta, Ga., 1983, pp. 1007–1032.
3. Moisture in building construction. Chapter 21 in *ASHRAE Handbook—1981 Fundamentals Volume*. American Society of Heating, Refrigeration and Air-Conditioning Engineers, Atlanta, Ga., 1981.
4. B. LINN: *Conservation of Old Buildings*. National Swedish Institute for Building Research, Stockholm, 1975.

K.M.K

VERMICULITE is the collective name of a number of minerals in the mica group with the general formula $(Mg, Fe, Al)_3\,(Al, Si)_4\,O_{10}\,(OH)_2 \cdot 4H_2O$. Like all micas they have cleavage planes; in addition they have the unique property of expanding in a direction perpendicular to the cleavage planes when they are heated to 100° to 220°C (212° to 428°F). This *exfoliation* results from the conversion of the water of hydration into steam.

Vermiculite has good THERMAL and SOUND INSULATION properties. It has excellent DURABILITY, and it has a low density. It is thus particularly suitable for the FIRE PROTECTION OF STEEL STRUCTURES, and it has been used extensively for this purpose, particularly since restrictions have been placed on the use of ASBESTOS for health reasons. It is also used for LIGHTWEIGHT CONCRETE.

The main producers of vermiculite are the United States and South Africa. Supplies are also obtainable from the U.S.S.R., India, Brazil, and Argentina.

The expansion coefficient of vermiculite varies widely. For South African vermiculite it averages 26, and for American vermiculite it averages 15.

References

1. J. J. SCHOEMAN; Vermiculite, in *Mineral Resources of the Repubic of South Africa*, 5th ed. Handbook 7, Geological Survey, Government Printer, Pretoria, S. Africa, 1976, pp. 431–432.
2. H. J. SCHROEDER: Vermiculite, in *Mineral Facts and Problems*. Bulletin 650, U.S. Bureau of Mines, Washington, D.C., 1970, pp. 1283–1291.
3. E. R. VARLEY: *Vermiculite*. Colonial Geological Surveys. Her Majesty's Stationery Office, London, 1952.

H.J.C.

VERNACULAR ARCHITECTURE. For centuries, vernacular buildings have been primarily constructed to provide shelter for humans, animals, and goods. Building techniques are largely dictated by the nature of available materials. The ways these materials are prepared for construction depends on the way labor is distributed and this, in turn, depends on the organization of the community. In most cases, there is a close affinity between preparation of building components and the practice of local handicrafts, such as pottery in hot, dry regions where earth is the basic material, and basket and mat weaving in warm and humid areas, where bountiful rain generates lush vegetation, providing a rich source of organic materials.

Development of building techniques is also influenced by their ability to cope with climatic extremes in regions that sometimes pose specific problems, such as seasonal floods or earth tremors. Vernacular buildings could be classified according to whether they are in hot-humid, hot-dry, or cold climatic regions (see also CLIMATE).

Hot-Humid areas, where temperatures remain mildly hot, between 26°C (80°F) and 32°C (90°F), and where humidity is extremely high throughout the year, occupy much of Equatorial Africa, the Amazon Basin, and most of Southeast Asia from the southern Indian coast to Papua New Guinea. There is no seasonal variation, and heavy rainfalls produce a wealth of vegetation, such as timber, grasses, reeds, leaves, canes, and bamboo. With the exception of timber, all building materials are relatively impermanent and still continue to be used in rural areas where pressures of population have not depleted their supplies.

The structures are light skeletal frames whose walls are covered with mats and whose roofs are thatched. They can withstand earthquakes, and their walls are ideal for cross-ventilation. Fabrication and later repairs are easy, and their reclamation value is fairly high.

Few of these materials, however, are able to survive the numerous agents of decay, such as weather and insects. The average life of thatch roofs and wall mats rarely exceeds 4 to 5 years. The fire hazard is usually great and, as a consequence, there have been incidents when whole villages have been reduced to ashes.

High volume and intensity of rainfall also cause problems of water penetration and flooding. Prolonged wetting delays drying of materials, and this promotes fungal growth, highlighting the need for shedding water and the provision of adequate ventilation and good drainage.

Hot-Dry climates, where summer temperatures are high, winters are temperate, and the diurnal range is large prevail in North Africa, West Asia, coastal Peru, Southwest America, and large parts of Australia. There is little rain, hence low humidity and sparse vegetation.

Here, particularly in rural areas, earth in one form or another constitutes the basic construction material, which is primarily used for building solid load-bearing walls and as a roof cover. In urban and semiurban centers, earth roofs are often replaced by light timber frames covered with galvanized iron sheets. The structure has a high heat storage capacity that enables it to remain cool during the hottest parts of the day and retain heat during cold winter nights. When interiors are uncomfortably hot, people utilize secondary areas, such as verandas and courtyards for evenings and roof tops for sleeping at night.

A major disadvantage of mud buildings lies in the lack of sufficient light and ventilation. The mud is slowly washed away by rainwater, which makes the buildings vulnerable in monsoonal climates, thus necessitating continuous maintenance. Deterioration is also caused by the abrasive action of dust and sand under windy conditions.

The **Very Cold** arctic and subarctic regions, for instance in Northern Canada, Scandinavia, and Greenland, as well as the mild and temperate regions of North America and Southern Europe, all have continuously cold winters and very little sun and suffer strong winds. The ice and fur-lined igloos or snowdomes of the Eskimos have only low heat storage capacity, but offer maximum stability with minimum surface in high winds. More sophisticated examples of vernacular buildings in cold climates can be seen in the rural areas of Norway and Switzerland where farmers use timber logs as a basic construction material.

Typical log cabins and chalets of the Swiss mountain areas, for instance, have a masonry base, walls constructed of logs laid horizontally one above the other, with their ends interlocking at the corners, and a shallow bridge roof. Use of logs imposes serious limitations on the building because their size cannot exceed the normal height of a tree, as splicing of the logs is difficult and weakens the stability of the structure. This restriction, along with the linear nature of this material, produces straight rectangular plans. It is also difficult to make these structures absolutely weathertight.

See also ADOBE, PRESSED-EARTH BLOCKS, AND RAMMED-EARTH CONSTRUCTION; MASONRY ARCHES, VAULTS, AND DOMES; PASSIVE SOLAR DESIGN; and SANITATION IN DEVELOPING COUNTRIES.

References

1. A. H. BRODRICK: Grass roots, *Architectural Review*, Vol. 115 (Feb. 1954), pp. 101–111.
2. J. M. FITCH and D. P. BRANCH: Primitive architecture and climate, *Scientific American*, Vol. 107, No. 6 (Dec. 1960), pp. 134–144.

3. P. OLIVER (ed.): *Shelter and Society*. Barrie & Rockliff; The Cresset Press, London, 1969.
4. AMOS RAPOPORT: *House Form and Culture*. Prentice-Hall, Englewood Cliffs, N.J., 1969.
5. BERNARD RUDOFSKY: *Architecture without Architects*. Museum of Modern Art, New York, 1964.

B.S.S.

VIBRATIONS IN TALL BUILDINGS come from vortex shedding and buffeting from the wind, ground motion from earthquake, harmonic impact from its own mechanical systems (pumps, fans, elevators, and the like), impact from its own occupants (footfall), and other sources. They include the vertical vibration of beams and girders, the vertical vibration of the entire building frame, with its columns acting as springs, the lateral and torsional vibration of the entire building frame, and the oscillation of elevator cables.

The first-mode lateral period of vibration for tall buildings can be estimated crudely as $0.1N$, where N is the number of floors in the building. Lateral frequencies of the entire building frame, then, range from about 0.1 Hz for very tall, flexible buildings to 10 Hz for low, stiff buildings. Typical floor beams vibrate with a natural frequency of about 5 Hz, and a 40-story building, with its columns as springs, also vibrates vertically at a natural frequency of about 5 Hz.

These vibrations can be detected readily by building occupants through body sensors. In addition, the lateral and torsional swaying motion of the entire building frame can be detected from the visual perception of hanging plants and from the squeaking sound of walls and partitions. Elevator cable oscillation resulting from wind or earthquake and induced by the swaying motion of the building frame can be detected by the slapping of elevator cable against shaft walls and by occasional elevator malfunction.

Vibrations couple from one mode to another and from one structural system to another. In a 40-story building, for example, dancing on the fortieth floor may set tea cups rattling 20 stories below, since both the floor beams and the building as a whole may have the same natural frequency.

Damping is an essential ingredient in controlling structure vibrations due to wind or earthquake. Reference 1 considers various damping devices for the reduction of transient vibrations in floor assemblies. It was, however, the design and the research effort associated with the twin towers of the World Trade Center (New York City) that first brought damping systems into the practical world of building construction.

Figure 301 Vibration damping system used in the Citicorp Center, New York City, erected in 1978.

In the World Trade Center, and later in Columbia Center (Seattle), viscoelastic devices were employed to reduce wind-induced building oscillations. For the World Trade Center, the damping system also provides for a modest reduction in transient oscillation of the floor system. For the Citicorp Center (New York City), the tuned mass damper was first put to the practical test of building oscillations. Here a block of concrete, vertically supported on oil bearings and laterally supported on nitrogen springs, is used to dissipate wind-induced energy, thus reducing building oscillations (Fig. 301). This system was later installed in the John Hancock Center (Boston) as a retrofit installation. Through the thoughtful selection of building shape, aeroelastic damping was used in the United States Steel Building (Pittsburgh) to control wind-induced building oscillation (Ref. 2).

References

1. B. J. LAZAN: *Shock and Vibration Handbook*. McGraw-Hill, New York, 1961.
2. *Proceedings of Seminar: Wind Loads on Structures*. National Science Foundation, Japan Society for Promotion of Science, Honolulu, 1970.

L.E.R.

W

WATER QUALITY AND TREATMENT. Water supplied should be good enough for any proposed uses. Drinking water must be esthetically acceptable (clear, colorless, and cool without objectionable odors), nontoxic, free of disease germs, and free of excessive amounts of chemicals that may cause undesirable effects either in the short term (e.g., magnesium sulfate that causes diarrhea) or in the long term (e.g., cancer-causing impurities). Washing water must form lather, not cause stains in clothes or fittings, and not cause scale in heating devices. Water should not be corrosive to pipes, and garden water should not contain too much salt.

The amount and type of treatment used depends very much on the nature of the impurities to be removed. The most common method of treatment for water from a stream that sometimes contains microscopic particles of clay and other minor impurities consists of a number of steps. First, the river water may be screened to remove floating leaves and twigs and to exclude fish. Coagulant chemicals then added to the water (e.g., alum) cause the minute particles to clump together during gentle stirring (flocculation) so that most of them can be settled out in a quiescent tank or settling basin. The particles that do not settle out in these conditions can then be removed as the water passes through a bed of sand that acts as a filter. The sand bed is cleaned at intervals by reversing the flow of water and sometimes by using air bubbles to loosen the accumulated impurities. After filtration, the water can be disinfected by the addition of a suitable chemical, such as chlorine. As a final step, the corrosiveness can be reduced by neutralizing some of the acidity with either lime, sodium carbonate, or sodium hydroxide. The finished water should contain sufficient disinfectant to ensure that it reaches the consumer free from disease germs.

Table 1 indicates some of the processes used for removing or inactivating several unwanted impurities. Figure 302 indicates some of the options that are sometimes used when treating water from lakes and streams for domestic use; the thick line indicates the most common sequence of treatments.

See also HYDROLOGY OF ROOF WATER COLLECTION and WATER SUPPLY.

TABLE 1
Typical Methods Used for Correcting Some Quality Deficiencies

Property	Treatment Processes Used
Color	a. Coagulation and filtration (with activated carbon) b. Sometimes oxidation by chlorine or ozone.
Turbidity	Coagulation and filtration (coagulation may be omitted if raw water has relatively low turbidity)
Tastes and odors	a. Activated carbon adsorption b. Coagulation and filtration c. Catchment and stream control d. Chlorine dioxide oxidation e. Aeration f. Chlorination g. Ozonation
Calcium and magnesium (hardness)	a. Precipitation as magnesium hydroxide and calcium carbonate on addition of lime and soda. b. Ion-exchange processes.
Iron and manganese	Treatment methods depend on other impurities present. In general removal is affected by oxidation and precipitation as hydroxide, but precipitation of iron as ferrous carbonate and removal by ion exchange are sometimes used.
Sodium, potassium sulfate, chloride nitrate	These anions and cations cannot be removed from water supplies by moderately priced processs. Expensive desalting is required.
pH	The water can be made more acid or alkaline by the addition of a suitable acid or alkali. Carbon dioxide, sulfuric and hydrochloric acids, lime, caustic soda, or soda ash are commonly used.
Phenolic substances	a. Oxidation with ozone or chlorine dioxide. b. Adsorption on activated carbon.

Extracted from Table 13.1, Ref. 2, by courtesy of Pitman Books Ltd.

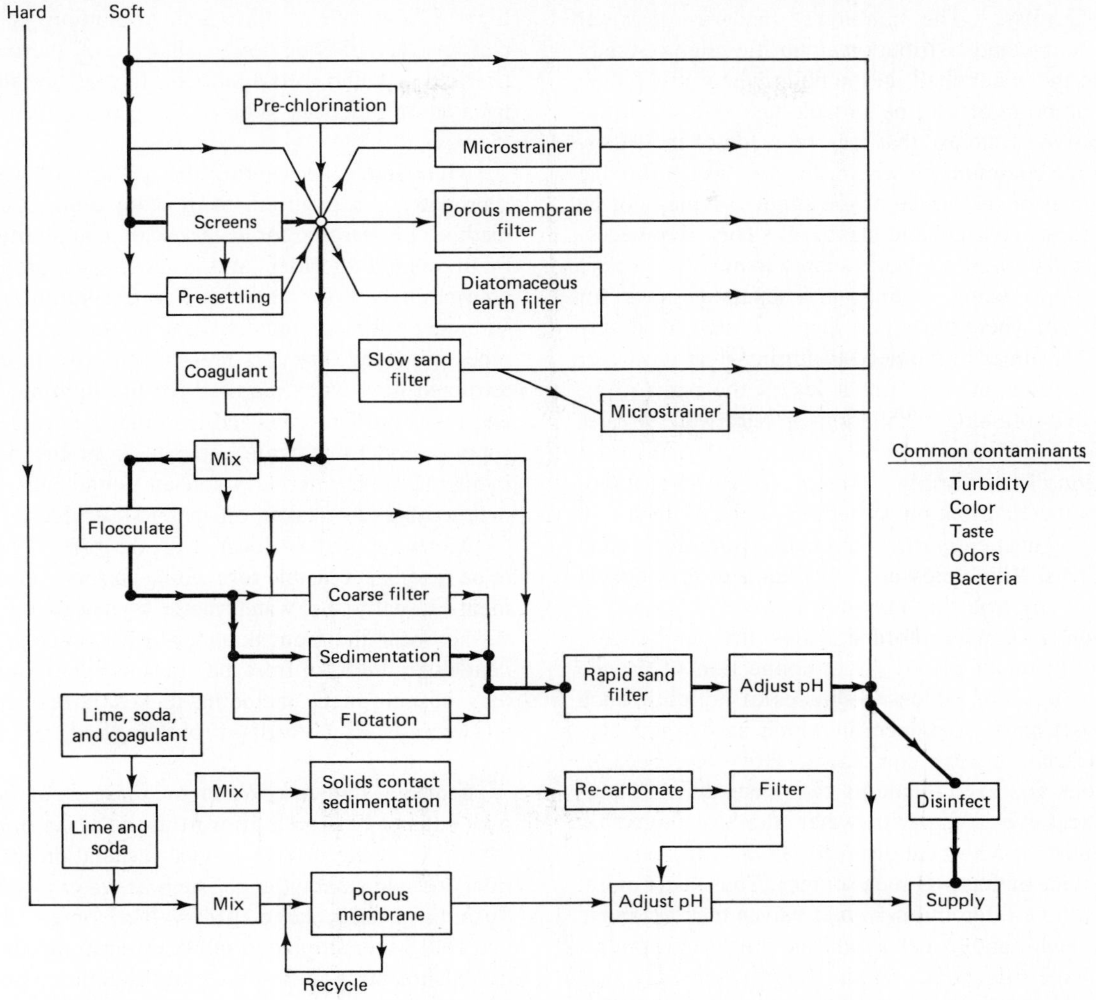

Figure 302 Some options for surface water treatment.

References

1. AMERICAN WATER WORKS ASSOCIATION: *Water Quality and Treatment*. McGraw-Hill, New York, 1971.
2. D. BARNES, P. J. BLISS, B. W. GOULD, and H. R. VALLENTINE: *Water and Wastewater Engineering Systems*. Pitman, London, 1981.
3. G. SMETHURST: *Basic Water Treatment*. Thomas Telford, London, 1979.

B.W.G.

WATER SUPPLY. The supply of water to buildings from the public system is so arranged that, throughout the water authority's distribution area, water can be delivered under pressure to levels at least three or four stories above the street. For taller buildings or buildings on elevated sites, it may be necessary for additional pressure to be applied by the use of pumps fitted within the property as part of the private water service installation.

Water supplied from the mains to properly designed and installed private services is available at rates of flow that are adequate for the usual domestic purposes, including drinking, washing, food preparation, sanitary flushing, and for the water carriage removal of waste matter in the building drainage and the public sewerage system.

Water may also be drawn off through faucets for garden watering and, in the case of industry, for manufacturing and other processes. Where suitable valves are provided, water may be drawn off through private services for fire fighting, although provision is also made in the form of hydrants at intervals along street mains for the use of water by fire authorities, whose hoses and mobile pumps may be connected to these hydrants so as to draw directly from the main. Street flushing using fire hydrants is permitted by some water authorities.

Water Quality. The minimum standards of water quality that supply and distribution authorities aim to satisfy are such that the water shall, at the point where it is drawn off for consumption or use, be suitable and safe for drinking, irrespective of the use that may be made of the water. There are some communities where, for part (i.e., from one or other source) or the whole of the supply, it may not be practicable to achieve potable standards. This may necessitate the installation of duplicate supply mains, the impure part of the supply being confined in a separate system for fire-fighting use; where the whole supply is affected, it may involve the consumer in the need to further treat the water for particular requirements after it leaves the main (e.g., domestic water softening installations in hard water areas).

Protection of the Supply. The pressure of water supplies within the distribution system is a major factor in safeguarding against entry of contaminants from the ground or other sources. While pressure is maintained, flow occurs *outward* from any leak that may develop.

Protection is likewise afforded against the supply being polluted due to unauthorized direct connection of private water service pipes or outlets to sources of pollution such as water closet pans, containers in which used water may lie, or industrial process equipment. However, because pressure drops and even vacuums can occur due to main-breaks or excessive demand for water, the best protection against pollution is a vertical distance (an *air gap*) between the water service outlets and such sources. Thus water closet cisterns should be filled through inlet valves that are above stored water level, and faucet spouts and other valve outlets should be above rim levels of fixtures or vessels they supply. Some authorities permit mechanical devices, such as vacuum-breaker backflow preventers, to be used in lieu of air gaps so as to enable apparatus that may contain pollutants to be operated under the water supply pressure that is essential to its operation (e.g., dishwashing machines using household detergents and water filtration plants using coagulants).

Some consumers may wish to use a second source of water supply (such as a private well, stream, or dam) in addition to the public supply to economize on water bills. In these cases it is essential to avoid polluting the public supply by keeping the two supplies entirely separate, necessitating separate service pipes for each and separate outlet fittings up to the point where water is discharged for use or for storage immediately prior to use. Of course, water from other than public sources should not be used in any application that requires water of potable standard unless it is known always to be safe for that purpose.

Installing the Service. The private water service installation between the main and the points of water usage must be of materials that are suitable to withstand water pressure and internal and external corrosive influences. Many local water authorities have plumbing codes or regulations that provide for installation of these services by licensed or authorized plumbers. In most the materials and installation practices to be used are prescribed.

Metering. Registering the quantity of water used on a property is a requirement of many authorities. Different methods of charging for water use or availability are applied by different authorities, and in most cases property owners are not charged for water used in fire fighting, if the measurement of the amount used can be separated from that for other purposes. For this reason, and also to ensure uninterrupted flow when required for fire fighting, commercial and industrial buildings and land may be provided with fire supply services that are separate from the domestic and industrial supply services and are unmetered, or with unmetered branches taken off metered services.

Meters should be located on the *private* service so as to be readily accessible for reading, inspection, and replacement. Usually the water meter is placed on the ground surface near the front boundary of a property; but where vandalism, theft, or frost may be a problem, other positions may be permitted, including locked boxes, covered pits, and recesses in the walls of buildings.

Storage Tanks. Water supplied from public mains may need to be stored prior to use in tanks or cisterns that form part of the private service installation. Such applications include small storage, such as water closet and urinal flushing cisterns, large flush valve storage, domestic hot and cold water storage in tall buildings, and, in some large establishments, water storage for fire-fighting purposes (Fig. 303). It is preferable, and with some applications essential, that the water stored be held in separate tanks, although offtakes at different levels in a tank can ensure that such needs as fire reserve storage will not be drawn down by the domestic demand taken off from the higher level. Automatic fire sprinkler storage is separate from other fire supplies so that at no time will the former be diminished except by the operation of the sprinklers.

Sanitary flushing supplies to water closet and urinal flush valves should not be stored in tanks holding other domestic supplies because there is a direct connection to the sewerage service through the flush valves that could constitute a severe pollution risk.

With tall buildings it may be necessary to provide water storage for the various requirements not only at the highest or the lowest levels, but also at one or more intermediate levels to enable adequate flow rates to be maintained throughout when water is drawn off or to reduce the static head to that for which pipes and fittings have been designed.

Hot-Water Supply. The supply of water to hot-water systems is usually furnished by means of a branch from the metered service, with delivery either to a cold-water feed

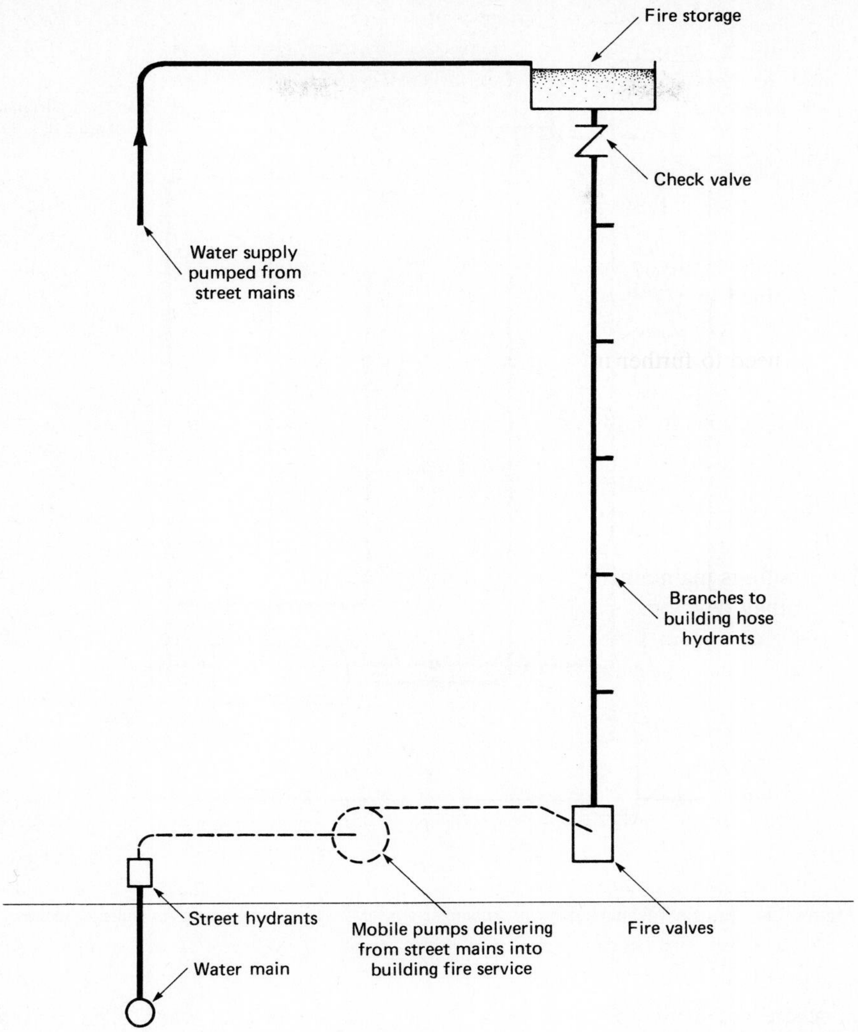

Figure 303 Fire services for tall buildings, with provision for boosting by mobile fire pumps.

tank or direct into the hot-water system. With the latter, unless a pressure reducing valve is installed (as may be done where water pressures are extremely high), the entire hot-water system is subjected to the water pressure applied from the water main, while the former installation has on it the head produced by the level of stored water in the cold feed tank.

Safety Devices. The maximum temperature of hot water supplies is usually about 95°C (170°F) and heater installations should be furnished with thermal cutouts and thermostats to control temperatures, and with temperature and pressure-relief valves in the case of direct-connected systems to safeguard against dangerously high pressures being generated. With direct connections, some authorities require nonreturn valves or vacuum-breaker backflow preventers to be fitted so as to protect against the flow of hot water, steam, or pollution back into the cold-water service.

Where cold-water feed tanks are used, hot-water storage vessels at levels below the feed tank need to be furnished with expansion pipes as an alternative to pressure-relief valves.

Relief valve drains should not be connected directly to any sewerage service installation because of the risk of pollution of the water supply; where relief drains discharge into such plumbing installations, an air gap should be provided.

Water Heating. Modern commercial and domestic hot-water supply systems use either direct or indirect heating. Direct heating involves electrically heated surfaces or immersed elements or gas or oil-fire-heated surfaces. Indirect heating uses heat exchange from high-temperature hot water or live steam in an exchange vessel called a *calorifier* (Fig. 304).

Indirect heating may be preferred in hard water or very

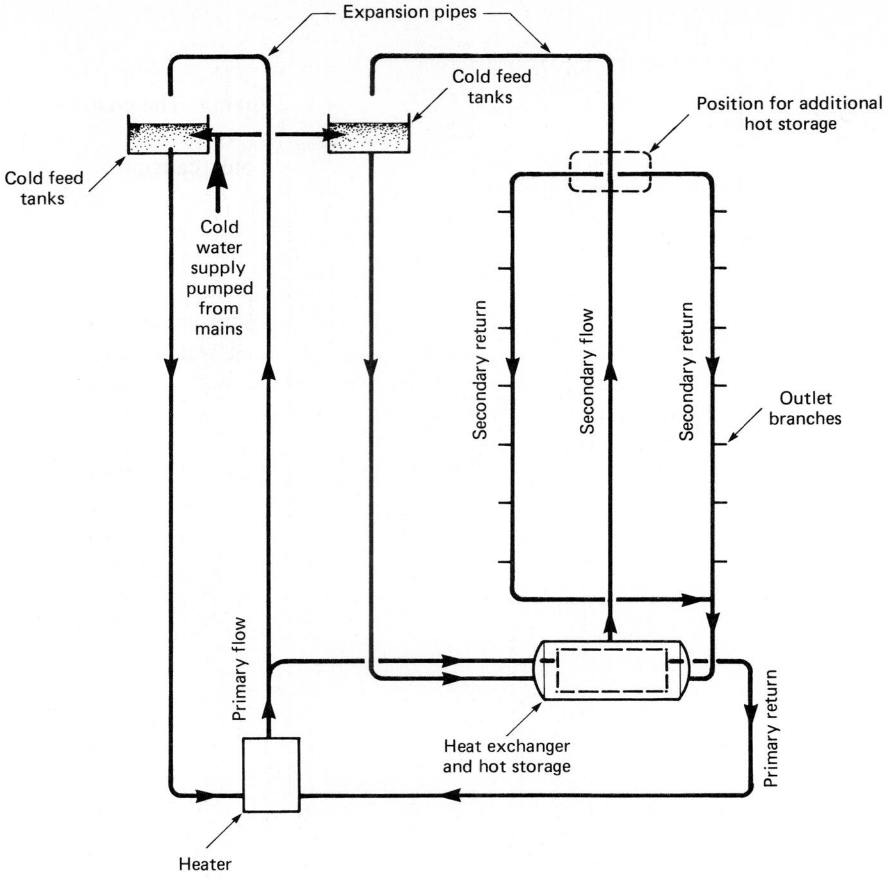

Figure 304 Principal features of heat-exchanger-type hot-water system with downfed branches.

soft water supply areas, where the incidence of scale deposition or corrosion in water heaters may be a problem. Indirect heating may also be preferred where the generation of high-temperature hot water or steam is necessary for other purposes, such as heating of buildings or for laundry or hospital applications.

As energy costs rise, solar heating of domestic, and to a lesser extent commercial, hot-water supplies becomes feasible, depending on the available sunlight and the latitude. SOLAR COLLECTOR panels may be incorporated in the roof structures or separately mounted to achieve the maximum extent of the sun's heat; they are connected to the hot-water storage tank by separate flow and return pipes to permit continuous circulation through the absorbers by convection, with supplementary heating of the stored water usually achieved by temperature-activated electric elements.

Supply of Hot Water to Outlets. This involves the provision of hot-water outlets at the usual domestic fixtures. Hot-water supply may be discharged at these outlets separately from cold or, by use of combination fittings, as a mixture with cold water. The mixing of hot and cold supplies can be achieved either by hand manipulation of two separate outlet faucets, of two valves on a combination fitting, or of a manual valve, or by the automatic functioning of a preset thermostatically controlled mixer valve that delivers water at the required temperature when the outlet faucet is turned on. Hot water may also be required at larger equipment, such as commercial dishwashing machines, laundry machines, and a variety of commercial and industrial process plants.

The delivery of hot water from its source to the outlets should be through pipelines of the shortest possible lengths and diameters to minimize heat losses. It is usual to insulate these pipes, except where very small diameters and very short lines are involved.

With commercial and large domestic installations, continuous circulation of hot water within the system is usually provided; this is achieved by the use of flow and return piping, with either downfeed or upfeed or a combined up- and downfeed system, having branch pipes to outlets taken off from either or both the flow and return pipes. This ensures that there will be the most equitable distribution of heated water throughout the building using convection currents to achieve circulation flow, assisted, in the case of extensive installations, by a circulating pump.

See also HYDROLOGY OF ROOF WATER COLLECTION; and WATER QUALITY AND TREATMENT.

References

1. L. S. NIELSEN: *Standard Plumbing Engineering Design*. McGraw-Hill, New York, 1963.
2. H. E. BABBITT: *Plumbing*, 3rd ed. McGraw-Hill, New York, 1960.
3. L. BLENDERMANN: *Design of Plumbing and Drainage Systems*, 2nd ed. Industrial Press, New York, 1963.

W.L.H.

WELDING, BRAZING, AND SOLDERING. There are over 40 welding processes available to join both similar and dissimilar metals used in building technology. The term welding denotes the process of connecting metals under the application of heat and/or pressure and either with or without the use of filler metal. Perhaps the most common welding process is arc welding, which refers to a wide spectrum of welding processes that use an electric arc to melt and join the parts. Brazing and soldering are welding processes that are similar to each other in that a filler metal is used to join the parts.

Figure 305 shows, in simplified form, the welding circuit basic to all arc-welding processes. A power source is connected on one line by the ground cable to the parts being welded and on the other line to the electrode. An arc is formed between the electrodes and the work parts.

Shielded Metal Arc Welding (SMAW) is shown in Fig. 306. The shield refers to the gaseous shield that surrounds the arc, gives stability to it, and prevents the molten metal from oxidizing. The coating of the electrode contains chemicals that perform many functions, such as providing the gaseous shield, causing the impurities to float to the surface as a slag, adding alloying elements to the weld metal as needed, and allowing the electrode to be designed to be compatible with the electrical characteristics of the power source. The shielded metal arc-welding process is used generally with manual welding, that is, for fitting up and for short tack welds used in assembly. Electrodes are available for use with a wide variety of steels and for aluminum, copper, and their alloys.

Submerged Arc Welding (SAW) uses a bare metal electrode, with the arc submerged under a mound of granular flux. The process is usually automatic or semiautomatic and includes an automatic wire feed for the electrode and a feed tube for the flux. Both electrode and flux material can be chosen to suit the particular work at hand. The process is suitable for a number of steels and nickel and nickel alloys.

Gas Metal Arc Welding (GMAW), formerly known as MIG welding for metal inert gas, and also known as CO_2 welding, uses a continuous mechanically fed consumable electrode with shielding from externally supplied gas, which may be either inert or reactive. The process is suitable for automatic or semiautomatic modes and for most steels, aluminum, and copper.

Gas Tungsten Arc Welding (GTAW), also referred to as TIG welding for tungsten inert gas, uses a tungsten nonconsumable electrode with shielding from externally sup-

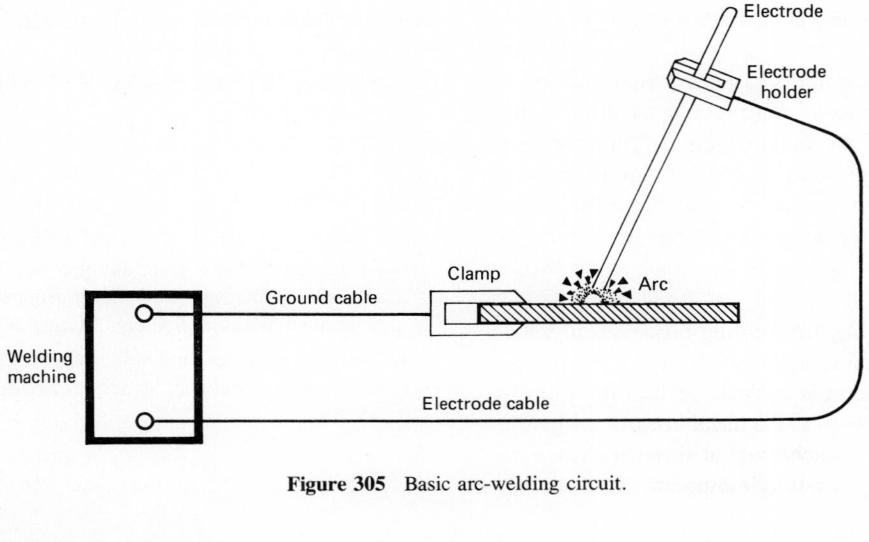

Figure 305 Basic arc-welding circuit.

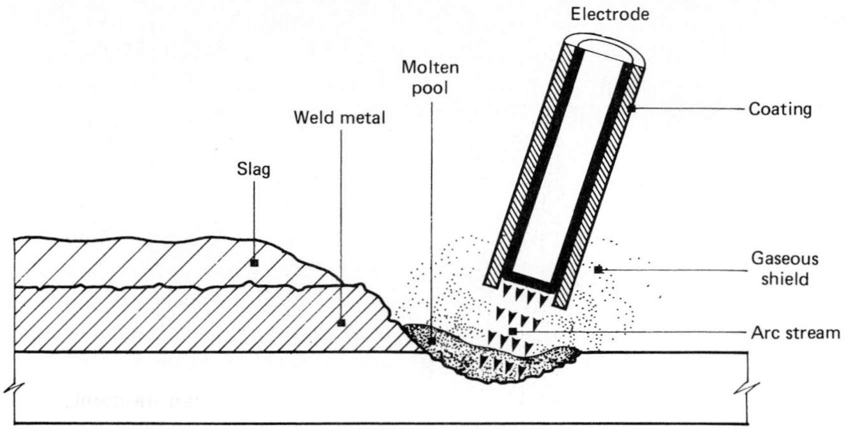

Figure 306 Shielded-metal arc welding.

plied gas. A welding rod may be used to supply filler metal if needed. The process may be automatic or semiautomatic and is suitable for most steels, aluminum, aluminum alloys, copper, bronzes, and brasses.

Flux Cored Arc Welding (FCAW) uses a continuous consumable electrode with flux within its core, with shielding supplied from the flux. Some processes use externally supplied gas. The welding process is usually used in semiautomatic form and is suitable for most steels.

Electroslag (ESW) and **Electrogas (GMAW-EG or FCAW-EG) Welding** are variations of the same process for single-pass vertical welding for plates. The welding process is suitable for most steels.

Stud Arc Welding is a specific arc-welding process for joining a metal stud to a plate or workpiece. Flux is contained at the end of the stud, which uses a gun to position the stud, create the arc, and apply pressure. Stud welding is available for most steels, aluminum and its alloys, and for various bronzes, brasses, and nickel–copper alloys.

Oxyfuel Gas Welding (OFW) uses a fuel gas mixed with oxygen and may use a welding rod as filler metal. Oxyacetylene welding is a typical example. The process is available for a wide range of steels and for aluminum, copper, and their alloys. Fluxes are needed for most metals other than steel to remove oxides from the surface of the base metal.

Brazing and Soldering are welding processes in which the materials to be welded together are heated to the proper temperature and filler metal is introduced into the joint by capillary action. *Brazing* employs filler metals with relatively high melting temperatures, while *soldering* uses filler metals with relatively low melting temperatures; in neither case is the base metal melted, but a metallic bond is formed through a solvent action between filler metal and base metal.

Most metals used in building construction, such as steel, aluminum, copper, and their alloys, are amenable to the brazing process, but care needs to be taken to choose the filler metal so that it melts before the base metal does, so that it flows properly into the joint, and so that the joint has the mechanical and physical properties compatible with those of the base metal. A variety of fluxes and gases are available. *Torch brazing* and *furnace brazing* are two of the most common brazing processes used in building technology. Other brazing processes are *resistance brazing* and *induction brazing*. The strength of the joint depends on the interaction between the filler metal and the base metal, the joint clearance, and the defects within the joint metal.

The soldering process may be used with most metals, and in many ways it is similar to the brazing process. *Torch soldering* is the most common process used in building construction; other methods are *resistance soldering*, *induction soldering*, and *wave soldering*. A variety of fluxes and solders is available, depending on the base metal and the properties desired in the joint. Most solders are tin–lead alloys.

See also JOINTS FOR STEEL STRUCTURES.

References

1. Welding processes—Arc and gas welding *and* Cutting, brazing and soldering, in *Welding Handbook*, Vol. 2, 7th ed. American Welding Society, Miami, Fla., 1978.
2. Welding, brazing, and soldering, in *Metals Handbook*, Vol. 6, 9th ed. American Society for Metals, Cleveland, Ohio, 1983.

L.T.

WIND LOADS on buildings are the result of air pressures acting on building surfaces that are different from the undisturbed atmospheric pressure. Shear stresses also cause forces but, on nonstreamlined building shapes, they are small compared to those caused by surface pressures and are not included in wind-load calculations. Pressures $p(x, y, z, t)$ vary strongly with time and location on a building due to increase of mean wind speed $\overline{U}(z)$ with height, turbulence, and flow disturbances caused by the building. The pressure is usually expressed as a one-half to one-hour time average $\overline{p}$ plus an instantaneous fluctuation from the mean p'; that is, $p(x, y, z, t) = \overline{p}(x, y, z, t) + p'(x, y, z, t)$. Integration of pressures over tributary areas for curtain-wall elements, roof sheathing, purlins, girts, studs, roof trusses, and so on, gives local wind loads $\vec{f} = \vec{\overline{f}} + \vec{f}'$. When integration is over the entire building surface, base shears $\vec{F} = \vec{\overline{F}} + \vec{F}'$ and base moments $\vec{M} = \vec{\overline{M}} + \vec{M}'$ are obtained. For a building with height to minimum width ratio in excess of 5, $\vec{F}'$ and $\vec{M}'$ may result in excessive dynamic response (deflections and accelerations) if not designed using information obtained from wind-tunnel tests and analysis.

Dimensionless Coefficients. A reference pressure $\rho \overline{U}_g^2/2$ is commonly used to form dimensionless coefficients (ρ = mass density of air, $\overline{U}_g$ = mean wind speed at gradient height or at top of wind-tunnel boundary layer). Coefficient definitions using the symbols shown in Fig. 307 are as follows:

$$\text{mean pressure, } C_{\overline{p}} = \frac{\overline{p}}{0.5\rho\,\overline{U}_g^2}$$

$$\text{peak maximum, } C_{\hat{p}} = \frac{p_{\max}}{0.5\rho\,\overline{U}_g^2}$$

$$\text{peak minimum, } C_{\check{p}} = \frac{p_{\min}}{0.5\rho\,\overline{U}_g^2}$$

$$\text{typical mean force, } C_{Fx} = \frac{F_x}{0.5\rho\,\overline{U}_g^2 WH}$$

$$\text{typical mean moment, } C_{My} = \frac{M_y}{0.5\rho\,\overline{U}_g^2 WH^2}$$

Mean wind speed at building height $\overline{U}_H$ and average mean wind speed between ground and building height $\overline{U}_A$ are sometimes used to form coefficients; therefore, care must be exercised when using coefficients for which the reference pressure is not defined precisely. Figure 308 illustrates the magnitude and variation of pressure coefficients around the periphery of an isolated block-type building at a single elevation. The radical variation of magnitudes and distributions with building height, building geometry, neighboring buildings, and wind direction can be determined accurately only by wind-tunnel tests on small-scale models. When a building does not have unusual geometrical form, a site with channeling or wake buffeting characteristics, or high dynamic-response characteristics, design wind loads may be determined from information given in building standards, such as Ref. 3.

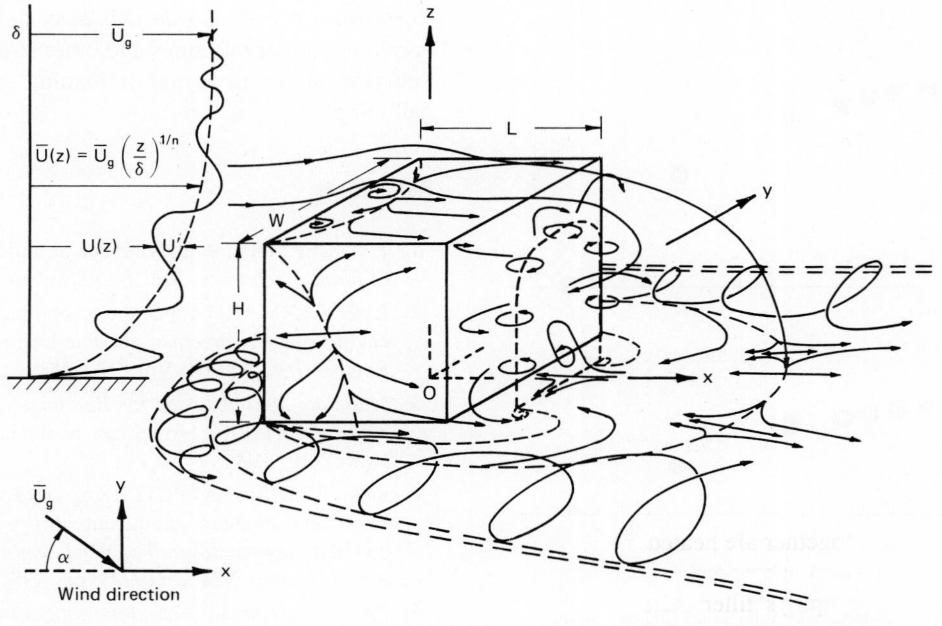

Figure 307 Flow pattern and definition for boundary-layer flow over an isolated block-type building.

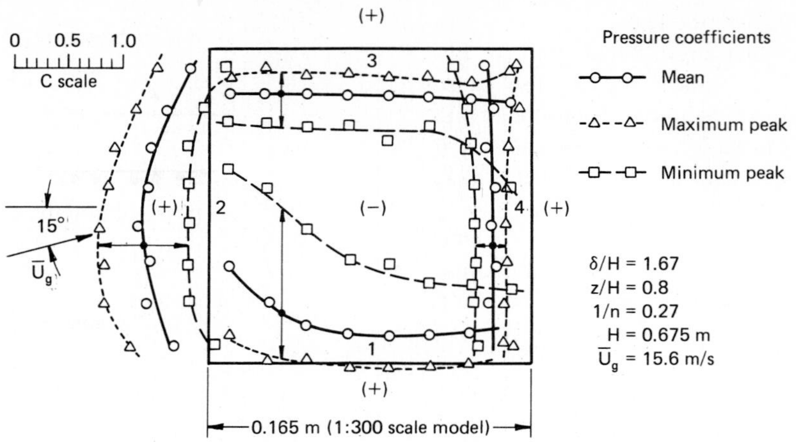

Figure 308 Pressure coefficient distributions around a 203 m (664 ft) high block-type building at a height of 162 m (531 ft) measured on a 1:300 scale model.

Of particular concern for the design of exterior curtain walls (glass, stone, and metal cladding and mullions) is the magnitude of $C_{\breve{p}}$, Near edges where flow separation occurs, $C_{\breve{p}}$ ranges from about -1.5 to -4.0. Vortex formation at discontinuities of vertical lines (setbacks, roof tops, intersection of a high-rise building with an attached low-rise structure, etc.) can result in values of $C_{\breve{p}}$ ranging from -4.0 to -10.0.

Examples of C_{Fx} and C_{My} (reference pressure $= 0.5\rho\,\bar{U}_A^2$) for an isolated block building in a boundary-layer wind field taken from Ref. 4 are shown in Fig. 309. Peak forces and moments may be approximated from these mean values by applying a gust factor of 1.3 to the *wind speed* $\bar{U}_A$.

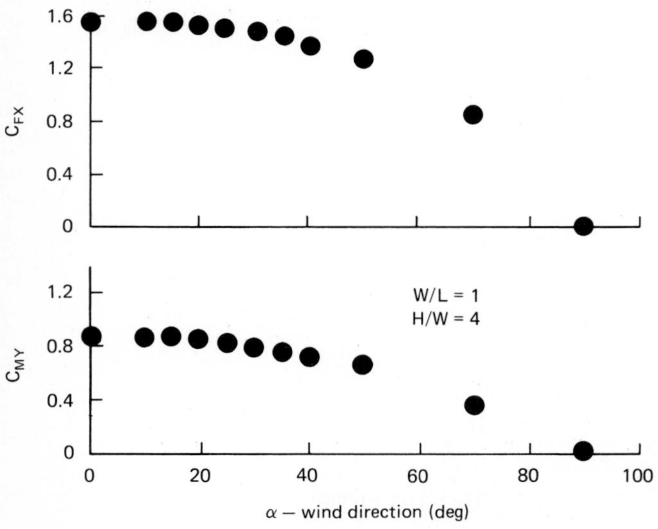

Figure 309 Typical mean force and moment coefficient variation with wind direction on a block-type building measured in a boundary-layer wind tunnel.

Wind-Load Determination. The increasing wind sensitivity of modern buildings (more extensive use of glass, greater flexibility of curtain-wall systems, less mass density of buildings, and reduced structural damping) has increased the need for accurate determination of wind pressure and loads. Small-scale models placed in a boundary-layer wind tunnel are commonly used for this purpose. Rigid (static) models of 1/100 to 1/500 scale perforated with 300 to 1000 pressure taps are used to obtain $C_{\bar{p}}$, $C_{\hat{p}}$, and $C_{\breve{p}}$. For buildings having motion that does not affect aerodynamic loading of the building, $\vec{\bar{F}}$, $\vec{\hat{F}}'$, $\vec{\bar{M}}$, $\vec{\hat{M}}'$, and their spectra are obtained by placing a lightweight rigid model on a high-natural-frequency base balance. Aeroelastic models are used to measure $\vec{\bar{M}}$, $\vec{\hat{M}}'$, mean and peak deflections, and peak accelerations for buildings and other structures with motion that may affect aerodynamic loading (vortex shedding or galloping).

References

1. J. E. CERMAK: Aerodynamics of buildings. *Annual Review of Fluid Mechanics*, Vol. 8 (1975), pp. 75–106.
2. J. E. CERMAK: Applications of fluid mechanics to wind engineering—a Freeman scholar lecture. *Journal of Fluids Engineering*, ASME, Vol. 97 (March 1975), pp. 9–38.
3. *Minimum Design Loads for Buildings and Other Structures*. ANSI A58.1-1982, American National Standards Institute, New York, 1982.
4. R. E. AKINS, J. A. PETERKA, and J. E. CERMAK: Mean force and moment coefficients for buildings in turbulent boundary layers. *Journal of Wind Engineering and Industrial Aerodynamics*, Vol. 2 (1977), pp. 195–209.
5. *Proceedings of the Fifth International Conference on Wind Engineering* (J. E. Cermak, ed.), 2 Vols. Pergamon Press, Elmsford, N.Y., 1980.

6. J. E. CERMAK: Wind-tunnel testing of structures. *Journal of the Engineering Mechanics Division*, ASCE, Vol. 103, No. EM6 (Dec. 1977), pp. 1125–1140.

J.E.C.

WIND TUNNELS have been developed with a capability for physical modeling of natural winds in the atmospheric boundary layer (ABL). Wind tunnels designed for this purpose are called boundary-layer wind tunnels (BLWTs) and may be of the closed-circuit or the open-circuit type. However, the open-circuit type illustrated in Fig. 310 is most common. These low-speed (subsonic) long-test-section wind tunnels enable accurate determination of wind pressure and forces on buildings, as well as transport of air pollutants in industrial and urban settings by measurements on small-scale models. A typical rigid (static) model of a building and surroundings used for measurement of $\overline{p}$, $\hat{p}$, $\check{p}$, $\vec{F}$, and $\vec{M}$ and street-level (pedestrian-level) winds is shown in Fig. 311. The meanings of the symbols used are explained in Fig. 310 and in the article on WIND LOADS.

Similarity Requirements. Essential requirements for physical modeling of strong-wind ABLs with neutral thermal stratification are (1) geometrical similarity of upwind surface geometry stratification (vegetation, buildings, and topography), (2) Reynolds number, $\text{Re} = \overline{U}_g L_x/\nu_0 > 2(L_x/K_s) \times 10^3$, where ν_0 is the kinematic viscosity of air, and (3) zero pressure gradient in the flow direction. For physical modeling of thermally stratified ABLs, the Richardson number, $\text{Ri} = (\Delta T_0/T_0)\,(gL_0/\overline{U}_0^2)$ must be equal for model and prototype, as well as the Prandtl number,

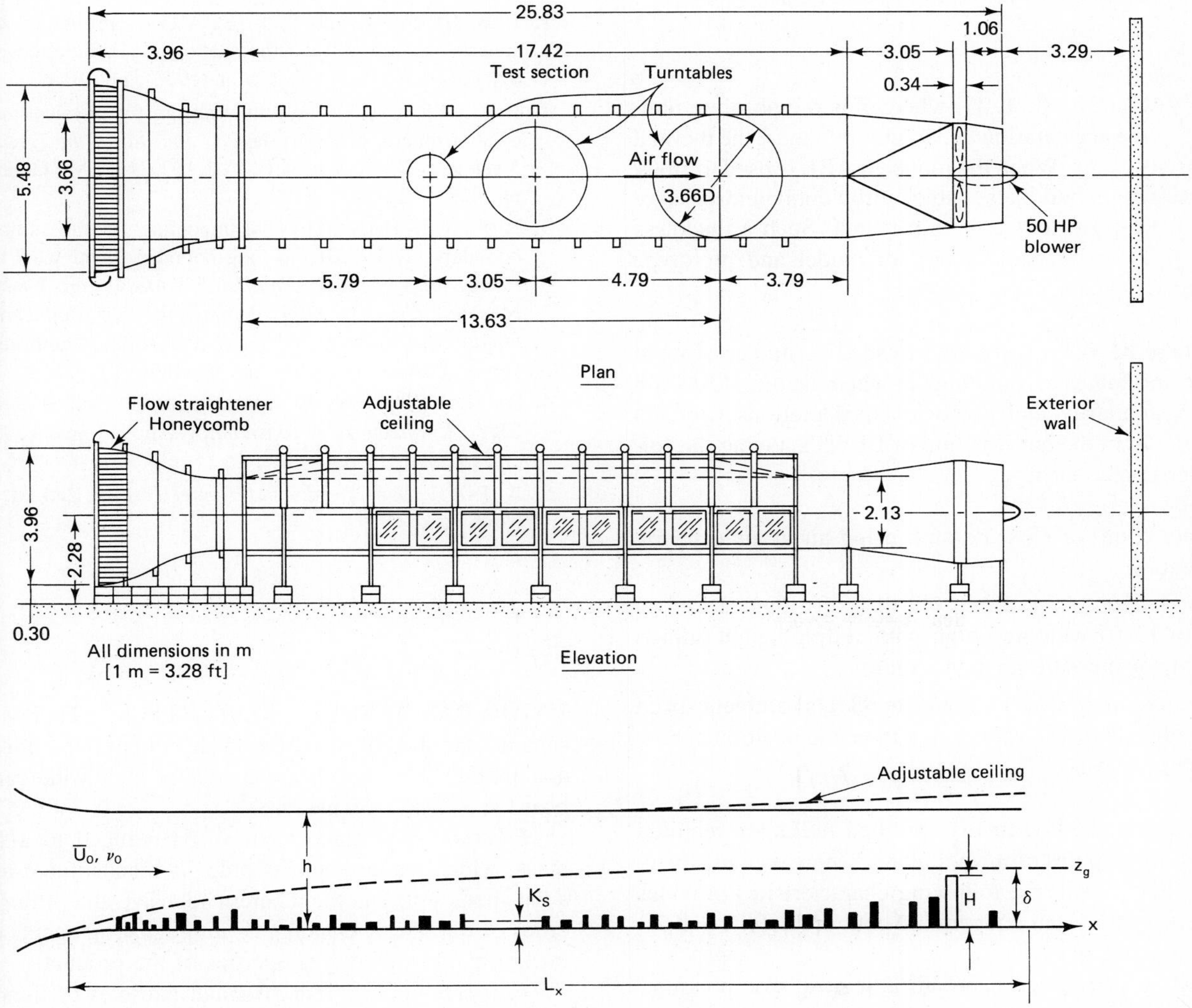

Figure 310 Typical open-circuit boundary-layer wind tunnel for simulation of neutral atmospheric boundary layers. (Courtesy Fluid Dynamics and Diffusion Laboratory, Colorado State University, Fort Collins, Colo.)

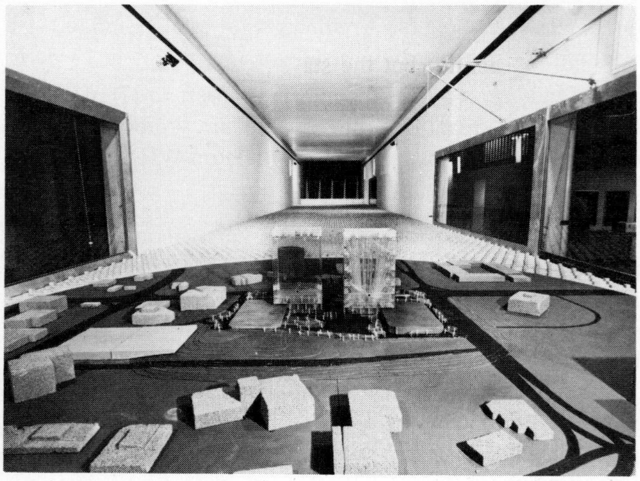

Figure 311 Rigid (static) building models (1:300 scale) prepared for pressure and pedestrian-level wind measurements in 3.1 m (10 ft) wide BLWT. (Courtesy Wind Engineering Laboratory, Cermak/Peterka and Associates, Inc., Fort Collins, Colo.)

$\text{Pr} = \nu_0/(k/\rho C_p)_0$ (Ref. 1), where T is the absolute temperature, g the acceleration due to gravity, and k the thermal conductivity of air. When the modeled ABL is used to study wind pressure on buildings, a thick turbulent boundary layer (0.5 to 1.3 m; 1.6 to 4.2 ft) is required. Such a thickness permits δ/H to be made equal for model and prototype without excessive scale reduction.

Typical BLWT Characteristics. For studies of wind pressure and forces on buildings in which neutral ABLs are required, similarity and practical considerations (Ref. 2) have resulted in the construction of BLWTs having the following characteristics:

1. *Type:* open- or closed-circuit using air at atmospheric pressure.
2. *Test section:* 12 to 20 m (39 to 66 ft) long, 2 to 4 m (7 to 13 ft) wide and high with ceiling-height adjustment for pressure-gradient control.
3. *Flow features:* 2 to 30 m/s (7 to 98 ft/s) ambient speed with less than 1% variation in time and position at test-section entrance.

When simulated thermally stratified ABLs are required (primarily for studies of air-pollutant dispersion and advection), a BLWT with the following characteristics provides the necessary flow and temperature control (Refs. 1 and 2):

1. *Type:* closed-circuit using air at atmospheric pressure.
2. *Test section:* 15 to 30 m (49 to 98 ft) long, 2 to 4 m (7 to 13 ft) wide, 2 to 3 m (7 to 10 ft) high with adjustment for pressure-gradient control.
3. *Flow features:* 0.5 to 30 m/s (2 to 98 ft/s) ambient air speed with less than 1% variation in time and space at test-section entrance.
4. *Thermal features:* ±75° to 100°C (±167° to 212°F) mean temperature difference between test-section floor and ambient air.

Boundary-layer characteristics for a particular BLWT depend on roughness of the test-section floor and the type of spires, vortex generators, and/or fences placed at the test-section entrance to increase boundary-layer thickness. Data on the effect of these flow control measures has been reported (Ref. 3).

See also ANEMOMETERS.

References

1. J. E. CERMAK and S. P. S. ARYA: Problems of atmospheric shear flows and their laboratory simulation. *Boundary-Layer Meteorology*, Vol. 1 (1970), pp. 40–60.
2. J. E. CERMAK: Wind tunnel design for physical modeling of atmospheric boundary layers. *Journal of the Engineering Mechanics Division*, ASCE, Vol. 102, No. EM3 (June 1981), pp. 623–642.
3. J. E. CERMAK: Physical modeling of the atmospheric boundary layer (ABL) in long boundary-layer wind tunnels. *Proceedings of the International Workshop on Wind Tunnel Modeling Criteria and Techniques in Civil Engineering Applications*, Gaithersburg, Md. (April 1982), Cambridge University Press, Cambridge, pp. 97–125.
4. R. M. AYNSLEY, W. MELBOURNE, and B. J. VICKERY: *Architectural Aerodynamics*. Applied Science Publishers, London, 1977.
5. A. POPE: *Low-Speed Wind Tunnel Testing*, 2nd ed. Wiley, New York, 1984.

J.E.C.

WOOD-BASED SHEET MATERIALS. There are two separate families of wood-based sheet materials, fiberboard and particleboard; both are available in a wide range of board sizes and thicknesses.

Fiberboard is made from wood reduced to a fibrous mass, which, as an aqueous pulp, is reconstituted to form a flat sheet, utilizing the inherent adhesive and felting properties of the fibers. Bonding or impregnating agents can be incorporated to modify properties of the product.

Particleboard consists of small particles of lignocellulosic material (generally wood) coated with a bonding agent; sheet material is formed by the application of heat and pressure.

History. In England in 1898 Sutherland hot-pressed layers of waste paper and produced a material akin to present-day hardboard. Softboard was produced in Canada 10 years later by mechanically grinding solid timber to a wet pulp; this felted together under light pressure to form a semirigid board. In the United States, Mason invented the explosion process of fiber production and in 1926 commenced production of Masonite, a hardboard formed under high pressure. Since Asplund developed the defibrator process for producing fibers in Sweden in 1935, fiberboard production and consumption have grown apace. The majority of fiberboards are made by the wet felting process, but in recent years a dry process employing adhesives has also evolved.

Particleboard manufacture became possible with the introduction of synthetic resins (see PLASTICS) in the 1930s; it was first produced commercially in Germany in 1941. Manufacture expanded after 1945, and it spread throughout the world in the next two decades. The production of particle board is now an important industry, because it can more easily and economically be tailored to meet a specific requirement than any other wood-based material.

Board Types. There are two types of fiberboard. *Insulating boards* are noncompressed low-density boards, tiles, or planks for thermal or acoustic insulation. Their thickness ranges from 9 to 25 mm ($\frac{3}{8}$ to 1 in.) in a wide range of panel sizes. Acoustic products generally have stopped holes or fissures to increase their sound-absorbing capacity. Impregnation with bitumen (asphalt) increases resistance to water.

Hardboards are highly compressed boards. They have at least one smooth surface; the reverse may be smooth or have a mesh pattern. Their thickness ranges, according to type, from 2 to 13 mm ($\frac{1}{10}$ to $\frac{1}{2}$ in.); however, medium-density fiberboard (MDF) may be up to 35 mm ($1\frac{3}{8}$ in.) thick. Densified (tempered) hardboard, the heaviest and hardest type, has an average density of 1100 kg/m^3 (70 lb/ft^3); it is suitable for exterior use. Standard hardboard has an average density of 800 kg/m^3 (50 lb/ft^3) and a variety of uses in buildings. Slightly less dense are medium board and medium-density fiberboard (MDF). MDF contains a resin bonding agent, enabling surfaces and edges to be cleanly profiled. It is used increasingly in place of solid wood for small section components.

Wood is the most widely used raw material in *particleboard* manufacture; but the stalks of seasonal crops, flax, bagasse, hemp, and cotton can be used. Coniferous timber has technical advantages; but hardwood is also utilized. The first stage of the manufacturing process is the production of particles. Their geometry and size determine the properties of the end product. Particles are coated with synthetic resin, commonly urea formaldehyde, although phenol or melamine resins may be used to give better resistance to moisture. They are then consolidated under heat and pressure to produce a mat of layered construction, frequently with a low-density core and dense outer layers of fine, closely packed particles. This gives high bending strength relative to weight and provides a good substrate for laying facing materials or for painting.

Extruded board is made by an alternative process, by forcing the particles between parallel dies. Two types of mat-formed board, waferboard and Oriental strand board (OSB), have a bending strength and stability approaching that of plywood. In normal particleboard, particle direction, and therefore fiber orientation, is entirely random; but in Oriental strand board particles are laid with their longitudinal axes in a desired plane. Cement board has cement as the bonding agent; it is suitable for siding and other external uses.

Fiberboards and particleboards can be treated to be flame retardant, primed, painted, grain printed, faced with wood veneers, or with plastic overlays. Many other process variations are available.

References

1. Wood based panels in 1980's. *Proceedings of United Nations Economic Committee for Europe Symposium*. Finnish Paper and Timber Publishing Co., Helsinki, 1980.
2. T. M. MALONEY: *Modern Particle Board and Dry Process Fiberboard Manufacture*. Miller Freeman, San Francisco, 1977.
3. L. E. AKERS: *Particle Board and Hardboard*. Pergamon Press, Oxford and Elmsford, N.Y., 1966.

L.E.A.

Index to Contributors

The number(s) following each name are those of the first page(s) of a contributor's article(s).

J

K

L

M

N

P

R

S

T

V

W

Z

General Index

The numbers in ordinary type denote the pages on which the subject is mentioned, not necessarily using exactly the same words. The numbers in **bold type** denote the first page of an article specifically on that subject.
The encyclopedia has an extensive system of cross references in the text of the articles, which are denoted by CAPITAL LETTERS. These cross references are generally a quicker method for finding further information *on related subjects* than this index.

B

C

G

H

Q

R

S

Y

Z